INTERNATIONAL PHARMACEUTICAL SERVICES: THE DRUG INDUSTRY AND PHARMACY PRACTICE IN TWENTY-THREE MAJOR COUNTRIES OF THE WORLD

Richard N. Spivey, PharmD, PhD
Albert I. Wertheimer, PhD, RPh, MBA
T. Donald Rucker, PhD, MA
Editors

SOME ADVANCE REVIEWS

"This book is a health care almanac for those persons interested in the globalization of health care systems. The organization of the chapters permits quick comparisons of one country with another in the areas of national health, drugs, pharmacy practice, and special situations. Sections with generous numbers of tables and graphs make direct comparisons in narrow areas possible. A must reference for anyone involved at any level of pharmaceutical services."

Lawrence C. Weaver, PhD
Dean Emeritus, College of Pharmacy
University of Minnesota

"A much-needed addition to the pharmacy literature. This book will fill existing needs in graduate and undergraduate pharmacy education. Thorough, well-written, and timely, it will be widely used as a reference text both within and outside of the pharmacy community, nationally and internationally. It will have a major beneficial influence for leaders in the educational drug manufacturing and policy analysis arenas. A must-have book."

Jack E. Fincham, PhD
Professor and Associate Dean
School of Pharmacy and Allied Health
Creighton University

"At the present time, each country is looking for alternatives to provide the best health care system within their respective resources available. This [book] . . . provides very useful information with respect to the development of health care programs pertaining to twenty-three countries. . . . Through this book, the reader will be able to evaluate various health care programs and what has or has not worked in a given country."

Andrés Malavé, PhD
Dean, School of Pharmacy
University of Puerto Rico

NOTES FOR PROFESSIONAL LIBRARIANS AND LIBRARY USERS

This is an original book title published by Pharmaceutical Products Press, an imprint of The Haworth Press, Inc. Unless otherwise noted in specific chapters with attribution, materials in this book have not been previously published elsewhere in any format or language.

CONSERVATION AND PRESERVATION NOTES

The paper used in this publication meets the minimum requirements of American National Standard for Information Sciences — Permanence of Paper for Printed Material, ANSI Z39.48-1984.

International Pharmaceutical Services

The Drug Industry and Pharmacy Practice in Twenty-Three Major Countries of the World

PHARMACEUTICAL PRODUCTS PRESS
Pharmaceutical Sciences
Mickey C. Smith, PhD
Executive Editor

New, Recent, and Forthcoming Titles:

Principles of Pharmaceutical Marketing, Third Edition edited by Mickey C. Smith

Pharmacy Ethics edited by Mickey C. Smith, Steven Strauss, John Baldwin, and Kelly T. Alberts

Drug-Related Problems in Geriatric Nursing Home Patients by James W. Cooper

Pharmacy and the U.S. Health Care System edited by Jack E. Fincham and Albert I. Wertheimer

Pharmaceutical Marketing: Strategy and Cases by Mickey C. Smith

International Pharmaceutical Services: The Drug Industry and Pharmacy Practice in Twenty-Three Major Countries of the World edited by Richard N. Spivey, Albert I. Wertheimer, and T. Donald Rucker

A Social History of the Minor Tranquilizers: The Quest for Small Comfort in the Age of Anxiety by Mickey C. Smith

Marketing Pharmaceutical Services: Patron Loyalty, Satisfaction, and Preferences edited by Harry A. Smith and Stephen Joel Coons

The Honest Herbal: A Sensible Guide to the Use of Herbs and Related Remedies, Third Edition by Varro E. Tyler

Nicotine Replacement: A Critical Evaluation edited by Ovide F. Pomerleau and Cynthia S. Pomerleau

International Pharmaceutical Services

The Drug Industry and Pharmacy Practice in Twenty-Three Major Countries of the World

Richard N. Spivey, PharmD, PhD
Albert I. Wertheimer, PhD, RPh, MBA
T. Donald Rucker, PhD, MA
Editors

Pharmaceutical Products Press
An Imprint of The Haworth Press, Inc.
New York • London • Norwood (Australia)

Published by

Pharmaceutical Products Press, an imprint of The Haworth Press, Inc., 10 Alice Street, Binghamton, NY 13904-1580

Library of Congress Cataloging-in-Publication Data

International pharmaceutical services : the drug industry and pharmacy practice in twenty-three major countries of the world / [edited by] Richard N. Spivey, Albert I. Wertheimer, T. Donald Rucker.
    p. cm.
Includes bibliographical references and index.
ISBN 0-86656-906-5 (acid-free paper)
    1. Pharmaceutical industry. 2. Pharmaceutical policy. 3. Medical care. I. Spivey, Richard N. II. Wertheimer, Albert I. III. Rucker, T. Donald.
HD9665.5.I577 1991
338.4′76151 — dc20   91-6794
                                                                CIP

CONTENTS

ABOUT THE EDITORS

Richard N. Spivey, PharmD, PhD, is currently Director, Worldwide Regulatory Affairs at Parke-Davis Pharmaceutical Research Division of Warner-Lambert Company, Ann Arbor, Michigan. He has held various positions in regulatory affairs and clinical research in the pharmaceutical industry. Prior to joining the industry, he conducted drug policy related research, was on the clinical pharmacy faculty at a major university, and practiced hospital pharmacy. Dr. Spivey earned his doctorate in Pharmacy Administration from the University of Minnesota and his PharmD degree from the University of Southern California.

Albert I. Wertheimer, PhD, RPh, MBA, is Dean of the Philadelphia College of Pharmacy and Science. Previously, he was the chairman and driving force for the creation of the Department of Social and Administrative Pharmacy at the University of Minnesota and the co-director of the Kellogg Pharmaceutical Clinical Scientist Program at the University of Minnesota. He has served as the president of the Academic Section of the International Federation of Pharmacy and has held numerous positions in various pharmaceutical organizations. A prominent researcher and recognized authority on social pharmacy, international pharmacy, and international health care, he has served as a consultant to the World Health Organization. Dr. Wertheimer is the author of several hundred publications and more than a dozen books. He has taught graduate and undergraduate courses in social and administrative pharmacy and is frequently invited to speak at international symposia. His areas of research emphasis include social pharmacy and sociobehavioral aspects of drug use. He received his PhD in Pharmacy Administration from Purdue University and his MBA from SUNY Buffalo.

T. Donald Rucker, PhD, MA, is Professor Emeritus, College of Pharmacy, University of Illinois, Chicago. Dr. Rucker has held several academic positions at major universities. Prior to joining academia, Dr. Rucker was Director of the Drug Studies Branch of the Social Security Administration in Washington, DC. He is a recognized authority on drug reimbursement issues and is widely published in this field. Dr. Rucker received his doctorate in Economics from Syracuse University.

List of Contributors

ARGENTINA

MARCELO JORGE VERNENGO, Rep. Arabe Siria (ex-Malabia) 2711, So. Piso-1425-Buenos Aires, Argentina

AUSTRALIA

MR. ROBERT DAVIES, Pharmacy Guild of Australia, 14 Thesiger Court, P.O. Box 36, Deakin, A.C.T. 2600, Australia

DR. MICHAEL TATCHELL, Director, Health Economics Division, Pharmacy Guild of Australia, 14 Thesiger Court, P.O. Box 36, Deakin, A.C.T. 2600, Australia

BRAZIL

DR. JOSE N. CALLEGARI LOPES, Universidade de Sao Paulo, Rua do Anfiteatro, 295, Cidade Universitária, "Armando de Salles Oliveira," CEP 05508, Sao Paulo, Brasil

DR. ANGELO J. COLOMBO, Faculdade de Ciencias Farmaceuticas, Universidade de Sao Paulo, Cidade Universitária, "Armando de Salles Oliveira," Caixa Postal 30.786, Sao Paulo, Brasil

DR. RENATO BARUFFALDI, Faculdade de Ciencias Farmaceuticas, Universidade de Sao Paulo, Cidade Universitária, "Armando de Salles Oliveira," Caixa Postal 30.786, Sao Paulo, Brasil

CANADA

DR. JOHN A. BACHYNSKY, Dean, Faculty of Pharmacy and Pharmaceutical Sciences, University of Alberta, 3118 Dentistry/Pharmacy Center, Edmonton, Alberta T6G 2N8, Canada

PEOPLE'S REPUBLIC OF CHINA

DR. HU TING-XI, China Pharmaceutical University, 24 Tong Jia Xiang, Nanjing, Hiangsu, People's Republic of China

DENMARK

MR. PETER KIELGAST, Danmarks Apoteker forening, Bredgade 54, P.O. Box 2181, DK-1017 Copenhagen K, Denmark

JENS POVELSEN, Danmarks Apoteker forening, Bredgade 54, P.O. Box 2181, DK-1017 Copenhagen K, Denmark

ARAB REPUBLIC OF EGYPT

DR. YEHIA M. DESSOUKY, Professor of Pharmaceutical Chemistry, Faculty of Pharmacy, Cairo University, P.O. Box 352, Heliopolis, Cairo 11757, Egypt

FRANCE

MR. JEAN BRUDON, Conseil National de L'ordre des Pharmaciens, 4, Avenue Ruysdael (VII), France

MR. GEORGES VIALA, Conseil National de L'ordre des Pharmaciens, 4, Avenue Ruysdael (VII), France

FEDERAL REPUBLIC OF GERMANY

DR. HANNELORE SITZIUS-ZEHENDER, Bundesvereinigung Deutscher, Apothkerverbande-ABDA, Beethovenplatz 1-3, 6000 Frankfurt/Main 97, Germany

BERTRAM DERVENILCH, Bundesvereinigung Deutscher, Apothkerverbande-ABDA, Beethovenplatz 1-3, 6000 Frankfurt/Main 97, Germany

FRANK DIENER, Bundesvereinigung Deutscher, Apothkerverbande-ABDA, Beethovenplatz 1-3, 6000 Frankfurt/Main 97, Germany

GERD M. FOH, Bundesvereinigung Deutscher, Apothkerverbande-ABDA, Beethovenplatz 1-3, 6000 Frankfurt/Main 97, Germany

HUNGARY

DR. ZOLTAN VINCZE, Hungarian Pharmaceutical Society, Hogyes Endre u.7, Budapest 1092, Hungary

ITALY

DR. MARCELLO MARCHETTI, Instituto Chimica Farmaceutica,
Viale Abruzzi u. 42, Milano 20131, Italy

DR. PAOLA MINGHETTI, Instituto Chimica Farmaceutica, Viale
Abruzzi u. 42, Milano 20131, Italy

JAPAN

DR. MINORI TATSUNO, Department of Hygiene and Preventive
Medicine, School of Medicine, Osaka University, 2-2 Yamada-oka,
Suita-shi, Osaka, Japan

DR. KIYOSHI MURAOKA, Department of Hygiene and Preventive
Medicine, School of Medicine, Osaka University, 2-2 Yamada-oka,
Suita-shi, Osaka, Japan

DR. TSUNEJI NAGAI, Pharmaceutics Department, Hoshi University,
Ebara, Shinagawa-Ku, Tokyo, Japan

MEXICO

DR. FELA VISO GUROVICH, Piramide de la Luna 104-401, Colonia
Avante, Delegacion Coyoacan, C.P. 04460, Mexico, D.F., Mexico

CARMEN GIRAL BARNES, Callejon del Olivio Number 10, Antes
84, Col. Florida, Delegacion Alvaro Obregon, 01030, Mexico, D.F.,
Mexico

NETHERLANDS

HUBERT G. LEUFKENS, Department of Pharmacoepidemiology,
University of Utrecht, Croesestraat 79, 3522 AD Utrecht,
Netherlands

PROF. ALBERT BAKKER, Department of Pharmacoepidemiology,
University of Utrecht, Croesestraat 79, 3522 AD Utrecht,
Netherlands

NEW ZEALAND

MS. JENNY CADE, Pharmaceutical Society of New Zealand,
P.O. Box 11-640, Wellington, New Zealand

NIGERIA

DR. OLANREWAJU OGUNLANA, LanPharm Laboratories
& Scientific Services Limited, 3rd Floor, Lapal House, 235,
Igbosere Road, Lagos, P.O. Box 52830, Falomo, Ikoyi, Lagos,
Nigeria

NORWAY

MR. BJORN JOLDAL, Directorate of Health, Pharmacy Department,
P.O. Box 8128, Oslo 1, 0032, Norway

BJORG STROMNES, The Norwegian Association of Proprietor
Pharmacists, P.O. Box 5070 Majorstua, Oslo 3, 0301, Norway

PANAMA

DR. JERONIMO AVERZA C., P.O. Box 4793, Panama 5, Panama

SPAIN

DR. JOAQUIMA SERRADELL, School of Medicine, Clinical
Epidemiology Unit, University of Pennsylvania, Philadelphia, PA
19104

MR. EUGENI SEDANO, Department of Pharmacy Services,
Government of Catalonia, Barcelona, Spain

MS. JOAN SERRA, Department of Pharmacy Services, Government
of Catalonia, Barcelona, Spain

SWEDEN

MR. TOMMY WESTERLUND, Pharmacy Manager, Apoteket Staren,
Blakullagatan 5, S-25457 Helsingborg, Sweden

TAIWAN, REPUBLIC OF CHINA

DR. WENG F. HUANG, Bureau of Drugs, Department of Health,
Executive Yuan, 11F, 100 Ai-kuo E. Road, Taipei, Taiwan 10726,
Republic of China

UNITED KINGDOM

DR. PETER R. NOYCE, Department of Pharmacy, University
of Manchester, Oxford Road, Manchester M13 9PL, United
Kingdom

JEANNETTE A. HOWE, Department of Health & Social Security,
Eileen House, Room 112, Newington Causeway, London SE 6EF,
United Kingdom

UNITED STATES

DR. KATHLEEN A. JOHNSON, School of Pharmacy, University
of Southern California, 1985 Zonal Avenue, Los Angeles, CA 90033

FOREWORD

MR. GEORGE GRIFFENHAGEN, American Pharmaceutical Assn.,
2215 Constitution Avenue N.W., Washington, DC 20037

EDITORS

DR. RICHARD N. SPIVEY, Parke-Davis Pharmaceutical Research
Division, Warner-Lambert Company, 2800 Plymouth Road,
Ann Arbor, MI 48106-1047

DR. ALBERT I. WERTHEIMER, Dean, Philadelphia College
of Pharmacy and Science, Woodland Avenue at 43rd Street,
Philadelphia, PA 19104

DR. T. DONALD RUCKER, Professor Emeritus, College of Pharmacy,
University of Illinois-Chicago, Chicago, IL

Foreword

We are living today in a shrinking world where the problems facing us and the profession of pharmacy in our country have become increasingly the concern of pharmacists around the world.

This description of the importance of international understanding, which I wrote in December 1962 for the *Journal of the American Pharmaceutical Association*, is as timely today as it was 30 years ago. As editor of the same journal, I edited and published from time to time various studies of pharmacy and health care systems in countries around the "shrinking" world.

One such study, reported by Richard A. Hall, then pharmacist director of the U.S. Public Health Service, appears in the December 1969 issue of the *Journal of the American Pharmaceutical Association*. Already, more than 20 years ago, over 80 nations throughout the world had established compulsory social insurance programs that provided health services to individuals for sickness or maternity or helped finance such services. Pharmaceutical benefits under these health insurance and government medical care programs were universally limited to prescribed drugs and appliances. With but few exceptions, nonprescribed (OTC) drugs were not included in the benefits provided. Hall went on to point out:

> Once the costs of a segment of medical care are assumed on a social basis, a social responsibility is created to help assure that the money is spent wisely. As a result, the use of drugs in such programs were circumscribed with various restraints.

Over 20 years later, we have a current assessment of these restraints, as well as the practice of pharmacy, the pharmaceutical industry, and the national health care system of some 20 countries around the world in this volume. The authors of each chapter—all experts in their countries—are variously executives of national pharmaceutical associations, faculty members of schools of pharmacy, or government officials. The national health care system of each country is reviewed, including history and background, social and political environment, public and private financing, facilities, health manpower, and population coverage. The pharmaceutical industry overview of each country includes a description of research and development of new drugs, manufacturing, pharmaceutical patents and licensing, approval process and government regulations, advertising and promotion, price regulations, competition (brand and generics), exports, and imports. Pharmacy practice covers dispensing, record-keeping, prescription drugs and OTCs, formularies, reimbursement, primary care, drug insurance programs, patient education, nonpharmacy outlets, and pharmacy technicians.

In addition, unique features in individual countries are described, including use of traditional remedies, differing healing systems, and non-Western medicine. Most importantly, an assessment of the strengths, weaknesses, and trends of health care in each country is presented.

Unfortunately, provincialism characterizes a good part of America. There are still those who refuse to admit that an understanding of pharmacy and the health care systems in other countries can assist us in charting our own future. A careful review of this invaluable resource should provide convincing proof that a study of international pharmacy and health care systems around our shrinking world offers approaches for both professional growth and the well-being of mankind.

George Griffenhagen

Preface

It is not satisfactory for us to be unaware of activities beyond our national borders. Yet, it is difficult, oftentimes, to obtain information about many topics on an international basis. The provision of pharmaceuticals, their distribution, payment schemes, relevant laws, and organizational characteristics are the major focal points of this book. We are convinced that knowledge about practices in distant places can help us in our own professional and scientific endeavors.

We recognize that the production, distribution, promotion, and sales of pharmaceutical products differ throughout the world based upon the specific characteristics of a nation's political system, its history, geographic characteristics, economic orientation, wealth, and history/tradition. Through this diversity, it is possible to consider features present in the drug use process in other lands that could be beneficial to readers in diverse countries and regions. Large, developed countries have no monopoly in ideas and original concepts. In the following chapters, we will encounter vastly different pharmacy schemes from all continents, from nations representing differing levels of development and political orientations.

We contacted colleagues and friends in these countries, providing them with a general outline of topic areas that should be included in their respective chapters, but due to the enormous diversity in the role of governments, development of a domestic pharmaceutical industry, the power and influence of regulatory structures, and the relative sizes of the public and private sectors in the provision of care, it became obvious to us that all chapters could not be uniform.

We enjoyed this opportunity to work together and to learn more about pharmacy and drug issues, policies, and practices in the four corners of this world. We hope that this book is of value to our colleagues and that it can aid in improvements that will bring higher quality and more efficient services to patients around the world. All suggestions and comments would be welcomed by us from interested readers. And finally, we need to add a caveat. We believe the material contained in the following pages was accurate in late 1989 and early 1990 when the book was in preparation, but events change daily due to changes in governments, budget crises, scandals, and other internal and external forces, so that we are unable to warrant the total accuracy of this material in the future.

Richard N. Spivey
Ann Arbor, MI

Albert I. Wertheimer
Philadelphia, PA

T. Donald Rucker
Chicago, IL

Acknowledgements

We are grateful for the assistance of many persons who helped this project survive to completion. Despite the usual complaining about international postal services, they do function, albeit sometimes slowly. We are grateful to the inventors of the facsimile machine and of the telephone. Naturally, this work could not have existed without the generous contributions of time and the sharing of knowledge and experience of our chapter authors. In addition, we are grateful for the cooperation and help of James Ice and Linda Cohen of The Haworth Press, and of our colleague Dr. Mickey C. Smith, Haworth's pharmacy area editor. And we want to say thank you to Thomas Benson and Holly Schoonover, who typed the final manuscript, and to Connie Vo and Sophia Pham, who assisted with clerical efforts and proofreading.

Chapter 1

Argentina

Marcelo Jorge Vernengo

A. THE NATIONAL HEALTH CARE SYSTEM

1. History and Background

Until the 1940s, private medical care was available only to a fraction of the population, and public hospitals provided free care to the rest of the people. There were a number of foreign community organizations (Italian, Spanish, French, British, German, and Jewish communities) administering nonprofit hospitals in large urban areas such as Buenos Aires. A small number of hospitals and care centers were established by some trade unions, mutuals, and beneficent organizations.[1]

Initial attempts during 1945-1955 were toward development of the capabilities of the federal government within the jurisdiction of the Ministerio de Salud Publica (created in 1949 following the Departamento Nacional de Higiene [1880-1944], the Direccion Nacional de Salud Publica 1944-1946, and the Secretaria de Salud Publica [1946-1949]). Control of endemic diseases was conducted at the federal level with significant success in the case of malaria. Medical care and preventive activities were centered in small health units.

The social security system began in the late 1940s, based mainly on trade unions that established health care systems through contracts with private providers. Initially this system covered only workers of the public sector, including those in state corporations (railways, electricity, petroleum, ports, etc.), but it was slowly extended to all trade unions and also covered family members.

Decentralization of the federal government system, with transference of its hospitals to the different provinces, began before 1960, and by 1985, only a small number of health care facilities remained under federal jurisdiction. In 1958, the Escuela Nacional de Salud Publica and a Public

Health School at the University of Buenos Aires were established. In 1964, Law Nos. 16.462 and 16.463 providing for drug regulations at the economic and technical level were enacted, and the Instituto Nacional de Farmacologia y Bromatologia was created to take charge of some drug control activities. In 1970, Law No. 18.610 provided for the reorganization of health agencies of the social security system. Universal coverage for all dependent workers and employees was set up, uniform compulsory salary deductions were established for all workers, and the Instituto Nacional de Obras Sociales (INOS) was put in charge of surveillance and promotion of the system.

The private sector expanded its activities in the 1970s, introducing the use of new technologies and extending inpatient treatment. Private providers increasingly contracted with the Obras Sociales, promoting commercial interests. Prepaid health care organizations became common. In 1974, a national health bill was proposed to unify the different organizations under an integrated national health system (Sistema Nacional Integrado de Salud [SNIS]) based upon the public health sector.

During 1976-1983, the policy of transference of health units to local and provincial authorities was completed. The Federal Health Authority remained in charge of planning, standardization, control, and training activities and provided budgetary contributions for programs such as rural health, maternal and child care, and communicable diseases.

The Government Organization Act of 1981, Law No. 22.450, took away from the Ministry of Health almost all authority for policy planning and coordination and diluted any possibility of an integrated health care system. In December 1983, a new Ministerio de Salud Publica y Accion Social was established. By the end of 1988, legislation was enacted establishing a National Health Insurance Scheme and regulating the activities of the Obras Sociales and other NHI agencies.

The National Health Insurance Scheme will be based on common standards and central coordination, while local and provincial authorities and intermediate organizations (including the Obras Sociales) will be in charge of all health activities. A new agency, the Administracion Nacional del Seguro de Salud (ANSSAL), will be in charge of central activities. The Insurance Scheme will be financed as before by salary deductions, contributions from employers, and by fiscal revenues devoted to specific segments of the population, as those without direct coverage or resources. A new Fondo Solidario de Redistribucion was established for promotion of equal access to health services among the beneficiaries of the NHI agencies.

2. Social and Political Environment

Argentina is the eighth largest country in the world and the second largest in South America, with a population of 31,536,000 in 1987 and an estimated doubling time of 54 years.[2,3] The population density is extremely variable. Metropolitan Buenos Aires contains approximately 35% of the population, with a population of nearly 10 million. All other major towns are under the one million mark, although the urban population reached 85% in 1985. Children under 15 years of age accounted for 30.5% of the population. Those who were 15-64 years old composed 60.9% of the population, and those 65 and older accounted for 8.6%. Sex distribution is almost equal, with 50.4% being female. The population is mostly of European ethnic origin (85%); the predominant religion is Catholicism.[3]

The total land area of 2,795,000 square kilometers includes the Territory of Tierra del Fuego and the disputed Malvinas and South Atlantic Islands. The Antarctic Region comprises 964,250 square kilometers. Argentina is situated on the eastern part of the South American land mass. The country spans over 3,800 kilometers north-south and 1,423 kilometers west-east. To the east, it covers the extended fertile pampas and most of the population, economical activities, and communications resources; to the south, the area known as the Patagonia; to the west, partially desert limited by the Andes and the neighboring Republic of Chile; and the north, the less developed region near Bolivia. The north-east region near Paraguay and Brazil is much richer in natural resources.

The climate is temperate although the range varies from − 16 °C in the south to 48°C in the north. The largest part of the population experiences regular and no extreme temperatures. Median temperature in Buenos Aires is 17°C with 22°C in summer and 5°C in winter and high levels of humidity. It rains predominantly in summer and autumn.

Argentina is a federal republic, with the head of state and the government being the President of Republica Argentina. It has two legislative houses (the Senate, with 46 members, and the Chamber of Deputies, with 254) and a judicial system headed by a Corte Suprema de Justicia. The executive power under the President is directed by eight ministries and several autonomous or semiautonomous agencies.

The country is composed of 22 provinces (states), a Federal District (Buenos Aires), and a federal territory. Legislation has been enacted to move the capital from Buenos Aires to Viedma in northern Patagonia, but it has not yet been implemented. It could reverse both the historical tendency toward political and economic development centered in Buenos Aires and population movements toward big urban centers.

The provinces are organized along similar lines. Local power resides in district, municipal, or city governments. The Federal Constitution, enacted in 1853, limits the power of the federal authorities, leaving almost all internal matters to the provincial authorities. Implementation of a number of policies related to education, health, etc., depends on agreements between different levels of authority. The Federal District is under the jurisdiction of the federal government, and the intendente (mayor) is appointed by the President, although the city is administered as an autonomous organization and in health matters participates at the level of the other provincial authorities.

The official language is Spanish, spoken by an overwhelming majority of the population. The currency is the austral.

Education is compulsory up to the end of the primary school, and in official establishments all levels of education are free. More than 40,000 educational units are in operation, three-fourths of which are state run. The literacy rate, according to the 1980 National Census, was 94.9% for the population over 15 years. The percentage of the population aged 25 and over with no formal schooling was 6.0%, with primary and secondary level almost 55% and higher education 6.9%. Primary school enrollment makes up approximately 95% of the school-age population.[1]

Natural population increase rate was 14.3 in 1980-85, with a tendency toward stability or slight decrease. The birth rate was 23.0 per 1,000 for 1980-85, with a legitimate rate of 70.2%, a fertility rate of 3.4, and a marriage rate of 6.0. Infant mortality rate per 1,000 live births was 29.5 in 1987; death rate in 1985 was 8.7 for 1980-85; and life expectancy at birth was, in 1985, 68.6 years for males and 75.0 for females.[3] Major causes of death were circulatory diseases (371.9 per 100,000), cancers (148.8), respiratory ailments (51.8), and accidents (42.8).[1,3,4]

Morbidity and mortality owing to infectious diseases are declining. Acute respiratory infections are responsible for around 3% of all deaths and 10% of infant deaths. Acute diarrheal diseases are still among the five leading causes of morbidity and mortality for children under five years, and together with malnutrition, account for a significant number of public hospital discharges. Other diseases under extensive control activities are Chagas' disease, malaria, Argentine hemorrhagic fever, hydatidosis, and leprosy. Social changes and increasing urbanization are exacerbating the importance of noncommunicable diseases (cardiovascular ailments and cancer, etc.).

The daily per capita caloric intake is estimated as 3,195, including 31%

animal products, and is about 119% of the recommended minimum requirements. Studies conducted in 1982 of children under 5 years of age led to development of food supplement programs for them and for relief for families facing acute food shortages.

By the end of 1983, 93% of the urban population had adequate systems for excreta disposal, but only 37% of the rural population has adequate latrine systems. Coverage of disposal services was over 22 million people or 84%, approximately. Waste collection is extended to 80% of the population. Sanitary landfills for final disposal were used, up to 1980, by only six provinces.[1]

Federal-directed food control programs based on common standards are in effect. Provincial and municipal governments have responsibility for executing those programs.

There is a shortage of housing facilities. Rural dwellings are characterized by sanitary deficiencies related to water supply, excreta disposal, environmental conditions, and overcrowding.[1]

3. Public and Private Financing, Health Costs

Lack of complete information is mainly due to the complex structure of the health system.[5] Different types of organizations coexist with widely different characteristics and heterogeneity of financial and accounting procedures. A complex flow of financial transactions occurs within the system. There are three subsectors: public, social security, and private.

In 1985, expenditures in health corresponded to 8.2% of the gross national product. This level of health expenditure is relatively high, but health expenditure was only one-third of the per capita expenditure of countries like the United Kingdom.[6]

The government subsector comprises all health units under the Ministerio de Salud y Accion Social, the 22 different provinces, and all the municipalities. A large proportion of the government budgets are devoted to maintenance of health units, provision for salaries, and implementation of specific programs. In 1985, public or government health expenses were about 22.7% of total health expenditures, of which the federal government was responsible for 4.2%.

The social security subsector is composed of around 300 organizations subdivided into about a dozen different types. The Obras Sociales do not have, in general, health facilities (only about 4% of the units and 5.4% of hospital beds in the country), and consequently, they contract services to provide services. Expenditures reached, in 1985, 39.2% of the total health expenditures in Argentina.

Private health expenditures in 1985 amounted to 38.1% of total expenditures.[6]

There is a tendency towards a slight percentage increase in private expenses and a decrease in government expenditures. Financial difficulties of government and social security health organizations are promoting both private disbursement of funds with utilization of cost-sharing schemes and restriction of expenditures.[1]

4. Facilities (Hospitals, Clinics, etc.)

In 1980, there were 9,642 health units, of which 3,186 had inpatient facilities. The public subsector had 5,123 units, of which 1,334 had inpatient means. The Obras Sociales had 374 units, 114 with admission facilities, and the private subsector had 4,145, with only 1,738 inpatient facilities. In general terms, data reveal a similar proportion of public and private facilities, although since 1969 there was a proportional increase in available beds in the private subsector.[1,7] There was a total of 150,000 hospital beds (1 per 186 people), of which 63.2% belonged to the public facilities, 5.4% to the Obras Sociales, and 31.4% to the private subsector.[1]

The government units are used mostly for infectious diseases and mental (chronic) disorders, as well as for rehabilitation purposes and pediatrics, while the private hospital units are devoted to surgery, obstetrics, geriatrics, psychiatry, ophthamology, and pediatrics. While the private subsector is oriented towards diagnostics and treatment of acute pathologies, the public sector is more prone to be responsive to the spontaneous demand of the population and treatment of chronic ailments. This situation has increased the difficulties of establishing care guidelines, information systems, and rationalization of facilities.

5. Manpower

Estimations for 1984 were 80,100 physicians, 13,809 nurses, 16,696 nursing auxiliaries, 6,416 dentists, and 620 nutritionists. The ratio of doctors to population is very high, but for nurses it is very low.[3] In 1979-80, 46.8% of the staff employed in public health units were professionals, 14.4% technicians, and 37.8% assistant staff, not including administrative, maintenance, and general service personnel. Medical doctors and dentists represent 80% of the professional staff, and only 681 pharmacists were employed, although there were 763 pharmacy positions. Nurses and assistant nurses constitute the largest proportion of the technical and assistant staff. Total staff in public health units numbered about 210,000 (2.1% of the economic active population).[1]

Physicians are distributed irregularly and without much connection to population density and medical needs, especially in rural areas and small towns.

6. The Social Security System

This system is based on Obras Sociales belonging to trade unions and other organizations within the different economic sectors of production, commerce, and services. Health services differ widely in terms of resources, membership, and distribution of units and facilities. Very powerful trade unions, like the Railway Union and the Banking Union, have a very extensive network of health units or services with a wide coverage and high level of medical complexity. Other unions, like the commerce and industry manager association, are small in membership size but have substantial resources and can rely on contracting services of private care organizations to provide medical care of the highest level to their members. Conversely, there are a number of Obras Sociales with small membership but widely dispersed in a number of regions and with few financial resources. Nearly 75% of the population (22 million people) is covered.[7]

The social security system is structured in a very complex form, resulting from the consolidation of the trade union-based system in the 1940s and 1950s. A distinctive characteristic is the implementation of other health-related activities such as housing, sports, and recreation. The social security system is essentially a financing system for health activities and services rendered by third parties such as private health organizations, health units, clinical laboratories, and medical doctors.

A few Obras Sociales are also financed by special contributions obtained from participation in revenues of the corresponding economic sector, such as banking, insurance, and energy. The Fondo Solidario de Redistribucion included in the new legislation will be used to promote similar services in all Obras Sociales, regardless of their structural differences.

7. The Medical Establishment: Entry, Specialists, Referrals

Available data of medical care in Argentina are indicative of a high level of utilization of medical services. A national health survey in 1970 showed 7.1 yearly medical contacts per inhabitant in the metropolitan area of Buenos Aires, 5.2 in the city of Cordoba, 6.3 in Mendoza, 5.2 in

Rosario, and 5.4 in Tucuman. In rural areas and small towns, those numbers were much smaller (on the order of 4.0). Data for 1980 showed a slight decrease in these figures.[5]

Levels of hospital admissions are not that high in comparison with data from Canada, the United States, or other developed countries. In 1969, Buenos Aires had 51 per 1,000 people who had been hospitalized for more than a night in the last 12 months and in 1980, 66 admissions per 1,000 inhabitants. Similar figures were obtained from other urban areas.

A 1970 study completed before the inception of the Obras Sociales regime (Public Law 18.610) showed the high propensity of the Argentine population towards perception of morbidity and procurement of medical services. Economic problems produced moderate restraints in using medical services.

About 66% of medical contacts were performed at private medical offices, constituting the most important way of entry. Only 25% of first contacts were made at public health units and around 10% through the Obras Sociales. These figures do not bear any direct relationship to how medical care was compensated. About 20% of the physician visits required laboratory tests, and 11% required radiological examinations.

There are no precise figures about the importance of general practitioners or family doctors versus specialists regarding first medical contacts and the functioning of referral systems. People, in general, use the services of internists or narrower specialists on a random basis depending on their perception of illness or cultural bias.

8. Payment and Reimbursement

Public hospitals, in general, do not provide for outpatient drug needs, while the rest of their medical services are free of charge. During 1985-87, a special fund (Law No. 23.102—Fondo de Asistencia de Medicamentos [FAM]) permitted the provision of a limited number of essential drugs to public health units at all levels, distributed according to estimations of families under poverty lines and selection of pathologies. It allowed for the implementation of rigorous measures of drug procurement concerning quality control and good manufacturing practices inspections.

Medical, laboratory, and drug services in the Obras Sociales are generally provided through private medical offices and private commercial outlets. Partial payments of services are made directly by the beneficiary to the physician, laboratory, or pharmacy owner, and reimbursement of the rest of the fees is made by the Obras Sociales to providers. There are a number of different administrative procedures concerning this partial payment and reimbursement system, but, in general terms, 70% to 80% of each medical contact and about 50% to 60% of drug expenses are paid for by the Obras Sociales, and the beneficiaries have to bear out direct payments of the differences.

9. Coverage

Data of the Instituto Nacional de Obras Sociales from 1983 revealed a social security health coverage of about 75% of the population, including all types of Obras Sociales (also those not within the NHI system). There is no known coverage system for 25% of the population. One quarter of the people are not beneficiaries of the Obras Sociales or attended by public or nonprofit hospitals or by the private subsector, including prepaid organizations. It is very difficult to estimate the proportion of each of these subgroups although quite a number of them may be users of public facilities, taking into consideration the fact that in 1980, 22% of the economically active population was self-employed.[1]

B. PHARMACEUTICAL INDUSTRY AND DRUG DISTRIBUTION

1. Structure of the Industry

The pharmaceutical industry in Argentina has been characterized as (1) using a high percentage of imported raw materials; (2) having no large sums for research and development; (3) devoting a great proportion of its resources to distribution and sales efforts, and (4) expending substantial amounts for payment of royalties and services.[8]

The domestic Argentine pharmaceutical companies, in contrast with those of many third world countries, have been successful in controlling a larger proportion of the market. In 1987, their market share was 54%.[9]

Industry sales in 1987 were U.S. $1,210,000, slightly lower than in the previous 2 years, reflecting both the economic recession and the continuous devaluation of the national currency. Prescription drug sales amounted to 70% of the units sold and 88.5% of the market, and imports of finished pharmaceutical products were nearly 5% of the market.[9]

One hundred and forty-two nationally-owned companies which account for almost all the nationally held market are characterized by a greater diversification across the different submarkets or therapeutic classes than foreign-owned firms (53 firms).[10]

Sales concentration and reduction of the number of enterprises are in progress. In 1987, the first 20 companies represented 53.4% of sales, and the first 50, almost 90%. In the last few years, a number of transnational firms have closed their plants, and others have transferred their marketing activities to other international or national firms. The American group still remains the most important, with a 21% share of the market, although the European firms have a larger proportion of their world market in Latin America.[9]

Official procurement is still limited ($274.6 million, or less than 25% of the total market in 1987), but there is growing tendency to increase official acquisitions.[9]

2. Research and Development, Imports, Exports, Patents, Licensing

National firms devote few efforts and resources to research activities. A few others have set up small research and development groups, including some related to bioengineering products. A substantial amount of effort is dedicated to galenical development and to synthesis of already patented drugs for local production. The Secretaria de Ciencia y Tecnica of the Ministerio de Educacion in connection with the Ministerio de Salud y Accion Social is also promoting research activities. Incentives for locally-owned firms were limited to general application of reduced tariffs for the importation of machinery and equipment.

By Decreto No. 1856/88, the Government signed an agreement with a section of the nationally-owned enterprises that will have a number of incentives for finished products registration and price fixation. Opposed by professional associations and members of Congress, the "Convenio General para el Desarrollo de la Farmoindustria" will give an important voice to a sector of the industry in planning and organization of government health research and regulatory policies. A joint group of government and industry representatives will supervise industry obligations related to research investment and expenses. For drug substances and finished products research and development, if locally produced, companies must invest 5% of net local sales excluding sales to government. A special tariff will be imposed on imports affecting these products. Fixed levels of exports have to be attained for companies to be eligible for incentives and adjustment of prices.

The problem of transfer pricing of drug substances is also a matter of government concern in Argentina. Prices were found in 1984 to be 143% to 3,700% higher than could be obtained from other sources. Overpricing was estimated at $80 million. Negotiations with industry in 1984-85 resulted in savings of $30 million (87% from transnational firms). Imports of 15 local firms and 39 transnational subsidiaries, 14 of which are American, 6 German, and 3 Swiss companies, were detained.[11] Raw materials imports remain stabilized at $270 million, and local production at $60 million. Taking into account price differentials, local production amounts to almost 20% of industry needs.

Argentina's pharmaceutical patent law recognizes only protection for production processes, not products. The courts went even further by permitting imports of drugs produced elsewhere with processes patented in Argentina.[10] International transfer of technology is regulated in Argentina by Law 22.426 of 1981. The pharmaceutical sectors rounded approximately 42% of contracts between 1982 and 1986 with $41.4 million (estimation of royalties to be paid). Panama was the leading seller of contracts, showing that contracts are used as an international transfer of capital rather than as an authentic means of commercial and technical cooperation. A small number (7%) were related to licensing local production of drug substances, a larger number (42%) for production supervision, and 32% for use of brand names.[12]

Exports of human and veterinary pharmaceuticals in 1987 amounted to $8.7 million, opotherapic products to $6 million, gonadotropins to $4 million, and heparins to $3.4 million. In 1983, 73% of exports were of foreign companies to related companies in other countries.[11]

3. Manufacturing

Argentina is one of the developing countries with an advanced pharmaceutical industry like Brazil, Egypt, India, Mexico, and Korea. The pharmaceutical industry in Argentina produces 99% of the marketed drugs. The degree of technological development achieved by local firms is high, and although many interpretations about its development could be put forward, there is no doubt that it is due to the open competition in terms of new products, launching, quality of products, and qualifications of producers.[10]

4. Advertising and Promotion, Price Regulation

Pharmaceutical companies in Argentina devote a large share of their expenses to promotion. More than 25% of sales are spent on promotional activities by both foreign firms and domestic ones, including sales representatives and distribution of free samples. The major pharmaceutical firms in Argentina have always followed strategies giving priority to the continuous launching of new products. New products are generally duplicative of or combinations of known products or new dosage forms with little therapeutic benefit for consumers. Product innovation is originated almost exclusively in the desire of increasing or maintaining sales. Promotion is consequently a weapon used for improving market share. Regulations concerning drug promotion in Argentina are very simple but not generally enforced. Advertisements of nonprescription drugs are permitted freely, although they should be approved or supervised by health authorities. Proper surveillance has not been enforced, and pharmaceutical firms are free to use their promotional methods. Concerning prescription drugs, informational activities should be limited to scientific and professional journals and to activities of medical representatives, but a series of promotional actions are made through medical congresses and other means. Every pharmaceutical firm is required to submit promotional materials at the time of drug registration, but there is no surveillance to see whether they are actually used.[10]

Median price per unit (at pharmacy level) was $3.24 in 1987-88 and per capita consumption was $47. Enforcement of price regulations was very flexible and more related to industry strategies for product differentiation. Different accounting systems were used by the government in the last 25 years to evaluate industry costs and to determine prices. The pharmaceutical firms responded with strategies of developing new presentations, new formulations, new dosage forms, transference of products, changes of brand names, discontinuation of sales of products of scarce profitability, etc.[11]

Prices are fixed according to Regulation No. 328 (1984) of the Secretaria de Comercio, which classifies products within one company whether they are new, similar, or me-too type. Wholesaler markup is 13% of factory prices, and retail pharmacies' markup is 25%. Drug prices must be printed on packages.

Pharmaceutical companies associated with the newly signed Convenio General para el Desarrollo de la Farmoindustria will have monthly adjustments of prices following special formulae. Prices for new products will be calculated with an admitted profitability of 15% before taxes, and the government has accepted a mechanism of price recomposition for older products.

5. Drug Approval and Government Regulation

Regulatory actions are based on Law 16.463 of 1964 and Decreto No. 9673/64. The Subsecretaria de Regulacion y Control is in charge of all regulatory actions concerning drugs, foods, cosmetics, medical devices, etc. The Direccion Nacional de Medicamentos, Alimentos y Farmacia deals with registration, inspection, and enforcement actions, and the Instituto Nacional de Farmacologia y Bromatologia performs evaluation activities and laboratory studies. Regulatory activities are seriously affected by the dispersion of responsibilities.

According to the federal law, federal regulations and surveillance are exercised when drug production and marketing are held in the Federal District of Buenos Aires or national territories, when interprovincial or interjurisdictional commerce is involved for importation and exportation activities, and when provincial governments require federal cooperation within their own areas. There are a number of provincial regulations concerning plant licensing or even product registration.

All pharmaceutical activities can be carried out only with official authorization in firms and premises previously inspected and licensed under the technical direction of a professional. Products can be marketed if they are in accordance with the National Pharmacopoeia or any other known standards accepted by the health authorities. Legal responsibility lies with the legally appointed company officer and with the technical director of the company. Third-party production, control, packaging, or related activities are only allowed in special cases and under official authorization. Nevertheless, third-party activities are always authorized, and the requirement for previous authorization is not really strictly enforced.

All pharmaceutical products must be registered. Registration is granted when scientific, therapeutic, technical, or economic advantages are demonstrated. Requirements for registration and evaluation procedures are described in Decreto 9763 and a number of regulations and are similar to those established in western countries. There are regulations concerning clinical investigations, but there are no provisions for previous clearance of investigational drugs, and surveillance of clinical investigations is very lax. Research could be conducted in human subjects without any real control by health authorities, research directors, or institutional ethics groups.

Evaluation of applications is made by the Instituto Nacional de Farmacologia y Bromatologia, and registration is granted by the Subsecretaria de Regulacion y Control. The Direccion Nacional de Medicamentos, Alimentos y Farmacia is charged with formal examination of applications and maintenance of the Drug Registry. Registration is granted for five-year periods. Product registration can be withdrawn at any time, based on scientific or medical evaluations. Registration procedures related to the Convenio de Desarrollo de la Industria Farmaquimica, including classification of applications in terms of new chemical entities, new combinations, and new dosage forms, were recently approved, with specific time limits for evaluations reflecting the permanent interest of industry for rapid registration of products.

Federal authorities are also empowered to make factory inspections and analytical surveillance of marketed products. A complete, good manufacturing regulation is still lacking. Compliance with GMPs is consequently lax, and enforcement of inspections findings is still weak in spite of previous experiences of regulatory actions. For instance, in 1985-87, procedures for drug procurement under the Fondo de Asistencia de Medicamentos (FAM) required manufacturers to comply with detailed and specific production and control requirements.

Regulations contain a few provisions in relation to adverse reactions. Side effects, contraindications, and precautions must be stated and described in drug applications, on labels, and in advertising materials. Deficiencies of regulatory actions, lack of a system for adverse effects data collection, and lack of drug information centers have allowed the marketing of prohibited or restricted products in other countries. The medical profession as a whole is not well informed about drug characteristics, and is unaware of the importance of adverse effects characterization and evaluation, and of the importance of epidemiological and utilization studies.

6. Competition: Brand and Generic

In general terms, the Argentine pharmaceutical market is based on branded products and branded generics mostly marketed by national enterprises, but also marketed by international firms. Very few products are marketed as truly generic products, and market share of those products is very small.[10]

The Argentine market is characterized by intensive nonprice competition for multiple source products sold by a number of suppliers. Such products include some licensed or obtained from another international supplier or locally produced.

7. Cost-Containment Activities

In 1985, total expenditures in medicines at the consumer level was about $1,610 million, 2.4% of the gross national product, $52.67 per capita, and 29.6% of total health expenditures. The estimated amount includes homeopathic drugs, medicinal herbs, formulated products, and data from commercial outlets not computed by industry.[6]

This level of drug expenditures in relation to total health expenses is high, like in many developing countries (in Latin America, 21.1%), and much higher than in industrialized countries. Per capita consumption of medicines in Argentina is higher than in any other Latin American country.

Public expenditures in drugs amounted in 1985 to 6.3% of total drug expenses; the Obras Sociales, 26.6%; and the private sector, 67.2%. Since the Obras Sociales covers only 25% and the official subsector only 5% of total drug expenses, any official action for cost containment would have a limited effect.

Analysis of the situation should not be restricted to quantitative aspects, but rather to qualitative consideration of the Argentine pharmaceutical market and the physicians' and consumers' approach to drug utilization. This includes such aspects as therapeutic value of marketed products, overprescription and overutilization of medicinal products, automedication practices, the almost unrestricted sale of prescription drugs, the unavailability of essential drugs to unprotected groups of the population, and the absence of educational measures to promote a rational use of drugs.

C. PHARMACY PRACTICE

1. Dispensing a Prescription: Where, What, Payment Formulary

Physicians are the only ones entitled to prescribe medicines to patients. Pharmaceutical products are scheduled in four classes: (1) psychotropic drugs prescribed in official formularies in triplicate (kept by the prescriber, by the dispenser, and the third is used for pharmacy and drug surveillance programs), (2) other psychoactive drugs and antibiotics, which are dispensed with presentation of prescription retained by the pharmacist, (3) regular prescription drugs, and (4) over-the-counter drugs. A proposal for the reexamination of classification of products with the aim of increasing the number of over-the-counter drugs not subjected to price regulations was included in the recent Convenio para el Fomento de la Industria Farmaquimica.

Medicinal products for hospitalized patients at public health units are free. A number of provinces or local authorities use drug formularies. A national formulary was introduced but is not really in use. There is a very strong and permanent industry pressure to avoid implementation of the national formulary or any other restricted list of drugs (under generic names) for public sector procurement or as a general guideline for hospital management or prescription orientation. Outpatient departments in public hospitals do not generally provide drugs, and prescriptions have to be filled normally in private pharmacies.

Drug payment in the Obras Sociales, both for inpatient and outpatient treatment, is borne by the beneficiary and the health organization, in different proportions according to each organization and the type of treatment. Drugs for hospitalized patients are bought by each contracted hospital, while medicines for outpatients are dispensed at private pharmacies or outlets belonging to the Obras Sociales. There is a great deal of variation in health services among the different Obras Sociales and, consequently, there are great differences in drug coverage, participation in drug benefits, and reimbursement of drug expenses, as well as in cost-containment systems. Per capita expenditures in the social security sector in 1985 varied from $2.44 to $37.15, depending on the health organization, with an average of $19.

2. Record Keeping

No record is kept of patients' use except the ones maintained by doctors in their private practices. No analysis of drug utilization is performed on a routine basis at public or private health units. Record keeping in pharmacies is restricted to prescription of scheduled drugs and the preparation of special formulae. There are no provisions for patients records, and, in general, pharmacists are not trained for that activity. Wholesalers and pharmacies do not participate in programs to detect drug quality problems.

3. Patient Education

Argentina is an overprescribed and overmedicated society with a great deal of irrational automedication of prescription drugs. There are very few patient educational efforts. Only recently, professional pharmacist organizations of the Federal District and the biggest province (Buenos Aires) have become engaged in public media campaigns to induce a rational approach to medication. In a number of public hospitals, physicians and pharmacists are also involved in educational activities, and the Faculty of Pharmacy of the University of Buenos Aires is studying the organization of a drug information center to keep doctors and consumers aware of the newest data regarding drugs.

4. OTCs and Other Classes and Schedules of Drugs, Other Nonpharmacy Outlets

Although regulations do not allow for sales of drugs outside pharmacies, a number of over-the-counter drugs are usually sold at supermarkets and retail shops. Free sale of prescription drugs is allowed, and government, professional organizations, and pharmacists are not taking any stand in this regard.

5. The Use of Pharmacists, Technicians, and Others

The 1980 National Census registered 10,734 pharmacists, of whom 56.6% were women. Data available from 1947 show that the number of private pharmacies remained stable at around 3,200-3,500 people per outlet, with a total of 8,916 pharmacies in 1983. In 1980, there were 2,387 pharmacies or drug deposits in 3,186 public and private health hospitals, clinics, and small units; at federal (29), university (6), provincial (1,015), and municipal (259) units; Obras Sociales (144), or at the armed forces services (173), while the private sector had a total of 698 pharmacies both at profit or beneficent institutions.[13]

Pharmacists employed at those units numbered 668 in 1969, 919 in 1978, and 765 in 1980: 71.5% at public units, 16.0% at the Obras Sociales, and only 12.5% at private clinics, a percentage that remained stable through the years. It is relevant to mention that only 455 health hospitals and clinics had pharmacists in 1980 in spite of the larger number (2,387) with pharmacies or pharmaceutical services of some sort.

A large proportion of the pharmacists were employed by or owners of private pharmaceutical outlets, which employed about 25,000 people in 1983 (around 3 per pharmacy). A number of pharmacists work at clinical laboratories or third-party organizations or are employed by the pharmaceutical industry at the production, quality control, and managerial areas.

The above data reveal the scarce utilization of pharmacists at the institutional level of medical care and probably the utilization of nonprepared assistant staff to carry out pharmaceutical activities.

Hospital pharmacy and related activities, such as clinical pharmacy, are still underdeveloped in Argentina and may require intensive efforts for integration with other health team actions.

6. Pharmacy and Primary Care

No information exists regarding utilization of pharmaceutical elements in the development of primary care activities in Argentina, nor about formulary development and pharmaceutical administrative procedures to be used in small health units devoted to medical attention in rural and small communities.

7. Drug Insurance and Third Parties

In previous sections, references were made to pharmaceutical services and drug payment and reimbursement systems in social security. An average of 50%-60% of drug expenses are either paid directly by the health organizations or reimbursed to beneficiaries, while at public hospitals, inpatient drug requirements are free, and outpatients normally have to get and pay for medicines in private pharmacies.

Regarding the private sector, a predominant outlet for financing drug expenses lately has been prepaid health organizations. A few of these organizations (*mutuales*), are under the supervision of the Instituto Nacional de Accion Mutual. There are no data concerning medical care activities, population covered, and pharmaceutical services by the *mutuales*, although they have about 5 million members, many important health activities, and a long tradition.[6]

Private, commercial prepaid health organizations consist of health insurance systems that generally do not have facilities but contract services with hospitals, clinical laboratories, pharmacies, and physicians. Coverage may be around 2 million people, or 12% of the population. Financing may come from individual membership fees, companies' payments for employees' coverage, and contracts with small but powerful trade unions or Obras Sociales. Estimation of drug expenditures of this group is about 4%-7% of total expenses, or $8.7 million. Beneficiaries are entitled to a variety of drug benefits (from 50% reimbursements to almost free drugs), depending on the insurance company.

D. UNIQUE OR INTERESTING FEATURES OR SPECIAL SITUATIONS IN ARGENTINA

The ethnic population of Argentina is almost completely of European origin, and the growing urban society has discouraged the use of traditional remedies based on nonwestern medical practices. There are still a number of herbal remedies in use. Free automedication practices and economic problems have also promoted the use of traditional herbal drugs. There are a substantial number of herbal sales outlets. Since its inception, the Instituto Nacional de Farmacologia y Bromatologia has been involved in laboratory actions devoted to identification of plants used in popular medicine and the finding of common adulterations.

Homeopathic medicine and homeopathic preparations are also widespread. There are a substantial number of manufacturers of these products and homeopathic pharmacies, although it is difficult to report specific figures to establish percentage participation in medical care and pharmaceutical services.

In the last few years, the use of organ lisates prepared by small production units has become more common, some of them, according to reports, without proper habilitation or registration of products. There are estimations of sales of about 100 million.

E. CONCLUSIONS

In spite of a number of program weaknesses, medical care and drug needs are generally well provided in Argentina. Public, social security, and private attention cover a large majority of the population. There is a need for supplemental programs for the nonprotected population.

Drug consumption and expenditures in Argentina are high due to a variety of coverage and reimbursement systems, industry promotion activities, and the high propensity of the population towards drug utilization and automedication.

Government policies regarding pharmaceutical industry development (research incentives; patents, licensing, and price regulations; generic products; etc.) should be oriented by health priorities.

Pharmacological and clinical studies and programs for medical training and education in drug utilization should be encouraged.

Government surveillance of industry activities (registration of drugs, inspection of facilities, sampling and laboratory analysis of marketed products) needs to be strengthened. Postmarketing actions should be incorporated into official programs (surveillance of promotional activities, adverse effects recollection, and epidemiological studies, etc.).

There is a lack of appropriate pharmaceutical services and pharmacy manpower at all levels of the care system, and adequate actions should be encouraged, such as education of patients and consumers, organization of hospital pharmacies, prescription and dispensing policies, record keeping, etc.

REFERENCES

1. Ministerio de Salud y Accion Social, Organizacion Panamericana de la Salud, eds. Argentina: Descripcion de su Situacion de Salud. Buenos Aires. 1985.

2. Centro Latinoamericano de Demografia, ed. America Latina en el Ano de los 5.000.000.000. Santiago de Chile. 1987.

3. World Health Organization, ed. 1988 World Health Statistics. Geneva. 1988.

4. Organizacion Panamericana de la Salud, ed. Los Servicios de Salud en las Americas. Analisis de Indicadores Basicos, Cuaderno Tecnico No. 14. Washington. 1988.

5. Neri A. Salud y Politica Social. Hachette. Buenos Aires. 1982.

6. Gonzalez Gines G, Abadie P, Lovet JJ, et al. El Gasto en Salud y en Medicamentos. Centro de Estudios de Estado y Sociedad. Buenos Aires. 1988.

7. Mera JA. Politicas de Salud en la Argentina. Hachette. Buenos Aires. 1988.

8. Comision Economica para America Latina y El Caribe, ed. Estudios e Informes de la Cepal 65, La industria farmaceutica y farmaquimica: desarrollo historico y posibilidades futuras. Naciones Unidas. Santiago de Chile. 1987.

9. Industria Farmaceutica Latinoamericana, Alifar-Mercados, Argentina, Ano VII No. 13, pg. 2, Junio de 1988.

10. United Nations Centre on Transnational Corporations, ed. Transnational Corporations in the Pharmaceutical Industry of Developing Countries. United Nations. New York. 1984.

11. Groisman S, Katz J. La Industria Farmaceutica en Argentina. Periodo 1983-1988 in: OMS/OPS/Secretaria de Salud de Mexico, eds. Memorias I Conferencia Latinoamericana sobre Politicas Farmaceuticas y Medicamentos Esenciales. Mexico. 1988.

12. Wetzler G, Ravizzini L, Tedesco AM, et al. Contratos de Transferencia de Tecnologia de la Industria de Productos Farmaceuticos en los ultimos anos. Unpublished report. Instituto Nacional de Tecnologia Industrial. Buenos Aires. 1987.

13. Confederación Farmaceutica Argentina, ed. El Farmaceútico y la Oficinade Farmacia en la Argentina. Buenos Aires. 1988.

Chapter 2

Australia

Robert Davies
Michael Tatchell

A. THE NATIONAL HEALTH CARE SYSTEM

1. History and Background

In 1988, Australians celebrated the 200th anniversary of European settlement. The nation, as we know it today, began in January 1788 when 11 small ships of the First Fleet landed at Sydney Cove. Three-quarters of the first settlers, who numbered some 1,000, were convicts from Britain's overloaded penal system.[1,2] The continent they came to settle dates back to the earliest known land forms on the planet. It housed a rich and diverse aboriginal culture which may have begun 40,000 years before the beginning of the Christian era.

The early history of the colony tells of a struggle for survival in the new land, much of which was either harsh desert or dense scrub. As the Europeans gradually expanded their influence through exploration and the establishment of new settlements, the traditional aboriginal way of life rapidly disappeared.

The first half of the nineteenth century saw a change in the pattern of European settlement. As the transportation of convicts was phased out between 1840 and 1868, immigration of free settlers was encouraged. The discovery of gold in the 1850s led to a dramatic increase in population from 430,000 to almost 1.2 million in 1861.

2. Social and Political Environment

The Commonwealth of Australia came into being on January 1, 1901. One of the main purposes of the federation was to bring the trade and commerce of the previously independent and often quarrelsome colonies into a single trading system. All six colonies were retained as states with individual governments, with the separate federal government taking control over matters affecting Australia as a nation.

The Commonwealth of Australia is now comprised of six states and two internal territories. It is the least populated country on earth and the most highly urbanized. Approximately 25% of the 16.5 million inhabitants, who occupy 7.7 million square kilometers, are immigrants or the children of immigrants.

Responsibility for the provision of health services in Australia is divided between the commonwealth and state governments.[3,4] The system itself is an amalgam of the British and American systems. Though links with the founding British institutions remain strong, medical practice is organized along North American lines, with the predominant mode being fee-for-service practice.

The Australian Constitution grants the commonwealth government powers over certain specified activities, with all other authority vested in the states. Health is not one of these specified areas. However, in practice, the commonwealth's control over the major revenue source—income tax—together with a 1946 constitutional amendment allowing it to provide a range of hospital, medical, dental, pharmaceutical, and other grants to the states on such terms as it thinks fit, enables the commonwealth to influence health policy significantly.

At the national level, the commonwealth government is primarily concerned with the formation of broad national policies. It influences policy-making in health services through its financial arrangements with the state and territory governments, through the provision of benefits and grants to organizations and individuals, and through the regulation of health insurance.

The primary responsibility for planning and provision of health services is with the state and territory governments, with the commonwealth government providing financial assistance to their public hospitals.

3. Public and Private Financing, Health Costs

The health service systems that have developed in the states mix public and private service delivery and public and private finance. Health care is also delivered by local government, semivoluntary agencies, and profit-making nongovernmental organizations.

Health expenditure information for 1986-87 estimates that more than $20 billion was spent on health care in Australia (both current and capital expenditure).[5,6] This is close to $1,300 per head and represents about 7.9% of the gross domestic product (GDP). While the amount spent on health in recent years has risen substantially in dollar terms, when expressed as a percentage of GDP there has been little change over the last decade (Figure 1). Also, relative to spending levels in other comparable nations (e.g., United States, 11%; Sweden, 9.5%; and Canada, 8.5%), Australia's health spending does not appear to be unduly high.

Health services in Australia are funded from three major sources: commonwealth government, state and local government, and the private sector. Principle private sector sources include the various health insurance funds, as well as payments by individual patients.

Since the 1980s, the proportion of health expenditure funded by the commonwealth government has increased, while that funded by the private sector has decreased. This was due to the introduction in February 1984 of Medicare, a universal system of health insurance which provides basic coverage to all Australians for medical and hospital treatment. Medicare is described in detail in Sections A.8 and A.9. In 1986, the commonwealth funded 38% of health expenditure, state and local government 34%, and the private sector 28%.[5]

General hospitals and medical services account for nearly 60% of all Australian health expenditures.[6] The remaining large items are psychiatric hospitals, nursing homes, and out-of-hospital prescribed drugs. Psychiatric hospitals are almost entirely state owned and provide free care. By contrast, long-term care in nursing homes is provided largely by private organizations (of which about half are profit-seeking), supported by a

FIGURE 1. Total health expenditure ($millions) and as a percent GDP 1971-1985, Australia.

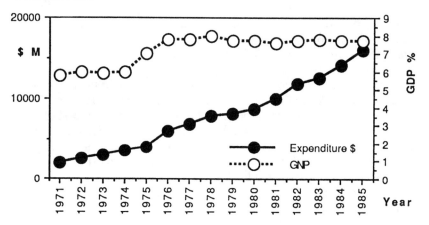

commonwealth government subsidy scheme which is, in effect, an extension of the age pension system.

4. Facilities (Hospitals, Clinics, etc.)

In 1986, there were 1,050 acute care hospitals in Australia, facilities providing short-term surgical and medical services as distinct from longer-term nursing homes and psychiatric institutions.[7] Acute care hospitals provided 84,129 beds, corresponding to a bed supply of 5.3 per 1,000 population. Private hospitals provided 23% of available acute care beds. A further 5.6 long-stay beds per 1,000 population were provided by psychiatric hospitals and nursing homes.

The overall supply of acute care beds in Australia has been falling in recent years, from 6.2 per 1,000 in 1982 to 5.3 per 1,000 in 1986. This reduction, the result of deliberate government policy, has occurred entirely in public hospitals. The supply of private hospital beds has remained almost constant over this period.

The average length of stay for patients in public hospitals has fallen from 10.3 days in 1966 to 6.7 days in 1986. Over the same period, it is estimated that length of stay in private hospitals has fallen from 8.1 days to 5.2 days.

5. Manpower

According to the 1981 census, almost 274,000 persons (3.15% of the total work force) were employed in health occupations.[8] This figure includes medical practitioners, dentists, nurses, pharmacists, and miscellaneous other health workers.

In 1986, there were more than 33,000 doctors in Australia, a ratio of 1 doctor to 466 people. Forty-four percent of these were general practitioners, 34% were specialists and 22% were registered medical officers in hospitals.[9]

The medical profession is one of the largest groups in the health work force. In many Australian cities, there are signs of an oversupply of doctors, which has implications for quality of care and the total cost of care. At the same time, there are shortages in a number of isolated country regions. There is also concern over the high numbers of doctors seeking to train in the more popular specialties while shortages persist in specialties such as geriatrics and mental health.

There has been a declining demand for dentists in Australia for several years, largely because of the effectiveness of fluoridation. In 1986, there were 5,677 dentists in Australia.[9] Like the medical profession, there is still a shortage of dentists in country areas.

Nursing is the largest group of health professionals in Australia. At the time of the census in 1981, there were 160,500 nurses employed in various locations throughout the community, including hospitals, community health centers, medical practices, and dental surgeries.[9]

In 1988, there were approximately 15,000 pharmacists registered with the various state and territory pharmacy boards.[10] Of all pharmacists employed either full-time or part-time, some 85% work in community (retail) pharmacies, 10% in hospital pharmacies, and 5% in other categories such as manufacturing, wholesaling, education, armed forces, or administration. According to 1985 figures, 20.5% of community pharmacy proprietors were women. There are twice as many women as men in hospital pharmacy, while industry attracts approximately equal numbers.

There are currently 84 doctors, all situated in remote country areas, with approval to dispense Pharmaceutical Benefit Scheme prescriptions in Australia.

Pharmacy is a degree course available in each state. It consists of three years of full-time study plus one year of practical training after qualification and before registration. Some 60% of graduates from pharmacy courses are women. If current trends continue, pharmacy will be a female-dominated profession by the year 2000.

Continuing education is not mandatory in Australia, but it is heavily promoted by various associations, particularly by the Pharmaceutical Society of Australia.

6. The Social Security System

The commonwealth government, the state governments, and voluntary welfare organizations all provide social welfare services.[11]

Age pensions are payable to men and women who have reached the ages of 65 and 60, respectively. Payment of these pensions is subject to residence qualifications, an income test, and an assets test. There were 1.3 million age pensioners in Australia in 1987 who were paid a total of $6.3 billion.

Invalid pensions are payable to persons between 16 years of age and pension age who are permanently incapacitated for work to the extent of at least 85%, or permanently blind. Invalid pensions are also subject to residence qualifications, an income test, and an assets test. There were over 289,000 invalid pensioners in Australia in 1987 who were paid a total of $1.9 billion.

Various other welfare programs and social security benefits are available, including unemployment and sickness benefits, widows' pensions, supporting parent's benefits, and family allowances.

7. The Medical Establishment: Entry, Specialists, Referrals

The structure of the medical profession in Australia, in occupational terms, is complex, as it is differentiated at the postgraduate stage into more than 30 specialties. The two principle occupational groups in the medical labor force are general practitioners and medical specialists.[12]

General medical practice is characterized by the provision of primary and continuing medical care to patients. This takes place mainly in the community and is particularly concerned with providing family-oriented health care. Medical practitioners can enter general medical practice on completion of the general requirements for registration. This typically includes six years of full-time study at one of Australia's ten medical schools, followed by two years of internship at a large public hospital.

Most general medical practitioners are in private practice and charge a fee for service.

More than one-third of the medical practitioners in Australia are specialists. These are doctors who have completed four to six years of postgraduate training and have passed the appropriate exams to become physicians, surgeons, anesthetists, obstetricians, psychiatrists, or pathologists. With increasing specialization, most physicians and surgeons are subspecialists within these fields.

The majority of specialists are in private practice and charge a fee for service. To obtain the Medicare subsidy for specialist consultations, patients must have a referral from a general practitioner. This mechanism acts as a filter, as most patients go to their general practitioner before visiting a specialist.

A significant proportion of medical practitioners are found not in the traditional community-based generalist or specialist practice but in full-time salaried occupations, mainly in hospitals throughout Australia.

8. Payment and Reimbursement

Medicare, Australia's universal system of health insurance, was introduced in February 1984 and is partly funded by an income tax.[13] This levy was set at 1% of taxable income until December 1986, when it was increased to 1.25%. All permanent Australian residents and any persons with approval to remain in Australia for more than six months are eligible for Medicare benefits.

Medicare provides:

— Automatic entitlement under a single public health insurance fund to medical and optometry benefits for all Australian residents (for medical services, Medicare covers 85% of the schedule fee)
— Free shared-ward accommodation and inpatient and outpatient treatment at public hospitals
— Subsidies for private treatment in public hospitals.

Private health insurance in Australia is currently confined to private hospital and ancillary insurance, the latter covering such items as dentistry, physiotherapy, and nonsubsidized pharmaceuticals. It is provided by a number of nonprofit private health funds, one government-owned fund, and some for-profit companies. The commonwealth government monitors the operations of the funds, approves contribution rates, and stipulates the amounts which funds must keep in reserve.

9. Coverage

Medicare provides basic coverage for medical and hospital treatment to all permanent Australian residents and any persons with approval to remain in Australia for more than six months.

In addition to the benefits provided under Medicare, a significant proportion of the population chooses to take out additional insurance with one of a number of health insurance funds. In March 1986, some 44% of the population was covered by health funds for private hospital and ancillary costs.[14] A further 3% had been insured against ancillary costs only. The remaining 53% relied on public hospitals for inpatient treatment under Medicare.

B. PHARMACEUTICAL INDUSTRY
AND DRUG DISTRIBUTION

1. Structure of the Industry

The Australian pharmaceutical manufacturing industry is a high-technology industry consisting of more than 150 companies involved in the manufacture and/or supply of pharmaceutical products to the Australian market. Some of the major manufacturers are wholly Australian owned; most are subsidiaries of overseas-based parent companies and are primarily engaged in the formulation and packaging of imported constituents.

Apart from providing products for Australians, international companies make a contribution to the country through transfer of capital, technology, and research and development, and by providing employment. The local subsidiary has access through its parent to new technologies in the manufacturing and packaging of pharmaceutical and in the construction of new production facilities. Skills are transferred to Australia in areas such as specialized building design and applications in engineering, quality control, and analytical chemistry.

The industry employs a high proportion of graduates who have qualifications in disciplines such as medicine; pharmacology; pharmacy; chemistry; microbiology; chemical, mechanical, and electrical engineering; commerce; economics; business administration; computer science; and marketing. Local direct employment numbers approximately 7,000, of whom 16% are graduates. This compares favorably with 4% for manufacturing industry generally.[15]

2. Research and Development, Imports, Exports, Patents, Licensing

Australian research and development expenditure currently amounts to $25 million per annum, and although limited, involves clinical testing of new drugs.[15] Clinical trial research, highly respected internationally, is usually conducted in teaching hospitals.

After extensive research and development, a new product is subjected to a lengthy evaluation and approval process before general marketing. The average approval time of between 91 and 115 weeks is via the Therapeutics Division of the Commonwealth Department of Community Services and Health. It includes product information to be used in conjunction with marketing.

Australian Bureau of Statistics data on the export and import of medical and pharmaceutical products show that Australia incurs a substantial balance-of-payments deficit for these products.[15] The major export markets for Australian pharmaceutical products are New Zealand, Hong Kong, United Kingdom, U.S.A., and Japan.

The total cost to Australia for pharmaceuticals has been falling as a percentage of gross domestic product and now accounts for approximately 8% of total recurrent health expenditure.[5]

3. Manufacturing

More than three-quarters of the pharmaceuticals supplied to the Australian market are manufactured locally. However, less than 15% are manufactured using ingredients produced locally.

Production facilities in Australia are mainly used for the formulation and packaging of drugs. Apart from a limited number of products, the necessary active ingredients are imported from overseas, generally from a local subsidiary's parent company. Other materials and containers are usually purchased locally.

Most companies within the Australian pharmaceutical manufacturing industry earn an average gross margin of less than 40% of sales for human-use pharmaceutical products.[15] Gross margins are at their lowest in the sector relating to the Pharmaceutical Benefits Scheme (PBS).

There has been some decline over the past decade in local manufacturing activity, with the closure of manufacturing capacity by a number of companies. Most manufacturing facilities were built during the 1950s and 1960s when PBS profitability was higher. Low PBS profitability during the 1970s and 1980s has undermined the growth potential of the industry and left it with some excess capacity.

4. Advertising and Promotion, Price Regulation

The advertising and promotion of pharmaceutical products in Australia is subject to a variety of commonwealth and state legislation. These controls are complex and depend upon the schedule into which the drug falls, whether the drug is imported or locally manufactured, and whether it is an originator-brand drug or a generic.[16]

Commonwealth and state legislation prevent the advertising of prescription drugs and Schedule 3 drugs (drugs that can be obtained only from a pharmacist or a medical, dental, or veterinary practitioner) on television, in the lay press, on posters, and in public places. However, no direct controls exist over the content of advertisements in professional journals.

Advertising restrictions do not apply to Schedule 2 items (these are generally sold through pharmacies but may also be supplied by a medical, dental, or veterinary practitioner or a licensed poisons dealer) and non-scheduled products. All advertisements for radio and television are subject to prior approval by the Secretary of the Department of Community Services and Health. A voluntary code for the advertising of therapeutic goods also exists.

The system of price regulation in Australia is dominated by the Pharmaceutical Benefits Scheme (PBS). This is described in more detail in Section C. It is important to note, however, that close to 90% of all prescriptions dispensed in Australia are for products whose prices have been determined under the price regulations of the PBS.

At the time of the introduction of the PBS in the early 1950s, the Department of Health negotiated drug prices with pharmaceutical manufacturers for those items listed under the PBS. United Kingdom prices were used as a guide to the prices that should apply in Australia. This system operated virtually unchanged until the mid-1970s, after which there was a move away from the U.K. reference price system to a cost-based pricing system. Where agreement on product price was not reached between the Department of Health and the manufacturer, a new product would be confined to the private prescription market or not made available in Australia.

The former pricing bureau — the Pharmaceutical Benefits Pricing Bureau — was replaced in January 1988 with an independent Pharmaceutical Benefits Pricing Authority (PBPA). The factors to be considered by the PBPA in setting prices include prices of competing products, cost of manufacture, prescription volumes, prices in overseas markets, and other relevant factors. In addition, the new authority is required to take account of the level of activity being undertaken by the company in Australia, including new investment, production, and research and development.

5. Drug Approval and Government Regulation

Marketing and approval of drugs in Australia is subject to a variety of government regulations. Because of constitutional powers, the commonwealth's control over the marketing of drugs is largely based on its control over imports and its control over which products are listed under the PBS.

The regulatory process is based on the evaluation of human-use pharmaceuticals from the point of view of each product's chemistry, toxicity, safety, and efficacy. The objective of the Department of Health's policy is to ensure "the quality, safety and efficacy of products."[16,17]

The Minister for Community Services and Health has the power to grant marketing approvals, taking into consideration the recommendations of the Australian Drug Evaluation Committee (ADEC). ADEC makes its recommendations on the basis of information provided by companies and the evaluations undertaken by the department. The committee has regard to the safety and efficacy of the new drugs and advises on the conditions under which new products would be marketed.

The approval process has been subject to considerable delays in various stages of the approval route and is a source of major complaint by the industry.

6. Competition: Brand and Generic

Generic products have never enjoyed a large market share in Australia. This is because of the overwhelming propensity of medical practitioners to prescribe brand name products encouraged by extensive advertising and detailing by manufacturers. In addition, pharmacists are prevented by law from substituting generic alternatives when a doctor prescribes by brand name unless the pharmacist obtains the prescribing doctor's permission. Generic substitution can only take place in hospitals where a limited range of drugs is stocked and where supply is commonly by tender.

In an effort to encourage greater use of generics, the government introduced a new pricing arrangement in December 1988 which took effect on April 1, 1989. This required branded products with generic equivalents to be priced within 20¢ of the generic product to retain listing on the Pharmaceutical Benefits Scheme. The new pricing policy is part of an industry package announced by the government, which also includes longer patent terms (now 20 years) and higher prices on other products.

7. Cost-Containment Activities

The procedure for determining drug prices in Australia which has been in existence since the introduction of the Pharmaceutical Benefits Scheme in the early 1950s, has been very successful in keeping down the overall cost of pharmaceuticals. This success has been due to the commonwealth's monopsony control over drug prices and the listing of products on the PBS. The effect of this control is illustrated in Table 1, which shows expenditure on pharmaceuticals expressed as a proportion of total health expenditures

TABLE 1. Expenditure on pharmaceuticals as a proportion of total health expenditure, 1960 to 1985.

YEAR	PERCENT
1960-61	22.5
1969-70	19.1
1974-75	13.5
1979-80	8.4
1984-85	7.8

Source: 5, 6, 18.

since 1960.[18] The proportion of health expenditures spent on pharmaceuticals has fallen from 22.5% (1960-61) to 7.8% (1984-85).

The end result of this sustained period of drug price control has been that drug prices in Australia have for some time been among the lowest in the world. While this result has been of great benefit to consumers and the government, the effect on the pharmaceutical manufacturing industry has been severe. Low drug prices in Australia relative to those in other countries have provided little or no incentive for research and development, and have done little to encourage the development of a viable pharmaceutical manufacturing sector.

Because of a number of factors—including the introduction of higher priced new generation drugs, changing prescribing habits of doctors, a substantial devaluation of the Australian dollar in 1985 and 1986, and a more flexible approach taken by the Health Department in its price negotiations with pharmaceutical manufacturers—drug prices in Australia during the past several years have been rising at a significantly faster rate than previously.

Average prescription prices for the patient have doubled in the past six years, from $5.67 in 1982-83 to $10.38 in 1987-88.[19] The rate of increase in drug prices has changed from an average of 7% per annum during the 1970s to an average of 11% per annum during the 1980s. In 1987, prescription prices increased by 25% and in 1988 by 18%.

The impact of rising drug prices on the cost of the Pharmaceutical Benefits Scheme has been dramatic. The total cost of the scheme, both government and patient contributions, has doubled in the past six years from $549 million in 1982-83 to $1,083 million in 1987-88.

The government's response to this expenditure increase has been to introduce a number of measures aimed at containing the cost of the PBS. These include:

— Introduction of the revised generic pricing policy
— Introduction of controls on the prescribing of high priced items—the "authority" system
— Doubling of allowed maximum quantities on a wide range of prescription items available under the PBS
— Reduction of the margin available to wholesalers as part of the manufacturer's price.

At present, these measures appear to be having the desired effect. Government expenditure estimates indicate that the cost of the PBS in 1988-89 is close to or below the budget figure.

The share of the pharmaceutical dollar going to manufacturers, wholesalers, and pharmacists has changed markedly as a result of the government's recent policy initiatives. In 1988, the proportions were 55% (manufacturers), 39% (pharmacists) and 6% (wholesalers). In contrast, the equivalent 1985 proportions were 45%, 47%, and 8%.

C. PHARMACY PRACTICE

1. Dispensing a Prescription: Where, What, Payment Formulary

Pharmaceutical Benefits Scheme

The first comprehensive government-subsidized prescription scheme in Australia, the Pharmaceutical Benefits Scheme, was implemented in 1950 and expanded in 1960. Introduction of the scheme has had a great impact on the practice of community pharmacy in Australia.

The scheme now covers about 110 million prescriptions annually. In 1987-88 their total value, including copayment, was about $1,182 million. The cost to the government is some $1,046 million, excluding administration costs.[20]

The list of pharmaceutical benefits initially made available as a result of commonwealth legislation was composed of lifesaving and disease-preventing drugs. Subsequent changes have added a wide range of drugs for pensioners and extended the list to the nonpensioner population. The initial list of 139 drugs available under the scheme has progressively increased to more than 1,200 and now accounts for about 90% of prescribed drugs.

Further modifications to the scheme took place in 1960, when the government introduced a patient contribution (copayment).

In 1983, a new concessional charge for each prescription was introduced for health care card holders (mostly unemployed and low income earners), those pensioners who, as the result of a means test, are not entitled to free health care, and the dependents of both groups.

In 1986, the scheme underwent the most extensive modifications since 1960. Safety net arrangements were introduced to protect the chronically ill members of the community. Under the new scheme, general and concessional users contribute toward the cost of prescription drugs up to a maximum agreed price, which is currently $11 and $2.50, respectively. For prescriptions costing more than the maximum patient contribution, the government meets the balance. Pensioners continue to receive items on the scheme free of charge.

Under the safety net arrangement, each family or individual pays for the first 25 dispensed Pharmaceutical Benefit items. All subsequent Pharmaceutical Benefit items are free for both general and concessional users.

A scheme similar to the PBS, the Repatriation Pharmaceutical Benefits Scheme, is conducted for veterans, who receive listed items for no charge.

An independent body, the Pharmaceutical Benefits Advisory Committee, advises the commonwealth minister responsible for administering the scheme on matters concerning the listing of drugs and medicinal preparations as pharmaceutical benefits.

Pharmacists' Remuneration

Pharmacists' remuneration for dispensing pharmaceutical benefits prescriptions is determined through twice yearly hearings by an independent body, the Pharmaceutical Benefits Remuneration Tribunal. The main parties to the tribunal's inquiries are the Pharmacy Guild of Australia and the commonwealth government.

Pharmacists' remuneration derives from four main components:

- The "approved price to pharmacist" of the medication (the price agreed between the manufacturer and the government)
- A mark-up of 25% applied to that price
- A professional fee
- The net effect of surcharges and discounts applied by suppliers to the "agreed price to chemist."

An increased professional fee applies to a small number of extemporaneously-prepared items, and a dangerous drug fee is paid where applicable. Pharmacists in remote areas whose income is below a certain level receive an isolated pharmacy allowance, a small subsidy of $2,000-4,000 paid by the commonwealth.

Pharmacists obtain payment for national health prescriptions through the lodgment of claims, generally on a monthly basis. Payment is usually received within 20-30 days.

The commonwealth's payment system, known as Pharmpay, has been in operation since mid-1984. It has changed significantly in that time with the implementation of a computerized claims transmission system (CTS). Currently, more than 2,800 pharmacies (close to 50%) submit claims on computer diskettes or directly via modem link to the Commonwealth Department of Community Services and Health. Computer software allows pharmacists' claims to be compiled automatically in the dispensary computer. All claims have to be accompanied by the lodgment of all relevant prescription forms.

Pharmacy Numbers

The estimated population of Australia on June 30, 1988 was 16,541,600, and the number of pharmacies was 5,609, giving a population-to-pharmacy ratio of approximately 2,949:1. A relatively stable number of pharmacies in recent years, combined with a steady increase in the population, has meant that the population-to-pharmacy ratio has been rising since the mid-1970s.

Care should be taken when comparing population-to-pharmacy ratios between nations. Such comparisons should take into account the many significant differences that exist between countries in such factors as population density and distribution, licensing requirements, ownership regulations, and alternative distribution arrangements such as dispensing doctors.

Australia, for example, is a vast continent whose population is concentrated for the most part in coastal regions. Sparsely populated rural and outback regions receive their pharmaceutical services from country pharmacies and a small number of dispensing doctors. In 1988, 90 doctors were given approval to dispense Pharmaceutical Benefit prescriptions to patients in isolated areas who have no access to pharmaceutical service from a pharmacy.

Community Pharmacy in Profile

During 1987-88, 100.9 million national health service prescriptions, 9 million prescriptions for Repatriation Pharmaceutical Benefits, an estimated 15 million safety net prescriptions, and 19 million private prescriptions were dispensed.

Total sales through pharmacy currently exceed $2,500 million. Of these, approximately 47% are through the dispensary and 53% through retail sales, including nonprescription therapeutic products. Of the dispensary items, 76% (by sales value) were Pharmaceutical Benefits Scheme prescriptions, 8% Repatriation Pharmaceutical Benefit prescriptions, and around 16% private and safety net prescriptions.[21]

In 1987, the average pharmacy was open 52 hours a week, with the proprietor working a 50-hour week. On average, there are 1.1 proprietors per pharmacy. The average pharmacy occupies an area (excluding storage area) of 119 square meters and dispenses 28,185 prescriptions per year, or close to 100 per day.

The ownership of pharmacies is regulated in each state and territory through pharmacy acts and ordinances. Pharmacy ownership throughout Australia is restricted to pharmacists. The limits on the number of pharmacies in which individual pharmacists may have an interest varies from state to state.

In each state, pharmacy boards (or council in the case of Western Australia) are responsible for administering the state's pharmacy acts. The boards are also responsible for registering and disciplining pharmacists as well as safeguarding the public interest.

Marketing groups offering various levels of services and costs to members have a considerable impact on Australian pharmacy. The groups offer services such as promotion, advertising, store identification, uniforms, staff training, and negotiation with suppliers. There is no unity among groups; they present a varied picture of the pharmacy profession to the public.

Hospital Pharmacy

There are pharmacy departments in most large hospitals in Australia, providing a diverse range of services and employing a total of some 1,500 pharmacists. Many of these pharmacists practice clinical pharmacy, ensuring that patients receive the optimum dose of medicine in the most appropriate manner, that untoward effects and drug interactions are identified and resolved, and that patients and prescribers are presented with easily understood information and support.

Tasks performed by clinical pharmacists include provision of drug information, drug product selection, drug therapy monitoring, patient counseling, education, research, and distribution of drugs throughout the hospital using impress, unit dose, or modified unit dose systems. Other services include nonsterile and sterile manufacturing of pharmaceuticals, including the preparation of intravenous preparations, solutions for total parenteral nutrition, cytotoxic drugs, radio-pharmaceuticals, and other specialized sterile products. Most large hospitals also provide outpatient dispensing services.

Conclusion

Today, community pharmacy in Australia is in a position to deliver a comprehensive basic health care service covering a range of contemporary health issues. It has demonstrated its ability to respond to the demands and pressures of a vast country with a relatively low population density in both urban and rural areas.

The way ahead lies in the maximum utilization of the extensive professional and decentralized health care delivery system that the network of community pharmacies provides across the country. The extent to which this can be achieved depends to a great extent on the ability of representatives of the pharmacy profession and the government to work together towards the common goal of providing economically viable and effective health services to all Australians.

2. Record Keeping

Pharmacists throughout Australia are required to keep a record of the dispensing or supply of prescribed medicines. Such records, which must be maintained at the pharmacy from which the medicine is dispensed, must be made at the time of dispensing or supply.

The pharmacist is also required to endorse and sign the original prescription with other information, including the date of supply, the name and address of the pharmacy, any alterations made to the prescription, and other information regarded as being necessary or which is legally required. A wide range of other records is required to be kept for drugs of addiction and restricted substances under relevant poisons legislation.

The advent of computers has meant that much information previously recorded in writing is now stored in the computer. Close to 75% of community pharmacies in Australia are now computerized. In addition to meeting the normal record keeping requirements, most pharmacists are now able to monitor drug usage by patients and to develop patient profiles. It is also now possible to provide computer-generated drug information leaflets to patients.

3. Patient Education

Pharmacists occupy a unique position in Australia's health care system as the most accessible of all primary health care professionals. As such, they are well placed to provide counseling and advice on drug therapy, on minor ailments and injuries, and more broadly on ways of achieving and maintaining better health.

Patient education by Australian pharmacists takes a number of forms:

— Advice on preventive health and healthy living
— Advice in relation to minor ailments
— Advice on the appropriateness of over-the-counter (OTC) medicines
— Counseling the patient on prescribed drugs:
  • How and when to take the medication

• Drug actions, drug interactions, the explanation of possible side effects, and precautions to be observed when taking the drug
— Advice on diet and dietary habits, nutrition, vitamins, and so on.

The Pharmaceutical Society of Australia has recently introduced a self-care program throughout Australia aimed at encouraging the public to view pharmacists as the primary source of advice and guidance on health promotion, good health maintenance, and self-care. As part of the program, backup material on health and lifestyle matters is available from pharmacies, and referral to other health professionals can be made where appropriate.

4. OTCs and Other Classes and Schedules of Drugs, Other Nonpharmacy Outlets

All pharmaceutical products in Australia are assigned to a Poisons Schedule prior to marketing. The classification of a drug to a particular schedule determines the restrictions to be placed on its availability and advertising together with labeling and packaging requirements. The schedules are based on the Standard for the Uniform Scheduling of Drugs and Poisons and are enforced by state legislation. Due to interstate differences in poisons scheduling legislation, classification of drugs under the various schedules can differ between states.

The Standard for Uniform Scheduling of Drugs and Poisons contains eight components:

Schedule 1. Dangerous substances (e.g., belladonna herb) that can be obtained only from medical practitioners, dentists, veterinary surgeons, pharmacists, or licensed poisons dealers.

Schedule 2. Substances that are dangerous to human life if misused or carelessly handled (e.g., Benadryl Expectorant cough/cold product). These goods are generally sold through pharmacies (by a pharmacist or a person under his/her supervision) but may also be supplied by a medical, dental, or veterinary practitioner or a licensed poisons dealer. Direct customer access to these goods is restricted in some states.

Schedule 3. Substances for therapeutic use that can be obtained only from a pharmacist (or a person under his/her direct supervision) or from a medical, dental, or veterinary practitioner (e.g., insulin). Any person who supplies these substances is required to keep them in a separate part of the premises to which customers do not have access.

Schedule 4. Substances that can be obtained from pharmacies only with an authorization (prescription) from a medical, dental, or veterinary practitioner (e.g., amoxycillin).

Schedule 5. Commonly used substances of a hazardous nature which require caution in handling, use, and storage.

Schedule 6. Substances of a poisonous nature that are commonly used for domestic, agricultural, pastoral, horticultural, veterinary, photographic, or industrial purposes or for the destruction of pests.

Schedule 7. Substances of exceptional danger that require special precautions in manufacture and use and for which special individual labeling and distribution regulations may be required.

Schedule 8. Substances that are addiction-producing, or potentially addiction-producing, including those so classified by the United Nations Organization or its agencies (e.g., morphine).

The range of goods and services provided by Australian pharmacies has followed the British and, to a lesser extent, American models.

Goods range from nonprescription drugs, sickroom supplies, toiletries, cosmetics, and photographic supplies to a diversity of commercial products. Diversification has been accelerated in the last decade as pharmacists seek to replace lost markets and maintain incomes. The abandonment of chemist-only policies by major manufacturers in the late 1970s caused a significant loss of market share from pharmacy to other retail outlets in a number of major product categories such as analgesics, hair care, and vitamins.

In Australia, nonprescription medication is largely restricted to sale through pharmacy. In most states, the more potent nonprescription medications are not freely accessible to the public and require the pharmacist's involvement in their supply and recording.

5. The Use of Technicians and Others

Employment of pharmacy technicians in Australia is restricted to hospitals, where strict guidelines govern the activities that technicians are permitted to perform. Generally, these are functions that involve no judgment on the part of the technician.

The regulations applying to dispensing activities in community pharmacy have prevented the widespread use of technicians in that setting. The law requires pharmacists to dispense all medicinal products in community pharmacy, leaving little opportunity for technicians legally to do so.

Unqualified personnel may carry out nonjudgmental tasks related to dispensing but must do so under the actual personal supervision of a pharmacist. The New South Wales Branch of the Pharmaceutical Society of Australia conducts a one-year course for dispensary assistants. However, the certificate for this course has no legal standing.

6. Pharmacy and Primary Care

Recent years have witnessed a growing emphasis on primary health care, health promotion, and illness prevention in Australia. Much of the reason for this change in direction is due to the ever-increasing cost of institutional care. Hospital services continue to account for the bulk of the resources expended on health care in Australia, and the proportion spent on institutional care is growing. Nevertheless, the services most often used are noninstitutional—doctors, specialists, community pharmacists, and other nonhospital professionals.

Community pharmacists in Australia are the most accessible of the primary health care professionals. Results from the 1983 Australian Health Survey show that:

- Seventy percent of the population took some form of action relating to their health during the two weeks prior to the interview.
- Of these persons, 94% reported taking medicine, 25% reported having a consultation with a doctor, and 9% reported a consultation with a health professional other than a doctor or a dentist.
- The most frequently consulted health professionals, after doctors and dentists, were community pharmacists.[22]

These results underline the importance of the community pharmacist as the most accessible point of first contact in the primary health care system.

7. Drug Insurance and Third Parties

Payment arrangements for pharmaceuticals are dominated by the Pharmaceutical Benefits Scheme, under which most prescription medications are provided free of charge or at a subsidized rate for the whole population.

Under the recently introduced safety net scheme, general patients pay up to a maximum of $11 per item. When more than 25 prescriptions have been dispensed for a family unit in any 1 year, prescriptions for the remainder of the year can be obtained free of charge; the cost is born entirely by the Commonwealth.

As indicated in Section A.8, private health insurance in Australia is currently confined to private hospital and ancillary insurance, the latter covering such items as dentistry, pharmaceuticals, physiotherapy, etc. Typically, private insurance protection for pharmaceuticals covers only those prescriptions not included under the Pharmaceutical Benefits Scheme (i.e., private prescriptions). Health insurance funds reimburse their members for any private prescription costs in excess of $11 per item. Most funds set an upper limit per prescription and an annual limit per person.

D. UNIQUE OR INTERESTING FEATURES
OR SPECIAL SITUATIONS IN AUSTRALIA

Australia is unique for its size and the distribution of its population. Australia is an island continent of vast proportions. It has a land area of 7.7 million square kilometers. It is the lowest, flattest, and, apart from Antarctica, the driest of the continents.

Australia's population of 16.5 million is concentrated in capital and other major cities, mainly on the south and east coasts of the continent. Nearly three-quarters of the population live in the state capitals, the national capital (Canberra), Darwin, and four other major cities of 100,000 or more persons.

This unusual pattern of population distribution has created problems for the efficient location and operation of health services. Providing health services to remote areas and small country towns has lead to the development of unique arrangements such as the flying doctor service. Ten percent of all pharmacies in Australia service small country towns in remote areas. In those areas where it is not viable for a pharmacy to exist, a small number of doctors are given approval to dispense prescriptions.

E. CONCLUSIONS

Australia can claim to be one of the healthiest nations in the world.[23] By 1986, life expectancy at birth for males had climbed to 72.77 years, and for females to 79.13 years, placing Australia about one year behind the populations of Iceland and Japan, which have the longest life spans. Australia's infant mortality rate has continuously declined from 103.6 per 1,000 live births at the turn of the century to a rate of 8.8 per 1,000 in 1986. The world's lowest rate is in Japan: 6 deaths per 1,000 live births.

Death rates from leading causes are falling, in some cases quite dramatically. Between 1972 and 1986, deaths per 100,000 population fell as follows: ischemic heart disease by 22%, stroke by 36%, motor vehicle accidents by 30%, other accidents by 33%, with all causes falling by 15%. In the case of cancer, chronic obstructive airways disease, AIDS, and mental disorders, however, the rates have increased in this period.

Despite these improvements, major challenges persist in the need to improve the health of all Australians. Australia's national vital statistics camouflage large variations in health status within the population. Inequalities exist in the prevalence and incidence of many diseases; the distribution of major disease risk factors such as smoking, hypertension, and obesity; the practice of health-related behavior, such as good nutrition or having a Pap smear; and most importantly, the social and economic circumstances that predispose people to these risk factors and behaviors.

No greater contrast in the extremes of health status can be found in this country than that between aborigines and other Australians. Any complacency about Australia's health being "good enough" seems unwarranted. The health status of aboriginal adults in 1988 appears to be worse on many standard indicators than that of any other population group in the world for whom records are available. The crude mortality rate of aborigines is about four times higher than that of nonaborigines. Aboriginal life expectancy is up to 15 to 20 years less than that of other Australians.

REFERENCES

1. Lloyd G. Pharmacy in Australia. International Pharmacy Journal. 1988; 2:100-102.

2. The Pharmacy Guild of Australia. Pharmacy in Australia. Canberra: PGOA, 1988.

3. Deeble JS. Unscrambling the Omelet: Public and Private Health Care Financing in Australia. In: McLachlan G, Maynard Q, eds. The Public/Private Mix for Health. London: The Nuffield Provincial Hospitals Trust, 1982: 425-465.

4. Sax S. A Strife of Interests – Politics and Policies in Australian Health Services. Sydney: George Allen and Unwin Australia, 1984.

5. Australian Institute of Health. Australian Health Expenditure 1982-83 to 1985-86. Information Bulletin No. 3. Canberra: AGPS, 1988.

6. Australian Institute of Health. Australian Health Expenditure 1979-80 to 1981-82. Canberra: AGPS, 1986.

7. Mathers CD, Harvey R. Provision and Utilization of Acute Care Hospitals in Australia, 1985/86. Paper Presented to Conference of Public Health Association of Australia and New Zealand. Sydney: 1987.

8. Australian Bureau of Statistics. Persons Employed in Health Occupations and Industries. Canberra: ABS, 1984.

9. Grant C, Lapsley HM. The Australian Health Care System, 1986. Australian Studies in Health Service Administration No. 60. Sydney: University of New South Wales, 1987.

10. The Pharmacy Guild of Australia. Pharmacy Manpower in Australia. Canberra: PGOA, 1988.

11. Australian Bureau of Statistics. Year Book Australia 1988. Canberra: ABS, 1988.

12. Commonwealth Department of Health. Handbook on Health Manpower. Canberra: AGPS, 1980.

13. Health Issues Centre. Medicare: A Double-Edged Sword. Melbourne: Health Issues Centre, 1987.

14. Australian Bureau of Statistics. Health Insurance Survey Australia March 1986. Canberra: ABS, 1986.

15. Parry TG, Thwaites RMA. The Pharmaceutical Industry in Australia — A Benchmark Study. Sydney: APMA, 1988.

16. Industries Assistance Commission. Pharmaceutical Products Report. Canberra: AGPS, 1986.

17. Commonwealth Department of Health. Therapeutics in Australia — An Information Handbook. Tasmania: Government Printer, 1986.

18. Tatchell PM. The Importance of Total Cost Containment — Do Expensive Drugs Really Exist? Australian Journal of Hospital Pharmacy. 1988; 18:48-54.

19. Tatchell PM. The Future for Community Pharmacy in Australia. Paper presented to the Biennial Conference of the Pharmaceutical Societies of Australia and New Zealand. Sydney: 1988.

20. Commonwealth of Australia. Budget Statements 1988-89. Canberra: AGPS, 1988.

21. The Pharmacy Guild of Australia. Community Pharmacy in Australia 1988. Canberra: PGOA, 1988.

22. Australian Bureau of Statistics. Australian Health Survey, 1983. Canberra: ABS, 1984.

23. Australian Institute of Health. Australia's Health — First Biennial Report. Canberra: AGPS, 1988.

Chapter 3

Brazil

Jose N. Callegari Lopes
Angelo J. Colombo
Renato Baruffaldi

A. THE NATIONAL HEALTH CARE SYSTEM

Brazil, the largest of the Latin American countries, is more than 8.5 million square kilometers in area, and the population forecasts are 147 million inhabitants for 1989 and 179 million by the end of the century.* The demographic distribution is very heterogeneous (Table 1) through its different regions, and a little predominance of females (0.2%) can be observed.

As Table 1 shows, more than 40% of the population lives in the four states of the southeastern region (Espirito Santo, Minas Gerais, Rio de Janeiro, and Sao Paulo), which represents a little more than 10% of the total area. Also of interest is that the northern region (where Amazonia is) represents more than 40% of the Brazilian territory and has little more than two inhabitants per square kilometer; that is, almost 30 times fewer than in the southeastern region. These two examples show the discrepancies in the distribution of the Brazilian population.

Also to be considered is the migration of the rural population (Table 2), which now represents only 25% of the Brazilian population and by the middle of the century amounted to 65%; this fact also becomes evident when we examine the increase in the number of inhabitants in the largest capital cities (Table 3). Considering that the population in 1985 was almost three times that in 1950, only Rio de Janeiro and Recife did not equal or exceed this proportion.

---

* Forecast according to the 1980 demographic census. In this census the settled population was 199 million inhabitants. Source: IBGE.

TABLE 1. Brazil's territorial and populational data.

Region	Areas (Km²)		Forecast of the Settled Population (1000 Inhabitants)[1]		Demographic Index (inhab./Km²)	
	Total	Relative	1989	2000	1989	2000
Northern	3,581,118	42.07	8,640.2	11,489.7	2.4	3.2
Northeastern	1,548,672	18.20	42,062.1	50,182.1	27.2	32.4
Southeastern	924,935	10.86	64,274.0	78,150.7	69.5	84.5
South	577,723	6.79	22,348.6	26,792.5	38.7	46.4
Center-Western	1,879,455	22.08	10,079.4	12,781.5	5.4	6.8
Brazil	8,511,965	100.00	147,404.3	179,486.5	17.3	21.2

1 Source: IBGE

TABLE 2. Relation between rural and urban populations (1,000 inhabitants).

YEAR Population	1950	1960	1970	1980	1989[1]
Rural	33,161	38,767	41,054	38,566	37,707
Urban	18,782	31,303	52,084	80,436	109,696
Rural/Urban	1.77	1.24	0.79	0.48	0.34

1: Forecast. Source: IBGE

Analysis of the age groups shows that the mean ages are significantly increasing when data from the last decades are compared with the 1990 forecast (Table 4). Average birth rate from 1984 to 1986 was 32.9 per 1,000 inhabitants, while the mortality rate was 6.2. Life expectancy was 60.1 years (1980), and the infant mortality and birth was 87.9/1,000 from 1970 to 1980.

Living conditions vary significantly throughout the different regions of the country, as well as in the cities themselves. As a consequence of the rural migration, a significant part of the population lives on the periphery of the cities without water and sewer systems. Thus, in 1970, only 55% of

TABLE 3. Population in the largest state capitals exceeding one million inhabitants in 1985 (× 1,000).

City YEAR	1950	1980	1985[1]
Sao Paulo	2,198	8,493	10,063
Rio de Janeiro	2,377	5,090	5,603
Belo Horizonte	353	1,780	2,114
Salvador	417	1,501	1,804
Fortaleza	270	1,307	1,582
Recife	525	1,203	1,287
Brasilia	---	1,176	1,567
Porto Alegre	394	1,125	1,272
Curitiba	181	1,024	1,279
Belem	255	933	1,117

1 Estimated settled population according to IBGE.

the urban dwellings were served by water systems and 22% by sewer systems. In 1983, 84% had water systems, but the sewer systems were inadequate. However, in absolute numbers, these data show that 1.3 million dwellings in the northeastern and 1 million in the southeastern regions were not served by the water system.

Some diseases also have significant incidences (1984). Malaria (380,000 cases) is mainly observed in the Amazonia, Chagas' disease (more than 2 million) in the rural areas, and schistosomiases (more than 6 million) mainly in the northeastern and southeastern regions, besides malnutrition (only 30% of the population has adequate caloric diets) and the infectious diarrheas. The differences in the infant mortality rate, which exceeds 120/1,000 in the northeastern region, is 60/1,000 in the southeastern region, and doubles among the low income population, can thus be explained.

Similarly, life expectancy, approximately 70 years among the high income population and less than 55 among the low income population, can also be explained. Average life span is 56.3 for males and 62.8 for females, 12 years less for the northeasterners.

Public health services were significantly altered during the last decades as far as organization was concerned. Until the 1950s, there were federal public institutions that accounted for retirement (pensions) and health ser-

TABLE 4. Settled population according to age groups ($\times$ 1,000).

Year of Census Age Groups*	1900	1950	1960	1970	1980	1990§
00 - 04	3,001.5	8,370.9	11,193.4	13,811.8	16,423.7	18,963.0
05 - 09	2,622.5	7,015.5	10,158.4	13,459.5	14,773.7	17,734.0
10 - 14	2,062.3	6,308.6	8,561.0	11,859.1	14,263.3	16,280.0
15 - 19	1,862.8	5,502.3	7,174.9	10,253.3	13,576.0	24,847.0
20 - 24	1,573.1	4,991.1	6,237.9	8,285.8	11,513.2	13,823.0
25 - 29	1,453.3	4,132.3	5,245.8	6,504.1	9,442.2	13,485.0
30 - 39	2,040.0	6,286.1	8,466.4	10,754.3	14,039.1	21,294.0
40 - 49	1,350.0	4,365.4	5,950.7	8,082.3	10,377.3	13,883.0
50 - 59	771.3	2,650.3	3,753.0	5,228.7	7,250.1	9,536.0
60 - 69	355.2	1,451.5	2,190.6	3,007.6	4,474.5	6,383.0
70 or more	203.2	753.9	1,140.4	1,708.6	2,741.5	4,230.0
Not Known	143.2	116.6	99.0	184.0	120.0	---
Total	17,438.4	51,944.4	70,191.4	93,139.0	19,002.7	150,366.0

* Years
§ Forecast
Source: IBGE

vices; they were the Institutes for Retirement and Pensions (for factory workers, businessmen, bank and railroad employees, etc.). All of them were joined to constitute INPS (National Institute for Social Welfare), which was turned into INAMPS, bound to the Ministries of Health (MS) and Welfare and Social Work (MPAS). On the other hand, state and municipal institutions significantly meet the need in those areas.

The new constitution foresees a unified and decentralized health system (SUDS) with allocation of funds aiming municipal attendance.

The allocation of funds for the public health system is mainly provided by the federal government (60%); the states provide 30% and the municipalities 10%. The resources come mainly from the contributions of users of the welfare and social work systems.

Also significant is the private health service, as far as medicines dispensation is concerned. Whereas patients interned in clinics belonging to the public health system receive free medication, a great part of the population buys medicines in pharmacies (drugstores), nearly always without medical prescription, and pays the total price for them.

Frequently, nonqualified personnel are in charge of dispensation, since the number of pharmacy graduates who dedicate themselves to dispensation is relatively small. Interestingly, the word "pharmacist" is commonly and unduly attributed to the individual who sells drugs in the pharmacies.

In 1986, 21.5% of the people over seven years of age were illiterate, and there were 187,273 elementary schools. Universities offered 430,482 vacancies, most of them in private colleges, and 404,115 students were enrolled in the universities (64.3% in the private ones).

In 1987, 42 institutions offered degrees in pharmacy, some with more than one option of professional formation, according to the working area: public pharmacy (drug dispensation), clinical analysis, pharmaceutical industry, and food industry.

Brazil had 28,972 health institutions in 1985; 58.9% were government-controlled, and the remaining were private. Among the public institutions, 92% provided general attendance, 1,002 were hospitals, and 9,670 were health centers providing medical and sanitary assistance. Among the 11,896 private institutions, 5,132 were hospitals and 6,136 polyclinics. All these health institutions offered 532,282 beds; that is, one bed per 255 inhabitants.

Government provides free assistance. The private system is paid by the patient him/herself, by government assistance institutions, or by the health insurance companies.

In 1985, health professionals working in hospitals were distributed as follows:

GRADUATE PROFESSIONALS IN HOSPITALS

	PUBLIC	PRIVATE	TOTAL
DOCTORS	95,220	103,109	198,329
NURSES	16,049	7,893	23,942
DENTISTS	14,875	10,955	25,830
PHARMACISTS	2,793	2,343	5,136
BIOCHEMIST/			
CLINICAL PATHOLOGIST	1,614	1,854	3,468

It is worthwhile to observe that doctors, *biomedicos* (professionals who work in clinical and pathological analysis) and biochemist-pharmacists are named biochemists, since they work in the areas of clinical analysis and clinical pathology.

In 1989, 36,097 pharmacists were enrolled in the CFF (Federal Council of Pharmacy). Table 5 shows the number of pharmacists enrolled in the regional councils as well as the institutions where pharmacists were supposed to be working.

B. PHARMACEUTICAL INDUSTRY AND DRUG DISTRIBUTION

1. Dispensation and Prescriptions

In Brazil, the commercialization of drugs, medicines, and pharmaceutical raw materials only can be carried by companies and establishments define in Law 5991 of December 17, 1973. The commercialization of related items such as equipment, accessories, diagnostic and analytical products, toiletries, cosmetics, and perfumes, can be done by pharmacies and drugstores and also by specialized establishments.

Dietary products not containing drugs can be marketed by commerce in general. However, the dispensation of medicines is the exclusive right of pharmacies, drugstores, medicine posts, riding units, and medicine dispensatories. The riding units are vehicles that go to distant and less inhabited places. In Amazonia, the means of transportation is boat.

Hotels and establishments of the same sort can provide their guests anodyne medicines in accordance with the list provided by the federal sanitary institution. Included in this list are, basically, the analgesics, the antipyretics, and first-aid articles.

TABLE 5. Number of professionals and establishments enrolled in the various regional councils of pharmacy.

Regional Council	Pharmacists	"Provisionados"	Pharmacy	Drugstore	Hospital Pharmacy	Lab. of Clinical analysis	Farmaceutical industry	other Related	Postos Medicamentos
CRF-1 (Para)	684	19	755	150	16	95	06	15	23
CRF-2 (Ceara)	1,174	44	949	131	182	148	07	01	124
CRF-3 (Pernambuco)	1,203	23	1,486	237	79	58	22	-	-
CRF-4 (Bahia)	1,420	21	1,756	02	21	172	12	13	299
CRF-5 (Goias)	1,140	80	602	632	49	248	23	04	288
CRF-6 (M.Gerais)	4,845	190	1,688	2,118	148	673	125	-	544
CRF-7 (R.Janeiro)	3,849	113	1,592	1,077	11	116	147	92	55
CRF-8 (S.Paulo)	9,119	1,167	3,232	5,281	380	345	298	-	12
CRF-9 (Parana)	2,984	222	2,282	47	206	397	18	84	-
CRF-10 (R.G.Sul)	3,038	177	1,520	702	332	674	51	48	04
CRF-11 (S.Catarina)	1,458	103	1,013	23	293	355	11	25	220
CRF-12 (Maranhao)	632	05	463	253	13	74	12	01	-
CRF-13 (Piaui)	219	05	44	414	38	44	08	02	158
CRF-14 (R.G.Norte)	753	05	346	271	23	109	09	15	43
CRF-15 (Paraiba)	1,103	10	630	55	04	106	04	02	-
CRF-16 (Alagoas)	80	12	188	192	26	12	04	-	37
CRF-17 (Sergipe)	76	18	345	07	03	37	02	01	91
CRF-18 (E.Santo)	580	48	382	251	10	129	06	-	91
CRF-20 (M.G.Sul)	523	32	340	222	52	93	02	06	03
CRF-21 (D.Federal)	510	05	32	380	46	33	05	-	30
CRF-22 (Amazonia)	317	-	31	360	04	50	-	12	07
CRF-23 (M.G.Norte)	390	33	70	900	50	110	-	05	50
TOTAL	36,097	2,332	19,746	13,705	1,986	4,078	772	326	2,079

The dispensation of medicinal plants can be exclusively effected by pharmacies and herbalists, with special attention to packing and botanical classification.

Commercialization, dispensation, representation, or distribution and importation of drugs, medicines, pharmaceutical raw materials, and related items can be performed only by companies and establishments licensed by the competent sanitation institution of each state.

In order that an establishment can operate, a one-year license is issued. This license remains valid in spite of transference of property, alterations in the social contract, or changes in the name of the establishment; however, moves to different places demand new licenses. Suspension of a license will be accomplished through a documented dispatch emitted by the competent authority. The right of defense is asserted by an administrative process to be instituted by the sanitary institution.

A prescription is dispensed when some requirements are met concerning presentation, such as legible and correct identification of the physician, and his/her number of registration and signature. When the product contains narcotics or causes dependency, more requirements have to be met. Although legislation demands that dispensation is accomplished after a prescription, that is not always the case.

The above mentioned rules concerning narcotics come from a series of annexes based in international acts. Annex 1 is based in List IV from the Narcotics Unique Convention, which took place in Vienna in 1961, and in List 1 from the Psychotropic Substances convention in New York in 1971, and corresponds to the narcotics and psychotropics of prescribed use in Brazil. Annex 2 corresponds to List I of the Vienna Convention. Annex 3 considers the substances included in List II and mentions the supervision of narcotics. Annex 4 contains the substances in List III and also deals with the control of narcotics. The remaining are based in Lists from the New York Convention dealing with Psychotropic Substances as follows: Annex 5 corresponds to List II, Annex 6 to List III, and Annex 7 to List IV.

There are other directives that determine the relationship among pharmaceutical specialties that contain narcotics and psychotropic substances and directives that create the notification of prescriptions, standardizing the official prescription book.

2. Registers of Information

The prescriptions containing narcotics and similar substances have their own standardization and will be registered in a specific book. Besides that, the entries of these substances and the filled prescriptions are controlled.

The prescription of magistral and official medicines prepared in the pharmacies are entirely registered in a prescription book, with no erasures.

Code prescriptions can be filled in private pharmacies when prescribed by professionals bound to a hospital unit.

All pharmacies and drugstores must have printed labels for the packages of the filled products. Also necessary are prints with instructions such as: "external use," "internal use," "shake before use," "veterinary use," and "poison."

When the dosage of the prescribed medicine exceeds the pharmacological limits or incompatibilities are detected, the responsible pharmacist must ask the prescribing doctor for confirmation.

The official institution in charge of supervision set rules about:

— Standardization of the registers of stock and sale or dispensation of medicines under special sanitary control
— Minimal stocks of particular drugs, observing the local nosocomial setting
— Medicines and materials reserved for emergencies, including the prophylactic sera.

Brazilians are not educated to adequately use medicines. They often medicate themselves since it is easy to buy medicines without prescription, and it is difficult to obtain a prescription through medical assistance.

3. Homeopathic Pharmacies

This branch of pharmacy is developing significantly in this country, so much so that an edition of the Homeopathic Pharmacopoeia is already available. The homeopathic pharmacy is authorized to manipulate offical and magistral formulae, following the homeopathic pharmacotechnique. The manipulation of formulae not included in the pharmacopoeia depends on ministry approval.

A homeopathic pharmacy can sell medicines and related products in their original packages. A prescription will be necessary to dispense homeopathic medicines containing concentrations of active substances in their maximum pharmacological dosage.

Localities were homeopathic pharmacies are not available can have a post of homeopathic medicines, or their dispensation can be made by the allopathic pharmacies.

4. Responsible Professionals

In Brazil, pharmacies and drugstores must have a responsible professional enrolled in the Regional Council of Pharmacy (CRF). This professional, according to Law 5991, can be a pharmacist in any specialization or any other duly licensed professional who is the responsible professional for the establishment.

The minimum curriculum proposed by the Federal Council of Education predicts the basic formation of a pharmacist in 3.5 years. After the basic information, the professional has two options: industrial pharmacist or biochemist-pharmacist. The curriculum for industry has disciplines aiming at the pharmaceutical industry. The curriculum for the biochemist-pharmacist has two areas: clinical analysis and food industry. The names are self-explanatory, so the curriculum for the former (Option A) is devised for students who want to work in clinical analysis, and the latter (Option B) aims at production, control, supervision, and research in the food area.

According to Table 5 there are more than 2,000 *provisionados* (professionals licensed to be the responsible professionals) registered in the CRFS. The responsible professional must be present during the working hours of the establishment. In case of absence, there must be a substitute, who does not necessarily have to be a licentiate. In spite of that, the substitute should be another pharmacist.

To allow *provisionamento* the law demands that the professional prove that he/she was working in a pharmacy when the law was promulgated and that he/she had been working for ten years in this or any other pharmacy. Another requirement is that he/she is enrolled in the CRF and has a proficiency certificate issued by the sanitary institution in charge of supervision.

To assist the populations in distant localities and to respond to interest on the part of the public, if the need of a pharmacy or drugstore is confirmed and it is not possible to have a pharmacist, the local sanitary institution has licensed establishments under the responsibility of the *practico de farmacia*, *oficial de farmacia*, or other professionals enrolled in the CRF.*

Pharmacies and drugstores will only be allowed to function for 30 days without the assistance of a responsible professional. During this period,

---

* Formerly there were courses to form this kind of professional. When they worked in a pharmacy for more than 10 years, they could be the responsible professional.

magistral and official formulae will not be filled, and medicines under special control shall not be sold.

Injections can be administered by skilled professionals in pharmacies and drugstores when a proper place is available. All Brazilian pharmacies and drugstores provide this kind of service.

The pharmacies can also maintain a clinical analysis laboratory in a separate room when the responsible professional is a biochemist-pharmacist, for the disciplines that characterize the clinical analysis are exclusive of the curriculum of the option "biochemistry."

The riding unit, as well as the medicine post, need not have professional assistance. Each pharmacist can be responsible for two pharmacies, one commercial and one hospital pharmacy.

The *praticos* and *oficials* who were working in pharmacies or were owners or co-owners of a pharmacy until November 11, 1960, will be *provisionados* by the Federal Council and by the CRFs to assume the responsibility of the pharmaceutical establishment. However, they will not be allowed to handle other activities exclusive of the pharmacists.

5. Pharmacy and Primary Care

The pharmacies refer the patient to the health centers of the municipality. SUDS (Unified Decentralized Health System) is being implemented, but the pharmacies are not integrated into this new structure. Depending on the region, some pharmacies provide primary care. On the other hand, SUDS is supposed to provide pharmaceutical care in its units.

Some pharmacies are required to give their help to administer vaccines in the vaccination campaigns.

Pharmacies and drugstores have duty hours and work in shifts to give continuous assistance to the community, obeying rules of the states, federal district, territories, and municipalities.

The internal areas of the pharmacies or drugstores must not be used as consultation rooms or for any other purpose.

In Brazil, there is no insurance for the use of medicines and drugs.

6. Supervision

Sanitary supervision of drugs, medicines, pharmaceutical raw materials, and related items will be performed by the competent institution in the states, federal district, and territories.

When an infraction of the laws or rules is observed, the responsible professionals will be subjected to the sanctions of the penal and administrative legislation, with no loss of the disciplinary action of the judicial regimen to which they are submitted.

Physical analysis is periodically made through sampling of products and materials from pharmaceutical establishments.

If an infraction is detected due to nonobservance of ethical-professional precepts, the supervisors will communicate the fact to the CRF of the jurisdiction.

C. PHARMACY PRACTICE

In Brazil, the pharmaceutical industry is essentially of the transformation type, since research is almost nonexistent, and most raw materials are imported. The lack of research can be explained by a number of factors, among which costs play the most important role. Research is very expensive. The multinational companies have a worldwide field of action, and losses in one country are compensated by profits in other regions. Brazil does not recognize patents in this area, and this hinders transference of technology or authorization for use, since only the multinational companies use their processes.

One must emphasize that the approximately 90 companies that work with foreign capitals in the Brazilian pharmaceutical market account for 85% to 90% of the market, while the almost 400 companies working with national capital share the remaining 10% to 15%.

A very important aspect for the development of this market is advertisement. According to Bruno Cunha, each foreign pharmaceutical industry would have an average expense of $850,000 per year, whereas the national ones would spend about $40,000. These figures clearly show the perspectives of the companies.

A rather negative aspect of the advertisement of medicines is the use of images of athletes or artists.

Another important issue is the number of medicines in the market, contributing to the complexity of the pharmaceutical market. According to the bibliographic references, the Ministry of Health has registered 7,000 pharmaceutical specialties, adding up to 15,000 presentations. Notwithstanding these figures, some authors state that these figures exceed 30,000.

Economically, this sector represents about 1.1% of the internal gross product. Price control depends on the CIP-Interministerial Price Commission (Price Control Agency). Considering that the country is under an inflationary process, with relatively high rates, this commission has been authorizing frequent and practically monthly rises in prices.

To attend the poorer populations, the implementation of the CEME-Center of Medicines was proposed. A series of technical and political moves were made toward the nationalization and institutionalization of an official system to produce and control the quality and distribution of medicines, as well as to support the scientific and technological development of the sector and the improvement of human resources, all necessary to implement the proposition. Originally, a directional plan for medicines was proposed, aiming at adoption of specific moves to support the genuinely Brazilian chemical and pharmaceutical industries considering the condition and the needs of the country. To put the directional plan into operation, three subsystems were instituted: production (integrated by three government industries), quality control (integrated by 11 reference industries), and distribution (integrated by 47 warehouses). The National List of Medicines (RENAME) was instituted as a rationalized form of production and consumption of pharmaceutical products through use of the public service system, offering the people free medicines.

It can be seen that, apart from the existence of the propositions, if the directional plan is considered, the benefits to the population are meaningless, since CEME does not have adequate firms. National raw materials are little used, technical and scientific development has little support, improvement of human resources is not encouraged, and distribution depends on the multinational companies. One can conclude that CEME did not produce the expected results.

As far as production is concerned, one must mention FURP (Foundation for Popular Medicine). FURP produces medicines for public beneficent institutions, asylums, and labor unions. All of FURP's medicines have two basic characteristics. First, they integrate the National List of Essential Medicines (RENAME) with approximately 350 products sufficient and adequate to attend about 90% of the health requirements. Of this list, FURP manufactures 130 items. The second characteristic is that FURP's and RENAME's medicines do not have trademarks. They are known by the active principles they contain to prevent automedication and commercialization.

Law 6360 of September 24, 1976 and Decree 79094 of January 5, 1977 refer to medicines, drugs, sanitary products, pharmaceutical raw materials, related items, cosmetics, perfumes, products for esthetic correction, and other items related to health care, establishing rules concerning manufacture, importation, storage, and packing. Companies that have activities referred to in this legislation must have an authorization by the competent sanitary institution of the Ministry of Health to operate. The states, the federal district, and the territories can create supplementary legislation to license establishments, considering among other aspects:

— Adequate localization, out of the urban areas, of plants that produced biological products or other items that can offer risks of contamination
— Installation of equipment to treat water and sewage in industries working with pathogenic organisms
— Adequate equipment against environmental pollution.

To manufacture products related to health, the assistance of a responsible professional who is legally authorized is compulsory. The companies also must have qualitative and quantitative legally authorized personnel to adequately supervise the several branches of production.

The law and the decree carry definitions for drugs, medicines, pharmaceutical raw materials, related items, dietary products, nutrition, hygiene products, perfumes, cosmetics, sanitary products, additives, raw materials, semielaborated products, labels, packages, fabrication, register of products, authorization, license, report, name, trademark, batch, quality control, quality inspection, purity, previous analysis, control analysis, physical analysis, competent sanitary supervision, institution, official laboratory, company, and establishment.

The Ministry of Health is in charge of approving the pharmaceutical products, as well as authorizing the operation of the establishments. Establishments belonging to the public administration do not need a previous license to operate; however, they must attend to the exigencies concerning installations, equipment, and technical responsibility.

The importation of medicines, drugs, and pharmaceutical raw material also depends on the approval of the Ministry of Health. The importation of products not submitted to a special control regimen, to be used in individual dosages, and with no selling or marketing purposes does not demand authorization.

Imported medicines, except the ones whose commercialization depends on medical prescription, must have in their packages or labels information written in Portuguese about composition, indications, and use, as well as contraindications and precautions, when necessary.

None of the products mentioned in this legislation, including the imported ones, can be industrialized, sold, or consumed before registration in the Ministry of Health for a period of five years. After this period, the register can be renewed successively. Validity for dietary products last two years.

The register of medicines, drugs, and pharmaceutical raw materials, due to their sanitary, medicinal, or prophylactic, curative, palliative, or diagnostic characteristics must also meet the necessary requirements concerning security, efficacy, activity, quality, purity, and innocuousness.

In addition to meeting those requirements of the local legislations, imported drugs, medicines, or raw materials must have been registered in their countries of origin.

Drugs, medicines, and pharmaceutical raw materials containing narcotics or causing physical or psychological dependency are subjected to specific legal exigencies.

Legislation exempts from registration:

— Products whose formulae are registered in the Brazilian Pharmacopoeia, in the codex, or in formulary(ies) accepted by the Ministry of Health
— Homeopathic preparations constituted by simple associations of tinctures or incorporated into solid substances
— Concentrated solutes aiming at extemporaneous obtainment of pharmaceutical preparations
— Products similar to the official whose formulae are not in the Pharmacopoeia or in the formulary(ies) but were approved and authorized by the Ministry of Health.

New products exclusively for experimental use under medical control are exempt from registration. In this case, exemption will last up to three years.

Related materials, equipment, instruments, and accessories used in medicine, dentistry, and related activities, as well as physical education, beauty, and esthetic correction, can only be manufactured, imported, consumed, or sold after decision of the Ministry of Health about the necessity of registration.

Toilet articles, perfumes, and other related items, products for external use or ambient utilization, according to their esthetic, protective, hygienic, or odoriferous purposes are registered as cosmetics if they do not cause harm or skin irritation. One must emphasize the possibility that some cosmetics have therapeutic action, in which case the same exigencies as for medicines are made.

Cosmetics and hygiene products for pediatric use cannot be presented in spray form. They must be exempt from caustic or irritant substances, and the packages must not have bruising parts.

The formulae of insecticides must be elaborated, considering the precautions necessary to manipulate the product. Exigencies are made concerning the form of preparation, the mode of application, toxicity, metabolic alterations in mammals, and indication of antidotes.

Besides the above-mentioned products, dietary products must also be considered, as well as their various purposes, such as meeting special dietary needs, supplementing or enriching the food, and lessening hunger.

Labels, tags, package inserts, and other printed material referring to medicines and cosmetics that contain active substances, or referring to disinfectants whose active agent has to be mentioned by the chemical nomenclature, will present the indications of the substances in the formula, the scientific names of the components, and the quantities expressed by the decimal metric system or by international units. The tags, labels, package inserts, and printed materials must show:

- Names of product, manufacturer, manufacturing company, and address
- Number of registration and initials of the Ministry of Health institution
- Number of the batch and date of manufacture
- Purpose, use, and application
- Way of preparing, when necessary
- Precautions and information about risks of handling, when necessary
- Name of the responsible professional, number of inscription, and initials of the institution.

The labels on packages containing medicines, dietary products, and related materials that only can be sold after medical prescription must exhibit a red stripe with the inscription "Sold under prescription."

Labels of packages containing drugs and medicines must exhibit therapeutic indications, contraindications, side effects, and precautions (when necessary), dosage, use, administrations route, and validity period. Labels of packages of medicines containing narcotics or causing physical or psychological dependency must exhibit a black stripe with the inscription "Sold under prescription." The labels of homeopathic medicines must exhibit the inscription "Brazilian Homeopathic Pharmacopoeia," the pertinent scale, the administration route, and the pharmaceutical presentation.

Advertisement of medicines, drugs, or any other products depending on prescription by physicians or dentist-surgeons, can only be aimed at these professionals, by means of specific publications. For over-the-counter products, the advertisement does not depend on previous authorization of the Ministry of Health, since some requirement are met, as follows:

—The text and figures must not generate false interpretations concerning composition, purpose, use and origin, and must not proclaim unsubstantiated therapeutic properties

—Indications, contraindications, precautions, and warnings must be declared.

Approval by the Ministry of Health of packages internally covered by substances that may alter the effects or cause harm is compulsory.

When the medicines contain narcotics or substances causing physical or psychological addiction, the packages must fit the standards approved by the Ministry of Health.

If special conditions for storage and protection are necessary, transportation must be made with equipment that guarantees purity, security, and efficacy.

The action of sanitary vigilance also includes inspection of the products; the manufacturing companies; distribution, storage, and sale; and vehicles used for transportation.

The responsibility for the inspection belongs to the sanitary vigilance institution of the Ministry of Health when the product passes through the states, when the product is imported or exported, and when sampling is necessary to provide physical and control analysis.

When the products are produced and consumed in the area of jurisdiction, the responsibility for the inspection belongs to the health institutions of the state, federal district, and territories.

The Ministry of Health is in charge of publishing, among other, the lists of:

—Raw materials whose importation depends on previous authorization

—Substances and medicines under special control

—Innocuous substances that can be used in cosmetics, perfumes, toilet articles, and other articles

—Additives for manufacture

—Propellants for sprays

—Allowed or forbidden substances to manufacture packages.

D. UNIQUE OR INTERESTING FEATURES OR SPECIAL SITUATIONS IN BRAZIL

Nontraditional forms of treatment also exist in Brazil, most of them illegal. Among them are:

—*Curandeiragem* performed by lay persons who prescribe poultices containing infusions and mixtures of herbs

—*Pajelanca*, similar to *curandeiragem*, consisting of exotic rituals

—*Benzecao*, in which substances are not used, but which consists of prayers to keep bad spirits away.

Evidently, these forms are most commonly found among the less informed and, consequently, low income population.

Among the nontraditional methods recognized by medicine is acupuncture.

E. CONCLUSIONS

Improvements that can be introduced are:

—Education of the patients to use medicines and to avoid all kinds of self-medication

—Existence of a client register in the pharmacies that, when associated with the doctor's register, would result in a personalized attendance. This should improve the health social work.

—Development of an industry with national characteristics aimed at the Brazilian nosocomial setting

—Give the users of the social welfare system special conditions to buy medicines with varied discounts, according to the necessity (permanent or temporary use) of the medicines. This also would result in a better health social service.

Chapter 4

Canada

John A. Bachynsky

DESCRIPTION OF CANADA

Canada is the largest country in the western hemisphere and the second largest in the world, with almost 10 million square kilometers. A very large proportion of the country has no permanent settlement, with the population located on only about 10% of the land area, predominantly along the border with the United States. Thus, the population is in a very long, thin strip stretching 3,000 miles in a band across North America.

Canada's population of 26 million people live, for the most part, in built-up metropolitan areas having a population of 100,000 or more. There is a rural population of approximately 20% of the population.

The growth of population in Canada has been largely due to immigration from other countries. As a result, there is a diversity of cultures and languages within the country. There are two official languages, English and French. The number of Canadians reporting English as their first language is approximately 62% (1976), while those reporting French is approximately 26%.

There are approximately 300,000 people registered with the federal government as Indians or Eskimos. In addition, there are an equal number of persons of mixed blood. These persons have special status within Canadian society and have a number of programs to deal with their needs; this includes their health needs.

Government of Canada

Canada is a country with a federal government. A central government, the federal government, has legislative jurisdiction over matters of national concern and over those matters not assigned to the provinces. The ten provincial governments are assigned specific areas of jurisdiction.

This division of responsibilities was set out in part by the British North America Act of 1867 and was the basis for confederation, the beginning of Canada as a nation. More recently, in 1981, a constitution was developed for Canada which further clarifies the division of responsibilities.

Canada differs from many other countries in that there is a fusion of executive and legislative powers. Formal executive power is vested in the Queen, whose authority is delegated to the Governor General, her representative. It should be noted that the Queen in this context is the Queen of Canada and not the Queen of England and that her representative, the Governor General, is appointed by her on the recommendation of the Cabinet. Legislative power is vested in the Parliament of Canada, which consists of an appointed upper house (the Senate) and the elected lower house (the House of Commons), as well as the Governor General representing the Queen.

The Prime Minister is the leader of the political party requested by the Governor General to form the government. This really means that he represents the party with the strongest representation in the House of Commons. His position is one of exceptional authority. It is his responsibility to choose a Cabinet. The members of this Cabinet represent the governing committee of the country.

The Senate is an appointed body responsible for the protection of the various provincial, minority, and regional interests in Canada. The members are appointed based on the principle of equal regional representation. The function of the Senate is to act as a court of revision by reviewing Commons bills and by recommending amendments to them.

Canada consists of ten provinces, each with its own government. The form of government is similar to that at the federal level, with the Queen's representative, the lieutenant governor, appointed by the Governor General. The lieutenant governor acts on the advice and with the assistance of his ministry or executive council in the province, which is, in fact, the Cabinet of the provincial government. The legislature of each province is unicameral, consisting of the lieutenant governor and a legislative assembly. In addition to the ten provinces, there are two territories: the Yukon Territories and the Northwest Territories. Each is in the process of moving towards self-government, and each has an elected legislature with powers similar to those in the provinces.

The provinces have acquired a substantial amount of power in the past few decades. This has come about through a number of changes in the Constitution of Canada redefining authority. It has also come about because the areas of education and health are under the control of the provinces, and these have grown very quickly over the past few years. In fact, health expenditures represent 20%-30% of provincial government expenditures.

A. THE NATIONAL HEALTH CARE SYSTEM

1. Public and Private Financing, Health Costs

In Canada, health expenditures almost tripled in the period 1975 to 1985 on a per capita basis. Health expenditures as a proportion of gross national product were 8.62% in 1985. This reflects a continued increase in the proportion of resources devoted to health care. In 1960, health expenditures represented 5.53% of gross national product. In 1970, it had grown to 7.13%, and in 1981 it had grown to 7.73%.

In the period 1975 to 1985, there was an average growth rate of approximately 11% in health care expenditures. The rate of growth in the hospital sector was the least rapid, at 10.2%, in this period. In contrast, the category of drugs, which include prescription and nonprescription drugs, grew at an average annual rate of 12.8%. Physician services grew at 11.3% and dental services at 12.6%.

In terms of the expenditures in the public sector as opposed to the private sector, Canada went through a period in which the proportion of expenditures in the public sector grew from just over 40% during 1960-1975, to 76% in 1975, and it remained at that level until 1985. In 1985, hospitals accounted for 40.4% of total health care expenditures while the category drugs accounted for 10.2% and physician services accounted for 15.7%. This represents a decrease for hospital expenditures, which in 1975 represented 44.5% of expenditures, drugs 8.9%, and physicians 15.7%. The total bill for health care expenditures in 1985 was approximately $40 billion with an average per capita expenditure of $1,568. (See Tables 1 and 2, Figures 1-5.)

2. Pharmacy Manpower

There are approximately 17,000 pharmacists in Canada, which results in a ratio of 1 pharmacist for every 1,500 persons. Although the ratio of pharmacists-to-population has been increasing steadily over the past few decades, there is a shortage of pharmacists in Canada.

Approximately 80% of the pharmacists are employed in community pharmacy, with 15%-20% in hospitals and a small additional number in government, drug programs, industry, and wholesale. The demand for pharmacists is increasing in each of these areas.

TABLE 1. Health expenditures as percentages of the gross national products, Canada and United States, 1960 to 1985.

Percentages, selected years

Year	Canada	United States	Difference (U.S. minus Canada)
1960	5.53	5.22	-0.31
1970	7.13	7.39	0.26
1975	7.24	8.30	1.06
1979	7.22	8.55	1.33
1981	7.73	9.40	1.67
1982 [a]	8.61	10.22	1.61
1985	8.62	10.63	2.01

(a)   Provisional

Source: National Health Expenditures in Canada 1975-1985; Health and Welfare Canada, 1987, p. 6.

TABLE 2. Total health expenditures per person, Canada and United States, 1975 to 1985.

Expenditures per person, selected years

Year	Canada		United States	
	Can. $ per person	% change per year	U.S. $ per person	% change per year
1975	539		590	
		11.0		11.7
1979	817		920	
		15.7		13.6
1982	1,264		1,347	
		7.4		8.5
1985 [a]	1,568		1,721	

(a)   Provisional

Source: National Health Expenditures in Canada 1975-1985; Health and Welfare Canada, 1987, p. 8.

FIGURE 1. Health expenditures as a percentage of gross national product, Canada and United States, 1960 to 1985.

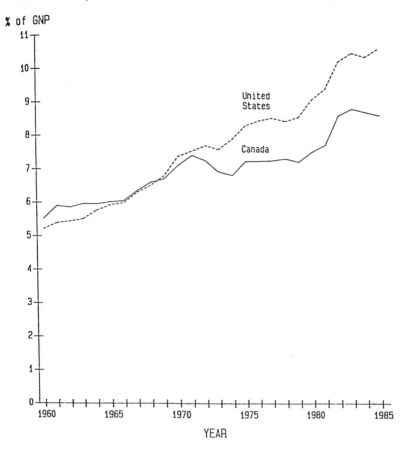

Canadian statistics for 1984 and 1985 and United States' statistics for 1985 are provisional.

Source: National Health Expenditures in Canada 1975-1985; Health and Welfare Canada, 1987, p. 7.

FIGURE 2. Health expenditures per person, Canada and United States, 1975 to 1985.

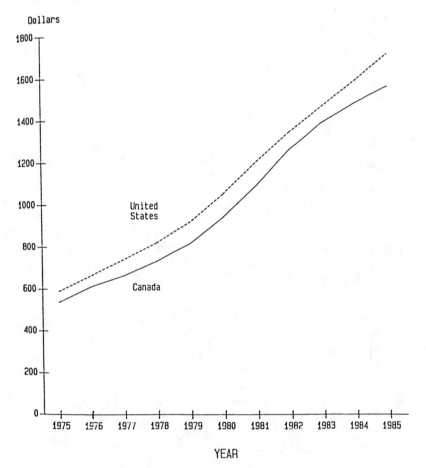

Figures for Canada are in Canadian dollars; figures for the United States are in American dollars. Canadian statistics for 1984 and 1985 and United States' statistics for 1985 are provisional.

Source: National Health Expenditures in Canada 1975-1985; Health and Welfare Canada, 1987, p. 9.

FIGURE 3. Selected categories of health expenditures as percentages of total, 1975 to 1985.

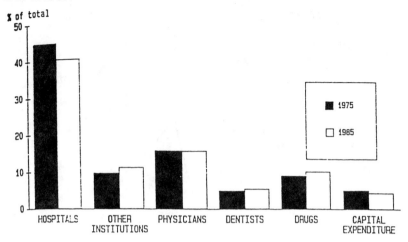

Source: National Health Expenditures in Canada 1975-1985; Health and Welfare Canada, 1987, p. 13.

There are approximately 5,500 pharmacies in Canada. Over the past decade, there has been a rapid growth in the number of chain stores or franchise stores. More recently, a number of grocery outlets have incorporated pharmacies, and these have kept the same long hours that are normal for the grocery store. The result of these changes is that each pharmacy now requires more pharmacists to cover the shifts. There has been an absorption of a large number of pharmacists into these new pharmacies, while the existing pharmacies have continued in business. Growth in the size and volume of pharmacies reflects the growth in prescription volume in Canada, presumably as a result of the aging of the population and the introduction of new products to treat chronic conditions.

In hospitals, the number of pharmacists in hospital pharmacy has almost doubled over the past decade as a result of the more intensive therapy used in acute care hospitals and the expansion in the number of chronic care beds. Some of the new forms of therapy, such as total parenteral nutrition, centralized intravenous admixtures, and unit dose distribution systems, have added a large number of pharmacists to the hospital staff. There is still a large potential for growth of services, however, as less than one-third of the hospitals now have unit dose, and one-third do not supply

FIGURE 4. Percentage distribution, by category, of the total increase in health expenditures from 1975-1985.

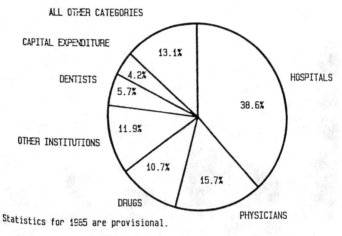

ALL OTHER CATEGORIES

CAPITAL EXPENDITURE

DENTISTS

OTHER INSTITUTIONS

DRUGS

HOSPITALS

PHYSICIANS

13.1%
4.2%
5.7%
11.9%
10.7%
15.7%
38.6%

Statistics for 1965 are provisional.

Source: National Health Expenditures in Canada 1975-1985; Health and Welfare Canada, 1987, p. 13.

clinical services. It is in the area of hospital pharmacy that the most pronounced shortages are seen, particularly in small and rural hospitals.

The pharmaceutical industry is growing rapidly in Canada and is searching for more pharmacists for technical and marketing positions. As the number of clinical trials increases, there is a particular need for pharmacists as clinical research associates.

Supply of Pharmacists

The nine pharmacy schools in Canada graduated 732 students in 1988. Over the past five years, the proportion of female pharmacy graduates consistently has been two-thirds of the total number. As a result, the numbers of male and female pharmacists in Canada are roughly equal, with a higher proportion of females in the younger age group and a higher proportion of males in the older age group.

There are approximately 17,000 pharmacists in Canada (1986), with about 13,500 in community pharmacy, 3,000 in hospital pharmacy, and the balance in industry, government, education, etc.

There is very little immigration of pharmacists into Canada from other countries. Recently, some of the chain stores have been recruiting for

FIGURE 5. Public shares of total health expenditures, Canada and United States, 1960 to 1985.

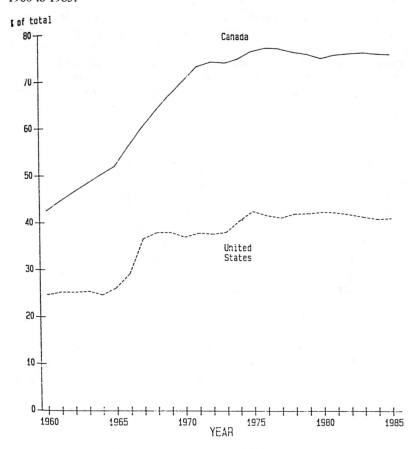

Canadian statistics for 1984 and 1985 and United States' statistics for 1985 are provisional.

Source: National Health Expenditures in Canada 1975-1985; Health and Welfare Canada, 1987, p. 23.

pharmacists in Britain, and the total number of pharmacists coming to Canada from all countries would be about 100 per year.

With the shortage of pharmacists that exists, there is now more emphasis on pharmacy technicians and on pharmacy technician training programs. There are approximately five college-level programs for training technicians in Canada, with the majority of the graduates employed in hospitals. A number of other educational institutions are now looking at developing various pharmacy technician programs, some related to secretarial training programs.

Education Level

The pharmacy schools in Canada have a requirement for a four-year degree program. Over the past decade, a number of the pharmacy schools have introduced a requirement for at least one year of postsecondary education prior to the four years of pharmacy. Currently, six of the pharmacy schools have a requirement for at least one year of postsecondary education prior to entering pharmacy. The other schools are in various stages of meeting this requirement. There are no pharmacy schools in Canada that provide the PharmD degree. A number of schools have clinical masters programs, however, and some have graduated PhDs in the clinical science area.

There is a shortage of pharmacists with advanced degrees in the areas of clinical pharmacy, pharmacy administration, pharmaceutics, and radiopharmacy. The first graduate of a program of graduate studies in the pharmaceutical sciences was in 1961. Now, virtually all pharmacy schools have graduate programs. Most academic appointments are now made from graduates of these programs.

3. Hospital Facilities

The universal coverage of the hospital insurance system in Canada has resulted in a high demand for hospital services and a growth in hospital beds during the period 1965-1985. There have been efforts to reduce the number of acute care beds and to develop more chronic care and nursing home beds as the aging population requires more care.

According to the 1988 *Canadian Hospital Directory* of the Canadian Hospital Association, for 1986-87 there were 1,004 public hospitals with 156,132 beds in Canada, of which about 100 hospitals had over 300 beds. There are virtually no privately owned acute care hospitals in Canada. These hospitals have available to the population approximately 6.08 beds per 1,000 population. It is generally believed that 4 to 5 per 1,000 would be adequate. There was an occupancy rate of 83.1 and an average length of stay of 10.85 days in 1986-1987. Drug expenditures amount to approximately 3% of the total expenditures of these hospitals.

In 1986-1987, there were 2,917 long-term care centers in Canada (*Directory of Long-Term Care Centers in Canada 1988*), of which one-third had under 25 beds. Ownership varies, with private, lay, religious, municipal, provincial, and federal ownership. There are a large number — perhaps half — of the nursing homes owned and operated as private organizations for profit.

4. Social Security System

In keeping with the Canadian system of federalism, most of the social assistance and related programs are the responsibility of the provinces. To assist the provinces in cases where there is a demonstrated need, the federal government will provide assistance. One of the earliest forms of assistance was the conditional grant program, which allowed for cost-sharing for provincially administered programs. One of the first such programs was for old age assistance. In 1937, cost-sharing programs were widened to include assistance to the blind. In 1940, the federal government initiated an Unemployment Insurance Act to deal with the widespread problems of unemployment which had occurred during the Depression. It was at this point that unemployment insurance and income insurance for the elderly would be a federal responsibility and the social services would remain a matter of provincial jurisdiction. A universal Old Age Security program, which provides a pension to all persons over the age of 70, was initiated in 1952. It was complemented by means-tested pension for those age 65 to 70.

Through the 1950s, a number of joint programs were initiated, such as the Blind Persons' Allowance, the Disabled Persons' Act, and the Unemployment Assistance Act. Their unique feature was that the federal government coordinated the social security system by agreeing to pay one-half of the cost of the programs, provided that certain common standards were established in programs throughout Canada. These social arrangements provided a blueprint for the introduction of hospital insurance in 1958, which was the first major cost-sharing arrangement not in the area of income security.

In the 1960s, the federal government introduced a youth allowance to families with children. This provided a monthly allowance for each child. The allowance was at several levels so that older children received a larger amount.

In the 1960s, there was also a good deal of discussion to seek a consensus on a national pension scheme. This resulted in the Canada Pension Plan, which came into effect in 1966, and an amendment to the Old Age Security Act, which would provide direct federal delivery of the Guaranteed Income Supplement. The Guaranteed Income Supplement was an additional payment over and above the Old Age Security payment and was subject to a means tests; that is, people who had limited incomes could request the additional amount if their incomes were below a certain level.

An important administrative change was enacted into law in 1965. This was the Established Programs Act, which permitted the provinces to acquire a share of the federal income tax rather than account for their total expenditures under each of the programs and then claim reimbursement on the basis of the cost-sharing formula. This is now the major mechanism for the financing of expenditures on health and postsecondary education due to a major change in 1976.

Another major social program is the Canada Assistance Plan Act, which was enacted in 1966. It provided for federal cost-sharing toward provincial social service expenditures. Its major feature was that it provided for a needs test rather than a means test as a program requirement to obtain federal cost-sharing. Administration costs were also shared. This funding resulted in a rapid expansion of provincial social services and social assistance programs in the late 1960s.

By the 1970s, the social programs in the provinces had expanded beyond the programs that were cost shared, and provinces were funding programs in areas such as day care and homemaker services, which were not shared. In the past decade, Canada's system of social security has become more complex and less integrated.

In summary, there are social assistance programs and programs of social services for those persons who are in need, normally through unemployment or disability. For children, there are family allowances or youth allowances that help provide income for families. There is also a child tax credit which operates through the income tax system and provides a payment to low income families as a form of negative income tax. For the elderly, there is a universal Old Age Security payment on reaching age 65, with a guaranteed income supplement for low income persons. In addition, there is a Canada Pension Plan which is a mandatory contributor insurance program for all persons who are employed. Finally, there is an Unemployment Insurance Act which provides unemployment benefits to those persons who become unemployed. This is one of the more rapidly growing areas of expenditure and has been subject to a number of changes due to escalating costs and allegations of abuse.

5. The Medical Establishment:
Entry, Specialists, Referrals

There are roughly equal numbers of medical specialists and general practitioners in Canada. Families are encouraged to have a general practitioner to look after their general needs and to have referrals to specialists as the occasion occurs. Specialist services are on referral for episodic health needs. Where there is a need for continuing care, the specialist assumes direct control of treatment. Both specialists and general practitioners have hospital privileges. Referrals may also be made to general practitioners or specialists by pharmacists, dentists, chiropodists, and nurses.

Physicians are normally in a private practice with an office that may be alone or in a group or clinic. These are widely distributed and are easily accessible. Few physicians providing primary care have their offices in a hospital. Hospital-based physicians tend to be specialists in certain areas such as radiology, pathology, or psychiatry. Teaching hospitals will also have specialists of various kinds with offices in a hospital. Overall, these tend to be a very small part of the licensed physicians.

6. Payment and Reimbursement

Provincial governments are responsible for all matters dealing with physicians: licensing, payments, supply, education, and scope of practice. This is accomplished through several mechanisms. First, in each province, a medical act is in place that sets out the regulation and licensing of physicians as a self-governing profession. Disciplinary procedures and penalties are set out in the regulations of these acts, and the regulations are set by the provincial government. In the context of payment and reimbursement, fraudulent billings are dealt with by the physicians' licensing body.

Payment levels in the form of a schedule of fees are negotiated by the provincial medical association. The rate for each service is established, and negotiations take place on an annual basis. The predominant method of payment is fee-for-service, negotiated between the province and the medical association. Other forms of payments such as salary, sessional fees, and contractual arrangements (capitation), also exist but are not significant. Physicians have the choice of participating or not participating. If they choose to participate, they must accept the fee levels as full payment. If they operate outside the program, they can set their own fees but will not be reimbursed either directly or indirectly by the provincial plan. Very few physicians operate outside the provincial plans.

7. Coverage

Virtually all persons living in Canada are covered by health insurance. The Canada Health Act of 1984 sets out the criteria of Canada's health program.

Public Administration. Provincial health programs must be administered by a public authority accountable to the provincial government.

Comprehensive. Provincial programs must cover all necessary hospital and medical services and surgical-dental services rendered in hospitals.

Universality. All residents must be entitled to insured health services on uniform terms and conditions, with residency as the sole criterion for entitlement.

Portability. There is no waiting period for persons moving province to province. Waiting period for new residents may not exceed three months.

Accessibility. Reasonable access to insured services is not to be impeded or prevented by charges or other mechanisms. There are penalties to provinces that allow extra billing by physicians and dentists for insured health services.

B. PHARMACEUTICAL INDUSTRY AND DRUG DISTRIBUTION

1. Structure of the Industry

The pharmaceutical industry in Canada consists of firms that market products based on their own research or research they have acquired through a license, firms that make standard products or manufacture products under a compulsory license, and newer firms in the biotechnology area that produce diagnostic products or other forms of biological products. The firms that are research-based are primarily represented by the Pharmaceutical Manufacturers Association of Canada (PMAC). A majority of the firms are subsidiaries of large multinational corporations with headquarters abroad, primarily in the United States. There are relatively few Canadian-owned firms in this group, and they tend to be smaller firms. The Pharmaceutical Manufacturers Association of Canada has a number of committees and activities and is extremely busy in the setting of standards for drug quality, conducting negotiations regarding regulatory procedures, and acting as liaison with government and universities.

The firms that make standard products or products under a compulsory license are largely grouped in the Canadian Drug Manufacturers Association (CDMA), which has approximately 20 firms. These firms are all Canadian-owned firms, and some of them are quite large in size, with sales over $100 million. The major area of growth has been in the marketing of products manufactured under compulsory license.

A number of firms have emerged in the area of biotechnology and biological products in Canada, predominantly in the area of diagnostics. There are two major Canadian firms in the manufacture of biological products: l'Institut Armand Frappier and Connaught Laboratories. This group has recently formed its own organization of biotechnology firms, Industrial Biotechnology Association of Canada, which has a loose association with PMAC.

There are a number of other firms in related areas that are part of the pharmaceutical industry. They produce fine chemicals as active ingredients or as components of the dosage formulation. Other firms do custom manufacturing or packaging. Some are in the area of drug testing, while others are in the area of medical devices that deliver drugs into the body. The estimated size of the Canadian market for ethical pharmaceuticals at the manufacturers' level is approximately $3 billion in 1989. The total number of pharmaceutical companies in Canada is approximately 140, some of which make nonprescription drugs. Eighty-five percent of the sales are through multinational firms operating in Canada, with Canadian firms selling so-called generics having about 10% of the total market. There are roughly 14,000-15,000 employees in the industry.

The biological and diagnostic companies have sales in the range of just over $100 million and are primarily Canadian-owned operations.

2. Research and Development, Imports, Exports, Patents, Licensing

Total pharmaceutical research in Canada is approximately $90 million per year, which is approximately 5%-6% of sales. The major part of the research has been clinical trials, with only about ten firms having major research activities in Canada involving laboratory facilities.

Research as a percentage of sales has been fairly stable and declined slightly in the period 1969 to 1987 due to the weak patent protection given to pharmaceuticals in Canada. In 1986, the government of Canada announced changes to the Patent Act that would restore a greater degree of patent protection for new pharmaceutical discoveries. In return for this, the pharmaceutical industry has made a commitment to increase the amount of research to 10% of sales by 1996. This is a doubling of research and will result in $3 billion in research over the period 1986 to 1996.

Some examples of the commitments to research that have been made by individual firms are as follows: Merck Frosst has committed approximately $200 million to new research over the decade; Astra Pharmaceuticals has committed $33 million to clinical research over the next 5 years; and Glaxo has a commitment of $50 million on basic and clinical research over the next 5 years and will be building a new $5 million research center.

The rapid expansion of clinical research and fundamental research in Canada will improve the international stature of biomedical research in Canada. The impetus from the pharmaceutical industry will have a substantial impact on universities and university research. Graduates in the biological sciences will have a larger number of career openings in the pharmaceutical industry. The Medical Research Council has initiated funding for joint industry-university projects, and a number of these have been approved.

Patents and Licensing

In Canada, under Section 41 of the Patent Act, it is possible for a firm to apply for a compulsory license. This compulsory license can be used to import the raw material from abroad or to manufacture the active ingredients in Canada. In the period 1969 to 1986, several firms actively pursued a course of obtaining compulsory licenses and importing active ingredients for manufacture of dosage forms in Canada. This was done for a select group of high-volume products and resulted in the so-called generic firms generating a very rapid increase in sales (The products were not sold under the generic name of the drug but under the brand name of the firm so that the firms were not generic drug-producing firms in a strict sense of the word.)

The compulsory licensing provisions resulted in innovative firms marketing a product, achieving a fairly high sales volume, gearing up for the manufacture of that volume in Canada, and then suddenly losing one-half or more of their market to products made and sold under compulsory licensing. In the period 1980 to 1985, the length of time during which firms could develop a market for a product and receive a return on investment rapidly shrank as compulsory licenses were issued. This meant that a firm might have as little as two years on the market before a competitive product was introduced. Firms reacted to this high-risk situation by setting their initial prices at a higher level to recoup some of their investment. In turn, this discouraged investment in Canada in physical facilities to manufacture products. The decrease in Canadian manufacturing and increased imports of dosage forms, particularly newer dosage forms, made Canada less self-reliant in pharmaceuticals so that by 1985, imported pharmaceuticals were approximately one-third of the total manufacturers' sales.

It was the decline in investment, manufacturing, and research and the appearance of a drug lag that initiated concern in the biomedical community. At the political level, representatives from countries that had multinational firms operating in Canada also made formal protest to the Canadian government, particularly the U.S. government. To deal with these problems and to stimulate research in Canada, the government initiated amendments to the Patent Act that would allow up to ten years of exclusivity for those firms that conducted substantial research in Canada. This new legislation did not alter the compulsory license provision. It simply delayed the marketing of drugs under a compulsory license.

Another cost of the increased patent protection was the establishment of a Patented Medicines Price Review Board. This board was set up and authorized to review the price at which patented pharmaceuticals were sold in Canada. There is a requirement that the price increases for pharmaceuticals not exceed the consumer price index of all items. This price review board has very strong powers and can rescind the patent on a product that is sold at an excessive price and can, as a penalty, remove the patent protection of one other product of the same firm. This procedure has placed a major constraint on the pricing of pharmaceuticals, even the entry price. Prior to 1988, there had been a very rapid increase in the price of pharmaceuticals as measured by the Consumer Price Index, and this mechanism is seen as an important method of controlling the price of pharmaceuticals in Canada.

3. Manufacturing

The pharmaceutical industry in Canada prepares over 90% of the dosage forms that are used in the country. There are state-of-the-art manufacturing facilities operated by most of the major firms in Canada. Over the past decade, there has been a trend by firms toward the import of products in whole or in part as a result of patent legislation. It will take some time for this situation to reverse itself.

The manufacturing facilities must meet good manufacturing practice standards (GMP), and they are inspected periodically by the Food and Drug Inspectors of Health Protection Branch. The facilities meet high international standards, and there are very few problems due to quality of the product or to contamination of products.

4. Advertising and Promotion, Price Regulation

Ethical pharmaceuticals are promoted to the health professions through journal advertising, direct mail, sales representatives, and audio- or video-tape messages. Any promotional material indicating that drugs are to be used for a certain purpose can only advertise these claims in the context of the product monograph which has been cleared through the Health Protection Branch. Firms also face the requirement that all advertising is screened by the Pharmaceutical Advertising Advisory Board, an independent Board that must approve advertising before the medical journals will accept it for publication. This is a mechanism that was voluntarily implemented by the pharmaceutical industry to ensure that the advertisements were accurate, in good taste, and met the ethical standards of the industry.

The Pharmaceutical Manufacturers Association of Canada adopted, in February 1988, a code of marketing practices which sets out guidelines for various kinds of promotional activities, including samples, continuing education, convention or clinic displays, and service-oriented items.

These measures have resulted in a system of pharmaceutical promotion that generates much less negative comment and which appears to be working well. There is still some irritation with respect to samples; however, physician signatures for samples has been in existence for some time, and the flagrant abuse that once occurred is rarely seen.

5. Drug Approval and Government Regulation

Any drug to be marketed in Canada must first be submitted to the Health Protection Branch for review. The review process will examine the documentation as to the safety and ethicality of the product, and if the claims are supported and the drug appears to be reasonably safe, a notice of compliance will be issued. This indicates that the drug complies with the regulations. As in other countries, this process requires a very large volume of documentation and tends to be lengthy.

For changes in a product monograph, a firm would submit supplemental new drug applications that would document and request certain changes. A similar process exists for investigational drugs. Where a drug has been marketed for some time, a copy of the drug can be marketed with a simplified procedure in which the manufacturer need only show documented evidence of bioequivalence and a suitable dosage form in terms of formulation. This process has increased the rate at which generic copies of products are marketed. There is a first-in, first-out system in operation in which the submissions are handled in the order they are received. As a result, the submissions for copies of existing products are given the same priority as new innovative compounds that are advances in therapy. Currently, no fast track system is in operation.

Regulation of drugs by the federal government is achieved by assigning them to various categories, such as narcotics, controlled drugs (amphetamines, barbiturates, etc.), and prescription drugs. The classification of drugs into these various schedules is used by the provincial pharmacy licensing bodies to control the sale and distribution of these products within each province. In effect, each province then has legislation with respect to drug controls which parallel the federal legislation.

6. Competition: Brand and Generic

Pharmaceutical products are introduced to the marketplace by the innovative drug manufacturers under a brand name, and the product is promoted by this name. Where the product has high sales volume and a dosage form that can be easily copied, generic firms will seek a compulsory license and sell their own brand product in competition with the originator's product. What is significant is that these alternate brands become available quite soon after a product is marketed, and there is competition based primarily on price due to the structure of the market.

Within each of the provinces, legislation has been enacted (beginning in 1962) which authorizes the pharmacist to dispense brands other than those prescribed. This is fostered by third-party repayment programs (primarily governmental), which will reimburse pharmacists only at the price of the lower-cost items. In the case of hospitals, a bidding process is used, and the lower-priced products tend to be purchased for a group of hospitals. These procedures, when taken in company with the introduction of compulsory licensed products, have forced down the price of many popular products in Canada. The leading 30 or 40 products are sold in a price-competitive environment. The physicians, to some extent, have also become involved in this, and many of them exert their prerogative to write in their own handwriting "no substitution," which will enable them to have the particular brand they desire dispensed to the patient with full coverage by the third party.

7. Cost-Containment Activities

In most provinces, the government itself is a major payor for benefits to the elderly, the social assistance recipients, government wards, and persons in certain disease categories. To contain costs, some provinces have insisted on repayment programs based on the lower-cost products and, in some cases, a requirement that the pharmacist must select a lower-priced product. Group purchasing by the federal and provincial governments has also been encouraged to reduce expenditures on pharmaceuticals.

C. PHARMACY PRACTICE

1. Dispensing a Prescription: Where, What, Payment Formulary

The prescriptions written by physicians for patients in Canada are normally dispensed by pharmacists in community pharmacies. It is now firmly established that the patients have the right to take their prescription order to whatever pharmacy they wish. It is considered unethical for physicians to steer patients to a pharmacy or to dispense medication to them for a profit.

The patients will normally take prescriptions to a pharmacy located near their homes. There are pharmacies distributed throughout Canada in all the areas with a significant population. As a result, no individual has to go very far to find a pharmacy at which pharmaceutical services could be obtained. There is a general tendency for people to deal primarily in one pharmacy and to get virtually all of their medication at that pharmacy.

Some pharmacies are located in buildings that contain a large number of medical practitioners. In these cases, many patients may have their medications dispensed in the building and have future prescriptions either dispensed there or transferred to another pharmacy closer to their homes. There is an easy process of moving prescription orders from one pharmacy to another at the request of the patient, provided the prescriptions are not for narcotics.

Each pharmacy keeps a wide selection of pharmaceutical products (5,000-10,000) and can dispense most of the prescriptions received from the stock available. Normally, the physicians in a particular area will have a group of products that they prefer, and this will form the usual products stocked by the pharmacy. Pharmacies generally have access to drug wholesalers where they can obtain medication within 24 hours.

When the prescription medication is obtained by the patient, the patient will normally pay for it and, if he/she has some form of third-party coverage, later claim reimbursement. For some individuals, particularly those covered by a government program, charges are made by the pharmacy directly to the government unit, and the patient either does not pay for the medication or pays only a portion of the cost.

Payment for prescriptions is based on a formula of the cost of the medication plus a dispensing fee. This system has now been in place for approximately 20 years and is generally accepted in Canada. Increasingly, the cost of the medication is the actual acquisition cost of the medication plus a fee that is negotiated with the provincial government in each province.

In each province, the government has differing involvement in the pharmaceutical services provided. In some provinces, there is a list of benefit drugs that can be provided to recipients in that province, while in other provinces, there is very little government involvement. One general trend that is emerging is that where various brands of a product are available, the government programs will normally reimburse the pharmacist at the lower-priced level.

2. Record Keeping

The key in record keeping is the original prescription order. In most cases, this is a prescription order written by the physician and carried by the patient to a pharmacy to have the medication dispensed. For most drugs, it is also possible for the physician to telephone the prescription drug order to the pharmacy where the pharmacist will make out an order and place it on file. This record is used primarily to account for medication dispensed, to ensure that pharmacists comply with the regulations, and to maintain a patient profile.

With the widespread use of computers, now thought to be in over 80% of all pharmacies, most of the record keeping is done on computer. This enables the rapid retrieval of information and a regular update of the patient's profile. Patient profiles are a requirement in most provinces, and they are now a standard feature on most computer software systems.

The Department of National Health and Welfare, which is responsible for the Narcotic Control Act in Canada, requires that pharmacists maintain a record of all narcotics received. This is used to check against the prescription issues of narcotics.

3. Patient Education

The pharmacist's role in counseling patients about their prescription medication is one of the key elements in patient education. It is generally accepted that pharmacists should counsel their patients on medication when a prescription is dispensed. This expectation is not always fully met, as some patients have had medication dispensed over a period of years and are well acquainted with the products and their use. In other cases, the pharmacist may be too busy or the prescription may be delivered, in which case the counseling is not provided.

In recognition of the importance of patient counseling and the difficulty of conveying accurate information to the patient in a timely fashion, there has developed the use of printed information about prescription drugs which is distributed to the patient. These are distributed either as patient leaflets that are developed by companies or pharmacy organizations and given out with the prescription or, as is more commonly done now, the computer will print out along with the prescription label some appropriate information for the patient.

The majority of pharmacies in the last few years have established patient education centers in the waiting area of the pharmacy. These usually consist of a rack that contains brochures and pamphlets dealing with diseases or forms of therapy for various diseases. There is often information on lifestyle health matters as well. Pharmacies will also carry health magazines and books for the benefit of their patients.

There is now a growing tendency for pharmacists to become more active in the community and to give talks to community groups or school children on matters dealing with drug abuse, the proper use of drugs, ways to maintain or improve health status, and diseases. The curricula in the pharmacy schools is oriented towards fostering this type of involvement in health promotion.

4. OTCs and Other Classes and Schedules of Drugs, Other Nonpharmacy Outlets

Drugs other than prescription drugs are referred to as nonprescription drugs. They are categorized into three groups. The first is a group of drugs that are sold in pharmacies but are not placed on self-service display and must be requested from the pharmacist or pharmacy staff. The next category of drugs are those that are restricted to sale in pharmacies only although they can be advertised and made available for self-service. The final category of drugs are those that have no restrictions on their sale and can be sold through any outlet.

These three classes of drugs tend to vary from province to province. To achieve a greater degree of uniformity, the Canadian Drug Advisory Committee has been formed through the Canadian Pharmaceutical Association in an attempt to achieve a more rational and consistent framework for classifying drugs. In this process, the federal government has little involvement. They do, however, assign two types of drug identification codes. One is referred to as the DIN (drug identification number), which is normally used for prescription and nonprescription drugs restricted to pharmacies. The other type of number is GP (general public), which is assigned to drugs available through all outlets without restriction. It is intended that, over time, the classifications of DIN and GP will fit the provincial schedules more closely and that there will be greater liaison between organized pharmacy and the federal government.

The movement of drugs from prescription to nonprescription status is strengthening the development of various schedules or categories of drugs. It is now generally accepted that any drug coming off prescription-only status would be available first in pharmacies under the direct control and supervision of the pharmacist for a period of one or two years and then move to an appropriate schedule based on the experience at that time. This has been the experience to date with the drugs hydrocortisone and ibuprofen.

The one contentious area with respect to drug schedules involves the classification of nutritional products. Although vitamins are labeled for therapeutic use only at higher concentrations, their widespread sale is allowed due to the pressure of lobby groups and patients who believe that it is important to have access to these materials as nutritional substances. There are continual fads involving the use of high-dosage vitamins and there is continuing concern over the potential toxicity of some of these substances.

The use of the various schedules in Canada has enabled a wide variety of effective products to be made available to the public under the control of the profession. Some of the medications that are now available without a prescription include acetylsalicylic acid with codeine 8 mg, antihistamines, drugs for the treatment of pinworm, hydrocortisone, ibuprofen, and benzoyl peroxide.

5. The Use of Pharmacists, Technicians and Others

Pharmacies in Canada have become larger both in terms of the floor area and number of staff. Over a period of 25 years, the total number of pharmacies has grown appreciably, while the number of pharmacists and the general population has grown at a greater rate. As a result, the once common one-man pharmacy has virtually disappeared. Pharmacies are now staffed with several pharmacists and a number of support staff. Prescriptions account for 30% to 45% of total sales volume, on average. This percentage has increased year by year even though the total volume of sales has increased, as has the product mix.

As the prescription volume has increased in each pharmacy, there has been an increase in the number of pharmacists employed and also an attempt to improve productivity by the use of computers and nonprofessional staff, particularly technicians.

The provincial licensing bodies have set out the duties and responsibilities of the pharmacist. These are responsibilities that cannot be delegated to technicians or other staff. There are other duties that can be delegated. There is also an attempt to control the use of nonprofessional staff through having ratios of the number of support staff that can be employed in the dispensary. In most cases, this is a ratio of one-to-one or, in some cases, one pharmacist to two nonprofessional staff.

Pharmacy technicians are normally trained on the job; however, there are five formal education programs in Canada: four in Ontario and one in the province of Alberta. These formal training programs are normally approximately one year in duration, and there are several shorter courses offered by business schools. There are now attempts to have various kinds of distance education to provide a higher level of instruction to nonprofessional staff.

With larger volume pharmacies and a greater variety of products, there is an increasing move towards specialization for some of the staff in the pharmacy. In some stores, cosmeticians with formal training are employed. The area of home care products requires staff with special skills. In very large stores, there is often a merchandising manager — a nonpharmacist — who is responsible for the general merchandise in the pharmacy.

6. Pharmacy and Primary Care

There is a high degree of trust in pharmacists by the general public, and there is a greater use of pharmacists in providing primary care as a result of several factors. First, there is a wide variety of effective products that are available to the public without professional control in the form of prescribing. These products are available to the public either on a self-serve basis or in consultation with the pharmacist. A second factor is the growing acceptance and desire on the part of the public to treat their own conditions. This self-treatment is encouraged by the pharmacy profession, the government, and a good many lay publications. A third factor is the ease of accessibility to pharmacies due to their widespread geographical locations and their extended hours of service. In contrast, it is increasingly difficult to obtain physician services, as an appointment is necessary, and several days may elapse before a patient has access to the physician. Once an appointment is made, it is not unusual for a patient to wait several hours to see the physician and this time plus the travel time imposes a substantial time commitment on the patient. There is very little financial barrier to physician services in Canada, and this may account for the high level of use of physician services as well as the difficulty in obtaining an appointment within a short time period. One result of the difficulty in getting in to see physicians on short notice is the tendency of the public to seek emergency health in hospitals for a number of nonserious conditions.

Pharmacists are playing a larger role in the monitoring of the patient's condition and assisting the patient with obtaining optimal drug therapy. For example, in the areas of hypertension and diabetes, pharmacists will work with the patient to evaluate his/her level of control over the disease and to regulate the dosage or medications used to obtain the best possible results. This will often involve pharmacists taking the blood pressure of the patient during a visit to the pharmacy or selling the patient a glucose monitor so that the patient can measure blood glucose and adjust the dosage of the drug accordingly.

One measure of the extent to which pharmacists are involved in primary care is the rapid growth in the use of nonprescription drugs relative to the use of prescription drugs. The vast majority of nonprescription drugs (75%-80%) are sold through pharmacies and often involve discussion with the pharmacist. This is particularly the case where patients are taking several prescription medications and require some advice for guidance on the use of the products. In a 1988 study asking people the purpose of their last visit to a pharmacy, 23% stated that it was to buy a nonprescription drug, 50% to obtain prescription medication, and 33% to buy a nondrug item.

7. Drug Insurance and Third Parties

The forms of drug benefits that are most influential in Canada are the provincial drug benefit programs for the elderly and those on social assistance or who are government wards. Stemming from concern over the very high cost of drugs relative to the incomes of the elderly through the 1950s and 1960s, a number of drug benefit programs were begun in the early 1970s. A common feature was the negotiation of a fee with the pharmacy associations for the provision of pharmacy service along with some definition of the cost of medication. This has set a pattern in which the provincial pharmacy associations now must negotiate the maximum charge that can be made for their prescriptions. In most cases, the negotiated fee is the maximum that can be charged on any prescription, although in some cases, it represents only the maximum that can be charged on a government reimbursed prescription.

In each province, there is a program that provides benefits to the elderly and prevents the older person from having catastrophic drug expenditures. In most cases, this is a deductible form of coverage, with the government paying 80% of the cost over some amount, normally an amount in the region of $100. In other provinces, there is a vendor payment program in which the vendor will charge the government directly for prescriptions for the elderly. These programs can account for 25%-30% of the prescriptions in the province, and it is for this reason that they play such a dominant role. In many cases, those on social assistance or government wards will participate in similar programs. In almost no instance does the provincial government provide direct patient benefits itself; rather, these benefits flow through community pharmacies under a variety of arrangements.

Drug benefit programs are a popular fringe benefit in a number of collective agreements, and a number of insurance companies and third-party carriers have established drug benefit programs throughout Canada. Blue Cross is a major force in this area, and there are another five to ten firms that also compete actively.

It is estimated that about 15% of the population have no form of coverage and are solely responsible for paying the full charges for all of their prescribed drugs during the year. In theory, this is not a problem, since under the Canada Assistance Act, if these individuals show that they have a need for this medication which they cannot meet out of their own income, they could be eligible for drug benefits. In practice, this does not happen.

Drug programs use a variety of means to reduce the cost of the program to government. These will include an insistence on lower cost medications being provided, a drug benefit list or formulary, a requirement that a 100-day supply be provided on a prescription rather than a restriction on quantity to the normal 30-day supply, a deductible copayment feature, a reduced level of payment to pharmacies for prescriptions in excess of a certain amount (e.g., 20,000 prescriptions per year), and definitions of cost that are more beneficial to the province rather than to the pharmacist. Recently in Ontario, the acquired cost has been defined as one that excludes any payment to the wholesaler for the wholesaler function. This is now a focus for lobbying.

D. UNIQUE OR INTERESTING FEATURES
OR SPECIAL SITUATIONS IN CANADA

1. Drug Caution Code

The Drug Caution Program consists of a written code of statements about nonprescription medication on a poster that is visible to the consumer in the pharmacy. These codes correspond to bright green alphabetical labels on the nonprescription medication. The written codes provide the consumer with sufficient information to talk comfortably with the pharmacist about self-medication. All material is available for take-home

use, and its eye-catching, bright green color is in legible print. It assigns a
code letter (A, B, C, D, etc.) for a variety of potential problems with
nonprescription drugs. When the nonprescription products are priced in
the pharmacy, the caution code on a green sticker is also placed on the
container. The patient, when purchasing this product, would compare the
code letter against a large sign indicating the potential problems for the
product. For example, it may say that this product is not to be used in
conjunction with antihistamines or cough and cold remedies. The intent is
to have the patient then discuss with the pharmacist the potential problems
before obtaining medication. There are attempts to formalize this program
across Canada and to have the manufacturers place the designated codes
on the labels. At this point, the program is still in a transitional stage, and
approximately 75% of the pharmacies in the provinces of Manitoba, Saskatchewan, Alberta, and British Columbia participate to some extent in
the program.

2. On-Line Third-Party Billing

In one of the provinces, Saskatchewan, a drug benefit program had
been in operation in which every prescription had part — the major part —
of the cost reimbursed to the pharmacist. This government program had a
number of interesting features. Due to the escalating costs of the program,
however, the benefit level was reduced through the use of a copayment
program with a deductible component.

While the universal coverage program was in effect, the data on drug
use for a population of one million provided a unique database. In an
effort to maintain this database and to reduce the inconvenience to individuals and pharmacists, a new third-party payment system was introduced.

A plastic health services card with a magnetic strip was issued to each
adult. Using this technology from the credit card and banking industry,
the pharmacists are linked to the Prescription Drug Plan computer. Each
claim is immediately assessed and the costs calculated. This enables the
pharmacists to know when the patient has paid the deductible portion
($125 for a family), and the charge to the patient is then 20% of the
prescription price, with the balance automatically billed to the government
program.

The link between the government computer and the pharmacist's computer has allowed a single entry in dispensing that also serves as the billing
process. There is a weekly billing by the program, so the pharmacists
receive reimbursement within two to three weeks of dispensing a prescription.

E. CONCLUSIONS

Pharmacy plays an important role in Canada's health care system. High quality pharmaceuticals are readily available, and the great majority of the population has coverage against high drug expenditures.

The Canadian population has a comprehensive health program that provides excellent care. It is a source of pride to Canadians and is an important political issue. The costs of the program, however, have been growing much faster than the economy of the country, and health now takes a growing proportion of government expenditures.

There are an adequate number of well-educated pharmacists distributed across Canada, with a greater concentration in urban areas. The majority of the pharmacists are in community pharmacy.

Hospital pharmacists are developing clinical programs, but the major focus remains drug distribution. There has been a rapid growth in the number and quality of hospital pharmacists. This could lead to more patient-oriented programs.

Drug benefit programs have a very strong influence on pharmacy practice, and this is seen to be an issue, as the programs have more interest in controlling costs than in improving the scope and quality of pharmaceutical services.

The public is satisfied with pharmaceutical services and rates the performance of pharmacists highly. Growing emphases on counseling and patient care will maintain the high level of public support.

Over the past few decades, the issue of high drug prices has cast a shadow over the pharmaceutical manufacturers and pharmacists.

The high prices of newer products have continued. The public has been concerned over drug prices, and this, in turn, has led to political charges of profiteering. In this context, the negotiation of higher dispensing fees and expanding services has been difficult.

The future of pharmacy is bright in Canada, and continuing progress in patient care seems assured.

SELECTED READINGS

Canada Year Book 1988 Statistics Canada, Ottawa, 1987. Chapter 6, Social Security.

National Health Expenditures in Canada 1975-85, Health and Welfare Canada, Ottawa, 1987.

Malcolm G. Taylor, ed. Health Insurance and Canadian Public Policy, Institute of Public Administration in Canada, McGill-Queen's University Press, Kingston and Montreal, 1978.

1988 Canadian Hospital Directory, Canadian Hospital Association, Toronto.

Directory of Long-Term Care Centers in Canada 1988, Canadian Hospital Association, Toronto.

Pharmacy in a New Age, Report of the Commission on Pharmaceutical Service, Canadian Pharmaceutical Association, Toronto, 1971.

The Report of the Commission of Inquiry on the Pharmaceutical Industry, Canadian Government Publishing Center, Ottawa, 1985.

What Your Customers Think of You, Canadian Pharmacy Services Study 1988, The Upjohn Company of Canada, Toronto, 1988.

Spyros Andreopoulos, ed. National Health Insurance: Can We Learn From Canada, John Wiley and Sons, New York, 1975.

Chapter 5

People's Republic of China

Hu Ting-Xi

A. THE NATIONAL HEALTH CARE SYSTEM

1. History and Background

In selecting a social system, constructing the country, deciding domestic and foreign policies, and developing industrial and agricultural production, the historical background and present reality should be taken as a basis. Countless facts in historical events and present practices tell us that ignorance of these will end in failure. Some misunderstanding, confusion, and absurd actions by both Chinese and foreigners in dealing with Chinese affairs were also arising from their failure or refusal to see the Chinese historical past and the present situation.

China, one of the largest countries in the world, is located in the eastern part of Asia. It covers a total area of 96 million square kilometers, divided administratively into 22 provinces, five autonomous regions, and three municipalities directly under the state council (central government). The statistics on the first half of 1989 show that the total Chinese population has reached 1.1 billion, of which 93.3% belong to the Han nationality. China has 56 nationalities.

Here are some basic aspects of Chinese historical features and present reality:

China is a country with a long history and a brilliant civilization. As early as the Han (206 B.C.-220 A.D.) and Tong (618-907 A.D.) dynasties, China was in the lead in the world in culture, education and science. While Europe was in the long dark period of the Middle Ages, China was thriving and prosperous in science and the culture of feudal society. China had made some outstanding scientific contributions to mankind, mainly with respect to astronomy, geology, traditional medicine and pharmacy, mathematics, and agriculture. The prominent achievements in practical

93

technique were the four major inventions of the compass, type print, paper, and black explosives.

Chinese medicine and pharmacy are a great treasure house. Most subjects in science and technology in ancient times were replaced by Western ways when Western modern science was transmitted and spread over China. Only Chinese traditional medicine developed independently and gave a full play of its vital role in the Chinese national health care system. It even gradually transmitted and spread in the foreign countries instead of being replaced by Western medicine. Distinguished physicians and pharmacists and well-known medical works are quite common in Chinese history.

China is a country with the largest population, but limited arable land. The country has entered its third baby boom since the founding of New China in 1949, and the total population is expected to exceed 1.15 billion by the end of 1989. Some experts say the Chinese population may top 1.3 billion by the turn of the century. During the forty years since the founding of the People's Republic of China, the population of China has increased by 600 million, which is almost 2.5 times the American population. Such a big population means a heavy burden for the economy and social development, and poses many problems in food supply, housing, medical care, traffic facilities, and education. China is a little larger than the U.S. but has greater extremes of climate and larger areas that are unproductive and uninhabitable, and cultivated land area is much smaller than that of the U.S. China has richer natural resources, but the per capita distribution of farmland and water resources is very low.

China is a developing country with a low level of social productive forces and with backwardness in science and technology in comparison with Western advanced cultures. Although China today has successfully solved the problem of feeding one-fifth of the world's population with only 7% of the world's arable land and has made substantial progress in industrial, agricultural, and scientific development, it cannot be denied that China is still in very obvious economic and scientific backwardness. The Chinese total output in grain, cotton, coal, and many other products is in the lead all over the world. However, in per capita GNP and per capita outputs of steel, fuel, grain, electricity, transportation, telephone, and many other products, China falls behind Western advanced countries (e.g., the U.S.), by 30, 40, 50 years, even by a century. In addition, it is well known that there is still a long way to go in catching up with the advanced countries in science and technology.

2. Social and Political Environment

China is a country with socialism as its social system. Socialism is a social system that persists in taking public ownership as the main economic body and persists in "to each according to his work" as the major form of distribution. China persists in the socialist course, but at the same time, it firmly carries out the principles and policies of openness to overcome the shortcomings and incompleteness in the socialist system. The reform and openness are conducted on the basis of the four Cardinal Principles: upholding the socialist road, the people's democratic dictatorship, the leadership of the Communist Party, and Marxism-Leninism and Mao Zedong thought. Under this system, a mechanism of the socialist commodity economy is suited to China's conditions and is capable of combining itself with regulation through the market. Under this system, some people are permitted to become prosperous first through honest labor and legal means, but the goal of prosperity for all should be kept uppermost. Under this system, China maintains the system of People's Congress, the multiparty cooperation led by the Communist Party, and the system of political consultation.

3. Public Health Administration System

The Chinese national public health executive agency is the Ministry of Public Health under the state council (central government).
The main functions of the Ministry of Public Health are:

1. To develop, stipulate, and promulgate public health policies, regulations, rules, and various technical standards; to devise development programs and a working plan of health care; to enforce the various laws, regulations, and rules of public health; and to sponsor and organize the exchange of experience
2. To perform education with the policy of "putting prevention first;" to carry out national patriotic public health campaigns; to develop a plan for and an approach to preventing and treating disease (especially those diseases seriously affecting the health of the people); and to conduct national borders' quarantines to protect against importation or exportation of infections and epidemic diseases
3. To control environmental hygiene, industrial hygiene, drinking water hygiene, food hygiene, school hygiene, and radioactivity protection; to devise hygiene standards and organize hygiene inspection and surveillance
4. To allocate properly the medical and health agencies at all levels and

to establish the approaches for controlling their medical practice; and to devise the regulations of professional title, promotion, and salary standards for medical and health personnel

5. To carry on and strengthen traditional medicine; to accomplish government's policies on traditional medicine and promote the combination of western medicine with traditional medicine

6. To conduct maternity and child medical care and health; to protect against diseases seriously affecting the health of women and children; and to perform the introduction and adoption of contraceptive technology

7. To develop the training of medical and pharmaceutical professionals at high and middle levels; to devise a plan for medical education and continuing education; to compile teaching plans and programs to organize the writing of teaching material; and to directly supervise some top medical universities and colleges

8. To develop a national medical research plan to sponsor; and to organize the evaluation of results of medical and pharmaceutical research and promote academic exchange

9. To stipulate the quality standards of drugs and biological products; to enforce drug laws; and to organize the manufacture of biological products and blood products

10. To be responsible for international communication and cooperation of medical, pharmaceutical, and public health science.

In provinces, autonomous regions, municipalities, provincial cities, and counties, there are also departments of public health (at the provincial level), bureaus of public health (at the city level), and divisions of public health (at the county level). These are government public health agencies at different levels, which have responsibility for accomplishing the functions of the ministry in their own provinces, cities, and counties. In the government agencies under the county level (such as township, village), there are some appointed personnel who are responsible for public health administration.

4. Medical Service (Health Facilities and Manpower)

Since New China was formed in 1949, the country has made substantial progress in medical service. In reality, a radical change has occurred all over the country in all respects of health care. Before 1949, there were only 3,670 medical establishments, 80,000 hospital beds, and 541,000 health care professionals, including 50,500 technical personnel. In 1985, medical and health care establishments increased to some 200,000 in number, of which 60,000 were hospitals, 127,000 clinics, 3,410 health and epidemic prevention stations, and 2,724 maternity and child hospital centers. A nationwide medical and health care network at all levels has been formed. Hospital beds were increased by 27.9 times, for a total number of 2,229,000, averaging 2.2 per 1,000 population. The number of health care and medical professionals (including doctors and assistant doctors of both Western and traditional Chinese medicine) reached 1,413,090, averaging 1.41 per 1,000 population.

Organization Structure of Ministry of Public Health

1. General Office	Secretary Division Policy Research Division Administration Division Division of Correspondence And Personal Interview Record Division
2. Personnel Department	Office Division of Administrative Cadres Division of Technical Division of Labouring And Salary Division of Cadres working in The Ministry
3. Planning and Financing Department	Planning Division Statistics Division Material Supplies Division Division of Capital Construction Finance Division
4. Medical Administration Department	Urban Medical Care Division Rural Medical Care Division Minority Medical Care Division Agency Medical Care Division Senior People Medical Care Division
5. Hygiene and Quarantine Department	Acute Infections Disease Division Parasit-infections and Chronic Infectious Disease Division National Bordering Quarantine Division Labouring Hygiene Division Environmental and Radioactivity Protection Division Food Hygiene and School Comprehensive Standard Division
6. Science and Education Department	Division of Higher Medical Education Division of Secondary Medical Education Division of Continual Medical Education Division of Science and Technology Planning Division of Scientific and Technological Results Division of Scientific and Technological Communication
7. Chinese Traditional Medicine Department	Office Division of Scientific Research of Traditional Medicine Division of Medical Care of Traditional Medicine Division of Education of Chinese Traditional Medicine Division of Combination of Western Medical with Traditional Medical
8. Maternity and Child Public Health Department	Women Hygiene Division Child Hygiene Division Division Of Family Planning Technology

9. Drug Administration

Division of New Drug Registration
Division of Controlled Substance
Division of Traditional Medicines
Division of Bio-drugs
Division of Drug Inspection and Regulatory Affairs

10. Bureau of
    Foreign Affairs

Office
Liaison Division
International Communication Division
International Cooperation Division

Medical Establishments, Hospital Beds, and Medical and Health Care Professionals

Total number of medical establishments and hospital beds in 1949 and in 1988:

Item	1949	1988
Number of medical establishments	3,670	205,988
Number of hospital beds	84,625	2,794,926

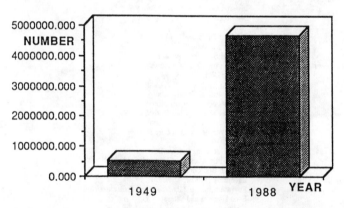

Number of health care professionals working in medical establishments at all levels in 1949 and 1988

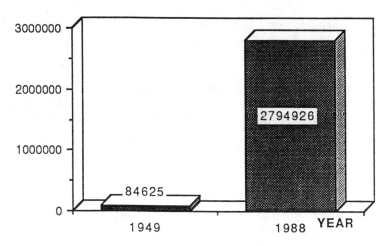

Number of Hospital Beds in 1949 and 1988

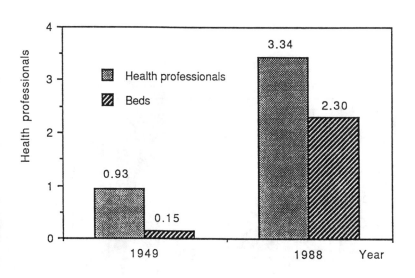

National Average Number of Hospital Beds and Health Care Professionals per 100 persons in 1949 and 1988.

*Medical Establishments in the Counties
and Rural Areas, Their Hospital Beds and Medical
and Health Care Professionals*

Since 1949, the medical service in large or middle-sized cities has greatly improved in both quantity and quality. The health care in the counties and rural areas (communes, townships, and villages — in China, there are 2,000 counties and 50,000 communes, which are now restoring the name of township) has also developed greatly. Each county has at least one general hospital, and each township has at least one clinic.

Total Number of Hospitals at county level, hospital beds and health care and medical personnel all over China in 1949 and 1988.

	1949	1988
Hospitals	1,437	2,256
Beds	11,224	383,429
Personnel	13,202	459,783

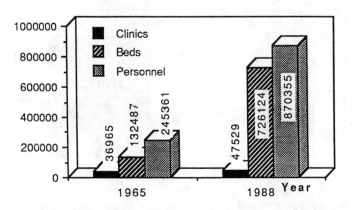

Total Number of Township Clinics, Sick Beds, and Medical and Health Personnel in 1965 and 1988

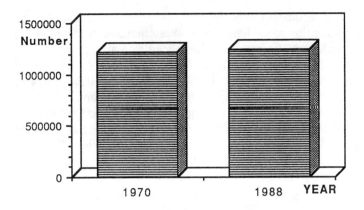

Total Number of Rural Doctors and Health Workers
in 1970 and 1988

Public Health and Epidemic Prevention Stations

Prevention is the guiding principle of the Chinese health service, and Chinese medical workers always remember the health care policy of "putting prevention first."

Thanks to the patriotic public health campaigns conducted all over China, infections and endemic diseases have been efficiently controlled. In the process of fighting these diseases, a network of public health and epidemic prevention stations at all levels (central, provincial, city, county, and township) has been established.

The Total Number of Public Health and Epidemic Prevention
Stations at All Levels and Public Health Professional in 1952
and 1988

	1952	1988
Stations	19	1932
Hospitals	224	155492
Personnel	306000	361882

In China, both in medicine and in pharmacy, there are two lines, i.e., Western and traditional. Since 1949, the Chinese government has paid great attention to the development of Chinese traditional medicine. Chinese traditional medicine has been greatly developed, strengthened, and perfected. In addition to the hospitals of Western medicine, there are established hospitals of Chinese traditional medicine at all levels. Many chronic diseases that are incurable or treated with difficulty by Western medicine can be relieved and cured by Chinese traditional treatment or the combination of Western treatment with the traditional method. Traditional medical services have enjoyed a very good reputation and have gained a wide acceptance among all walks of Chinese people, especially in counties, towns, and townships. In recent years, the traditional medical treatment is also supported and warmly welcomed by foreign patients.

Total Number of Hospitals of Chinese Traditional Medicines, Their Sick Beds, and Medical Professionals of Chinese Traditional Medicine in 1952 and 1988

	1952	1988
Hospitals	481	5303
Personnel	20504	221669

Maternal and Children's Medical Service

During the past 40 years, the maternal and children's medical service has expanded greatly, both in quantity and quality, all over the country. The prenatal physical examinations for pregnant women in urban areas average five exams per pregnancy, the examination rate having reached 98.4%, and those for pregnant women in rural areas average 2-3 exams per pregnancy, the examination rate having reached 70%.

An extensive survey of the most common diseases affecting women (such as uterine tumors, various inflammations, disorders of monthly cycle, cervical erosion, various injuries, etc.) are conducted regularly. Special medical care of women during their monthly cycle and periods of pregnancy, puerperium, and nursing is given routinely both in urban and in rural areas. Encouraging progress has been made in the prevention and treatment of common female diseases.

The total number of maternity and children's public health hospitals and clinics, sick beds, and medical professionals in 1949 and in 1988 is as follows:

Item	1949	1988
Number of maternity and children's health hospitals	80	310
Number of hospital sick beds	1,762	30,226
Number of children's hospitals	5	30
Number of hospital beds	139	7,336
Number of maternity and children's public health clinics or stations	9	2,793
Number of midwives	13,900	60,991
Number of rural midwives	NA	466,974

Medical Education and Research

In the last 40 years, medical education has developed very rapidly, forming a complete, multilevel and multidiscipline medical education system. A comprehensive system of medical continuing education at advanced, middle, and elementary levels has also been established in China. Since 1949, the number of graduates of medical schools and colleges has reached 650,000, and those graduated from the secondary medical school has reached 1,400,000. Several million part-time medical and health personnel have been trained for the rural areas through various kinds of short-term training courses.

In the last 40 years, medical research has also greatly expanded, forming a more complete research network at the national and local levels. Medical researchers, physicians, surgeons, and pharmacists have made great advances in basic medical science, clinical science, preventive medicine, Chinese traditional medicine, drug and biological products manufacturing, medical device and apparatus production, etc. Some of China's medical achievements enjoy international recognition and good reputation.

The total number of medical colleges and schools and the number of students studying in the schools in 1949 and in 1988 is as follows:

Item	1949	1988
Number of medical colleges and schools	22	119
Number of students studying in schools	15,234	191,527

The total number of medical research institutes and medical researchers in 1949 and in 1988 is as follows:

Item	1949	1988
Number of institutes	4	332
Number of medical researchers	300	37,419

Life Expectancy

As a result of the improvement in prevention and a much reduced mortality rate (especially infant mortality) both in rural and urban areas, Chinese average life expectancy has significantly increased. National average life expectancy before 1949 and in 1985 is as follows:

Life Expectancy before 1949	Life Expectancy in 1985 (on the basis of sampling investigation of National Population Statistics Bureau)
Total: 35.0 years	Total: 68.9 years
Male: NA	Male: 67 years
Female: NA	Female: 71 years

Incidence of Infectious Disease and Death Rate

Since the founding of New China, the government has taken a series of effective measures to prevent infections and endemic disease, and the morbidity rate has been greatly reduced. Some typical dreaded maladies of the biological type (such as cholera, plague, smallpox, relapsing fever, typhus, kala-azar, etc.) either have been completely eliminated or basically controlled. National incidence of infectious disease and death rate in 1980 and 1988 is as follows:

Year	Incidence (per 100,000)	Deaths (per 100,000)	Death Rate
1980	2,076.17	3.07	0.15
1988	465.89	1.49	0.32

5. Health Costs and Public and Private Medical Financing

In China, most of the health costs are paid by the government and most of the medical and health facilities in the cities are under state ownership. Part of the facilities at the township level, and most of them at the village level are under collective ownership, but receive some financial subsidies from the government. In recent years, both in cities and rural areas, a small number of private clinics and private individual practitioners have been emerging and developing.

All the employees working in various levels of government agencies; military units; educational, scientific, cultural, and commercial institutions or units; or state-run or collectively-run factories in China, enjoy a completely free medical care system, i.e., all the medical expenses will be paid by the government except that recipients must pay a very small amount of the hospital or clinic registration fee, normally 0.1-0.3 yuan (RMB), which is equivalent to $.03-$0.08. All the expenses of diagnosis (including various medical examinations), medicines, and hospitalization (except the food in the hospital) will be paid by a third party. The third party can be divided into two categories: government agencies or units, state-run institutions, or commercial units, etc., that receive a definite amount of the government health budget based on per capita and health expense; and state-run works, plants, factories, and collectively-run manufacturers who receive a certain amount of medical funds from their own profits on the basis of the percentage set by the government.

The directly-related family members of the employees—parents, spouse, and children—who are not state employees also, to a different extent, enjoy complete or partial, free medical care.

Some of the peasants in the countryside (various percentages in different areas) join a cooperative medical care system, and regulary pay a definite amount of money to the Committee of Cooperative Medical Care, which is responsible for organizing and financing rural medical care. The committee will pay 60%-80% of medical expenses for its members when they go to the hospital or clinic for treatment of common diseases or ailments. For the serious cases or hospitalized patients, the committee usually pays 50% of medical expenses.

Some patients in rural areas will not join the Cooperative Medical Care system and will pay all the medical fees themselves. The private business people or jobless inhabitants in the urban area also pay all their medical expenses themselves, when they go to state-run, cooperative, or private hospitals or clinics to receive medical care.

With the introduction and development of the rural family responsibility system (peasants enjoy full freedom to arrange their productive activities), the rural medical care system is faced with some difficulties, and new approaches are being discussed and designed.

The free medical care system in urban areas has also encountered some trouble, since it sometimes results in drug waste and drug misuse or overuse. Some officers of public health agencies, the public, the people's congress representatives at all levels, and medical and pharmaceutical experts have made many suggestions for improving the present system.

China is are looking forward to more complete medical care financing, both for cities and rural areas, to be devised in the near future.

Governmental Expenditure for Public Health
(Over the Different Periods)

Time	National Total Budget	Health Care Budget	Percentage of Health Care in National Total Budget
	1	2	3
1. Restoring Period	5.59	366.56	1.52
2. First 5-year plan	14.55	1,345.68	1.08
3.Second 5-year plan	23.45	2,288.67	1.02
4. Adjusting Period	18.84	1,204.98	1.56
5. Third 5-year plan	38.35	2,518.52	1.52
6. Forth 5-year plan	65.62	3,919.44	1.67
7. Fifth 5-year plan	113.65	5,247.35	2.17
8. 1976	16.96	806.20	2.10
9. 1978	22.42	1,110.95	2.02
10. 1980	30.16	1,212.73	2.49
11. Sixth 5-year plan	215.32	6,951.93	3.10
12. 1981	32.74	1,114.97	2.94
13. 1982	37.66	1,153.31	3.27
14. 1983	41.95	1,292.45	3.25
15. 1984	48.16	1,546.40	3.11
16. 1985	54.81	1,844.80	2.97
17. Seventh 5-year plan (1987)	64.28	2,291.10	2.81

B. PHARMACEUTICAL INDUSTRY
AND DRUG DISTRIBUTION

1. Structure of the Industry

All the Chinese manufacturers of pharmaceutical and medical products are state owned, except for some plantation farms of medicinal herbs and animals, which are collectively owned. In 1949-1978, the manufacture, research, development, marketing, sale, distribution, import, and export of medical and pharmaceutical products were supervised by different ministries under the State Council, e.g., Ministry of Public Health, Ministry of Foreign Trade, Ministry of Chemical Industry, and Ministry of Business. Since 1978, for implementing unified management and supervision, an agency of national pharmaceutical industry, the State Pharmaceutical Administration of China, has been established under the State Council. SPAC is a government agency responsible for planning, adjusting, organizing, and coordinating the manufacture, research, distribution, and marketing of Chinese traditional medicines, chemical drugs, and medical instruments all over China.

Here it must be mentioned that in China, the national drug law enforcement agency is the Drug Administration under the Ministry of Public Health. SPAC is not a regulatory body, but a supervisory organization of the national pharmaceutical industry. Since the English translation of the name of the state pharmaceutical administration is somewhat confusing, it was wrongly considered by foreigners to be the counterpart of the FDA in China.

In SPAC, a few functional departments have been established to implement government policy and to accomplish the functions mentioned above. They are the Departments of Planning and Capital Construction; Financing and Pricing; Enterprise Management; Scientific Research and Education; Quality Standard; Material Storage, Transport and Supply; Personnel and Labor; and Foreign Affairs; and the Office of General Administration and the Office of Veteran Cadres Affairs.

In addition, for accomplishing special and more direct supervision and guidance of the pharmaceutical industry, the following six corporations were set up:

1. China National Corporation of Traditional Herbal Medicines, whose function and responsibility is to organize and coordinate the production and marketing of traditional herbal and patent medicines

2. China National Pharmaceutical Industry Corporation, whose function and responsibility is to organize and coordinate the production of bulk pharmaceuticals and their dosage forms
3. China National Medical Instruments Industry Corporation, whose function and responsibility is to organize and coordinate the manufacture of medical instruments and hygienic supplies
4. China National Commercial Corporation of Medicines Distribution and Marketing, which is responsible for the distribution and sale of chemical medicines and medical instruments
5. China National Medicines and Health Products Imports and Exports Corporation, which is responsible for handling the import and export business of medicines, health tonics, and hygienic supplies
6. China National Pharmaceutical Economic and Technical International Corporation, whose function and responsibility is to conduct economic and technical cooperation with foreign countries.

There are over 2,100 manufacturers under the SPAC system of pharmaceutical and medical industry, including over 900 pharmaceutical factories, over 500 pharmaceutical manufacturers of traditional and patent medicines, and more than 600 medical instrument factories.

Pharmaceutical administration bureaus at the provincial level (municipalities and autonomous regions) and at the city level are the local branches of SPAC. They are in charge of accomplishing the functions of SPAC because the manufacturers are distributed at all corners of China, especially in large and midsized cities.

Apart from those agencies described above, there are some research institutes, colleges, universities, and commercial companies and firms directly supervised by SPAC. These are discussed in the following sections.

2. Manufacturing

Before 1949, there was practically no pharmaceutical industry in China. Most chemical medicines were largely dependent upon either imports or reprocessing imported drugs. Medicines were so scarce and expensive that many ordinary people could not afford to buy them.

During the past 40 years, a more complete and efficient system of pharmaceutical manufacturing has been formed in China. The products made under this system not only meet the domestic basic need, but are also exported as bulk chemical pharmaceuticals to other countries.

The Chinese pharmaceutical industry provides the people with various kinds of traditional patent medicines, chemical drugs in various dosage forms, medical instruments, and hygienic supplies, which play an increasingly greater role in the prevention and treatment of disease and in people's health care and welfare.

The industry has become a large employer. In 1989, the whole medical and pharmaceutical industry system employed a total staff of over 1.1 million. In 1989, the total output value of the production in the pharmaceutical industry system had reached 26.86 billion yuan. The output value has increased almost 100 times since 1949. The effective tax rate, excellent economic advantage, and great social benefit for the pharmaceutical industry are among the highest for all industries in China.

Pharmaceutical Industry Today

Through the efforts of 40 years after the founding of the People's Republic of China, the pharmaceutical industry has developed into a more complete industrial system involved with both bulk drugs and their various dosage forms. Over 900 manufacturers have produced more than 1,300 kinds of bulk pharmaceuticals and 3,500 kinds of preparations, with about 4,300 various specifications. The total staff has reached 440,000 people. The industry has manufactured 26 categories of bulk pharmaceuticals, such as antipyretics, analgesics, antibiotics, sulfa drugs, vitamins, drugs for respiratory system disease, drugs for endemic disease, anthelmintic drugs, drugs for diagnostic aids, antituberculosis agents, cardiovascular drugs, hormones, contraceptives, anticancer drugs, etc. Annual output has reached over 100,000 tons, ranking second in the world market.

Among the above-mentioned 900 pharmaceutical manufacturers, there are a few dozen top pharmaceutical enterprises and a few joint ventures with the U.S., Sweden, Japan, Belgium, Switzerland, etc. The former top pharmaceutical enterprises have equipped themselves with modern, sophisticated equipment and instruments by importing from western advanced countries, and have become more competent both in product output and in drug quality.

A List of Top Enterprises

Factory Name and a Brief Description	Number of total staff	Products and Output
1. North China Pharmaceutical Factory. The largest antibiotic manufacturers in China. Shijiazhang, Hobei Province	Over 10,000	83 various antibiotics and other products. Antibiotic output amounts to 1/6 of total Chinese antibiotics output. Many products have been awarded silver and golden medals.
2. Northeast General Pharmaceutical Factory. The largest organic chemistry synthetic pharmaceutical factory. Shengyang, Liaoning Province.	Over 10,000	More than 60 kinds of antibiotics, sulfa drugs, vitamins and drug intermediates and their preparations. Chloromycetin, SD, Vitamin C and B1 are four major exported products from which are earned a large amount of foreign exchange for the country.
3. Shan Dong Xin Hua Pharmaceutical Factory. Founded in 1943, important production base for chemosynthetic drugs, Shang Dong Province.	Over 5,000	Over 100 varieties of pharmaceuticals, whose annual productive capacity for chemical raw materials is 6,000 metric tons, injections over 100,000,000; tablets 3.5 billion.
4. Taiyuan Pharmaceutical Factory. One of key and backbone pharmaceutical manufacturers, specialized in producing antimicrobial agents.	Over 3,000	Many products have won the national medals for their superior quality. The products are sold in 52 countries. The annual quantity for export has reached 350 tons.

5. Dan Dong Pharmaceutical Factory. One of key and backbone pharmaceutical enterprises in China.	Over 2,000	Over 100 kinds of bulk drugs, intermediates, and antibiotics and injections, tablets, powder, ointment, capsules. Annual output has reached about 1,000 tons.
6. Changzhou Pharmaceutical Factory. (One of top pharmaceutical enterprises).	Over 1,300	Over 120 varieties of preparations and over 30 kinds of bulk pharmaceutical in four major categories (cardiovascular drugs, diuretics, antipyretics, and analgesics, and semi-synthetic antibiotics).
7. Beijing Pharmaceutical Works, a large-scale back bone enterprise. Beijing, China	Over 4,000	Over 300 varieties of sulfa drugs, vitamins, antituberculosis drugs, cardiovascular drugs in different dosage forms.
8. Xian Pharmaceutical Factory, a large-scale top pharmaceutical enterprise. Xian, Shan Xi Province	Over 3,000	Various antibiotics and vitamins and over 100 kinds of their preparations. Erythromycin and Vitamin B2 have been awarded silver medals.
9. Minsheng Pharmaceutical Factory. Found in 1926, one of large-scale Chinese pharmaceutical manufacturers with a long history. Hangzhou, Zhejiang Province	Over 2,400	Over 200 varieties of chemical drugs, plant extraction products and fermentation products, such as drugs of xanthine, of hyoscyamine, anthelmintics, cardiac muscle stimulants, anti-cancer drugs, vitamins, etc.

10. Fuzhou Antibiotic Factory. The largest pharmaceutical enterprise in Fujian province.	Over 1,800	The annual production of antibiotic bulk pharmaceutics amounts to 300 tons. The main products include gentamycin, aureomycin, medecamycin, cefazolin sodium, and their preparations.
11. Harbin Pharmaceutical Factory. An important base of bulk production materials in Heilongiang Province. Harbin, Heilongiang Province.	3,650	The main products are antibiotics.

Production of Traditional Medicines

The Chinese government has always attached much importance to the planting of medicinal herbs and the manufacture of Chinese patent medicines. There are very rich natural resources of medicinal herbs in China. Over 5,000 kinds of Chinese medicinal herbs have been found and investigated, of which about 1,000 are frequently used in medical practice. To meet the increasingly greater medical need, most of the plants are cultivated rather than collected as wild herbs. The area of land for planting medical crops has been enlarged to over 4,800,000 square kilometers. Some rare Chinese medicinal herbs, such as ginseng, radix angelicae sinensis, and rhizoma coptidis, have been successfully planted, and their output has been greatly increased.

A more complete system of manufacturing Chinese traditional medicines has been formed. Under this system, over 500 factories with a total staff of 300,000 are producing various dosage forms of patent medicines by adopting new techniques and using advanced equipment and instruments. The dosage forms include tablets, capsules, injections, ointments, and sugar granules, as well as traditional pills, powders, and extracts. About 4,000 kinds of patent medicines with 40 dosage forms are being produced and used in clinical practice. Fewer side effects and less toxic action have made the Chinese traditional medicines more well accepted by the people, both at home and abroad, especially in the treatment of chronic disease.

Manufacture of Medical Instruments

Since the founding of New China, the medical instrument industry has developed very rapidly. A manufacturing system of over 600 medical instrument factories with a total staff of over 140,000 has been set up. Now 43 categories (2,300 varieties and 9,000 specifications) of radiation equipment for medical use, medical electronic instruments, nuclear instruments, optical instruments, biochemical analyzers, and surgical instruments can be produced in China. Some of the instruments are quite well thought of, for example, a fiber laser tract endoscope, a B ultrasonic diagnostic unit, medium and small X-ray units, and surgical scissors of stainless steel with molybdenum. These have gained good acceptance in the world market. The Chinese medical instrument industry not only meets domestic needs, but also exports 15% of its total output to foreign countries annually. China may also manufacture some sophisticated new medical instruments, such as a CT brain scanner, a gamma camera, and a medical standing wave accelerator.

To implement quality assurance, China has established National Quality Inspection Centers and a Medical X-ray Unit Center in Shanghai and Shenyang. In addition, numerous local inspection stations have been set up in some large cities so that quality inspection of medical instruments can be strengthened.

3. Research and Development

The rapid development of the pharmaceutical industry and the medical instrument industry is closely connected with the advances in development and research. During the 40 years after the founding of New China, pharmaceutical research has made great progress. A nationwide research network on drugs has been formed. Under the system of SPAC, there are now 54 professional research academies and institutes, with more than 11,000 researchers. What is more important, practically all large-scale or middle-sized pharmaceutical and medical instrument enterprises have their own research institutes or central research labs, with more than 10,000 research personnel. China has already achieved a large number of scientific advancements, such as developing a number of creative new drugs of high quality with highly curative effects (e.g., artemether, arteannuim) and successfully introducing and cultivating the Western safflower.

These research units, by improving their facilities through importing numerous modern scientific equipment and instruments, have achieved over 1,600 research results. Of these, 27 items have been awarded national invention prizes, and 51 items have been conferred national science and technology progress prizes.

In addition to the research units under the SPAC system, there are numerous research institutes under the system of the Ministry of Public Health and under the system of the Chinese Academy of Science, which also undertakes the investigation of drugs.

In the research of chemical drugs, the emphasis has been on searching for new chemical entities and new dosage forms, introducing new technological processes and new techniques, and providing new equipment and instruments.

From the beginning of the 1950s to the middle of the 1960s, the research stress was placed on developing antibiotics, sulfa drugs, some anti-infection drugs, antituberculosis drugs, drugs for endemic disease, antiparasitic drugs, widely used vitamins, antipyretics, and analgesics. In the latter 1960s and in the 1970s, the research was focused on steroid hormones, contraceptives, antitumor drugs, and cardiovascular drugs. In the 1980s, a greater effort has been made to investigate drugs for diagnosis, amino acids, semisynthetic antibiotics, new analgesics, and novel anti-inflammatory drugs.

In the research of Chinese traditional medicine, greater attention has been paid to digging out and improving traditional dosage forms, folk recipes, experienced recipes, and secret recipes by introducing modern techniques under the principle of persisting, traditional theory. Every effort has been made to modernize the manufacture of and enlarge the production scale of patent medicines and to isolate and extract the active principles from both medicinal plants and animals. The stress of research has also been placed on the plantation, cultivation, transplantation, and domestication of many useful and precious medicinal plants.

In the research of medical instruments, the focus has been on routine surgical instruments and apparatus, and on new equipment and new instruments, using advanced and sophisticated techniques so that the Chinese medical instrument industry may reach and keep abreast of the state-of-the-art level of the world.

4. Drug Distribution, Cost Containment, and Price Regulation

Like the manufacturing and research units in the pharmaceutical and medical instruments industries, all the commercial corporations of pharmaceutical and medical products are state-run and state-owned. The commercial corporations at the national level, provincial level, city level, and county level have formed an efficient supplying and purchasing network for all medical and pharmaceutical merchandise.

    a. Six Commercial Wholesale Corporations at the National Level (also called the First Level.)

      1. Shanghai Purchasing and Supplying Center of Medical and Pharmaceutical Products
      2. Beijing Purchasing and Supplying Center of Medical and Pharmaceutical Products
      3. Tienjin Purchasing and Supplying Center of Medical and Pharmaceutical Products
      4. Guangzhou Purchasing and Supplying Center of Medical and Pharmaceutical Products
      5. Shengyang Purchasing and Supplying Center of Medical and Pharmaceutical Products
      6. Shanghai Purchasing and Supplying Center of Chemical Reagents

    b. Over 360 commercial wholesale corporations at the provincial level (the second level, including 29 provincial commercial wholesale corporations of traditional medicinal materials) are distributed in numerous cities of provinces, municipalities, and autonomous regions.
    c. Over 3,000 commercial wholesale corporations at the county level (the third level) includes about 2,000 county commercial wholesale corporations of traditional medicinal materials.
    d. Over 50,000 retail shops for pharmaceutical and medical products (equivalent to U.S. community pharmacies) and over 5,000 retailing sales and purchasing shops below the county level are distributed in all corners of the country, both in urban and rural areas.

The commercial units mentioned above are sold and supplied more than 20,000 kinds and specifications of medical and pharmaceutical merchandise. The total sales of medical and pharmaceutical merchandise have increased from 278 million Yuan in 1952 to 23.1 billion Yuan in 1988. The annual average medicine consumption per capita has increased from 8 Yuan in 1980 to 21 Yuan in 1988.

In recent years, the pharmaceutical commerce has developed tremendously. In 1988, the sales values of six major categories of drugs reached 2.4 billion Yuan. The pharmaceutical commercial units at different levels have successfully organized, allocated, and provided various medicines and medical instruments to meet the requirements of the people both in the urban and rural areas.

The cost containment and price regulation of medical and pharmaceutical merchandise are controlled by government agencies at various levels. The medical and pharmaceutical service is mainly operating for social benefit rather than economic profit. In the past 40 years since 1949, the retail sale prices of drugs have dropped many times, and compared with the 1950s, they have decreased by 82%. In addition, some medicines, such as contraceptives and medicines for endemic diseases, are supplied to the people free of charge.

5. Import and Export

With implementation of the policies of reform and openness, communication and cooperation with foreign countries in the pharmaceutical industry have been developed very rapidly. China has imported some medicines that are safe and effective and which are not produced in the country. China also imported some drug intermediates that are very deficient in the domestic market and some medicinal plants of foreign origin. To improve the facilities and change the obsolete equipment of its pharmaceutical enterprises, in recent years, China has imported 242 items of important equipment and 1,802 sets of single machines and analytical instruments which are worth $500 million. China also signed over 200 projects for importation of technologies and equipment to satisfy the urgent need of technical innovation of pharmaceutical factories. The country imported about 300 kinds of medicines from foreign countries, and there are still 600 kinds of medicines under application for approval by Chinese drug law enforcement agencies.

The exportation of Chinese medical and pharmaceutical products has also been expanded. The export value of drugs has been increased from $461 million in 1979 to $700 million in 1988. Twenty-two kinds of Chinese drugs have been examined and approved by the FDA, and twenty-eight kinds are under application.

In the exportation of Chinese advanced technologies, there have been some encouraging achievements. For example, Vitamin C production technology has been sold to Switzerland, and whole sets of Vitamin A, Vitamin B6, SD, and sodium mathoxide production technology have been exported to Romania.

6. Drug Approval and Government Regulation

A comprehensive drug control law for China, consisting of 11 chapters covering control and inspection over drug research and development (NDA); over drugs and their manufacture; over drug distribution, labeling, packaging, and advertising; over drug imports and exports; and over controlled drugs and drug uses in hospital and community pharmacies, was promulgated in September 1984 and entered into force on July 1, 1985. Since 1985, a series of more detailed acts, provisions, and regulations, such as a controlled drugs act, provisions for new drug application and approval, provisions for drug advertising, provisions for GMP, detailed regulations for accomplishing drug control law, and detailed regulations for accomplishing GMP provisions, have been enacted one after another by the enforcement agency, the Ministry of Public Health.

The requirements of advertising, NDA, and GMP are more or less similar to those in the U.S. FDCA.

Thus far (up to 1989), no complete pharmacy law (like the U.S. State Pharmacy Act) has been promulgated, and no pharmacist license examination has been required in China.

Up to now, the people responsible for drug law enforcement have reached about 2,000 (including drug inspectors at various levels). A more complete national network of drug control and supervision has been formed, which is composed of 1,100 drug quality control labs at different levels (800 drug control labs at the county level) with a total staff of 13,000.

C. PHARMACY PRACTICE

1. Varieties of Pharmacy

The whole Chinese health care system can be divided into two lines — East and West: Western medicine and traditional medicine; Western drugs and traditional drugs; Western hospital and hospital of traditional Chinese medicine; hospital pharmacy of Western drugs and pharmacy of Chinese traditional medicines; and community pharmacy shop of Western drugs and drugstore of medicinal herbs.

There are over 50,000 retail sale community pharmacies all over China, 2,030 traditional medicinal materials corporations at the county level, and 5,000 stations below the county level for the purchasing and selling of Chinese medicines.

In China, there are now over 73,000 hospitals and clinics, in which are usually set up both pharmacy of Western drugs and pharmacy of Chinese traditional medicines. It is interesting to note that in many hospitals, two kinds of pharmacy are facing each other in separate rooms. In the pharmacy of Chinese traditional medicines, there are drawers containing dry roots, stems, rhizomes, leaves, flowers, and herbs, and dried portions of small animals residing on the big shelves. The prescriptions of traditional pharmacy are filled by pharmaceutical personnel of traditional pharmacy and are often processed into decoction by boiling with water.

2. Organizational Structure of Hospital Pharmacy

a. Three Categories of Hospitals in China

Category I.

—Hospitals at provincial level (both general and special hospitals)
—Hospitals at city level (both general and special hospitals)
—Hospitals affiliated with the medical university and college
—Hospitals for large industrial and commercial enterprise

Category II.

—Hospitals at the county level
—Clinics in the city area
—Institutional clinics

Category III.

—Hospitals at the township level
—Village clinics

b. Organization Structure for Category I Hospitals Pharmacy

Hospital Director

Hospital Committee of Pharmacy Affairs

Department of Pharmacy

Dispensing	Manufacturing	Drug storage
1. Dispensary for out - patients	1. Common Preparations Manufacturing Lab	1. Storage House for Western Drugs
2. Dispensary for inpatients (central and ward)	2. Sterile Preparations Manufacturing Lab	2. Storage House for Traditional Drugs
3. Dispensary for Emergency patients	3. Traditional Preparations Manufacturing Lab	3. Storage House for Dangerous Drugs
4. Dispensary of Traditional Medicines	4. Repackager	4. Refrigerating Storage House (or Refrigerating)
		5. Packaging Material Store House Equipment

Drug Analysis	Information Office	Research Lab
1. Chemical Analysis Lab		Clinical Pharmacy Lab
2. Instrumental Analysis Lab		
3. Animal Experiment Lab		

c. Organization Structure for Category II Hospitals Pharmacy

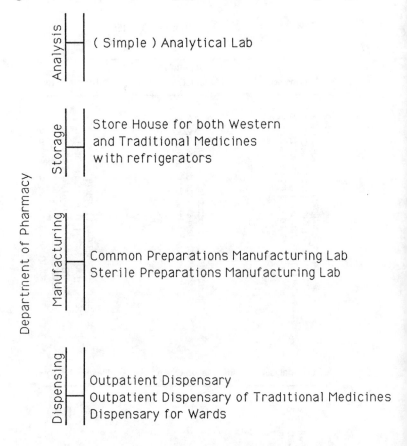

d. Organization Structure for Category III Hospitals Pharmacy

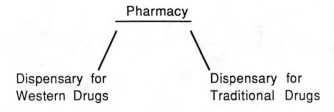

3. Pharmacists, Technicians, Assistant Technicians

According to the statistical data of 1982, the number of pharmacists with college degrees was 14,597. The number of technicians was 14,579, and the number of assistant technicians was 267,754. The pharmaceutical personnel specializing in traditional pharmacy numbered 35,922.

The latest accurate statistical data are not available. It is estimated that there are over 20,000 pharmacists with college degrees. Many technicians with secondary pharmaceutical school education and over ten years of practical pharmacy experience have recently been promoted to pharmacist.

There are six titles for personnel working in a pharmacy: chief pharmacist, vice-chief pharmacist, senior pharmacist, pharmacist, technician, and assistant technician.

a. Chief pharmacist. Very few in number, with college degree and more than 30 years of practical pharmacy experience and very profound expertise in pharmacy. Their function is to supervise the complicated and difficult dispensing, to guarantee the safety and efficiency of medicines, to inspect the controlled drugs, to introduce new drugs, to conduct research on new dosage forms and new techniques, and to teach pharmacists, technicians, trainees, and pharmaceutical students.

b. Vice-chief pharmacist. Much more than chief pharmacist in number. Both the qualification and function are quite similar to that of the chief pharmacist. Their theoretical basis and practical experience are not so outstanding as the chief pharmacist's.

c. Senior pharmacist. Pharmacist with college degree after 5-10 years of practice or a technician with secondary pharmacy school education after over 25-30 years of practice. Their function is to supervise technicians to dispense or manufacture preparations, to undertake drug quality assurance on the basis of state drug standards, to join in research on improving drug delivery, to help the chief pharmacist to inspect controlled drugs, and to undertake part of the teaching work.

d. Pharmacist. With college degree or with secondary pharmacy school education, but having practical experience of about 20 years in pharmacy. Their function is to undertake dispensing and preparation/manufacturing under the supervision of chief, vice-chief, and senior pharmacists; to conduct analysis, research, and controlled drugs delivery; and to teach or direct technicians, assistant technicians, and pharmacy students.

e. Technician. Usually graduated from secondary pharmacy school. Their function is to undertake prescription filling, to make purchasing plans, to take medicines from the storehouse, to purchase medicines, to fill in all kinds of forms, to recover and deliver medicines, to record statistics, and to do many more practical jobs under the guidance of various levels of pharmacists.

f. Assistant Technician. Usually graduated from junior or senior middle school and have received some pharmacy training in the three-to six-month short training course. Their job is to do the prescription filling and to be the operators of manufacturing common or sterile preparations under the direct guidance of the pharmacist or technician.

More pharmacists (including chief, vice-chief, and senior pharmacists) are working in Category I hospital pharmacies, a small percentage of pharmacists are working in Category II hospital pharmacies, and practically no pharmacists are working in Category III hospital pharmacies and community pharmacies (some community pharmacies specializing in new drugs sales are exceptions.). In the latter cases, the technician or even assistant technician is working and responsible for everything. It should be mentioned that in China there is not yet a pharmacist license examination system established, nor is there a separate state pharmacy law as in the U.S. It is expected that these will gradually be set up in the near future in the Chinese pharmacy profession.

4. Dispensing

The number of daily average prescriptions is 1,019-2,144 in Category I hospital pharmacies (the highest is in hospitals at the provincial level, and the lowest is in the special hospitals pharmacies). Pharmacists or technicians are responsible for distributing medications to outpatients, while ward nurses (who also receive some training given by the pharmacy department) are responsible for distributing medications to hospital inpatients under the direction of pharmacists who are working in ward-decentralized pharmacy. Dispensing also includes monitoring and directing the use of drugs. Now, more and more people are recognizing the importance of this system because in China, combination drug therapy is more common. In the process of filling prescriptions, over 99.9% of drugs are supplied in ready-made dosage forms, mostly made by manufacturers and partly by pharmacy manufacturing labs. The extemporaneous prescriptions have been reduced to the least, if any. The average number of drugs in each prescription is 1.92-3.05 in Category I hospital pharmacies. The average number of drug varieties is 360-735 kinds in Category I hospital pharmacies.

The frequently used dosage forms are (in order) tablet, injection, ointment, liquid dosage form, and infusion.

The prescriptions for traditional drugs still use multicomponent recipes, often in decoctions, but recently more ready-made dosage forms (tablets, pills, sugar granules) are used.

5. Manufacturing Labs of Hospital Pharmacies

Usually, Category I hospital pharmacies have manufacturing labs of larger scale, composed of common preparations lab, sterile preparations lab, and traditional medicines lab. They produce various dosage forms frequently used in clinical practice, such as injection, infusion, tablet, capsule, sugar granules, powder, solution, mixture, syrup, tincture, fluid extract, spiritus, glycerin, enema, eye drops, ointment, oculentum, suppository, paste, film, lotion, liniment, suspension, emulsion, plaster, etc. A few very large hospital pharmacies even produce tablets. All these products could only be sold within the hospital, and selling them to other pharmacies is illegal. In accordance with the new provisions enacted recently, GMP is also required for the manufacturing labs of hospital pharmacies.

In most large cities, there is an agreed formulations handbook, compiled by experienced pharmacists and sponsored by the local public health agency. The manufacturing labs of hospital pharmacies produce most preparations on the basis of this handbook. They also produce some preparation according to their self-designed formulations, which must be submitted to the public health agency for approval before manufacture. The quality assurance of preparations produced by the manufacturing labs is conducted by the drug analysis division of the pharmacy.

6. Patient Education

More and more Chinese professional people have begun to recognize the importance of patient education. Noncompliance with the physician's and pharmacist's directions by patients is very common in China. The common sense of medication remains to be popularized among the ordinary people. The pharmacy department is required to perform patient education. Many Category I hospital pharmacies are using a "blackpaper newspaper" to introduce some knowledge about correct use of drugs.

What pharmacy is presently doing in this respect is far from being required.

7. Record Keeping

Common prescriptions must be kept for one year in the pharmacy. Narcotics prescriptions are to be maintained for three years, and psychological drug prescriptions must be kept in the pharmacy for two years. Complete and accurate records of sales receipts, dispensing, deliveries, or other disposition of controlled substances (including narcotics, psychological drugs, toxic drugs, and radioactive drugs) must be maintained in good order, waiting for inspection by the drug enforcement agency.

Chapter 6

Denmark

Peter Kielgast
Jens Povelsen

A. THE NATIONAL HEALTH CARE SYSTEM

1. History and Background

The health service can be described as some service functions that aim to contribute to the maintenance and improvement of the nation's health. This description covers both the traditional functions of the health service, like pharmacies, hospitals, and doctors, as well as a number of prophylactic functions. The health service in Denmark has been in existence for several hundred years, during the course of which it has been subjected to repeated legislative amendment.

The health service is often divided into a primary and a secondary sector. The primary health service is defined as the places in the health service where the population can go with questions concerning illness and health care. The personnel of the primary health service are in close and ongoing contact with the population and work in the local community. As a rule, the complaints treated are not particularly grave. If this is not the case, patients are referred to the secondary sector. The secondary health service works mainly with a clientele that has been graded and referred to it by the primary health service. Hospitals are typical representatives of the secondary health service.

The social services in Denmark have been organized for several hun-

This report was drawn up by: Mrs. Bente Frokjaer, MSc (pharm), Head of Department, Steffen Kjaer, MSc (pharm) Strategic Planning, the Danish Pharmaceutical Association and, for Section B, by Miss Susanne Honore, Head of Department, MEFA (the Association of Danish Pharmaceutical Industry).

dred years. In the course of history, their legislation has been amended many times, but the underlying principle remains the same: that citizens shall have reasonable access to economic and practical assistance to enable them to cope.

2. Social and Political Environment

Denmark is a small country in terms of both population and land area. Its area is 43,000 square kilometers, and the population is 5.1 million.

Denmark has a constitutional monarchy and a one-chamber parliament, Folketinget, to which 179 members are elected for four-year terms. Political parties are represented in exact proportion to the percentage of the votes received in the general election, provided they have polled at least 2% of the total vote.

The Ministry of Health lays down and administers general legislation governing the entire health area. The most important laws in the field of pharmacies and pharmaceuticals are the Medicines Act of 1975 and the Pharmacy Act of 1984. The daily administration of acts and regulations governing the pharmacies and the pharmaceutical market is undertaken by the pharmaceutical department of the National Board of Health.

3. Public and Private Financing, Health Costs

The total expenditure on health care amounted to more than DKK 47 billion in 1988. Sixty-one percent of the expenditure concerns the hospital sector, 22% the primary health care sector, and 14% of the expenditure is financed by the patients themselves. During recent years, expenditure on the health sector as a percentage of the GDP (gross domestic product, DKK 735 billion in 1988) has tended to fall. The percentage fell from 7% in 1980 to 6.5% in 1988. As mentioned, expenditure on the hospital sector accounted for 61% of the total health expenditure. For comparison, this share amounted to 53% in 1970.

Expenditure on drugs has in recent years accounted for a relatively constant share of around 10% of total health costs, but in 1987, the figure was 11.1% and in 1988, 11%. (See Table 1.) Drug expenditure accounts for 4.7% of the total hospital expenditure, 17.6% of public expenditure on primary services, and 27.6% of citizens' own direct expenditure on primary health care.

According to the Organization for Economic Cooperation and Development (OECD), expenditure on the health sector as a percentage of the

TABLE 1. Total health expenditure divided according to the nature of expenditure, 1987.

<u>1987   DKK million</u>

Hospital sector	27,789
Primary health service	9,679
Total public operations costs	37,468
Citizens' own direct costs	6,569
Total operational costs	43,716
Construction costs	1,044
<u>Total Operational and construction cost</u>	44,702
Gross domestic product (GDP)	692,000

Source: Facts, Medicine and Health Care, Denmark. MEFA, 1989.

GDP varies considerably. The western industrialized countries and Japan spend, on average, 7.1% of the GDP on the health sector. According to OECD, 6.5% was spent in Denmark in 1986.

4. Facilities (Hospitals, Clinics, etc.)

All hospitals in Denmark are owned by the public authorities, either by the state, county, or municipal council. An entirely new departure, however, is the opening in 1989 of two private hospitals in Denmark. These have not been included in Table 2 regarding hospitals in Denmark.

The number of discharges from general hospitals rose from 916,000 in 1980 to 1,031,006 in 1988, corresponding to an average annual increase of about 2%. The average bed-day consumption was 7.6 days in 1987. During the same period, the number of treatments of outpatients rose from 3.3 million to almost 3.6 million, corresponding to an average annual increase of 1.5%. On average, every Dane is admitted to a hospital every 5 years and is treated in an outpatient clinic every 18 months.

TABLE 2. General hospitals and number of beds, 1987.

	No.	No. beds
General hospitals	97	27,680
Mental hospitals and nursing homes	1 7	4,645

Source: Facts, Medicine and Health Care, Denmark. MEFA, 1989.

5. Manpower

The total number of people employed full-time in the health sector in 1986 was about 121,000. More than 90,000 people work in the hospital sector.

6. The Social Security System

Welfare legislation in Denmark is aimed at protecting weaker population groups, for example, sick people and old-age pensioners. The state therefore subsidizes the cost of drugs and medical aids for the sick and handicapped. The public health insurance also covers consultations with general practitioners, hospital stays, subsidies for dental treatment, and specialist treatment if the patient has been referred by the general practitioner. Also, homes for the elderly are paid for by the public.

7. The Medical Establishment: Entry, Specialists, Referrals

There are in Denmark 2,956 general practitioners, 800 specialists in practice, and 8,482 doctors in the hospital sector (see Table 3).

The general practitioner may choose his or her own form of practice. Some prefer to have an individual practice, others to establish themselves in a cooperative practice (usually 2-4 doctors). The latter form gives economies through the shared use of premises, secretarial staff, etc., and it can be more stimulating professionally to work with other doctors.

In principle, the individual citizen has a free choice of doctor. There are, however, two groups within the national health insurance system, namely Group I and Group II. Most Danes are members of Group I. This means that once a year one has the right to choose a doctor working near

TABLE 3. Persons employed in the health sector. Converted into number of full-time employed (1986).

A. Hospital sector

1.	Doctors	8,482
2.	Others with university degrees	154
3.	Nurses	21,482
4.	Staff with other nursing training	18,091
5.	Staff with other health care training	11,114
6.	Other staff	<u>26,322</u>

A.	**Total**	**<u>85,645</u>**

B.	**Primary health service**	
7.	General practitioners	2,956
7.a.	Assistants	2,470
8.	Specialists	800
8.a.	Assistants	630
9.	Practicing dentists	3,570
9.a.	Assistants	4,600
10.	Pharmacy proprietors	312
10.a.	Pharmacists (employed at private pharmacies)	754
10.b.	Pharmacy assistants and trainees	3,302
11.	School dentists	1,246
11.a.	Assistants	2,747
12.	Health and sick care	4,463

B Total		**<u>27,850</u>**

C.	**Students, trainees, total**	7,239

A+B+C Total		**<u>120,734</u>**

Source: Facts, Medicine and Health Care, Denmark. MEFA, 1989.

one's home. Consultations with the doctor thus selected are free. On the other hand, one cannot go to a specialist without a referral from a general practitioner.

If one chooses to be a member of Group II, one has a free choice of doctor, and one can go to a specialist whenever one wishes. As a member of Group II, one has to pay part of the consultation fee but can have the sum refunded on application to a private health insurance company. Ninety-eight percent of the population are members of Group I, while 1.5% are members of Group II.

The general practitioner makes referrals to specialists and to hospitals. However, emergency admissions to hospitals can also take place.

8. Payment and Reimbursement

The largest part of the primary health sector is financed by the national health insurance system. Agreements concerning payments from the national health insurance system have been concluded with the following:

—pharmacy sector
—state-registered chiropodists
—chiropractors
—opticians
—practicing physiotherapists
—general practitioners
—practicing specialists
—practicing dentists

However, only the fees of general practitioners and specialists are fully covered by the national health insurance system. For the remaining areas of the primary health sector, the patient pays part of the bill, after which a bill for the remainder is sent directly to the national health insurance system by the health worker in question.

Private Insurance

In addition to the various public subsidies, a large number of Danes, about 1,000,000, have taken out private sickness insurance in the company Danmark. The most important area covered by this company is payment of medicine expenses for insurance subscribers. Grants are also provided toward dental treatment, optician fees, and fees to practicing specialists (medical doctors). The company provides a grant toward all medicines prescribed by a doctor at the rate of 25%, 50%, or 100% of the expenses paid by the patient, depending on the category of the drug and on the type of insurance the member has chosen.

Coverage. National health insurance expenditure on prophylaxis and treatment is shown in the table below (1987):

PRIMARY HEALTH SECTOR	DKK million
General practitioners	2,643
Practicing specialists (medical doctors)	895
Dentists	819
Pharmaceuticals	1,893
Physiotherapy	144
Chiropractors and other expenditures	121
National health insurance, total	6,515
Home nursing	1,069
Examination of children by health visitors and doctors	428
Midwifery, obstetrics	1,734
Prenatal care	214
School dental service	1,092
Medical officers and running costs of the Serum Institute	361
Primary Health sector, total	9,679
HOSPITAL SECTOR	27,789

As will appear from the following, special conditions apply to subsidies for medicine.

Subsidies Granted Under the National Health Insurance Act

By far the greatest contribution toward drug costs is paid under the National Health Insurance Act, which came into force in 1973. In addition, it is possible in certain cases to obtain individual subsidies.

Everyone has a sickness insurance card, and this gives adults who are permanent residents in Denmark the right to receive assistance toward the cost of medicine under the National Health Insurance Act. Children under the age of 16 are covered by their parents' sickness insurance cards.

In 1989, a self-payment minimum was introduced so that the individual patient has to pay the full cost of medicine, also of subsidized medicine, up to a limit of DKK 800, after which subsidies are paid according to the usual rules.

General Subsidies, Nonperson-Related

The general rule is that subsidies are granted to a range of pharmaceutical specialities, divided into specific therapeutic groups. In consultation with the National Board of Health the Ministry of Health determines what drugs and therapeutic groups should receive subsidies (the positive list) and also the rate of subsidy (50% or 75% of the price).

In general, the National Health Insurance Act operates with two rates of subsidy. For some pharmaceuticals the rate is 75%, for others 50%. The two rates are determined by medical reasons related to the seriousness of the disease. Drugs receiving 75% subsidy are considered products vital to life, while the 50% group is for more mundane, but often chronic, diseases. The 0% group contains drugs regarded as being of minor medical importance, for example, most psychopharmacological drugs, disinfectants for the urinary system, contraceptive pills, and a number of skin products.

B. PHARMACEUTICAL INDUSTRY
AND DRUG DISTRIBUTION

1. Structure of the Industry

The pharmaceutical industry in Denmark consists of 50 companies. Seven of the companies are Danish and organized in MEFA, the Association of the Danish Pharmaceutical Industry. The MEFA companies are all engaged in research and development of drugs and represent approximately 90% of the total pharmaceutical production and export in Denmark. Forty-three companies are organized in the Medicine Importers' Association, MEDIF. These companies represent all the major pharmaceutical companies in the Western world.

Of the total sale of drugs (in terms of money) on the Danish market, 29% is covered by the MEFA manufacturers, 58% by imported drugs (MEDIF companies), and 13% by pharmaceuticals produced by a company owned by the Danish Pharmaceutical Association (DAK Laboratories Ltd.). Ninety percent of the production of the Danish factories is exported. Production is primarily insulins, vitamins, antibiotics, psychotropics, diuretics, and sulfonamides.

2. Research and Development, Imports, Exports, Patents, Licensing

The seven Danish companies spent DKK 757 million in all on research for and development of new and better drugs in 1988. This corresponds to 13% of the companies' total drug turnover. A further DKK 147 million was spent on quality control and assurance testing of the products manufactured by the companies.

In 1988, the import of drugs was about DKK 2.8 billion, an increase of 9.1% compared with 1987. Ninety percent was imported from other western European countries. West Germany, U.K., Switzerland, France, and Sweden accounted for 61% of the total import.

In 1988, vitamins and antibiotics accounted for the biggest groups of imports, finished drugs accounted for 76%, and semi-products for 24% of the total import.

About 90% of the production of the Danish pharmaceutical industry is exported. Since 1980, the growth of drug exports has been much greater than that of other Danish industries or other products. With an index of 100 in 1980, drug exports had risen to an index of 292 in 1988. Denmark is the second largest per capita exporter of drugs in the world.

The main groups of drugs exported by Denmark are insulin, vitamins, diuretics, heparin, and antibiotics. The biggest markets are the U.S., West Germany, U.K., Sweden, and Japan. Sales to these five countries accounted for 42% of the total export in 1988.

The patent system has recently been changed from patents of methods to cover product patents as well.

3. Manufacturing

The total drug production in 1988 came to DKK 6.7 billion. The corresponding figure for 1980 was DKK 2.5 billion, an increase of 167%. The production was primarily in insulins, vitamins, antibiotics, psychotropics, diuretics, and sulfonamides.

The companies in Denmark differ very much in size. Even the largest Danish companies are small compared to companies in other countries. This is the reason why companies have a very specialized, high-quality production exported to the world market.

4. Advertising and Promotion, Price Regulation

The rules concerning advertising and promotion can be found in the Danish Medicines Act. Advertising directed toward doctors, pharmacists, dentists, and veterinarians is permitted for all medicines, while only non-prescription medicines may be advertised to the public. Furthermore, advertising toward the public is permitted only after approval by the National Board of Health and discussion in an advisory board concerning advertising. This board consists of members from all interested parties: the industry, the pharmacies, the consumers, the Ministry, doctors, veterinarians, and others. The role of the board is to assess whether the advertising is objective and factual.

Advertising to the public is allowed only for nonprescription medicine and only in newspapers and magazines. Advertising on television, the radio, and in public places other than pharmacies is not permitted. There is a tendency toward increasingly restrictive legislation governing the over-the-counter (OTC) market.

Advertising to doctors and other professionals must also be objective and factual but need not be approved in advance. Instead, the Medicines Act gives some details concerning the content of advertising. Advertisements must contain the name of the product and the names of any active substances. The names of the active substances must be printed in the same type and as conspicuously as the name of the product. Furthermore, indications, contraindications, side effects, dosage, dispensing form, reimbursement, and price must be stated.

To guarantee that the information given doctors and other professionals is objective and correct, the Association of the Danish Pharmaceutical Industry and the Medicine Importers' Association have set up their own regulatory body, the Danish Board of Drug Advertising. The board monitors all advertising and considers complaints and inquiries about the above-mentioned persons from authorities, drug producers and importers, and from other parties that may have a special and proper interest in a given case. The board may also take up cases for consideration at its own initiative.

In 1988, the number of advertisement units addressed to professionals was 2,487, corresponding to 73% of the total amount of advertising and information brochures to professionals. Of the total 3,392, there were complaints concerning 164. Forty-eight, corresponding to 29% of the complaints, concerned noncompliance with the advertising codex, i.e., exaggerated or unserious advertising. The rest referred to inadequate description of indications, documentation, generic name, composition, information on dosage or dispensing form, side effects, or lack of date.

As already stated, pharmacies in Denmark hold the exclusive right to retail pharmaceuticals. As a consequence, price levels are set by the government (the Ministry of Health) so that all drug retail prices are uniform throughout the country. The price of individual drugs is based on the price set by the manufacturers and importers. There are no separate price control measures applied by manufacturers and importers of drugs. The price set by the manufacturer or importer is determined by competition — the interplay of free market forces between equivalent drugs from different producers.

Pharmaceutical wholesalers in Denmark compete only in terms of service, not price. They maintain lists of product item numbers for all drugs registered in Denmark. At present, the gross profits of pharmaceutical wholesalers average 7.8% of the purchase price paid by the pharmacies (the range is about 4% to 17%). Drugs that are expensive to prepare and can be supplied at lower costs yield less than a 7.8% profit but are balanced out by higher profits on small turnover or low priced drugs that cost more to dispense, such as narcotics or large volume packages. In other words, wholesaler profits are cost related. Wholesalers submit a monthly report to the Monopolies Control Authority with information on their earnings and costs, so their profits are subject to close and continuous monitoring by that body.

Every two years, the Ministry of Health concludes a new agreement with the Danish Pharmaceutical Association to determine the pharmacy sector's gross profits for the next two years. The agreement is based on current figures and forecasts developments in the sector in the coming two years. The gross profit is the difference between pharmacy purchase prices and retail selling prices. The gross profits of the country's 310 pharmacies in 1988 totaled DKK 1.5 million, for a total turnover of DKK 4.6 billion.

5. Drug Approval and Government Regulation

As a member of the EEC and the Nordic Council, Denmark accepts applications in the format set out in the EEC Notice to Applicants and the Nordic Guideline on Registration. The statutory basis for the authorization system is the Medicines Act of June 26, 1975, with amendments as included in the promulgation of August 6, 1982. Furthermore, a number of ministerial decrees laying down the administrative procedure have been issued.

Medicines are defined in the act as follows:

Articles which are intended to be administered to human beings or animals to prevent, diagnose, alleviate, treat or cure disease, symptoms of disease and pain or to affect bodily functions. All medicines apart from the following must be registered:

1. Products made in pharmacies for a named patient (extemporaneous products)
2. Homeopathic drugs
3. Drugs made from only natural products, not up-concentrated and recognized as "safe"
4. Certain vitamins and all mineral drugs

Registration is granted for five years and is renewable.

The total number of pharmaceutical specialities registered on December 31, 1988 was 2,746 distributed over 4,951 dispensing forms corresponding to about 7,000 package sizes. However, only 2,100 of these were sold on the Danish market, of which 1,866 were for human use. Of these, 999 were imported, 400 were produced by DAK Laboratories Ltd. (owned by the Danish Pharmaceutical Association), and 476 were produced by the MEFA manufacturers. The specialities registered and marketed contained 1,119 different active substances.

Before being marketed, drugs must be registered by the National Board of Health. In 1988, the registration authorities handled 568 applications, of which 329 were approved. Most of these applications regarded reapplications and supplementary documentation (216); 78 were synonyms, and 26 concerned original products.

6. Competition: Brand and Generics

The Danish market is a highly competitive market. The reason for this is the free-price setting. The generic market is approximately 50% in terms of volume and about 25%-40% of turnover. About 60% of all original products in Denmark have a copy on the market. Apart from a large number of branded generics, there used to be production of generics at pharmacies. This production is now centralized to only a few pharmacies and the products registered with one company as holder of registration, the DAK Laboratories Ltd. Besides DAK's pharmacy production units, DAK has a production and filling plant. Today the DAK products are branded generics, OTC as well as prescription medicines.

7. Cost-Containment Activities

Danish doctors enjoy free prescription rights, which means that pharmacists cannot change a prescription or choose a similar drug as a substitute without contacting the doctor. Similarly, the authorities cannot compel doctors to prescribe specific pharmaceuticals on purely economic grounds.

Pharmacies are, however, frequently involved in local pharmacy and therapeutic committee work, in which they cooperate with authorities, including doctors, to try to achieve a more uniform prescription pattern. This work is usually carried out at meetings held every two or three months, typically to discuss a specific group of pharmaceuticals. These pharmacy and therapeutic committees are encouraged by the subsidizing authorities to keep down public health care costs.

With the purpose of saving DKK 600 million on the state reimbursement budget for 1989 and coming years, a yearly own-risk limit of DKK 800 has been introduced. For expenditure above DKK 800, the patient is reimbursed with 50% or 75% of his or her expenditure on reimbursable drugs. Some of the Danish political parties in Parliament are in favor of permitting pharmacies to carry out generic substitution for economic reasons. This proposal was put forward in 1989 by the Danish Pharmacists' Union.

C. PHARMACY PRACTICE

1. Dispensing a Prescription: Where, What, Payment, Formulary

Prescriptions can be dispensed at a pharmacy or branch pharmacy for immediate service, as there is always a pharmacist present during open hours. Prescriptions can also be phoned or telefaxed by the prescribing doctor to the pharmacy, which then dispenses the prescription. Finally, a number of pharmacy outlets (staffed by pharmacy assistants) and retailer's shops will often be attached to a pharmacy. Prescriptions handed in at these places are forwarded to the pharmacy, which dispenses the prescription and sends the medicine back. This delivery service takes place two to three times a day on weekdays.

In Denmark pharmaceuticals are defined as products that are intended to be administered to human beings or animals to prevent, diagnose, alleviate, treat, or cure disease, symptoms of disease, and pain or to affect bodily functions.

A prescription is defined as the authorization of a product by a physician, dentist, or veterinary surgeon for a specific person or an animal, a hospital, etc., or for use in the prescriber's practice, which is to be supplied by a pharmacy or hospital pharmacy.

Prescription Right of Doctors

Generally, doctors have the right to prescribe all pharmaceutical products listed in the National Board of Health's register of specialities. However, there are a few limitations; in these cases, a drug can be prescribed only by a medical specialist or a hospital. The doctor can also prescribe extemporaneous medicine produced especially for individual patients. Danish doctors thus have free right of prescription.

Prescription Right of Dentists

Dentists are subject to certain limitations in their right to prescribe medicine. In the first place, dentists can prescribe medicine only in connection with their activities as dentists. Dentists are not allowed to prescribe medicines that are included in a special list of narcotics. They are not allowed to prescribe extemporaneous medicine, apart from medicine containing fluorine, medicine for external use, and medicine for local treatment of the oral cavity.

Prescription Right of Veterinary Surgeons

Veterinary surgeons are subject to the same limitations as dentists in that they are allowed to prescribe medicine only in connection with their activities as veterinary surgeons.

Dispensing Groups

There are three categories under which medicine can be dispensed:

- Medicinal products requiring a prescription (A-narcotics, A and B)
- Nonprescription drugs on simple request (H)
- A few medicinal products sold outside the pharmacies (for example, certain pharmaceutical specialities for animals, vitamins, and anthelmihtics).

When registering a pharmaceutical speciality, the National Board of Health considers the element of danger represented by the product or special circumstances concerning its use. In accordance with its findings, drugs are classified in the different supply groups mentioned above.

The group classification determines what provisions apply to the issue of the pharmaceutical product. Products that are not explicitly mentioned in the list of dispensing groups can be sold without a prescription.

Dispensing Group A. Prescriptions for Group A-narcotics are no longer valid once the first supply of the drug has taken place, irrespective of whether or not the whole amount prescribed has been supplied. The same procedure is followed for products containing dextropropoxyphene. Pharmaceutical products in dispensing Group A can only be dispensed once on the same prescription. Examples of preparations included in Group A are all cytostatics and a number of individual substances such as indometacine, dopamine, nicotine, ketamine, and benzodiazepines. Special regulations apply to addictive pharmaceuticals. These are marked in Group A by an asterisk, and to receive such medicine, the patient's personal identification number must be shown on the prescription form for the purpose of possible central registration and statistics. One example is the group of benzodiazepines.

Dispensing Group B. Prescriptions for medicine in dispensing Group B must specify how many times the medicine may be supplied and possible time intervals. Drugs from this group may therefore be supplied several times on the same prescription. Examples of drugs included in Group B are antibiotics (for example, erythromycin), and diuretic preparations (for example, furosemide). The prescription is valid for a maximum of two years from the date of issue.

Dispensing Group H (nonprescription [OTC] Pharmaceuticals). If OTC preparations are prescribed on a prescription (as documentation for the reimbursement of pensioners and chronically ill persons), the prescription permits unrestricted supply as long as the prescription is valid (two years). Medicine subsidies follow the rules outlined in Section A.8. The patient pays his or her share of the price on receiving the medicine, while the pharmacy sends in a monthly account for the share to be reimbursed by the national health insurance system. The size of the subsidy is set by the Ministry of Health on the basis of a pharmacological assessment of the drug's importance. All prescription drugs are divided accordingly into three categories, which receive 75%, 50%, or 0% subsidy, respectively. Generally, OTC preparations do not receive public subsidy (only for pensioners and chronically ill persons, and only analgetics, and a few other therapeutic groups).

2. Record Keeping

Citizens in Denmark have a free choice of pharmacy and may change pharmacies as they wish. In Denmark, the registration of patients' consumption of medicine is forbidden by law.

The increasing use of computers in pharmacies affords greater possibilities for compiling general drug statistics on sales and consumption at local, regional, and national levels.

3. Patient Education

Patient education, understood as education of the patient concerning his or her illness and its treatment prior to discharge from a hospital, is not a routine occurrence in Denmark.

In 1989, a pilot project in patient education was started at a regional hospital. The project targets patients suffering from asthma and bronchitis.

4. OTCs and Other Classes and Schedules of Drugs, Other Nonpharmacy Outlets

Because of the pharmacies' sole concession, OTC drugs are sold only from the pharmacies and their supply facilities. Examples from this group are light analgesic preparations, such as acetylsalicylic acid and paracetamol, and various cough mixtures. In Denmark, the general view is that medicine is not harmless just because it can be sold over-the-counter, and this is why pharmacies remain the only sales outlet. In pharmacies, consumers can obtain correct and objective information from professionally qualified staff.

In addition to verbal information and guidance, OTC drugs must bear dosage directions and instructions. Moreover, at the National Board of Health's discretion, such drugs may require a red warning triangle plus a few warning sentences in cases where the medicine induces lethargy and can therefore affect driving ability.

The pharmacy system in Denmark contains four types of pharmacy units (figures as of December 31, 1988):

Pharmacies (307). At a pharmacy, there are a proprietor pharmacist (owner), employed pharmacist(s), pharmacy assistants, trainees, and unskilled personnel (cleaners, etc.). At a pharmacy, prescriptions can be dispensed, OTC drugs sold, extemporaneous pharmaceuticals produced, and, to a limited extent, actual pharmaceuticals are produced. As skilled staff are employed, qualified information can be given to the individual patient, to doctors, to hospitals, and to other customer groups.

Branch Pharmacies (31). A branch pharmacy is owned by and attached to a pharmacy, where the overall administration is located. The staff of a branch pharmacy will contain a pharmacist, pharmacy assistants, and unskilled personnel (cleaners, etc.). At a branch pharmacy, prescriptions can be dispensed, and OTC drugs are sold. As skilled staff are employed, qualified information can be given. No production of preparations takes place at a branch pharmacy.

Pharmacy Outlets (110). A pharmacy outlet is owned by and attached to a pharmacy, where the overall administration is located. A pharmacy outlet is staffed by one or two pharmacy assistants. At a pharmacy outlet, prescriptions can be handed in to be dispensed at the parent pharmacy, and the packaged medicine will later be handed over to the customer. Furthermore, OTC drugs are sold, and qualified guidance can be given.

Retailer's Shops (villages and rural areas) (930). A retailer's shop is a small pharmacy sales outlet, situated in a larger store, for example, a supermarket. The owner of the store has a contract with the proprietor pharmacist, who is responsible for the satisfactory functioning of the sale of medicines. At a retailer's shop, prescriptions can be handed in to be dispensed at the parent pharmacy, and the packaged medicine will later be handed over to the customer. Furthermore, a limited range, as determined by the pharmacy, of OTC drugs can be sold.

5. The Use of Technicians and Others

In Denmark, one can qualify as a pharmacist or a pharmacy assistant, and there are two educations leading to employment at a pharmacy and to qualifying diploma holders to dispense prescriptions. Members of the latter group are, however, only permitted to dispense prescriptions in collaboration with a pharmacist, as double control of prescription drugs is mandatory in Denmark.

It takes three years to train as a pharmacy assistant. The entrance qualifications are a high school diploma after 10 or 12 years of schooling. During training, students are at a pharmacy and receive instruction there. Furthermore, they attend three 5-week courses at the Danish Pharmacy Seminar College of the Danish Pharmaceutical Association.

The Pharmacy Seminar College administers the entire course of trainings, during which it sends assignments to students. The assignments are obligatory, and students have to write a number of reports on pharmacy tasks. The course is completed by an examination consisting of practical tests in customer handling/communication and production, written tests in pharmacology and the different laws regulating the work in the pharmacy and related topics, and an oral examination based on the written reports.

Calculated as full-time posts, the staff of the 307 pharmacies (including branch pharmacies and pharmacy outlets) consists of 307 proprietor pharmacists, 748 employed pharmacists, 2,644 pharmacy assistants, 407 pharmacy assistant trainees, 125 pharmacist trainees, and 911 other employees — in total, 5,141 full-time posts distributed among approximately 8,500 persons, most of whom are part-time employees. Unskilled staff are employed at Danish pharmacies only to deal with reception and delivery of articles and cleaning. They are not permitted to serve customers.

The education of pharmacists takes place at the Royal Danish School of Pharmacy. The education takes place at university level and lasts five years.

6. Pharmacy in Primary Care

The pharmacy sector's role in society is governed by the preamble to the Pharmacy Act of 1985: "that the population shall have reasonably easy and safe access to medical supply at a reasonable cost to society." The pharmacy is part of the primary health care sector. It is the part of the health services that is most easily accessible to the public, who can come straight in from the street and speak to a pharmacist, while it is necessary to obtain an appointment with or be referred to all other health professionals. In the local community, the pharmacy has a natural cooperation with other health workers, especially general practitioners, home nurses, health visitors, and the administrative departments for pensioners and social work.

As already mentioned, the pharmacies are involved in local pharmacy and therapeutics committee work in cooperation with general practitioners to introduce a more uniform prescription pattern, with emphasis on the interests of patients and social costs.

Cooperation with health visitors is widespread. It is important for the pharmacy to discuss with health visitors what recommendations should be made to young parents with respect to children, so that there is agreement on the guidance given. Similar cooperation has been established with home nurses. Several pharmacies measure out doses of medicine (dose dispensing) for the elderly, and this is done in collaboration with the home nurse.

Besides the already-mentioned forms of cooperation, the pharmacy's role in the primary health service can be described as follows:

—The pharmacy has a role to play in meeting the population's needs for health-promoting, healing, and support services. This can be a matter of showing special concern for risk groups, such as the elderly, pregnant women, children, and patient groups. The most important jobs are to give information about effects, side effects, and drug interactions and to give guidance in the correct use and keeping of pharmaceuticals.

—The pharmacy gives guidance concerning self-care. When the consumer comes to the pharmacy with symptoms, the staff will give guidance on choice of treatment and drug or advise the patient to go to a doctor. It is important in this connection that the pharmacy has a monopoly on the sale of pharmaceuticals, thus ensuring the consumer professional guidance.

—The preamble to the Pharmacy Act of 1985 states that the pharmacy has a duty to provide information about pharmaceuticals. There is a need to develop mechanisms that will ensure quality of treatment for the patient. This also applies to the pharmacy's supply of drugs and information on drugs. The pharmacy must ensure that such information is of a high quality. Studies on the use of pharmaceuticals are one way of improving the treatment of patients. Another possibility to improve the use of drugs is to raise consumers' consciousness concerning their use of drugs by establishing consumer groups, in which experiences of illness and drugs can be exchanged among consumers and passed on from consumers to the pharmacist.

—The pharmacy must collaborate in the collection of data on medicine consumption. Centrally, this is already taking place. Locally, the pharmacy can collect general data concerning the local population's consumption of medicine and pass on this knowledge to the other professions in the primary health service as a basis for a discussion of prescription habits, therapy, economy, etc.

—The pharmacy contributes to the secure disposal of dangerous waste and offers guidance with respect to the disposal of other types of waste.

—There is a general wish among the population for a healthier lifestyle. Here, too, the pharmacy can play an important role. Some pharmacies measure blood pressure and cholesterol, give guidance on nutrition and suitable physical activity, on medicine, alcohol and narcotics abuse, on contraception, and on AIDS, and hold courses to help people stop smoking.

7. Drug Insurance and Third Parties

The national health insurance system has been previously described in detail, namely in sections A.6, 7, 8 and 9. The private health insurance, Danmark, has also been described, namely in Section A8.

Besides these two forms of general insurance, there are many special rules for reimbursement for groups of ill people and pensioners. For example, diabetic patients receive 100% reimbursement for insulin, terminal cancer patients receive all medicines free of charge, and parents of chronically ill children get special reimbursement. Pensioners with low incomes are reimbursed by the local authority.

D. UNIQUE OR INTERESTING FEATURES OR SPECIAL SITUATIONS IN DENMARK

Despite the fact that the new Pharmacy Act was passed in Denmark as recently as 1985, there were in 1988-1989 a number of political initiatives aimed at fundamental changes in the distribution system for pharmaceuticals. Deregulation, liberalization, and debureaucratization are the key words in the various models that the Danish government is considering.

A report on "Prices and Competition in the Field of Pharmaceuticals" by a special government committee proposes that the pharmacy system should, by and large, be preserved. The most radical change in the ten-point plan suggests the total or partial abolition of the pharmacies' monopoly on the sale of nonprescription preparations. The committee recommends a reduction in the number of independent pharmacies and a rationalization of pharmacy work procedures, among other things, through the use of computer technology. Other proposals are connected with extended protection of documentation material for pharmaceuticals, public control of the industry's prices, liberalization of wholesale systems, and improved information and reduced VAT on pharmaceuticals. A proposed change in the rules for medicine subsidies has, as mentioned, already been implemented.

At the same time, the government is preparing a radical liberalization of the pharmacy system. A new working committee has been instructed to draft practical proposals for a reform of the present concession system.

In many countries, the Danish pharmacy system has long been regarded as an example for imitation, a system in which the privately owned pharmacies with their professional know-how and credibility were integrated into the public health system. A system of this kind costs money, and there is much to indicate that the government's initiatives are primarily motivated by economic arguments and by the fact that Denmark lacks a health policy.

During the past decade, pharmacies have invested many resources in becoming more than just places where the public can buy its medicine. Pharmacies are well on the way to becoming small and large know-how centers in which the public can receive guidance in the correct use of pharmaceuticals and advice concerning a healthy lifestyle from qualified staff. At the same time, pharmacies have established increasingly close cooperation with patient organizations and other consumer groups and have entered into many fields of interprofessional and prophylactic activity.

E. CONCLUSIONS

For generations, public control and insight have been fundamental principles in the field of pharmaceuticals. Generations of pharmacists have had the daily task of translating laws and rules into practice for the benefit of both society and the consumer. Through this well-developed practice, pharmacists assure the population not only that there is always reasonably easy access to pharmaceuticals at uniform prices, but also that the pharmacy and its highly qualified staff are a guarantee of safety, security, and impartial counselling.

In the 1980s, the main goal of the Danish pharmacy sector has been to "tear down the counters," to provide a unique professional expertise where it is most needed. This new role has received wide recognition in the health service and among the population. Danish pharmacists will be able to adapt to the demands for change in the next decade and to continue to be pioneers in the Danish health system.

Chapter 7

Arab Republic of Egypt

Yehia M. Dessouky

Egypt is located at the northeastern corner of Africa. Its total surface area is about one million square kilometers, of which about 4% is cultivatable, the remainder being mostly uninhabited desert. The population is concentrated in the delta and on both banks of the river Nile. The overall population density amounts to 48.3 persons per square kilometer (1986); in the inhabited area, however, it amounts to 1,170 persons per square kilometer, and in Greater Cairo, 28,259 persons per square kilometer (1986).

The net annual natural population increase amounted to 2.8%. According to the last census (1986), the population of Egypt was 50,455,049, of which an estimated 2.25 million reside abroad, mainly in the Arab countries. At the present growth rate, it is expected that the population will be 71.2 million in the year 2000. As a matter of fact, it was announced officially on March 16, 1989 that the total population of Egypt had reached 54 million. The population of Greater Cairo was 8,761,927, while that of Alexandria was 2,896,459 (both in 1986). Cairo and Alexandria are the two largest cities in Africa.

The country is administratively divided into 26 governorates.

A. THE NATIONAL HEALTH CARE SYSTEM

1. History and Background

Of all the nations of the ancient world, Egypt was the one where medicine and pharmacy flowered earliest and highest. By present-day standards, this flowering was to a great extent superstition-ridden fumbling, but men have usually fumbled before they have discovered. Greek doctors and pharmacists were later to catch the Egyptian torch and make it burn

147

clearer and brighter. The fact remains, however, that the torch, feeble as it may have first burned, was lit in the land of the pharaohs. A general acceptance of Egypts' medical and pharmaceutical preeminence threads through history. Here is great Homer speaking, around 950 B.C.: "In Egypt the men are more skilled in human medicine and drugs than any other human kind."

The Greeks, the Romans, and later, the Arabs, dug freely into the almost limitless mines of Egypt's papyruses. There they found prescriptions that to a great extent are still applicable. The English medical historian, Charles Singer, had estimated that a third of all herbal, metallic, and organic substances that serve even today's medical needs were known and used in Egypt. The Ebers papyrus alone lists 875 prescriptions containing countless ingredients and admixtures. It was discovered in Thebes in 1862 and was written by ancient Egyptians in 1550 B.C. Here and there, by luck or by observation, Egypt's pharmacists filled prescriptions that afforded their patients some good. Modern pharmacists would be lost without their balances, weights, and measures. Their counterparts in ancient Egypt had them too—surely not as fine, but effective enough.

Both public and personal hygiene were established at high levels in Egypt. Meat, for instance, was most carefully inspected before being offered for sale. No Egyptian ate tainted meat. Detectable spoilage forced it off the market. So did subtle contamination, such as tapeworm larvae, which the Egyptians even then were able to recognize as dangerous.

The Egyptians were proud of the purity of their drinking water, too, and made sure that no decomposing matter contaminated their wells. "They drink from cups of bronze which they cleanse out daily," wrote one bemused Egyptologist. Wealthy Egyptians even had rudimentary flush toilets. Remains of bathrooms and drainage systems, dating to about 1370 B.C., have been excavated at Akhetaton; these had crude but usable latrines with water jars hanging next to them for flushing.

The present policy of the national health care system of Egypt is based upon ensuring the welfare of the Egyptian population through increasing national production and upgrading the productivity of individuals. The government's policy takes into consideration the fact that man is the cornerstone of development in society. The government, acting in this line, is keen on providing services to the population to improve the standards of living.

Based upon the principles of the state's policy, health strategy and national health planning have been developed in a way that realizes the relevancy of health objectives to the main goals of the national development plans. All health services and patient care activities that can contribute to achieving the basic improvement of the health status of the country are encompassed by national health planning.

The health policy of Egypt concentrates upon attaining immediate objectives that ensure the fulfillment of the basic health needs in the form of basic minimum improvements in health services and patient care. In defining these needs, the health plan has taken into consideration the size of the health problems facing the country at the present and the availability of technical, administrative, and financial resources that can help meet the basic needs and at the same time permit the follow-up and evaluation of the progress made. Also, the demands on services as a result of popular pressure have been taken into consideration.

Outline of the National Health Program

It is the policy of the Egyptian government in general and the Ministry of Health (MOH) in particular to give priority to generalizing and developing the essential health care services in the rural areas as well as the urban areas. MOH supports maternity and infant health care services, as well as services for school pupils, dentistry, and curative and first aid services. This is achieved through 2,108 health units in rural regions, 96 general health units in urban regions, 250 centers for maternity and infant care, and 375 health offices, in addition to 300 school health units for the treatment of 9 million school pupils (1987). It is the goal of MOH to gather together all the services that are carried out by the health units in urban regions in general health centers in order to raise the standard of services (one general health center per 50,000 persons) to citizens in cities so as to lessen the pressure on hospitals and let them operate their facilities for the treatment of inpatients.

The health units in rural regions and the general health centers in urban regions are considered the first defensive line to preserve the health of all citizens. Therefore, it is the goal of the health sector to:

—Provide a rural health unit for every village that has a population exceeding 3,000 and at a distance of 3 kilometers from the nearest health services
—Establish a health center for every 50,000 people living in cities
—Connect these health units with higher levels by means of transportation, such as first aid ambulances equipped with telephones and other wireless means of communication, to assure good, efficient service
—Prepare technical staff to administrate and operate these units through special training programs.

Great effort and large investment are needed to achieve these goals. The goals are clear; this helps to reach them in stages because providing the necessary network of units will contribute to implementation of the preventive programs and make them more effective. This will have great positive results on the general health level at low cost to the majority of citizens.

Preventive Services

The goal of MOH is to expand preventive campaigns to:

— Avoid the spread of epidemics and contagious diseases through early detection, to take precautions necessary to stop their spread, and to follow up international epidemiological status
— Provide basic health services and raise their efficiency, as they are fundamental in preserving the health of citizens
— Eliminate, insofar as is possible, the diseases amenable to control by immunization and to attain self-sufficiency in the production of vaccines
— Continue to safeguard the country against the entry of epidemic and, in particular, quarantinable diseases
— Support basic services for food control to guarantee the protection and safety of citizens and to ensure the freedom of foods and drinks from contamination with infectious agents upon entry into the country and in local production and handling
— Provide thorough supervision of the health environment, such as drinking water, sewage, and treatment of waste and garbage
— Control vectors and rodents by using modern methods
— Provide necessary laboratories for diagnosing different diseases and supply them with modern means of diagnosis
— Develop specialized hospitals and supply them with modern facilities and intensive care units for certain diseases
— Establish specialized centers to study zoonoses for the protection of citizens.

Nursing

MOH seeks to improve and raise the standard of nursing performance in different health units according to the advanced development in medical sciences by:

— Implementing a plan to train nursing staff, centrally, in all specialized nursing fields such as intensive care, surgery, and emergency first aid
— Providing study missions in different fields to basic health care units in rural and urban regions.

MOH also hopes to overcome the deficiency already existing in nursing staff by:

— Increasing the number of candidates accepted in nursing schools to reach 7,000 graduates from 147 schools per year
— Encouraging opening schools to graduate male nurses, especially in governorates of Upper Egypt (schools were already opened Asyut, Kena and Aswan governorates)
— Revising, evaluating, and developing programs and updating the syllabi of nursing schools.

MOH plans to encourage all nursing staff and improve their financial status by increasing salaries, reasonable incentives, and fringe benefits for all nurses.

Emergency Health Care

It is very important to provide emergency services, as they are lifesaving duties, so the plan of MOH concerning this matter is to:

— Complete all the wireless network services in all 26 governorates of Egypt
— Provide a well-equipped first aid ambulance for every 15,000 citizens and one for every 5,000 citizens in remote desert regions
— Provide modern reception departments and establish new departments for burns and intensive care units in general and central hospitals
— Support blood banks with modern equipment and supply them with well-trained staff
— Encourage blood donation among citizens, apply the most recent techniques and tests to assure its safety, and have enough plasma to reach 25 blood units per 1,000 citizens.

Curative Health Care

MOH assures the right of every citizen to have medical treatment in all its hospitals and units, and also in hospitals of the public sector. Also, it assures that every citizen receives satisfactory medical services either free of charge or at a very reasonable cost compared to private hospitals and clinics. This is achieved through:

- Supporting general and central hospitals, specialized centers, and educational hospitals to offer excellent investigation, diagnosis, and treatment
- Increasing the number of centers for renal dialysis in all general hospitals throughout the country, manufacturing the filters and all the necessary appliances for the equipment to treat renal failure, and establishing centers for maintenance of such equipment
- Reaching the ratio of 2 beds per 1,000 citizens.

General Organization for Biological and Vaccine Products

MOH supports the production and supply vaccines of and sera not only to all the citizens of Egypt, but also to Arab and Middle Eastern countries, through employment of modern technology and research.

The Population Problem

This problem is rooted in many factors, and its solution entails the cooperation of several organizations and the integration of various efforts. At present, this coordination is carried on through the Supreme Council for Population and Family Planning. This council has developed a national plan to solve the population problem based upon three main elements:

- An integrated plan for social and rural development and concerns with projects to increase family income and to emphasize the role of women in the society
- Promotion of family planning services through coordinating the activities of MOH and their organizations and national societies
- An information plan depending on population information and personal communication to create awareness of the impact of the problem of the national economy and the standard of living.

Drug Policy

One of the major duties of MOH is to give full support to activities related to the production of pharmaceuticals so that it may face the increasing local demand for medicine. The government is very keen on securing medicines of the highest therapeutic efficiency, with prices compatible with various income brackets. The manufacturers of Egyptian drugs in the public sector are deemed to be a strategic necessity to the state. The government is committed to guaranteeing the effectiveness and soundness of locally produced drugs through the strictest control on production units and pharmaceutical research and development, and consolidation with the most modern technology. At the same time, the government encourages joint venture and private sector investment drug companies for the production of advanced modern medicinals.

2. Social and Political Environment

The social and political environment of the national health care system in Egypt is in line with the global health policy, which aims at achieving health for all by the year 2000. It calls for the promotion of primary health care services, satisfying the citizen's needs for drugs and family planning aids at suitable prices, expansion of health insurance, and subsidization of medical care. This policy has been endorsed at all official levels. The Egyptian Constitution provides for the right of every citizen to health care. Government health services, including those covering drugs, are provided either gratis or at reduced, government-subsidized prices.

The national health strategy reflects the health policy of the state. The aim is to achieve health for all through free government services, health insurance, and curative establishments, in cooperation with the joint venture and private investment sector. It is based upon well-defined principles. These are:

— To keep abreast of the socioeconomic and cultural development of the community
— To rely upon a clear political commitment for implementation
— To ensure the involvement of the various sectors within the community
— To encourage the role of the joint venture and private sector in complementing certain aspects of the health plans
— To apply decentralization widely, with emphasis on the role of local bodies
— To increase existing financial resources.

The objectives of the current five-year health plan (1987-1992) are the reduction of infant mortality and that of children below the age of five years.

The current health system has the basic characteristics of a primary health care based system, as determined by the Alma-Ata Declaration.

These are represented in the comprehensive nature of a system that contains all the elements of basic health care, accessible to all individuals, families, and the community. It is supported by other service levels, which provide training for personnel and guidance to communities, and which ensure coordination at the central level that provides the necessary expertise for planning, management, and training, as well as the financial and administrative support. Moreover, the health policy calls for expanding the provision of these services to all rural and urban regions, especially to new urban extension regions, in addition to providing them in an integrated form to urban regions. While this policy applies in full to the health system in the rural sector, where the integrated health care elements are provided at one service site and by one health team, it varies in the way it is provided in the urban sector. The primary health care has been provided through various health establishments, such as maternal and child health centers, school health units, and health offices. Indeed, MOH had, since the 1970s, begun to implement a plan to establish urban health centers which were to provide integrated health care.

MOH is headed by the Minister, who is assisted by a number of undersecretaries of health. It formulates health policies and plans and carries out organizational, supervisory, and training activities. However, the activities are implemented on a decentralized basis through local governments.

The present five-year health plan (1987-1992), as well as the previous one (1981-1986), stresses the importance of primary health care as a basis for the provision of health care services, with emphasis on preventive services. The national health system is composed of MOH and autonomous health institutions which fall under supervision of the Minister of Health (General Organization for Health Insurance, and Curative Establishments), universities (university hospitals), health institutions attached to other ministries (such as the Ministries of Defense, Interior, and Communications), and the private sector.

Managerial Process

Planning is undertaken annually through the preparation, in the various governorates or organizations, of projects to be put forward as a part of the national plan. The projects are compiled by the General Department of Planning, MOH, and the necessary modifications are made according to priorities. They are then submitted to the Ministry of Planning for discussion and approval at the national level within the comprehensive social development plan, prior to their submission for endorsement by the People's Assembly. A Department of Planning and Follow-up exists within each of the Directorates of Health in the governorates or in the health sector agencies. When finally approved, the plan is returned to the local governments for implementation, with periodic follow-up of implementation by the MOH.

A comprehensive supervisory, follow-up, and evaluation body has been established at all levels, from the service units, to the departments of health in the districts, to the directorates of health in the governorates, to the various technical departments, the Information and Documentation Center and the General Department of Statistics and Evaluation within MOH. The various activities are based on periodic progress reports, field supervision and follow-up reports, studies, and research and statistical reports. The evaluation results are referred from one level to the next, with information feedback to the former level.

Coordination takes place on a continuing basis among the technical departments of the health sector that are involved in primary health care. Periodic meetings, chaired by the Minister of Health, are held to coordinate the work of the technical departments of the Ministry and the Directorate of Health Affairs in the governorates. The chairmen of boards of directors of the various organizations attached to MOH attend these meetings. A number of committees have been formed, such as the higher committees for the control of communicable diseases, endemic diseases, and health education. Paramount attention is paid by MOH to the development of the managerial capabilities of health personnel at various levels through WHO-assisted training courses.

Social Environment and Service

MOH believes that the basic sociomedical service is to apply, cleverly, the art of serving the individual/patient to help him or her to overcome the difficulties and problems encountered from the environmental conditions and also to study the psychological, social, and economic conditions. When the reasons of the problem are discovered through social diagnosis, therefore, the treatment becomes possible.

Applying the art of group service helps meet individuals' needs, such as programs to improve their conditions, to help cure them psychologically, and to accommodate their new environment. They use modern pedagogical methods to solve their problems through routine visits by the medical social specialist. The increased effort offered to sociomedical care programs is considered one of the most important accomplishments of MOH.

Information about any social phenomenon related to diseases is acquired through field research, and through measuring and evaluating different opinions and ideas of beneficiaries of sociomedical services to improve the services.

3. Public and Private Financing, Health Costs

Local communities are involved through the People's Local Councils at the rural, urban, and governorate levels (each council has a health committee) and the boards of directors of health institutions (local communities are represented thereon). Local communities are involved right from the start, since health plans are initially prepared locally. Communities are, therefore, involved in the financing process, either through self-financing of certain health projects or by contributing to the service promotion funds at various health units or hospitals by collecting minimal fees for certain services.

Expenditure

Government investment in health care is always increasing. For example, the budget of MOH for the fiscal year 1986-1987 was as follows, in Egyptian pounds £E (1 £E = U.S.$0.42):

General Administration Office	85,064,900
Health Council	29,200
Health Governorates	331,638,000
General Organization for Health Insurance	182,458,600
General Organization for Biological and Vaccine Products	6,212,000
National Organization for Drug Control Research (NODCAR)	4,215,000
Curative Establishment in Cairo	25,416,000
Curative Establishment in Alexandria	2,916,700
Hospital Organization	28,359,000
Family Planning	2,558,000
Drug Organization for Chemical and Medical Appliances (DOCMA)	22,037,000
Total	£E 690,931,400

This total amount represents 2.5% of the total state budget of that year, but if, further, the expenditures of other ministries on such health-related activities as drinking water supply and waste disposal, environmental sanitation, education or provision of health care to their personnel are added, the percentage rises to more than 6% of the total public expenditure.

4. Facilities (Hospitals, Clinics, etc.)

The medical facilities operated by MOH are shown in Tables 1-8; also shown are the increase in the number of beds from 1952 to 1987 and the percentage of change from 1983 to 1987.

Aside from MOH, there are other facilities (e.g., hospitals) attached to the Ministries of Interior and Communication and military hospitals attached to the Ministry of Defense. There are also university hospitals, hospitals of public and private companies, and private hospitals and clinics which provide considerable health and medical services. The following graphs show the number of beds in hospitals.

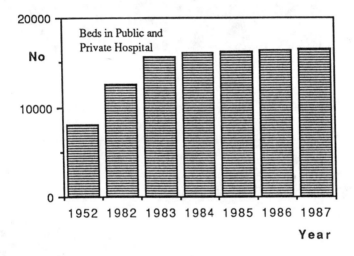

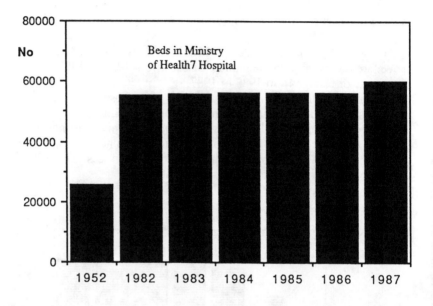

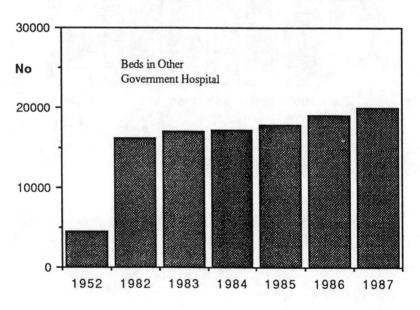

TABLE 1. Number of beds in hospitals, by type of hospital and year.

Type of Hospital	Year							Percent of Change 1983-87
	1982	1982	1983	1984	1985	1986	1987*	
All Types	35744	87453	90445	92003	92700	94354	98344	8.7
Beds in Ministry of Health Hospitals	25710	57470	57989	58867	58961	59912	62527	7.8
Beds in Other Governmental Hospitals	4076	16480	17173	17396	17951	18580	19632	14.3
Beds in Public Sector And Private Sector Hospitals	5958	13503	15283	15740	15788	15862	16185	5.9

TABLE 2. Number of treatment units and beds in towns, by type of unit and year.

Type of Treatment Unit	Unit	Year						Percent of Change 1983-87
		1982	1983	1984	1985	85/86	1987*	
All Types	No.	5579	5633	5741	5742	5849	6175	9.6
	Beds	59541	59872	60517	60788	61639	64325	7.4
General And District Hospitals	No.	331	330	329	324	330	335	1.5
	Beds	23025	23097	24129	24275	24768	25468	10.3
Eye Diseases Hospitals	No.	169	173	177	179	180	182	5.2
	Beds	2924	2952	2954	2975	2943	2940	-4.0
Endemic Diseases Hospitals	No.	160	164	163	162	160	161	-1.8
	Beds	918	953	823	827	850	852	-10.6
Chest Diseases Dispensaries	No.	137	133	144	143	138	121	-9.0
	Beds	395	365	417	417	400	123	-66.3
Chest Diseases Hospitals	No.	39	39	36	35	37	53	35.9
	Beds	7045	7046	7046	7130	7403	7498	6.4

TABLE 3. Number of health service units and beds in rural areas, by type of health unit and year.

Type of Service Unit	Unit	YEAR							Percent of change 1983-87
		1982	1983	1984	1985	1986	1987*		
All types	No.	2554	2559	2608	2596	2607	2740		7.1
	Beds	8961	9181	9070	8987	8890	9059		-1.3
Endemic diseases sections in health centers and combined units	No.	587	589	586	583	582	586		-0.5
	Beds	8961	9181	9070	8987	8890	9059		-1.3
Social centers and rural health units	No.	1967	1970	2022	2013	2025	2154		9.3

161

TABLE 4. Numbers of preventive services in the field of endemic diseases, by type of service and year.

Type of Service	YEAR						Percent of change 1983-87
	1982	1983	1984	1985	1986	1987*	
All Types	2158	2137	2186	2179	2136	2217	3.7
Bilharziasis inspectorates	123	117	119	119	114	118	0.9
Bilharziasis Centers	331	321	320	322	260	326	1.6
Bilharziasis Units	1250	1234	1265	1254	1242	1248	1.1
Malaria Units	420	432	449	450	485	489	13.2
Medical insect units and inspectorates	12	11	11	12	13	14	27.3
Mosquito units	22	22	22	22	22	22	0.0

TABLE 5. Number of services and beds in the field of endemic diseases, by type of service and year.

Type of Service Unit	Unit	Year						Percent of Change
		1982	1983	1984	1985	85/86	1987*	1983-87
All Types--------------	No.	2958	2966	3011	3004	3035	3154	6.3
	Beds	10480	10756	10524	10445	10373	10536	-2.1
Hospital and units for endemic diseases------	No.	587	589	586	583	582	5Ɛ6	-0.5
	Beds	8961	9181	9070	8987	8890	9050	-1.3
Endemic diseases sections in school centers---------	No.	244	243	2ᴛ0	246	268	253	4.1
	Beds	601	631	631	631	633	625	-1.0

TABLE 6. Number of centers for maternity and infant care, by type of center and year.

| Type of Center | Year | | | | | | Percent of Change 1983-87 |
	1982	1983	1984	1985	1986	1987*	
All Types	2796	2801	2853	2841	2864	3000	7.1
Maternity and infant care centers	242	242	245	245	257	260	7.4
Sections for maternity and infant care in health centers and combined units	587	589	586	583	582	586	-0.5
Sections for maternity and infant care in social centers and in rural health units	1967	1970	2022	2013	2025	2154	9.3

TABLE 7. Number of services provided by maternity and infant care centers, by type of service and year (in thousand, except for percent).

| Type of Service | Year | | | | | | | Percent of Change 82/83-86/87* |
	81/82	82/83	83/84	84/85	85/86	86/87*	
Post Pregnancy cases	1094	890	776	731	736	729	-18.1
New pregnancy cases	489	533	542	564	606	553	3.8
Deliveries in centers	408	451	414	481	417	445	-1.3
Infant visiting centers	3495	2795	2522	2914	2816	2547	-8.9
Circumcision for children	11	12	41	N.A	N.A	N.A	--
Small Pox Vaccinations	881	937	-	N.A	N.A	N.A	--
Diphteria vaccinations	1220	1042	1410	1677	1484	1471	41.2
Polio vaccinations	1437	1057	1383	1668	1569	1081	2.3

TABLE 8. Number of products of ministry of health labs by types of product and year (in thousand except for percent).

Type of product	YEAR						percent of change 1983-87
	1982	1983	1984	1985	85/86	86/87*	
D.T	164	1362	199	1326	2422	15048	708.2
DPT vaccine	3195	4693	3237	5201	5898	7770	65.6
Poliomylitics	8845	34800	29353	30000	25350	25758	-26.0
Scorpion bites vaccine	245	279	95	156	247	253	-9.3
Snake bites vaccine	2	2	3	5	14	33	1550.0
Cholera vaccine	59816	13783	15506	945	3035	2300	-83.3
D.T	520	395	7	1090	1530	487	23.3
T. vaccine	1363	2464	2030	1073	690	1158	-53.0
Dry B.C.G vaccine	2013	3500	122	2504	2996	3384	-3.3
Dry B.C.G vaccine for children	846	-	1840	1860	3825	2000	--
(R.V.) Rabies vaccine	753	597	549	28	31	36	-94.0
Anti haemophilia globul n	3	3	2	1	2	2	-33.3
	15	3	13	32	35	--	--
Anti haemophilia plasma	1	1	1	(1)	(1)	(1)	--

(1) Less than 1000

5. Manpower

Studies covering health services requirements until the year 2000 were prepared by the National Specialized Councils of Egypt. The manpower and material resources are being distributed to the various governorates on the basis of short- and long-term plans.

A plan for developing health manpower, particularly primary health care personnel, is being implemented. Physicians, nurses, and other categories are distributed in accordance with primary health care requirements at the governorate level. Admission to the technical health institutes is based on geographical distribution and on local needs. Moreover, programs of work, performance criteria, and responsibilities for each category of health worker have been formulated. The basic education at the faculties of medicine and nursing schools, and also at the higher institutes of nursing, is being developed in light of primary health care needs. Members of the health team receive training prior to taking up their duties; their training takes place locally in training centers in the various governorates. This is in addition to the continued training of these personnel. Each staff member attends a training course at least once every year throughout her/ his service at primary health care units. Such training covers all technical and managerial skills. It has also been found necessary, under the development plan, to provide each urban health care unit with a medical records technician.

MOH is keen on participating with the universities in studying the needs for health professionals for the present and for the future. Also, the Ministry is looking forward to an effective role for the universities in the development and implementation of continuing medical education programs and in planning, executing, and evaluating research studies aimed at improving the health status of the country. At the same time, it is the policy of MOH to encourage postgraduate studies in all specializations for physicians, pharmacists, dentists, and nurses.

Total Number of Registered Members
of Medical Professions in Egypt
1987

Physicians	86,778
Pharmacists	25,906
Dentists	11,926
Nurses	59,209*

*Only 2,043 are graduated from the High Institute of Nursing (i.e., are university graduates).

6. The Social Security System

The Health Insurance System

The health insurance system of Egypt started at the same time as the social security system, in the early days of the 1950s. The health insurance system of Egypt is under the direct supervision of the General Organization for Health Insurance, a major organization attached to MOH. It is considered the cornerstone of health services. It is the dream and hope of MOH to cover all the citizens of Egypt against disease and occupational accidents.

The policies and goals of the health insurance system of Egypt are to:

- Generalize and spread health insurance to cover all citizens of Egypt
- Develop the already existing insurance laws to suit all the citizens, including pensioners, widows, widowers, etc.
- Increase the number of health insurance polyclinics, both on the general practitioner and the specialized levels
- Continue to develop and modernize all their clinics
- Give extra care to cases of occupational accidents and diseases and establish special centers for such cases
- Cooperate with other institutes of MOH concerning programs of family planning and birth control
- Provide health care to all pupils in schools and also to all agriculture workers.

Health Insurance in Case of Occupational Accidents

The total number of those insured in occupational accidents was 6,877,000 in 1986. Occupational accidents are defined as:

- Injury due to any occupational disease
- Injury as a result of an accident during working hours or as a result of the work
- Injury as a result of fatigue because of work
- Any accident that occurs on the way to or from work.

Financing of the health insurance system for occupational accidents is through subscription by the employer (government pays 1%, public sector pays 2%, and private sector pays 3% of the total salary). The employee or worker does not pay anything when receiving any curative services, rehabilitation from any injury or accident for drugs.

Curative Services

All necessary medical services needed for the treatment of injured patients are free of charge for outpatient or inpatients, such as operations, laboratory investigations, X ray, rehabilitation services, physical therapy, and medicines.

7. Payment and Reimbursement

The following rules are applied:

- The injured person is entitled to financial aid during the whole period of injury (temporary disability), which equals 100% of salary on the day of the injury.
- Reimbursement for cases of permanent disability depends on each case. If the injury is less than 35% of total disability, the injured person receives reimbursement in the form of cash money, but if the injury is more than 35% of total disability, the disabled person retires and receives a monthly pension proportional to the degree of the disability.
- The pension for permanent disability or death due to occupational accident is 100% of the salary on the day of the accident or death.

Health Insurance Against Diseases

The total number of beneficiaries is 3,388,000 (1986). The General Organization for Health Insurance does hope to offer all necessary services to all workers, employees, civil servants, etc., in the government public or private sectors and similar establishments. Also, it is hoped that it will cover all members of their families.

Financing such cases depends on subscription through payment of 4% of the total salary of the insured person (1% deducted from the salary + 3% paid by the employer). This is the case of those who work for public or private sectors. Each beneficiary pays in the form of insurance stamps, £E 0.03 for curative service and £E 0.20 for a house visit. For those who work for government agencies and general establishments, insurance is the 2% of total salary of the insured person (0.5% deducted from the salary + 1.5% paid by the employer). Each beneficiary pays:

— £E 0.05 for general practitioner service
— £E 0.10 for specialist service
— £E 0.20 for a house visit
— £E 0.25 for the price of medicine (not to exceed £E 1.0),
— £E 0.25 for laboratory investigations, x-ray, etc. (not to exceed £E 1.0)
— 50% of the cost of any medical prosthesis equipment
— £E 0.25 to £E 0.50 per night in hospitals as an inpatient depending on the class of the room of choice.

Pensioners pay 1% of their pension and enjoy full health insurance services as outpatients. They pay only £E .03 in the form of an insurance stamp for every visit to either the general practitioner or specialist, and medicine is dispensed free of charge.

The General Organization for Health Insurance offers its services and activities in its general practitioner clinics, polyclinics, specialized centers, or hospitals. In rural regions where there are no such clinics or hospitals, beneficiaries receive their medical treatment at private doctors' clinics and general hospitals or specialized hospitals (such as fever, psychiatric, tumors, etc.), according to bilateral agreements.

B. PHARMACEUTICAL INDUSTRY AND DRUG DISTRIBUTION

1. Drug Policy

The national drug policy of Egypt is an integral part of its health policy to ensure an adequate supply of safe and effective drugs of good quality.

The goal of MOH concerning drug policy is to develop, within the resources of the country, the potential that drugs have to control common diseases and alleviate suffering, and also to supply the needs of the country with good quality drugs at reasonable prices. This is achieved by:

— Supporting activities related to local production of pharmaceuticals so that MOH may face the increasing local demand for medicines
— Limiting the role of importation as a complement to local production
— Encouraging locally manufactured drugs
— Exchanging information and specific cooperation with international and local scientific organizations and following up the requirements and rules of GMP, GLP, GVP, GSP, etc., for the sake of better production of high quality drugs
— Rationalization of drug consumption

— Avoiding sudden increases in drug prices and selling them at subsidized prices
— Supporting the role of NODCAR to assure the quality of drugs either locally manufactured or imported.

2. Structure of the Industry

The pharmaceutical industry in Egypt passed through three different stages during the last 55 years (1934-1989).

The First Stage: 1934 to 1959

This stage witnessed the birth of the Egyptian pharmaceutical industry, where either completely private or public limited companies and laboratories were dominating. They were the following:

— 3 large public limited companies
— 16 middle-size private companies
— 22 small private laboratories for the manufacture of simple pharmacopoeial preparation.

During that period, the authority of MOH was limited to approving registration and issuing licenses to locally manufactured as well as imported drugs. This was carried out by the General Administration of Pharmacy, MOH. In addition, some quality control activities were carried out by the central laboratories of this administration. There was no legislation to protect local production against the competition of this period. The volume of the national drug industry covered 30% of the total national drug requirements.

The Second Stage: 1960 to 1975

In 1960, drug trade was nationalized, followed by pharmaceutical industries in 1961. Since that year, all small laboratories that were unable to offer or use modern technical means for production were banned. At the same time, middle-size companies were merged to become bigger and economically stronger. From that time, all drug activities were under the direct authority and supervision of the Egyptian General Establishment for Drugs and Medical Appliances, MOH. This establishment completely dominated the process of obtaining raw materials, production importation, exportation, marketing, storage, and distribution through its various companies. The pharmaceutical body during that period consisted of 11 public sector large-size companies as follows:

- Seven companies for production of pharmaceutical preparations
- One company for both production of raw materials and pharmaceutical preparations
- Two commercial companies for import/export, storage, and distribution of pharmaceutical preparations, raw materials, and medical appliances
- One company for production of packaging appliances
- Joint-venture companies for production of pharmaceutical preparations
- A center for quality control and research (established in 1963).

At the end of that period, the volume of the national drug industry covered 74% of the total national drug requirements.

The Third Stage: 1976 to 1988-1989

At the beginning of this stage, the Egyptian government adopted the economic policy of open-door investment. As a result, the Egyptian General Establishment for Drugs and Medical Appliances was banned, to give full freedom to all pharmaceutical companies practicing in Egypt to plan, direct, and manage their own policy and business. Instead, the Drug Organization for Chemical and Medical Appliances (DOCMA), MOH was established in 1983 to practice only strategic planning, follow-up, and evaluation of the broad lines of the performance of all the pharmaceutical industries in Egypt. During that period, the following companies were established in addition to those which had been previously established and mentioned under the second stage:

- Two joint-venture companies for production of pharmaceutical preparations
- Six private companies for production of pharmaceutical preparations
- One private company for production of large parenteral solutions
- One private company for production of both raw materials and pharmaceutical preparations
- One private company for production of raw materials
- One private company for production of medical packaging appliances
- One private company for production of pharmaceutical glass
- One private company for production of gelatin pharmaceutical products (capsules)
- Two private companies for production of medicinal plants
- One private company for production of medical food products

— One private company for production of veterinary products
— One private company for production of veterinary products, animal feed, and food additives for veterinary use
— Three private companies for production of medical appliances
— One private company for production of hospital appliances.

These private companies are either purely Egyptian capital or foreign capital investment or a combination of both.

In 1976, NODCAR was established in place of the Center for Quality Control and Research and included the central laboratories of MOH. At the end of that period, the volume of the national drug industry covered 86.8% of the total national drug requirements, and the rest was covered through importation (13.2%).

3. Research and Development, Imports, Exports, Patents, Licensing

Research and Development

Research and development are an essential feature of Egyptian national strategies for health for all by the year 2000. The research priorities considered by DOCMA, MOH are to determine the needs of the drug industry in terms of its immediate production programs and the country's long-term needs. Research and development are to be focused on developing complete technological packages rather than fragmented ones, such as production of pharmaceutical raw materials using locally produced intermediates. The policy of open-door investment yields short-term and long-term benefits by leading to development of badly needed new drugs, improvement in existing ones, and a more rational use of drugs. Exploration and development research into Egyptian raw materials and traditional medicines and evaluation of their use is always the goal of the Egyptian pharmaceutical industry.

During the last few years, only 13 pharmaceutical chemicals (raw materials) were successfully synthesized through cooperation between research departments of some pharmaceutical companies and the National Research Center in Cairo. These compounds are produced nowadays for the first time on a semi-industrial scale and before the end of this year will be produced on an industrial scale. Very encouraging results had also been reached concerning another 18 pharmaceutical compounds that will be produced industrially within the next three years.

The seven Faculties of Pharmacy of Egypt are also engaged in various projects with some pharmaceutical companies to synthesize new compounds, develop new formulas of dosage forms, isolate and characterize active constituents from Egyptian medicinal plants, and carry out pharmacological and microbiological screening of new compounds and study their bioavailability.

Production, Importation, and Distribution Policy

Production by using pure Egyptian technology is tolerated in the form of license royalty agreements as long as it satisfies the standard contract issues by DOCMA, MOH, which necessitates that royalty is fixed as a maximum of 5% of the net sales for a maximum period of five years after being approved by DOCMA, MOH board.

Egypt, together with a few countries in South America, Asia, and East Europe, is classified by UNIDO/Vienna as a country that is able to manufacture most intermediates required and undertakes local research on the development of innovative products and manufacturing processes.

Local drugs are available for most sickness, but some drugs for treating important health problems such as diabetes and cancer must still be imported. Furthermore, some chemical raw materials for locally produced drugs, as well as bulk drugs and intermediates for drugs produced locally under license, have to be imported.

Importers

Owing to the foreign exchange control policy, imported life-saving drugs and baby milk receive preferential subsidized prices. In Egypt, the biggest importers of drugs, raw materials, and medical appliances are actually two public sector companies. They are:

— *The Egyptian Pharmaceutical Trading Company (EPTC)*
  This is the largest drug importer in the country. In 1988, it handled 86.565% of the drugs imported into Egypt, which is equivalent to 119.85 million Egyptian pounds (£E).
— *El-Gomhoria Pharmaceutical, Chemical and Medical Appliances Trading Company*
  This company is the exclusive importer of pharmaceutical raw materials and handles almost all raw materials and medical appliances consumed for the production of drugs by pharmaceutical manufacturers in Egypt.

The activities of these two companies are supervised by DOCMA, MOH. The smallest share of importation belongs to the private sector, which in 1988 handled 13.435% of the imported drugs in the country, equivalent to 18.60 million Egyptian pounds (£E).

Owing to the shortage of hard currency, the government of Egypt is encouraging and promoting local production of drugs and trying to reduce the number of imported ones. Since these drugs are quite important to health, smooth procurement and supply of imported drugs is one of the largest responsibilities of EPTC. It is observed that the number of imported drugs is high.

All imported drugs have to be approved by the Supreme Committee of Drugs, established within DOCMA, MOH, and each drug is assigned a rank of priority.

Exporters

MOH gives great attention to the activity of exportation nowadays and considers it one of the main goals of the sector of drugs. Private and joint-venture pharmaceutical companies have all the advanced technical facilities, well-trained qualified manpower, excellent production, and quality energies that make it easier to compete in the field of exportation of drugs.

The general policy of the country is to encourage opening new markets in Arab and African countries, through exporting quality drugs manufactured according to GMP regulations, to gain and keep a good name for the Egyptian pharmaceutical industry and its products. It is hoped that the amount of exportation will be increased more and more year after year. For example, total exportation of drugs during the last five years was as follows (in millions):

1983-1984	7.9	£E
1984-1985	8.6	£E
1985-1986	10.4	£E
1986-1987	12.6	£E
1987-1988	15.0	£E
1988-1989	25.0	£E (expected)

Special Requirements for Imports and Exports

Imports. An import permit is needed in all cases, and the applicant must hold a marketing authorization. However, importation of nonregistered drugs is allowed upon individual requests and on a physician's recommendation. Certificates are required proving authorization for sale of distribution within the country of export and stating price in the country of export, public sales price in country of origin and in countries where the drug is marketed, import price free on board (FOB), and the proposed public sale price in Egypt. In addition to quality standard references for the nonactive ingredients, certificates of analysis of the product and of the stated active ingredients(s), scientific references attached to the technical data, formulation certificates, statements of different trademarks, and certificates of validity are required. Antibiotics require a certificate of analysis for the active substance and for each batch of imported finished product. Vaccines require a certificate of analysis for each batch imported. Radioactive substances are imported by the specialized centers and universities and are not subject to registration. They are released from customs after approval from the Atomic Energy Authority on the responsibility of the importers. There are approved lists for colorings, flavors, and sweeteners. There are no arrangements for mutual recognition of inspection.

Exports. There are no specific requirements, but the producer in Egypt must be an authorized manufacturer, and certificates confirming this may be requested in accordance with the WHO Certification Scheme.

Distribution of Drugs

Drugs are incorporated in the welfare of the health care system of Egypt, so their influence on the public and the society is very strong. Any outbreak of disease is unpredictable, and demand is not constant either in time or location. Demand for drugs for influenza is seasonal, but on the whole, the demand for drugs is changing and is very unstable. Thus, necessary drugs, at the necessary time, in necessary amounts must be supplied through the well-adjusted distribution system.

In Egypt, drugs are distributed through a route of manufacturers, distributors, and dispensaries or pharmacies. The function of these distributors is to tie production to consumption by means of providing services through purchasing, collecting, storing, selling, sorting, and distributing drugs. Aside from these basic functions, distributors are responsible for providing constant supply, strict quality control, and accurate information.

EPTC, besides being the main and largest importer and distributor of drugs in Egypt, also distributes 25% to 50% of all drugs locally manufactured. By the end of 1988, distribution was carried out through its own distribution centers. There are seven main warehouses which supply 43 branches. These branches supply 27 networks of pharmacies spread throughout the country; they also supply hospitals and rural health centers and units with their pharmaceuticals and paramedicinals, either locally manufactured or imported. EPTC is increasing the number of distribution centers and outlets year after year, especially in remote rural regions. At the same time, pharmaceutical companies have the right to distribute a majority of their production directly to pharmacies, health centers, and hospitals. To ensure excellent, smooth, and efficient performance in the process of storage and distribution, EPTC adopted systems of automation and computerization.

It is the policy of MOH to assist all pharmacies of the country to obtain all their needs without any difficulties. The following table shows the distribution and number of pharmacies in the 26 governorates of Egypt:

Number of Pharmacies in 1987, By Governorates and ownership

Governorate	All Pharmacies	Individually Owned Pharmacies	private and Cooperative Pharmacies
All Governates	8962	8778	184
Alexandria	997	980	17
Aswan	63	59	4
Asyut	348	342	6
Behera	507	499	8
Beni-Suef	121	120	1
Cairo	1783	1727	56
Dakablia	812	810	2
Damietta	133	128	5
Fayum	178	174	4
Gharbia	530	520	10
Giza	845	825	20
Ismailia	94	89	5
Kafr-Ei-Sheikh	263	257	6
Kalyubia	400	395	5
Matruh	14	14	-
Menia	324	324	-
Munufia	339	335	4
New Valley	9	6	3
Port Said	93	89	4
Qena	234	227	7
Red Sea	11	6	5
Sharkia	485	481	4
North Sinai	17	17	-
Suez	59	54	5
Suhag	303	300	3
South Sinai	-	-	-

Patents

Protection of intellectual property is greatly considered in developing national drug policy in Egypt. Applicants for patent registration submit special application forms to the patent office of the Academy of Scientific Research and Technology of Egypt. Applicants should give a detailed description of the patent, including the most important elements needing protection.

Few Egyptian pharmaceutical companies have patents for completely new synthesized pharmaceutical compounds, formulas of dosage forms, or newly isolated, elucidated, or discovered natural products. Also, there are a number of patents belonging to the Organization of Atomic Energy of Egypt, for radioactive isotopes used either for medical treatment or diagnosis.

4. Manufacturing

Local Manufacturing of Drugs

The total local manufacturing of drugs in Egypt in 1987-1988 was £E 800 million and was approximately £E 900 million in 1988-1989.

Manufacturing is divided among public sector companies, which have the biggest share of the total production of Egypt, and the joint-venture and private companies.

In 1986-1987, the public sector manufactured 60.99%, the joint-venture sector manufactured 24.17%, and the private sector manufactured 14.84%. In 1988-1989, the public sector manufactured 55.8% (63% of Egyptian drugs, and 37% under license); the joint-venture 22.1% (all under license), and the private sector 22.1% (61% of Egyptian drugs, 39% under license).

All the needs of the country for locally manufactured drugs are provided in agreement with the general health policy of MOH and the actual needs of the country. The expected increase in drug consumption is always considered. For example, consumption of drugs had tripled since 1980. It was £E 6.38 per person per year in 1980 and reached £E 20.0 per person per year in 1988.

It is the role of DOCMA to coordinate all companies of the public sector, joint-venture, and private companies to guarantee smooth supply of the needs of the country for all groups of drugs, to be sure that every drug company is implementing and reaching its specific goals.

The goals of public sector companies are to implement the policy of the state in guaranteeing the manufacture of basic drugs needed by all citizens, and to ensure that these drugs are of good quality, safe, effective, and sold at reasonable subsidized prices. Their profit is just enough to cover their expenses and achieve an economic balance to assure the continuity of the process of manufacturing and production, the process of resetting and renewing their old lines of production, modernizing their laboratories, and paying salaries and incentives to all workers and staff.

The goals of joint-venture and private investment companies are to:

— Manufacture newly developed drugs by using advanced technology to gradually replace the imported drugs and, thus, contribute to the improvement of Egyptian trade balance and narrow the gap of balance of payment
— Manufacture raw materials, particularly those now imported, to supply the majority of the needs of pharmaceutical industries
— Export the excess of their production and thus increase hard currency in the country
— Cover the deficiency and shortage of some imported specialties which the public sector companies are not able to manufacture to satisfy the needs of the Egyptian market
— Produce medicinal plants and herbal preparations
— Manufacture medical appliances
— Manufacture intermediate medical commodities, such as X-ray films and surgical sutures
— Manufacturer pharmaceutical machinery used for production
— To manufacture packaging materials and appliances
— Apply specific high technology, such as biotechnology and genetic engineering, in drug manufacturing.

DOCMA always follows up all manufacturing and production plans of all pharmaceutical companies and tries to overcome all the problems which arise. Also, it has to review the drug manufacturing plan every three months according to the actual consumption bases of every product, and as a result, modify the plan if necessary.

5. Advertising and Promotion

All drug companies practicing in Egypt, as well as all importers, distributors, manufacturers, and exporters, depend on advertising and promotion of their products through:

— Medical/scientific offices and representatives
— Medical, scientific, and pharmaceutical journals, magazines, and newsletters
— Specialized periodicals
— Conferences, seminars, and symposia
— Professional exhibitions.

The medical staff depend on these five methods to get acquainted with new trends and discoveries of new drugs. Medical representatives usually visit doctors routinely in their clinics, health centers, or hospitals to advertise and promote their products through medical discussions and dissemination of unbiased and completely reliable, accurate, truthful, informative, balanced, up-to-date drug information. The representatives explain the therapeutic importance of their products, side effects, adverse reactions, etc. They usually support their discussions by showing independent scientific and medical literature such as booklets, pamphlets, brochures, catalogues, or pictures of cases before and after using the product(s). In most cases, they give the doctors free medical samples of prescription drugs for trials, but in modest quantities. Medical representatives play an important role because of the rapid dynamic development and the numerous discoveries of drugs.

The medical representatives should be qualified and have an appropriate educational background. They should possess sufficient medical and pharmaceutical knowledge and integrity to present information on products and carry out other promotional activities in an accurate and responsible manner. They have to abide by the Ethical Criteria for Medical Drug Promotion of Egypt, as their activities are medical/scientific rather than commercial.

The activities of these medical/scientific offices are under the direct supervision and control of CAPD, MOH to ensure that only complete medical, pharmaceutical, and scientific facts are practiced and also to ensure that promotion is in keeping with the national health policies and in compliance with the Egyptian regulations, as advertising and promotion of drugs may greatly influence their supply and use. Monitoring and control of both activities are essential parts of the national drug policy of the country. At the same time, all advertising materials such as labels, inner inserts, etc., should be scrutinized with a view to ensuring compliance with the legal requirement and must be agreed upon by CAPD. These materials should not contain misleading or unverifiable statements or omissions likely to induce medically unjustifiable drug use or to give rise to undue risks.

It is strictly prohibited by force of Egyptian legislation to advertise any medicine in daily newspapers, radio, or television, except for aspirin and oral rehydration therapy for children. In the case of advertisement in specialized journals or magazines, different regulatory provisions can be formulated to regulate different situations in which a manufacturer or distributor might advertise. In general, according to the state regulations, there must be a fair balance in presentation of information regarding merits and demerits.

The majority of medical/scientific offices practicing in Egypt are well equipped with audiovisual aids (e.g., movie films, videotapes, cassettes, small libraries, etc.) to demonstrate and help in advertising and promotion of their products to any interested member of the medical or pharmaceutical staff. Usually these offices arrange conferences, seminars, symposia, or lectures related to their products. Some of them issue their own newsletters and magazines and have their own exhibitions.

6. Cost Containment and Price Regulation

All drugs found in the Egyptian market are submitted to compulsory fixed prices. Control of drug prices is exercised by MOH, which fixes the prices of all locally manufactured and imported drugs according to the actual cost of production or importation. In the case of health insured citizens, sale prices are controlled by regulations governing social and health welfare reimbursement.

The current prevailing tendency in Egypt is for a price freeze and diminished profitability levels for essential drugs, as well as ensured tariff protection for locally manufactured products. The various mechanisms for such protection could be:

—High tariffs on imports of finished products
—Lower or no tariffs on imports of raw materials and bulk drugs
—Authorization of imports only for essential drugs that are not produced locally.

In addition, tariff policy offers special incentives to local, joint-venture, and private investment producers, such as special tax relief, a high profit margin, etc.

7. Drug Approval and Government Regulation

Registration and Licensing of Pharmaceuticals and Premises

Registration Policy

The registration of pharmaceuticals started in Egypt by 1930. The purposes of registration are: to ensure that safe, effective, and inexpensive drugs of good quality reach the consumer; to discourage the abuse and misuse of drugs; and to ensure fair trading practices. Registration requirements generally apply in respect to all drugs either locally produced or imported. The information for registration can be classified into two categories:

Administrative data. Include information relating to such matters as the name of the product and the details of the manufacturer, the status of production in the country of origin and other countries (for imported drugs), the content of labeling and advertising materials, etc.

Technological data. Include pharmaceutical, pharmacological, toxicological, therapeutic, and clinical data and give detailed information of a technical and scientific nature.

Three different institutions are engaged in the registration process and constitute the Egyptian licensing authority. These three institutions are subordinated to MOH. They are:

- *DOCMA:* performs medical, clinical evaluation and selection according to prevailing policies
- *NODCAR:* performs chemical, bacteriological, pharmaceutical, and pharmacological evaluation and stability testing of the selected pharmaceuticals
- *CAPD:* performs final evaluation and the administrative registration process until the registration is granted. The grant is valid only for 10 years, and the renewal of registration requires the same procedure as that for a new drug.

The sketch in Figure 1 shows the institutions, organizations, departments, and committees engaged in the registration procedure.

Requirements for Registration

A. Applicant

1. Applicant should be an authorized representative (of the local or foreign manufacturing company or medical/scientific office or agency representing one or more foreign companies), preferably a pharmacist or other member of the medical profession.
2. Authorization of the foreign manufacturer should be legalized from the Egyptian Embassy in the country of origin.

FIGURE 1

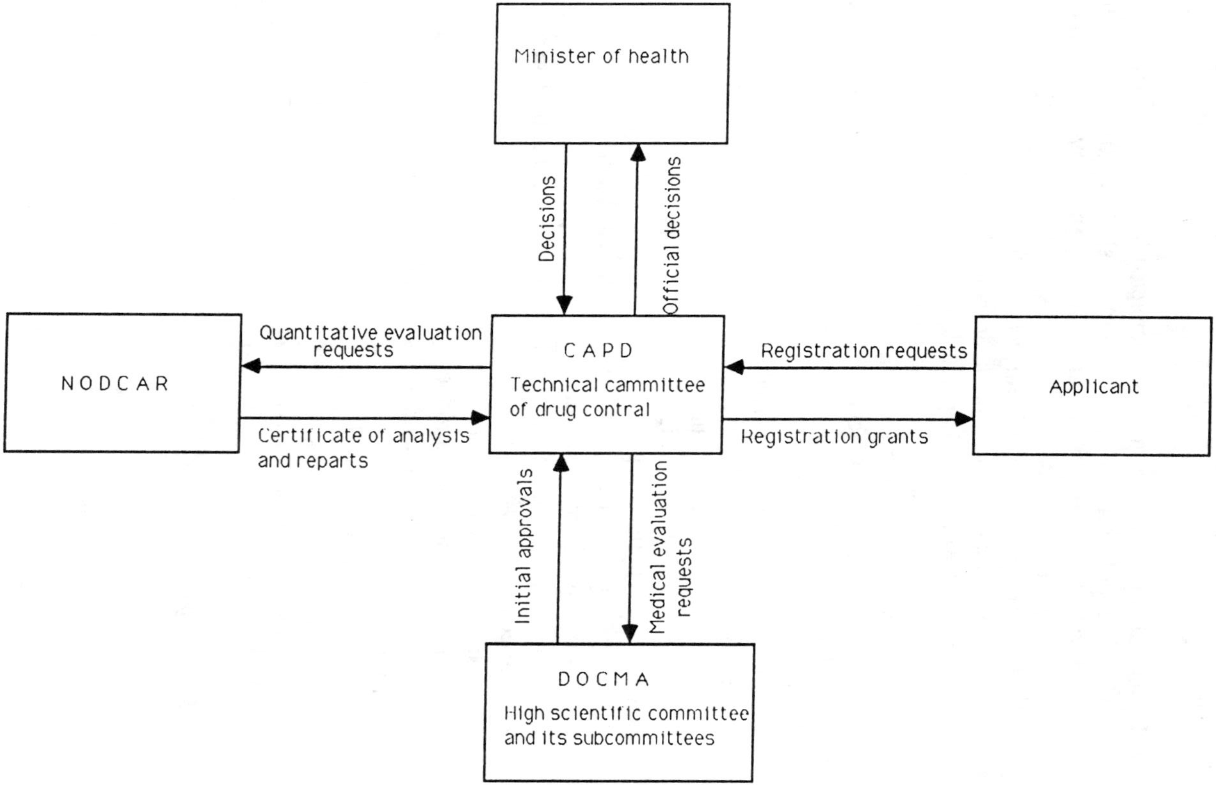

Minister of health
Decisions
Official decisions
NODCAR
Quantitative evaluation requests
Certificate of analysis and reparts
CAPD
Technical cammittee of drug contral
Registration requests
Registration grants
Applicant
Initial approvals
Medical evaluation requests
DOCMA
High scientific committee and its subcommittees

B. Forms

1. Special application form to be completed in English or Arabic
2. Certificate of origin (for foreign products) issued from the health authority in the country of origin and legalized by the Egyptian Embassy there, stating:
   - Name of product and pharmaceutical form
   - Name of manufacturer and address
   - Active constituents and percentage
   - Statement denoting that this product is sold under the same composition in the country of origin
   - Statement that manufacturing company follows and conforms to standards of GMP
3. Full details of constituents
4. Full details of methods of analysis
5. Certificate of analysis of the batches of samples presented (another certificate is needed for antibiotics)
6. Full scientific data, specially for new chemical entities, including chemical, pharmacological, toxicological, and biological data
7. *Labeling:* Copies of each inner and outer label should designate:
   - Name and form of product
   - Active ingredient(s) (preferably generic names) and percentage
   - Special precautions
   - Name of manufacturer and address
   - Manufacturing data (products are allowed five years maximum)
   - Expiration date (for antibiotics and biologicals)
8. Inner inserts
9. Expenses free on board (FOB).
   - Expected selling price in Egypt
10. Samples of the product having one and the same batch number for checking and analysis by NODCAR.

C. Biological Products

1. All sera, vaccines, and hormone products are subject to the same registration regulations as medicines.
2. All human plasma derivatives are subject to the same regulations as medicines, with extra certificates of AIDS- and Hepatitis B-free reports.

D. Diagnostics and Medical Devices

These are subjected to all previous consent, but at present, no registration is required; all medical devices for sterile handling are checked for sterility.

General Consideration for Drug Registration

All of the following articles must be registered before marketing:

- Pharmaceuticals for human and veterinary use
- Surgical sutures
- Contact lens preparations
- Food additives for veterinary use
- Sera and vaccines
- Cosmetics
- Domestic insecticides.

Registration Procedure

The applicant submits a package of three files to CAPD composed of:

- Main file
- Subsidiary file
- Scientific file.

Documents of the three files are revised by CAPD. The revised subsidiary and scientific files are to be sent to the Egyptian Drug Information Center (EDIC), a department of DOCMA, MOH.

Upon receipt of the files by EDIC, the Egyptian Drug Information File (EDIF) sheet document is revised to ensure the compliance of data and to add all WHO, FDA, and DHSS notifications and recommendations, as well as to select a number of similar substitutes available in the local market. Both files are submitted to the Technical Secretariat of the Scientific Committees, who refers them to the High Scientific Committee.

If medical and clinical evaluation studies performed on the product agree, the product is considered scientifically viable and will be granted an initial approval to continue the registration procedure. If the product is not approved, all files will be returned to the applicant; however, there is a possibility for the applicant to appeal against the disapproval decision, but reasonable recommendations for its reconsideration must be included.

High priority is given to the product that satisfies one or more of the following criteria:

- The product is a breakthrough, with a new mode of action, and is of vital therapeutic potency.
- The product is a new development of currently available products, but with modified effects.
- The product is similar to a currently imported product in the market with an expected lower public price but is on the same quality level.
- The product is similar to a currently locally produced one that is suffering from technical shortcomings leading to suspension or withdrawal of the product from the market.
- The product is similar to a currently locally manufactured product produced by foreign technology (i.e., under license), thus promoting the national technology.
- The product is an orphan drug.
- The product has no more than four locally produced competitors, thus encouraging competition and production of other items.

Once the initial approval is granted, the Technical Secretariat of the Scientific Committees sends the subsidiary and scientific files, together with the grant decision, to the Technical Committee for Drug Control at CAPD, where the samples and methods of assay will be sent to NODCAR to complete the chemical, bacteriological, pharmaceutical, and pharmacological evaluations. NODCAR then notifies the Technical Committee for Drug Control with the formal report, including all results and details of evaluation processes to get final evaluation and approval.

The Technical Committee for Drug Control at CAPD is the legal authority responsible for issuing the registration grant for licenses, and all decisions must be officially signed by the Minister of Health.

Licensing of Premises

There are basic sanitary requirements for pharmaceutical establishments which are always revised and amended.

Lately, owing to the wide expansion of the pharmaceutical industry in Egypt within both the public and private sectors, the need arose for more specific control. Through the concept of WHO specifications for GMP, efforts were directed to establish basic requirements and standards for licensing drug firms practicing in Egypt. A committee consisting of scientists, experts from the pharmaceutical industry, and technical officials from MOH work on this matter.

There is a ministerial decree specifying the basic technical and sanitary standards required to issue a license. Previously existing drug firms were given a transition period of three years to reform their situation in accordance with the decree.

8. Competition: Brand and Generic

Egyptian National Policy of Essential Drugs for Health Care

The participation in the selection and issuance of the list of essential drugs and implementation of the National Policy concerning this matter was started in Egypt in 1984 by CAPD, MOH, and the list was reviewed and updated during 1986-1987. It was approved, finally, as a model in 1988. Further updating and yearly revision will take place whenever needed.

Aim

1. Because of the growing number of drug products and competition under brand names that flood the Egyptian market (2,184 drug specialties, including 406 imported ones, in 1988), some of which are irrational, ineffective, inappropriate, and needlessly expensive and which, through misuse, may impair rather than improve health, the CAPD, MOH has undertaken to provide a list of essential drugs. The list is based on the selection of simple, high-quality, safe and effective drugs at a fairly reasonable cost.
2. The list covers prevalent diseases and meets real patients' needs according to different conditions, such as genetic, demographic, economic, treatment facilities, and available personnel, including general practitioners, specialists, pharmacists, and nurses.
3. The MOH will, in turn, establish a system for continuous, sufficient, and efficient supply of essential drugs for public health.
4. The local manufacturer will be encouraged to supply essential drugs under nonproprietary names to save money and to minimize drug importation.

General Information

- The list of essential drugs is classified into 29 pharmaceutical groups covering almost all basic health needs.
- The number of drug items is 227.
- International Nonproprietary Names (INN) (generic names) of active substances are adopted, with some references to trade names to

assist and familiarize prescribers, but with an emphasis on using INN.
— Individual active component drugs are mostly chosen, and combination products are avoided except when urgently needed.
— The list consists of drugs with defined dosage forms and strengths, with detailed clinical categorization.
— The list includes alternative drugs to provide a wider selection for different patients.
— The list does not imply that some drugs are not useful, but simply that, in a given situation, these drugs are the most needed by the majority of the population and should be available at all times in adequate amounts.

C. PHARMACY PRACTICE

1. Dispensing a Prescription: Where, What, Payment, Formulary

It is commonly agreed that the art of prescription writing originated thousands of years ago with the Ancient Egyptian priests, who practiced prescription writing because they engaged in treating the sick in addition to their religious duties. There are preserved in museums such Egyptian records as those cut on stone or written on papyrus that contain a variety of medical formulas. As a matter of fact, there are several formulas preserved in the British Museum which are said to date form the time of King Cheops, about 3700 B.C.

No one is allowed to prescribe a medicine and issue a prescription in Egypt unless she/he is a physician, dentist, or veterinarian (i.e., a member of the medical profession) directing the pharmacist to compound and dispense medication for a patient and usually accompanied by directions for its administration or use.

The prescription is written and signed by the prescriber. It is usually written in English and, in a few cases, in Latin or Arabic. It is the responsibility of the pharmacist to interpret the wishes of the prescriber, and thus she/he should be familiar with the typical form of the written prescription.

According to Egyptian legislation, every prescription should carry the name of the prescriber printed at its top, with title, specialization, registry number, address, and telephone number, if any. The following data should be mentioned in the prescription, which contains the symbol R/:

— Name of patient, address, age, and sex
— The inscription, which is the principal part of the prescription order, and which contains the names and quantities of the ingredients and usually employs the metric system, or which contains names and kinds of medicines and usually brand names
— The subscription, which consists of directions to the pharmacist for preparing the prescription (e.g., number of doses, injections, capsules, tablets, suppositories, etc., to be supplied)
— The Signa (i.e., the directions for the patient)
— The renewal instructions (sometimes)
— Signature of the prescriber and date.

Responsible directors of pharmacies in Egypt should be university graduates holding a minimum B.Sc. in pharmaceutical sciences. They should be registered licensed pharmacists. Usually the pharmacist is the person who personally receives the prescription from the patient or the person who presents the prescription for the patient. If any ambiguity, dangerous dose, incompatibility, or other features of the prescription are noticed, the physician must be consulted by the pharmacist before the prescription is dispensed.

The Egyptian laws do not allow the pharmacist to make any changes in the prescription that would affect its medicinal activity or alter the intent of the prescriber. Alterations or additions may be made only after consultation with the physician, either personally or, in most cases, by telephone.

Nowadays, contrary to the early days of the 1950s and before, the number of prescriptions that require compounding represents only a very small ratio compared to the total number of prescriptions dispensed in any pharmacy. It is very important to mention, therefore, that it is the policy and goal of all the pharmaceutical bodies of Egypt (Egyptian Syndicate of Pharmacists, Egyptian Pharmaceutical Society, and all Faculties of Pharmacy of Egypt) to encourage and convince all the other members of the medical profession to return to the old days of compounded prescriptions and prescribe them for their patients. Therapeutically, it will be more useful and effective, and rationalization of drugs could be achieved. Professionally, it is more prestigious to the pharmacist and the doctor and their professions.

Pricing of Prescriptions

Prices of drugs manufactured by pharmaceutical companies are fixed and controlled by MOH, but concerning compounded prescriptions or official preparations of the Egyptian Pharmacopoeia or Formulary, the price depends—and differs in a marked degree—upon the value that the pharmacist places upon his or her services, amount and quality of ingredients, time consumed, and the general economic condition of the community in which the pharmacy is located. Thus, in this latter case, personal evaluation and judgment are factors that play a major role in the setting of prescription prices and payment.

2. Record Keeping

The Egyptian laws require that the original prescription, if compounded, should be recorded in the record book of the pharmacy. Other prescriptions which contain only manufactured drugs do not need to be recorded. In both cases, the patient has the right to retain the prescription.

Prescriptions that include any drug classified as a narcotic should be compounded and dispensed within five days of the date of issue. In this case, it is obligatory for the prescriber to mention the full name of the patient, age, sex, detailed address, amount to be dispensed (within permissible limits), and number of doses. The prescription should be recorded in a special record book of narcotics of the pharmacy and retained. The patient has the right to have a stamped copy of the prescription. Renewal of such a prescription requires a new original one from the prescriber. Similar requirements are applied to prescriptions including hypnotics and other habit-forming drugs (e.g., tranquilizers, psychotropics, analeptics, CNS stimulants) and antibiotics, hormones, cortizones, etc., which are mentioned in special tables issued by CAPD, MOH.

The ordinary record book and the special record book of narcotics are subject to inspection by pharmacy inspectors.

3. Patient Education

Generally speaking, the community pharmacists of Egypt play a valuable role in educating their patients concerning drugs, as they are patient-oriented as well as drug-oriented, and in ensuring that medicines are taken in the proper way. Even when the directions are obvious and clearly understood by patients, additional instructions and advice from the pharmacists could result in better therapy and fewer problems.

It is very common for the pharmacists of Egypt to discuss with their patients whether they are taking other drugs, not listed in the prescription under discussion which may produce side effects. Pharmacists explain, sometimes, what the drug is intended to do and how to tell whether or not it is having the desired effect, how to recognize and what to do about possible adverse effects, and how to avoid or minimize problems by taking the drug in an appropriate way. Pharmacists are always ready to warn their patients against any adverse effects and their symptoms, stability of the dispensed drug under various conditions, and its toxicity.

The patients of Egypt are in a great need of such education, as the average one recognizes that physicians are always extremely busy, and patients do not feel free to tell physicians things they consider minor or of no importance. This is why the pharmacist is usually considered more accessible and, thus, it is possible for him or her to be a consultant giving the best advice.

The community pharmacists of Egypt, especially those who are practicing in remote rural regions, play a very important role as patient educators both in improving primary health care and in preventive medicine. They educate their patients about health matters including diet, life-style, the effect of environment, and drugs improve their health.

4. OTCs and Other Classes and Schedules of Drugs, Other Nonpharmacy Outlets

OTC drugs, or nonprescription drugs, play an important and ever-increasing role day after day in medical treatment in Egypt. Self-diagnosis and self-medication will continue to increase as the public becomes more and more knowledgeable about diseases and drugs. The pharmacist, who is always available and does not require an appointment, contrary to the physician, is the logical choice for an individual to be consulted.

Usually in Egypt the pharmacist, in handling a case that deals with self-diagnosis and OTC drugs, has the total responsibility for all professional decisions. The pharmacist first decides whether the patient's own efforts of self-treatment with home remedies are justifiable or hazardous, then decides whether the patient has accurately self-diagnosed his or her case, and accordingly decides if the patient should consult a physician or be recommended an OTC drug that suits his or her case.

When dispensing an OTC drug, the Egyptian pharmacists:

— Provide quick and effective relief of symptoms that do not require medical consultation
— Reduce the increasing pressure on medical services for the relief of minor symptoms, especially when resources and manpower are limited
— Increase the availability of health care to the population living in remote rural regions where access to medical advice may be difficult to obtain or sometimes does not exist.

The amount of OTC drugs sold in Egyptian pharmacies every day represents a very reasonable figure for the total sale, such as drugs for treatment of cold, cough, indigestion, constipation, diarrhea, pain, worm infestations, head lice, scabies, simple eye and nasal drops, hemorrhoids, burns and sunburns, reaction to insect bites, acne, dandruff, psoriasis, warts and corns, diaper rash, fungal infections, menstrual problems, feminine hygiene, obesity, poisoning emergencies, vitamins, minerals, and tonics. Herbal tea packages used as laxatives or for liver and bile function and for other simple symptoms are also sold as OTC drugs.

5. The Use of Technicians and Others

Usually there is a staff of technicians, and very few assistants working in Egyptian pharmacies. They hold high school certificates followed by two years in a special professional institute. They can handle simple drugs under the supervision of a qualified pharmacist, but they have no right to discuss drugs or give advice to patients.

6. Pharmacy and Primary Care

The community pharmacists in Egypt hold positions of the highest trust and responsibility within the community. They usually work within a time schedule that fits their patients rather than themselves, and the public always turns to them for advice because they are easily accessible. They play an important role in promoting primary health care, and believe that their professional responsibilities do not end at the pharmacy door and that they have equal professional responsibilities at the counter. They have come to believe that it is not a question of the pharmacist usurping the physicians's role or even of becoming the poor man's doctor. As community pharmacists are the most accessible of the health professionals, they are in an ideal position to provide information on a wide range of health topics.

In Egypt, pharmacists usually give advice, free of charge, on matters such as drug use and abuse, drug-drug interaction and drug-food interaction, adverse drug reaction, side effects, infant feeding, family planning, best way to store the drug to keep it stable, the proper adjustable diets (especially to obese patients), vaccination, potassium intake if patients are under diuretic treatment, etc. It is a well-known fact that pharmacists enjoy giving such advice, and they are very active in contributing to the primary health care program of the country.

D. UNIQUE OR INTERESTING FEATURES OR SPECIAL SITUATIONS IN EGYPT

By the end of 1988, the total number of drug specialties found in the Egyptian market was 2,184, including 406 imported ones. All the imported drugs are Western medicines, except for about 1% nonWestern medicines imported mainly from Hungary.

1. Traditional Remedies and Nonpharmacy Outlets

There are plenty of shops specializing in traditional remedies and medicine spread everywhere in Egypt and found usually in public poor districts and in remote rural regions. Sometimes they are called folk medicine, herbal medicine, or herbal health shops. These shops are usually directed by an herbalist who sells simple nontoxic or impotent herbs and medicinal plants. There is legislation governing and controlling the activities of these shops.

Herbal medicine is very popular among the natives of Egypt, especially the poor, the illiterate, and those who live in remote rural regions. The formulas dispensed by the herbalist are simple ones, containing either a single herb or a mixture of some dry herbs, medicinal plants, or natural products such as seeds, leaves, roots, barks, flowers, pods, fruits, volatile or fixed oils, resin, gum, etc. These formulas are either for internal or external use. They are usually administered in powder form or as a decoction or in the form of tea boiled in water.

2. A Different Healing System

Homeopathic prescriptions or preparations and the practice of homeopathy are not found at all in Egypt. As a matter of fact, this kind of practice is not known and has never been heard about in the country.

E. CONCLUSIONS

1. Assessment of Program Strength, Weakness, and Trends

The national pharmaceutical and health services policies are relevant to the goal of health for all, and all political and executive levels of Egypt are committed to this policy. The results currently available seem to indicate that both the strategy and efforts being made to implement them are in harmony, thanks to the wide distribution of health services throughout the country, local pharmaceutical production covering 86.8% of the local demand, suitable technical and administrative supervision designed to follow up the implementation of the policy and strategy, and the regular flow of information on the various activities.

The implementation for pharmaceutical and health service programs and strategy led to an increase in the utilization of health services. Among the main reasons for their strength and success are:

— Provision of the necessary manpower of all categories to man their services
— Implementation of strong pharmaceutical and medical legislation
— Encouragement of the policy of open-door investment in both the pharmaceutical and health sector
— The interest taken in training and continuing pharmaceutical and medical education
— Continued cooperation with other ministries, universities, and research institutes to strengthen primary health care and health-related areas, such as the pharmaceutical area
— The attention given to promote health awareness among the citizens.

Several programs and activities are being carried out to improve the quality of health care. Some proved to be very successful and gave very good results; some are on a trial basis in certain regions, with a view to generalize them once they prove efficient. These include programs of immunization, home visiting (which aims at providing health care to the family), schistosomiasis and diarrheal diseases control, eating habits of pregnant women and children, oral rehydration, family planning, and the program for strengthening and developing the referral system. The quality of service is assessed by analyzing the data and statistics received by the technical departments. The improvement in quality noted is due to:

— External assistance, which was used to upgrade the level of training
— Cooperation among the various departments and hospitals and the
  pharmaceutical sector in utilization of modern drugs and equipment
— Proper use of research results.

The effectiveness of these programs and strategies is evident from the
substantial change in the vital statistics during the last ten years, between
1978 and 1988. The crude death rate decreased from 10.5 (1978) to 9.0
(1988) per 1,000 population, the infant mortality rate decreased from 107
(1978) to 93 (1988) per 1,000 live births, and life expectancy for males
increased from 55.5 (1978) to 59.6 (1988) and for females increased from
58.7 (1978) to 62.4 (1988). Furthermore, no case of disease subject to the
International Health Regulations was noticed, while the incidence of and
death from infectious diseases, particularly those covered by the expanded
program on immunization, has decreased.

Production of pharmaceuticals has expanded tremendously and almost
tripled during the last two decades. The country depends nowadays on
only 13.2% imported drugs for its demand, compared to 70% 20 years
ago.

MOH is trying to overcome the difficulties encountered which slow
down reaching the planned goals and cause some weakness to pharmaceutical and health programs. For example, there is a shortage of funds allot-
ted to improve the quality of the service, and the amount of external assistance provided is insufficient to bring about the required improvement;
particularly, the rate of illiteracy is slowly decreasing, and health awareness among the public is slowly increasing. Due to the critical foreign
exchange conditions, the country is always in great need of hard currency
for establishing, expanding, or developing new pharmaceutical companies
and hospitals and importing advanced modern equipment and technology.
Also, as a result of this, unstable supply of raw materials represents serious problems. Another problem is that although many drugs are produced,
shipment to meet the trend of consumption still needs to be improved.

The socioeconomic and environmental factors that affect health, like
overpopulation, housing, clean drinking water, malnutrition, illiteracy,
and others, are not under the direct jurisdiction of health authorities. However, one cannot ignore their negative effect on health, and the only way
to deal with them is through multisectorial plans at the highest level of
administration and legislation.

It is the trend of Egyptian policy and strategy in general to encourage
high-quality production in all industries. The pharmaceutical sector is active in this matter and is also active as an exporter to Arab and African

countries, thus increasing the income of the country in hard currency to bridge the gap in the balance of payment. This policy is achieved by encouraging the joint-venture and private sectors in both the pharmaceutical and the medical fields, among other sectors, to practice with advanced modern technology to compete in foreign markets. But at the same time, there must be a balance between these two sectors and the improved developed public sector.

SELECTED READINGS

Statistical Year Book, Arab Republic of Egypt, Central Agency for Public Mobilization and Statistics, 1988.

Haub C., Kent M. Population Reference Bureau, Inc., Washington, DC, 1988.

The Golden Book 1936-1986, The Ministry of Health, 1986.

Medicine and Pharmacy, an Informal History, 1 — Ancient Egypt, Schering Corporation, New Jersey, U.S.A., 1955.

Egyptian Drug Information Center Brochure, Drug Organization for Chemical and Medical Appliances, 1988.

General Department of Statistics and Evaluation Bulletin, Ministry of Health, 1988.

General Organization for Health Insurance, Report of Activities, The General Administration of Statistics, 1986.

Herxheimer A., Davies C. Drug information for patients: bringing together the messages from prescriber, pharmacist and manufacturer. Journal of the Royal College of General Practitioners, 1982; 32:93-97.

Windle M., Moore R., Gourley D. et al. The community pharmacy as a health education center. American Pharmacy. 1981; NS21, No. 7: 22-25.

Anderson J., Bendush C., Chase G. et al. Remington's Pharmaceutical Sciences. 15th ed. Mack Publishing Company, Easton, Pennsylvania, 1975.

Lyman R., Sprowls J. Textbook of Pharmaceutical Compounding and Dispensing. 2nd ed. J.B. Lippincott Company, Philadelphia, 1955.

Jenkins G., Francke D., Brecht E. et al. Scoville's The Art of Compounding. 9th ed. The Blakiston Division, McGraw-Hill Book Company, Inc., New York, 1957.

Medicines in Egypt, Vol I, Drug Organization for Chemical and Medical Appliances, 1985.

Guidelines for Developing National Drug Policy, WHO, Geneva, 1988.

Chapter 8

France

Jean Brudon
Georges Viala

A. THE NATIONAL HEALTH CARE SYSTEM

1. History and Background

Nearly all people in France are covered by the national health care system, the Securite Sociale, against the following risks: illness, pregnancy, invalidity, old age, occupational health (disease and accidents), death, family allowances, and unemployment. But the act of October 4, 1945, establishing the Securite Sociale in fact created the *regime general*, a general system of coverage for salaried employees in commerce and industry and their dependents (the family), covering over 42 million people. The regime general is accompanied by other regimes: *regimes particuliers*, for civil servants, students, etc., who depend on the *regime generale* but have certain distinctive features; *regimes speciaux*, for specific categories of salaried workers (miners, railway workers, etc.); and *regimes autonomes* for farmers and the self-employed. State health insurance therefore consists of a whole array of different systems, all financed by compulsory contributions, which more or less completely cover the risks mentioned above.

The regimes are all base, except the one for the self-employed, on contributions from both employers and employees.

The regimes outlined above are enhanced by specific social legislation such as:

— Provisions for medical assistance which cover pharmaceutical services. This assistance is provided by local government (departments and communes) to people of limited financial means and is based on the idea of assistance rather than insurance.

— Provision for disabled servicemen and war victims whose injuries and illnesses are covered by the state (Ministry for War Veterans).

Lastly, one needs to include mutual insurance coverage, Mutualite, whose principles were laid down in 1898.

Mutual benefit society (friendly societies), *mutuelles*, are governed by private law. Members pay a fee. These companies act on behalf of their members to provide insurance, solidarity, and mutual assistance. These organizations act in conjunction with the Securite Sociale to cover the risks it does not provide for (this applies to the pharmaceutical charges not covered).[1,2]

Similarly, joint agreements signed between employer organizations and trade unions on working conditions may contain provision for additional health and pension insurance, *regimes complementaires*, on top of that provided by the Securite Sociale. This also applies to pharmaceutical company employees (industrial plant, retail chemists, and wholesale pharmacies).[3,4]

2. Social and Political Environment

The existence of the Securite Sociale has major effects on the whole social environment, particularly the following aspects:

— Demographic: fall in infant mortality and major increase in life expectancy
— Economic: consumption directed to health care, reducing the use of savings to cover personal risks, etc.
— Social: the elderly are no longer cared for by their children as was the case in the past. There has been a certain shift in resources to large and/or low income families.

The country's social expenditure (i.e., all resources directed to protection against social risks) largely exceeds public expenditure. The state will help balance the budgets of any branches of the Securite Sociale in financial difficulty, either by grants or injecting resources through taxation.

The French system could stay balanced as long as the postwar boom lasted through contributions mostly from salaried earnings. With the recession and growth in unemployment since 1975, resources have fallen while demand for services has increased or remained stable (e.g., unemployment benefit). (Unemployment is not one of the social risks covered by the Securite Sociale, but it is covered by a separate system of contributions from employers and salaried employees.)

3. Public and Private Financing, Health Costs

Ordinary health expenditure (i.e., all yearly health expenditure — total medical consumption, sickness benefit paid to people off work, public preventive health, research, teaching, health administration) rose from FF 452.515 billion in 1986 to FF 516 billion in 1988. Total medical consumption (care and medical goods including drugs) rose from FF 388.50 billion to FF 462 billion (an 8% increase since 1987; i.e., FF 8,270 per person, including FF 1,493 on drugs alone). These figures can be compared to expenditure on food (FF 12,000), housing (11,480), and transport and communication (FF 10,280). Forecasts by the Institut National de la Statistique (INSEE) predict that health expenditure will reach 20% of household expenditure by the end of the century.

Still, in 1988, expenditure on doctors' fees has grown considerably to FF 1,126 (13.2% up from 1987), while drug expenditure, after a short break, has continued to rise (13.1% up from 1987).

The state and, above all, the Securite Sociale have to some extent reduced their provision for expenditure on drugs. In the last eight years, the share paid by households has grown considerably, while that paid by mutual insurance companies providing additional coverage has risen the most, three times more in value terms and by five percentage points.[5,8]

In the field of drugs, there has been a trend toward self-medication, which is not reimbursed by the Securite Sociale. This has found favor with some politicians, health administrators, the pharmaceutical industry, and consumers themselves. Whatever the case, drugs accounted for 12.9% of health insurance (*regime general*) expenditure in 1988 and 3.4% of total Securite Sociale expenditure. In other words, although drug expenditure is a significant area, it is far from being the most expensive part of health care.[9] This position is taken by hospitals (53.1%), followed by medical and dental care (16.8%), services in kind (6%), and miscellaneous (11.3%). Lastly, in 1988, drugs accounted for 2.5% of total household consumption (2.3% in 1970) worth FF 3,402,291 million.

4. Facilities (Hospitals, Clinics, etc.)

On January 1, 1988, the public sector consisted of 1,065 facilities (i.e., 370,059 beds) compared to 2,691 in the private sector (i.e., 198,846 beds). Neither of the figures includes outpatient places or accommodations for the elderly. Public facilities have approximately 120,000 residential places in hospice units, old age homes, and annexes.

Public sector hospitals include 29 regional hospital centers which play an important part in training doctors through agreements with the universities. However, there has been a fall in the number of beds in public sector hospitals. Current provision is approximately 7 beds per 1,000 inhabitants. The majority of beds are medical (approximately 195,000), about double the number of surgical ones.

Over 400 private facilities collaborate by agreement with public utilities. The two largest categories are care facilities (over 1,300) and average stay facilities (about 770). There are also 20 cancer treatment centers. Private hospitals provide nearly 200,000 beds (30,400 medical, 63,400 in clinics, and 110,000 average stay), corresponding to 3.6 beds per 1,000 inhabitants.

Recently, there has been the development of alternative facilities to traditional hospitals:

- Day centers, started in psychiatry, but growing elsewhere
- Home-based treatment, particularly for hemodialysis patients, sufferers of respiratory insufficiency, etc.
- Hospitalization at home, decided after hospitalization or consultation in a public facility, which is run by private organizations (for patients suffering from severe cardiovascular disease, cancer, etc.).

A whole series of different coordinated care networks, modeled on the American HMOs, provide all the preventive care and treatment needed to maintain good health. Currently there are only a few such facilities.[10]

5. Manpower

On January 1, 1988, 62% (97,676) of the 157,527 doctors counted were self-employed. By the year 2000, despite the *numerus clausus* for students going into the second year of their studies, 180,000 doctors are forecast in France. This clause limits the number of professionals entering the next phase of training. At the same date, there were 39,075 dental surgeons, 34,651 of whom were self-employed (there is also a *numerus clausus* at the end of the first year). There were also 8,479 self-employed midwives (a much larger number work in hospitals).

There was a decline in the number of private medical biological testing laboratories run by doctors, pharmacists, veterinary surgeons, and scientists between 1980 (4,241) and 1987 (3,789), but the fall seems to have stopped for the moment.

The number of self-employed medical auxiliaries (mainly nurses and physiotherapists) has grown sharply, from 48,425 in 1980 to 64,525 on January 1, 1988 (there was already a total of nearly 300,000 nurses in 1985).[11]

It should be remembered that doctors qualify after studying for eight to ten years, depending whether they choose general practice or a specialty, while dentists qualify after five years. Since 1985, pharmacists can either study for six years (short retail chemist and industrial courses) or for nine years if they pass their internal pharmacy examinations and take a specialized diploma course in medical biology, hospital and institutional pharmacy, industrial and biomedical pharmacy, or specialized pharmacy.[12,13] Since 1980, the state doctor in pharmacy diploma (practitioner's doctorate) has replaced the former state pharmacist diploma.

Since 1981, a *numerus clausus* similar to that in medicine has been established at the end of the first year because of the excessive number of students (nearly 4,000 qualified students entering the employment market). Current numbers have been cut to 2,250.

There has been continuous expansion in the number of pharmacists over the last few decades, giving the following figures (all activities):

1976:	31,919 qualified pharmacists
1981:	40,099 qualified pharmacists
1985:	46,179 qualified pharmacists
1989:	51,390 qualified pharmacists
	(59.43% of whom are women)

Employment, like in medicine and dentistry, cannot continue expanding indefinitely.

6. The Social Security System

The largest system, the *regime general*, has the following main administrative features:

— Offices, *caisses*, handle different risks.
— Representatives of salaried employees make up the majority of their boards.
— The state enjoys a certain degree of power.[14,15]

Administrative Organization

— *Caisses primaires* — one per department (CPAM) — are responsible for registering their members and providing a benefits service for health, maternity, disability, accidents at work, and occupational illnesses.
— *Caisses regionales* (CRAM) have various responsibilities, which mainly involve running retirement pensions.

— The Caisse Nationale d'Assurance Maladie des Travailleurs Salaries (CNAMTS) is responsible for coordinating the activities of the *caisses* mentioned above. In practice, the pensions work of the *caisses regionales* is under a different *caisse nationale* from the CNAMTS. The *caisse* is responsible for balancing the budget and for medical control. It also has a role in preventive medicine and health and social education.

Lastly, there are family allowance *caisses* (one per department) responsible for paying the corresponding benefits (including to the self-employed) whose activities are, again, coordinated by an independent national *caisse*.

Generally speaking, the *caisses* are run by a board of directors and management bodies. The board consists of elected members representing salaried employees and employers, with the former enjoying a majority. The board is responsible for general administration and voting on the budget proposed by the management. It controls the management bodies. Members of the management bodies are appointed by the government for the *caisses nationales* and by the board, subject to approval by the Minister of Health and Social Security, in other cases.

It should also be realized that the state itself has control over decisions by the *caisses* through its regional health and social welfare managers. It has also established a welfare inspection department which is responsible for reports and surveys.

Securite Sociale organizations are subject to Finance Ministry and Treasury control.

Financial Organization

The Securite Sociale is confronted with financial difficulties. It is mostly financed through contributions that have been regularly raised over the years to cover its deficit. There is increasing discussion about fiscalizing the Securite Sociale, and there has already been recourse to taxation when the deficit has grown too high.

Nonetheless, the prevailing system is still based on contributions shared between employers and salaried employees, admittedly more so by the former than the latter. For instance, employees pay a 5.5% contribution for health insurance compared to 12.6% for employers, and the latter pay all family allowance and occupational accidents contributions. In some cases, contributions are based not on the full wage but only on a certain ceiling fixed by the government. This applies to pension and family allowance contributions but not to health insurance.

The self-employed, particularly pharmacists owning retail chemist shops, have to contribute to a special system, the *regime autonome*. Maternity and health insurance coverage is generally lower than in the *regime general*. In addition, the self-employed have their own pension fund divided into distinct professional sections. Thus, pharmacists have their own pension fund, which receives their contributions and pays their pensions. There is a certain amount of compensation between the various professional sections belonging to the Caisse Nationale d'Assurance Viellesse des Professions Liberales.

7. The Medical Establishment: Entry, Specialists, Referrals

Establishing a Medical Practice

Self-employed doctors, like other members of the medical profession wishing to set up their own practice, are not bound by any restrictions.[16] Practitioners with this status have a large amount of independence in their relations with patients. Subject to certain reservations stipulated by the code of medical practice, doctors are able to conclude and terminate health care contracts with patients.

This being said, they cannot treat medical practice as a business, and no forms of fee-splitting are allowed with colleagues. Patients are not transferable, but retiring doctors may sell lists of their patients to their successors, as well as their agreement not to reestablish under conditions that could compete with the purchasing physician.

Doctors must practice in person. They may not have their surgery managed on their behalf, although they are entitled to be replaced by locums. They may only be assisted by colleagues under exceptional circumstances. In recent years, alongside individual practice, other forms of group practice have developed which may extend to true professional status (fees are paid to the partnership and profits are shared according to the members' professional activity and status clauses).

Doctors' situations vary greatly according to whether they are under contract with Securite Sociale. If they are not, they are free to establish (within reasonable limits) their fees. If they are, they may either apply Securite Sociale rates or charge higher fees, although reimbursement by the Securite Sociale is based on the rates. Generally speaking, specialists' fees are higher than those of general practitioners.

Establishment of Retail Chemists

The situation of retail chemists is quite different. A retail chemist is considered to be a shop (Art. L. 570 of the Public Health Code). But contrary to the general principle of freedom for commerce and industry, its establishment is controlled. A retail chemist may only be opened under the conditions stipulated by the law (Art. L. 571 of the Public Health Code), which fixes a quorum of: 1 retail chemist per 3,000 inhabitants in communities with over 30,000 inhabitants, 1 retail chemist per 2,500 inhabitants in communities with 5,000 to 30,000 inhabitants, and 1 retail chemist per 2,000 inhabitants in communities with under 5,000 inhabitants. A special higher quorum exists in the departments of Alsace and Moselle and in French Guyana. However, even if the quorum is reached, the Prefet (the government's representative in the department) may allow an establishment by exception if it is really necessary for the permanent or seasonal community. In practice, and during periods of demographic stability, most establishments take place by exception due to population shifts and urbanization in certain areas. For instance, in 1988, 52 retail chemists were opened under the standard procedure (i.e., through application of the quorum) and 143 by exception.

It should be added that transfer of a retail chemist from one location to another within the same community must also be approved by the Prefet, who decides whether the operation is in the interest of public health.

On 1 January 1989, there was a total of 21,827 retail chemists in France. Some previous figures:

1963	15,099
1970	16,786
1980	19,709
1985	21,186

As can be seen, there has been steady growth, particularly in rural locations and the suburbs and new developments in large towns. There has been a slowing in growth since 1987.

It should be noted that alongside retail chemists, there are also pharmacies owned by public and private health care organizations where patients are treated (Art. L. 577) or by *mutuelles* (Art. L. 577a). These pharmacies are characterized by dispensing only for the internal use of the establishment (hospital, clinic, dispensary), i.e., for patients treated on the premises and outpatients. In the latter case, only members of the mutual benefit society or miners' society are eligible (the Securite Sociale for miners generally has recourse to this type of society, which is run on a mutual benefit basis).

On December 31, 1988, in addition to the 21,827 community pharmacies open to the general public, there were 68 mutual benefit pharmacies (accounting for about 2.5% of total retail pharmacy turnover) and 75 mining pharmacies serving, as indicated above, persons covered by the special miners' Securite Sociale. One should not ignore the effects of the expansion of such types of pharmacies as three-fourths of the French population belong to "friendly societies."

The only way that retail chemists can prevent the creation of mutual benefit pharmacies is to come to agreements with the mutual benefit societies to give their members equivalent advantages, which will be discussed later. It appears that some of these societies are striving to obtain satisfaction on both fronts (i.e., owning their own retail chemists and signing agreements granting advantages to their members with private pharmacies).

On the other hand, there are 3,500 health care establishment pharmacies of varying size. They could create problems for pharmacies open to the general public to the extent that, as home-based hospital treatment is growing, they would be increasingly involved, especially for certain powerful, costly drugs that only the health care establishment pharmacies are allowed to dispense. They are the only establishments entitled to dispense such drugs to patients who would otherwise have to go without.

The last feature common to pharmacies covered by Articles L. 577 and L. 577a is that they are not counted in the quorum.

8. Payment and Reimbursement

The principle of the Securite Sociale code is that the insured person advances payment and pays part of the cost, the *ticket moderateur*, or copayment, which is designed to help keep down costs. In practice, there is flexibility in these rules.

Advancing Payment and Reimbursement

Since 1986, the pricing principle for drugs has been that of fixing at all stages (manufacture, wholesale, retail), although this does not, in fact, apply to drugs reimbursed to Securite Sociale beneficiaries, where prices are fixed by the state (see Section B.4).

Customers first pay the pharmacist and are then reimbursed a fraction of their payment (Art. R.322-1 Securite Sociale Code). There is 100% coverage for high-priced special drugs considered irreplaceable. If this is not the case, reimbursement is at the rate of:

- 40% for specialized drugs mostly used for disorders and conditions that are not usually severe
- 70%, the standard rate, for all other specialized drugs, preparations made to prescription, and official drugs (i.e., listed in the Pharmacopoeia).

Recent trends are the following:

- Downgrading drug reimbursement rates from 100% to 70% and 70% to 40%
- No longer reimbursing some drug groups previously reimbursed, e.g., many vitamins. The same applies to preparations made to prescription and official drugs that do not meet well-defined criteria.

All specialized drugs carry a white self-adhesive label (if the *ticket moderateur* is 0% or 30%) or a blue one (if it is 60%). The retail price is given on the label. The insured person removes the label and sticks it on a claim form filled in by the pharmacists with the price of the supplies. This printed form has to be sent, together with the prescription, in any reimbursement claim. The label, as described, is a means of controlling the sale of the product.

Preparations are made according to a prescription and official drugs are sold by pharmacists at an official price, *Tarif pharmaceutique national* (see Section B.4).

Exceptions to the Rule of Participation by Insured Persons Advancing Payment

1. Insured persons may be exempted from payment of their portion of the cost of treatment for certain officially listed medical conditions, if they are long-term and require particularly expensive drugs. Thus, patients suffering from among 30 diseases, including AIDS, do not have to contribute anything for treatment.

Insured persons afflicted by a disease not on the list may be entitled to exemption if it involves long-term treatment with particularly expensive drugs. In December 1986, the government attempted to tighten up the conditions for this second type of exemption, which had become very common, but reversed itself in September 1988.

Victims of occupational accidents are also exempted from payment of their contribution to the cost of treatment.

Thus, in May 1989, for example, the medical insurance *caisses* reimbursed FF 1.903 billion to insured persons contributing to the cost of treatment (*ticket moderateur*) and FF 1.409 billion where they were exempted from such payment.

If the insured persons are also covered by a mutual benefit society or private insurance, paid out of contributions or by a premium, the insurance will cover the share normally paid by them.

2. By the system of agreements between the pharmacists (generally by a pharmaceutical association) with either the state health insurance *caisse* and/or mutual benefit society, the insured person may no longer have to advance payment for all or part of the supplies bought from the pharmacist.

9. Coverage

Insured persons have rates of coverage that vary with the product and service reimbursable (doctors' fees, hospital fees, laboratory tests, drugs, etc.) and their personal situation. The above discussion on drugs is a good example of how coverage rates can vary from one category of insured person to another.

Patients are also exempted from paying their portion of the cost of treatment in certain other situations, such as if they remain in a hospital for longer than 30 days, undergo particular treatments exceeding certain rates, and obtain major orthopedic prostheses.

B. PHARMACEUTICAL INDUSTRY AND DRUG DISTRIBUTION

1. Structure of the Industry

The pharmaceutical industry was mostly established in the last century in the back of retail chemists. This explains the large number of companies involved which, in the post-World War II period, have either merged or discontinued operations.

Degree of Concentration

According to figures supplied by the SNIP, in 1950 there was a total of 1,960 laboratories, 990 of which were part of retail chemists.[9] During the last 20 years concentration has gained pace, with a fall to 507 companies, of which 85 were laboratories and 27 were retail chemist laboratories. Lastly in 1988, SNIP figures were 345 companies and 13 pharmacies with production annexes. It specifies that the figure of 345 does not include companies doing work on behalf of companies with drug marketing licenses. Neither are laboratories included that only manufacture veterinary drugs.

The French pharmaceutical industry is still not very concentrated, even allowing for the reduction in the number of companies discussed above.[9] This can be seen by the percentage turnover of the main companies: Figures for 1987:

	Turnover
The top 5 achieved	15.1%
The top 10 achieved	24.8%
The top 20 achieved	40.0%
The top 50 achieved	67.5%
The top 100 achieved	88.5%

Turnover in 1987 was FF 55.7 billion, of which FF 10 billion was exported. The corresponding figures for 1988 were FF 63 billion and FF 10.5 billion. Furthermore, only seven companies held over 2% each of the market, and their combined contribution to it was 19.1%. At the same time, the five largest groups of companies took 31.3% of the market (a total of 32 companies), the top ten took 42.5%, and the top 20 took 58.4%. The largest group in 1988 was Rhone-Poulenc, which took about 10% of the market.

The order, according to REM-France in 1986, was as follows: Rhone-Poulenc, Sanofi, Hoescht-Roussel, Servier, Synthelabo, MSD-Chibret, Pierre Fabre, Ciba-Geigy, l'Aire Liquide, and Roche.[17] According to the same source, only 12 companies had a turnover of over FF 1 billion, and together they accounted for only 30% of production in the sector. The conclusion is that the pharmaceutical industry is far from being concentrated because of the wide variety of its products and techniques.

Human Factors and Location

Laboratories employed 60,500 people (all activities) in 1970. The figure rose to 72,143 in 1987 (including finishing activities). The percentages of these people employed by the top companies in the same year were as follows:

Top 5 companies	6.9%
Top 10 companies	12.9%
Top 20 companies	22.9%
Top 50 companies	48.1%

Only 11 companies employed over 1,000 people, 23 employed 500 to 1,000 people, and 57 employed 200 to 500 people.

One of the features of people working in the pharmaceutical industry is their level of training. Forty-five percent are workers and employees, 18.2% are management staff, and 19.3% are medical representatives. In this respect, there has been a trend to reducing the number of production workers (due to automation), while there has been an increase in numbers in medical information and research and development. Fifty-six percent of all employees and 46% of representatives are women, compared to under 30% in 1975.

The geographical distribution of the pharmaceutical industry is extremely uneven, with Ile-de-France first, followed by Rhone-Alpes, the Centre, Aquitaine, and Normandy.

Financial Items

As mentioned above, the turnover in 1988 of FF 63.9 billion had risen by over 15.5% since 1987. Does this mean that the industry is in a good economic state?

1. The price of reimbursable drugs (90% of the market) is strictly controlled, and the industry complains that the prices of its old products have not generally risen enough. In this respect, the industry maintains that French reimbursable drug prices are among the lowest in Europe and are similar to those in Greece and Spain. Generally speaking, if prices in France are 100, they are 137 in the U.K., 194 in Holland, and 205 in West Germany. However, French prices are likely to be gradually brought into line with those in other countries in Europe.

Another complaint of the pharmaceutical industry to the government, which relates to the previous one, is that French prices do not even keep pace with inflation. Thus, although the cost of living went up from a base of 100 in 1980 to 171.8 in 1988, the price index for reimbursable drugs only rose to 125.4. On the other hand, it is true that the index for nonreimbursable drugs increased from 100 to 193.4, which explains why there has been enthusiastic promotion of this category of drugs.

The low prices in France do not favor exports, as they encourage other, less desirable parallel exports and discourage the real exporters. French prices act as a sort of reference, and laboratories are unable to sell more expensively, which means that their margins are too low to finance promotion and conquer markets.

2. French manufacturers complain that the low profitability of their companies prevents them from making the investments needed to meet foreign competition. Physical investments accounted for only 2.7% of turnover before tax in 1987. Furthermore, profits after tax appear to be much higher abroad. Net profits of 3.1% of turnover before tax are well

below figures for the United States (13.6%), the United Kingdom (8%), Japan (5.7%), and Italy (5.8%).

Foreign Penetration

The Ministry of Industry study of January 1, 1985 shows a substantial foreign penetration in the pharmaceutical industry.

The penetration index for investments worth FF 525 million was 44.5%, while that for sales was 44.9% (FF 25.6 million) and that for manpower was 40.5%. These include most of the large American, European, and Japanese companies with various legal statuses.

In 1987, according to the Syndicat National de L'Industrie Pharmaceutique, firms with a majority French holding held 49% of the domestic market. Although this showed an increase in foreign penetration (36% in 1972 to 46.8% in 1980), the figure is similar in the other main European countries: West Germany, 44.7%; Great Britain, 66.2%; and Italy, 57%.[18]

The Specificities of French Pharmaceutical Law

French legislation is special because it stipulates that pharmacists must be employed in the industry. We shall not discuss the marginal case where the establishment belongs to the pharmacist. The Public Health Code (Art. L. 596) is mainly concerned with the case where the said establishment belongs to a company. It requires a pharmacist to participate in either the general management or running (depending on the case) of the company and specifies what office the pharmacist must hold. The pharmacist is furthermore *personally responsible* for applying public health regulations. The code defines the minimum responsibilities necessary to perform the pharmacist's functions. The pharmacist shall:

—Participate in the design of the research program.
—Organize and supervise pharmaceutical operations (manufacture, quality control, storage, release, and advertising).
—Sign product licensing applications.

Furthermore, it is stipulated that the pharmacist has authority over the company's assistant pharmacists. The legislation stipulates that the head pharmacist must have a certain number of assistant pharmacists at his or her disposal, their number depending on the personnel performing pharmaceutical operations.

This situation is not seen in other EEC countries, except Belgium, which has an industrial pharmacist. In the context of EEC directives, it is compatible with EEC directives and cannot be considered, specifically, as discriminatory, provided that qualified pharmacists from other EEC countries can be appointed to the posts discussed above.

2. Research and Development, Imports, Exports, Patents, Licensing

Both the manufacturers and government are aware of the importance of research. However, pharmacy is a field in which private funds are largely dominant. Thus, in 1986, the state financed only 0.3% of the pharmaceutical research and development (R&D) budget, compared to 47% for aeronautics and 36% for electronics. There has been considerable growth in the number of people employed in research, as can be seen by the following figures:

1980   5,090, including 1,916 researchers
1986   9,264, including 3,056 researchers

In terms of turnover, the percentage devoted to R&D grew from 11% in 1986 to 11.2% in 1987 and 11.9% in 1988.[19] Thus the pharmaceutical industry has the third highest research budget after aeronautics and electronics already mentioned. Even if the R&D budget appears to be honorable, it is smaller than in the United States and West Germany, for example, which also have larger turnovers.

Recent studies have shown that France fell behind during the period 1975-1986.[20,21] It was second in world drug research and has now dropped to fourth in the total number of drugs representing both a clinical advance and new chemical structure. If internationalization is taken into account as a criterion, France is seventh in expansion into the largest industrial countries.

Patents and Licenses

Although no statistics are available on patents granted, we can mention that France established its first patents system for drug production in 1944 and then in 1959 expanded it to cover their therapeutic applications. The present situation is one where drugs, with the exception of certain specific provisions such as licenses in the interests of public health, follow the general rules for patents. France has, of course, adhered to the main international patents conventions. Although French national patents do not require full examination before issuance, issue is preceded by an abbreviated written application procedure including a research report revealing any antecedence.

In the field of technical exchanges for patents and licenses, the most recent statistics published by the SNIP for 1987 show a gross deficit of FF 27 million compared to 1985 and 1986, where there were respective profits of FF 38 million and FF 94 million.[22]

Imports and Exports

There was growth in the value of exports in 1988 to FF 11.089 billion (this figure also incorporates veterinary drugs), compared to FF 10.837 billion in 1987, ranking France as the fourth largest exporting country. French exports traditionally go mainly to the Magreb and the Franc area — nearly 40% in 1986. This is followed by the EEC taking up 37%, while North, Central, and Latin America account for only 5%. The latest figures show a fall in exports to Algeria, while West Germany has become the largest customer for French exports.

There was a sizeable trading surplus of exports versus imports worth nearly FF 8 billion in 1988 for finished drugs (packed and bulk), sera and vaccines (exports: FF 11.089 billion; imports: FF 4.060 billion).[23] However, the deficit of FF 2 billion in active constituents for pharmaceutical use should be emphasized (imports: FF 8.09 billion; exports; FF 5.68 billion).[23]

3. Manufacturing

The operation of manufacturing plants is controlled by regulations that comply with EEC directives.[24] These plants (Art. L. 598 Public Health Code) have to be opened with a permission granted by the Ministry of Health for the preparation of pharmaceutical products specified after on-site inspection by the pharmacy inspectors from the aforesaid ministry. Any major modifications in the buildings, equipment, or drugs produced must be by prior approval from the same authority.

In France, in compliance with EEC directives, the aforementioned head pharmacist is responsible for the compliance of the product batches marketed with their specifications. The head pharmacist has to meet the EEC requirements of practical experience.

Any drug regularly manufactured by another EEC member state with a French product marketing authorization does not have to be tested again on entry into France. This does not apply to drugs manufactured in non-EEC countries.

Manufacturing must comply with good manufacturing and production standards (Art. R. 5115-9 Public Health Code) defined by the regulations of October 1, 1985. Pharmacy inspectors must verify at least once yearly that the regulations are complied with. The National Public Health Laboratory tests samples of the drugs (Art. R. 5140). If the plant's operations pose any risks to public health, ministerial approval may be withdrawn.

4. Advertising and Promotion, Price Regulation

Advertising and Promotion

At present, following the 1987 amendments to the Public Health Code (Art. R. 5045 et al.), the regulations are as follows:

— Any type of advertising to the general public has to be authorized by the Ministry, following approval by a commission.
— Advertising for the packaging must also be approved if it does not carry the regulatory drug labeling. It should be noted that the Ministry may refuse to approve the mention of certain therapeutic indications and side effects (known only to the medical and pharmaceutical professions).
— Advertising by the medical and pharmaceutical professions (still referred to as medical information) does not have to be approved beforehand. By contrast, all written advertising must be registered after release.[25]

Such advertising must be truthful, loyal, and comply with a summary of the drug's properties stipulated in the issue of the drug's marketing license. This summary is a rigorously objective information sheet. Unlawful or untruthful advertising may be punished by administrative sanctions, such as stopping advertising and publishing corrective statements, with the further possibility of penal or disciplinary sanctions (the latter by the Order of Pharmacists) against the head pharmacist.

The provision of medical samples is covered by certain restrictions and has to be by request of the practitioner. Any types of gifts or monetary advantages to prescribers or consumers are strictly forbidden. Only donations to teaching and research are permitted.

An EEC directive on advertising is likely to be published soon, probably due to the differences in the systems operating from one country to another. It is a point that should be considered.

Lastly, there is an old rule which is very strictly observed, namely that all direct advertising (see above) to the public means that the drug is automatically considered as nonreimbursable by the Securite Sociale.

Price Regulation

Taking the aforesaid into account (see Section A.8), drug pricing can be summarized as follows:[26]

1. Drugs reimbursed to insured persons: the price is (Art. L. 162-38 Securite Sociale Code and Departmental Order of August 4, 1987) fixed by the Ministries of Health, Social Security, and Economy in the course of the approval of the drug for reimbursement. This fixed pricing assumes the agreement of the laboratory. Some laboratories renounce projects as they are unable to obtain a satisfactory price in their negotiations with the pricing committee set up by the ministry departments involved. Any later price revisions must be approved by the ministries mentioned above. The price must be marked on the label required for reimbursement.

During the last ten years there has generally been a set annual price increase, which the laboratories are allowed to treat with some degree of flexibility, i.e, they can vary how they apply it to their products in accordance with those they most want to promote.

Having specified the maximum margins allowed between wholesale distributor and retail chemist, a drug price of FF 100 can be broken into:

Price at manufacture before tax	FF	59.54
Wholesaler's profit before tax	FF	6.40
Chemist's profit before tax	FF	28.85
V.A.T.	FF	5.21
	FF	100

(*N.B.* Retail chemist's margin in relation to manufacturer's price before tax: 48.46%; wholesale distributor's margin in relation to manufacturer's price before tax: 10.74%.)

2. Preparations according to prescription — official preparations: prices are determined in accordance with the French national pharmaceutical price list (Art. L. 593 Public Health Code). For components used, these are in practice calculated by multiplying the purchase price before tax by 2.15 and then adding handling charges and responsibility fees (for products covered by poisons legislation).

It should be noted that the prices mentioned in (a) and (b) are maximum prices (Art. L. 593). Pharmacists may therefore sell below this price, although they may not advertise any such discounts (Art. 377-7 Securite Sociale Code and Art. R. 5015-26 Public Health Code); furthermore, they must indicate the amount actually paid for the purpose of reimbursement. In practice, pharmacists charge the maximum price.

3. Prices for other drugs directly advertised to the public and for ethical drugs, which are not advertised and are only available on prescription, are not regulated. This is a consequence of the legislation already mentioned, of December 1, 1986, on freedom of pricing and competition. At most, the manufacturer may recommend a retail price to the public. Although legal, this practice is strictly supervised by the Ministry of the Economy, which means that laboratories use it less and less.

4. Other products: whatever their retail price these products are in principle unregulated. However, as some are reimbursable to insured persons (e.g., dressings, orthopedic equipment, etc.) the *caisses* reimburse a fraction of the official price (*Tarif interministeriel des prestations sanitaires*), which is generally lower than the pharmacist will charge (if only to avoid selling at a loss).

5. Drug Approval and Government Regulation

The Common Market is once again the source of aligning the 12 EEC countries on unified product marketing authorization standards and application procedures. However, mutual recognition of product marketing authorization has not yet been achieved. To sum up, a product marketing authorization (Art. L. 601 Public Health Code) is granted if the analytical, pharmacotoxicological, and clinical reports establish that:

— The product complies with its announced formula and is consistent in composition and quality once mass produced.
— The manufacturer's method of preparation and quality control comply with the objectives stated.
— The product is safe under normal conditions of use.
— It is a useful, i.e., an efficacious, medication.[27]

Once issued, the product marketing authorization is valid for five years and is then renewable unless any new unfavorable information about the product comes to light. It is likely to be withdrawn if any accidents occur or if there are any serious adverse effects (since 1976 France has had an official system of pharmacological supervision).

The possession of a product marketing authorization does not guarantee that the drug will be reimbursed by the Securite Sociale. A special procedure has been established, the pricing aspects of which have already been discussed (see Section B.4). This procedure involves the intervention of a special transparency commission whose function is to determine whether the product under consideration for reimbursement improves the medical service provided (in other words, therapeutic progress). This does not necessarily mean that the drug has to be a great innovation. Modest progress may be sufficient (e.g., if there is a new type of pharmaceutical preparation that is more suitable than ones already marketed). In situations where applicants are unable to establish that their product is an improvement, the only chance of obtaining reimbursement resides in pricing the product cheaper than other equivalent drugs already marketed (lower the cost of daily treatment).

Lastly, it should be noted that, alongside drugs that are reimbursed to insured persons (and to beneficiaries of other social legislation), there is a list of drugs solely approved for state hospitals (Art. L. 618 Public Health Code) that has now been extended to private care establishments. This list is considerably shorter, but includes drugs, which are sometimes very expensive, restricted for sole use in hospitals either because of government or laboratory policy. No price has been fixed for such drugs, and this results in competition to the extent that the hospitals either put out to tender or award contracts for this market.

6. Competition: Brand and Generic

All pharmaceutical specialties, i.e., ready prepared proprietary medicinal products, must carry special names (Art. L. 601 Public Health Code) and have specific special packaging. The special name may either be a pure invention or may have a standard common or scientific name, followed by the trademark or name of its manufacturer (demanded throughout the EEC).

The special name is mostly registered as a trademark (amended law of December 1964). The protection given by the special name is considerable and adds to the drug's prestige as a leading product which is a real innovation.

This being said, generics may obtain a product marketing license through a shortened product marketing license application procedure (through recourse to a bibliography for the pharmacotoxicological and clinical testing) and may then be admitted to the Securite Sociale at a lower price than the original drug. But one has to realize that the French procedure means that most are not used under their common name but have an invented name (which is perfectly legal).

One must also understand that retail pharmacists do not presently have any right to substitute. They must comply with the prescription of either the original drug or the generic one for two reasons:

— Their code of practice (Art. R. 5015-45 Public Health Code) forbids them to modify a prescription without the prior agreement of its author.
— Substitution of any product requested under its trademark is a penal offense for any retailer and doing so is considered counterfeiting. [28]

Retail chemists also supply drugs in their original packaging.

Lastly, French prices are generally lower than those of neighboring countries, a fact that does not encourage French retailers to use equivalent imported drugs. It is rather the opposite that happens (see CJ. CE. Schumacher, March 7, 1989, JO. CE. April 8, 1989).

7. Cost-Containment Activities

The suitability of the sale package for proper treatment without waste is an important consideration for the transparency commission in the application procedure for reimbursement by the Securite Sociale.

The transparency commission has to stipulate on the number of therapeutic units in the sale package. The laboratory may have to modify the contents of the aforesaid model.

C. PHARMACY PRACTICE

1. Dispensing a Prescription: Where, What, Payment, Formulary

The Prescription

A prescription is necessary if:

— The drug falls into the category of dangerous substances, currently in the process of amendment (decree of December 29, 1988).
— The drug (whether in the above category or not) is paid for by application of social or related legislation.

Only doctors have full ability to prescribe. However, they may have to comply with certain rules when writing prescriptions by virtue of the legislation concerning dangerous drugs. (Narcotics are given under strictly controlled conditions, and prescriptions must be written on a form taken from a counterfoil book given to doctors by their Order. The form must be kept by the pharmacist for three years.)

Dental surgeons and midwives have less extensive prescription rights than doctors (those of dentists have been substantially increased).

Dispensing a Prescription

Dispensing of prescriptions must be performed by retail pharmacies (Art. L. 568 Public Health Code). It has to be the property of a pharmacist or a group of pharmacists (co-owners or associates; Art. L. 575). A pharmacist or partnership of pharmacists may only own one retail chemist (same Article).

The great care (Art. R. 5015-23 Public Health Code) with which the prescriptions are dispensed implies that pharmacists check their authenticity, whether their authors are entitled to prescribe, and that the maximum doses fixed in the Pharmacopoeia or indicated by the drug's manufacturer are not exceeded. It the dose is exceeded, the pharmacist must notify the doctor and cannot dispense the prescription unless the doctor confirms it with the statement, "I state this particular dose." (And even in this case, dispensing a prescription carrying such a statement does not exempt the pharmacist from liability in case of accident.) The courts consider that the pharmacist must not only confine dispensing to properly written prescriptions but must also detect any mistakes (particularly concerning drug names). Lastly, the pharmacist must notify the customer, where appropriate, of any adverse effects, precautions, or contraindications, and the pharmacist is responsible for detecting any possible drug interactions.[29] For many years, medical prescriptions have been mainly for pharmaceutical specialties. Preparations according to prescription, although currently popular with some prescribers for various reasons, and official drugs (whose formulas are given by the French Pharmacopoeia or by the National Formulary, which completes it) do not account for more than 3% of a retail chemist's turnover.

As indicated above, insured customers generally pay the pharmacist, but as previously discussed (Section A.8), many insured persons could be exempted from advanced payments.

Pharmacists may not supply insured persons with more than one month's treatment at any one time, and the doctor may only prescribe for a maximum of six months (Art. R. 5148a Public Health Code; for contraceptives, read three months and 12 months).

Pharmacists obtain their supplies from two sources:[30]

—Wholesale distributors specializing in resale to retail chemists. (There are 26 such companies with about 200 sales points, i.e., a highly concentrated sector, where computerization has been particularly successful.)

Wholesale distributors are obliged to keep minimum stocks of pharmaceutical specialties (one month's supply in advance) and to supply their usual customers within 24 hours with the drugs they stock. They provide approximately 82.5% of supplies to retail chemists.

—Directly from the laboratories, which give discounts, or through agents.

Agents provide the same distribution function as wholesale distributors except that: (1) they represent the manufacturers and do not own their stock and (2) they do not have to comply with any specific stock obligations.

An original feature of French legislation is that the above wholesalers and wholesale distributors are covered by the same rules, including the employment of pharmacists, as in manufacturing plants (see Section B.3).

2. Record Keeping

The Public Health Code stipulates that retail chemists must keep records of:

—The formulae of preparations made according to prescription
—Dispensing of all medicines covered by the poisons legislation and their renewal if legally possible. Drugs based on dangerous compounds may be supplied without medical prescription below certain doses and concentration.

Special records must be kept of the purchase and dispensing of narcotics. Records have hitherto always been kept manually. Under the revised legislation on poisons and dangerous drugs, computerized record keeping will be allowed. However, manual records will still have to be kept for preparations according to a prescription and for narcotics.

3. Patient Education

Retail pharmacists play a role in both social and health education. A health and social education committee has been established by the Order of Pharmacists in collaboration with the unions, whose function is to provide assistance through: (1) news releases (leaflets, articles, etc.) and (2) information leaflets for the general public.

Other organizations, like the Union Technique Intersyndicale Pharmaceutique (U.T.I.P.), have also played a very effective part in informing pharmacists of the importance of the service that they provide to the community in the field of healthy living and the environment.

These considerations have been incorporated into university pharmacy courses.

4. OTCs and Other Classes and Schedules of Drugs, Nonpharmacy Outlets

A Monopoly Situation

In the interests of patients, health legislators (Article L. 512 Public Health Code) have confirmed pharmacists' monopoly of dispensing drugs. The monopoly is an institution dating back to the Royal Declaration in 1777 and even earlier.

This monopoly on drugs for human medicine extends to certain other products defined by the code:

– Dressings and items complying with the French Pharmacopoeia
– Insecticides and acaricides for human use
– Products used to clean and fit contact lenses
– Products and reagents in forms for sale to the public, which, despite not being medicines, are used in medical diagnoses or pregnancy testing
– Medicinal plants listed in the Pharmacopoeia, unless exempted by decree (34 medicinal plants are on unrestricted sale)
– Certain officially listed essential oils
– Baby milk products and baby foods for young babies defined by government legislation (few products concerned in practice).

Furthermore, pharmacists share with veterinary surgeons the supply of drugs for animals (Art. L. 610), but their share of the market is relatively small (5% to 6%). The law of December 27, 1967 on birth control grants them the exclusive right to sell oral contraceptives and contraceptive products and devices and other items, and, in collaboration with medical and surgical suppliers, the exclusive right to sell products and apparatus for abortion and syringes and needles for parenteral injection.[31]

As the spirit of the legislation is for pharmacists to devote their activities to their monopoly, they are simply authorized to sell in their pharmacy some other products and items listed by the Ministry of Health (Departmental Order of December 8, 1943) considered as parapharmaceutical items. There are, of course, no exclusive rights for such products, which in practice are mostly cosmetics, health care products, diet foods, and baby foods.

Consequences of the Monopoly

The monopoly on drugs means that they cannot be sold by other retailers. The only exceptions to the monopoly are:

— Doctors in areas without retail chemists experiencing supply difficulties (Arts. L. 594 and L. 595)
— Herbalists (Art. L. 659 Public Health Code) selling medicinal plants (herbalists constitute a disappearing profession as no diplomas have been issued since 1941).

Doctors still, however, require authorization by the Prefet of the Department and may only supply drugs prescribed in the course of their consultations (approximately 250 propharmacists). This authorization is withdrawn if a community pharmacy opens.

Despite the laws establishing the monopoly of retail pharmacists for the sale of medicines, large retailers (supermarkets and others) have recently started selling borderline products that pharmacists consider within their monopoly, like vitamins, capsules based on medicinal plants, etc. A great many litigations are currently underway in the courts.

Supermarket chains are well aware that the monopoly of pharmacists in France is more extensive than elsewhere in the EEC. They expect liberalization of the law in their favor. At present date, the EEC considers the monopoly a matter for the individual member states and that France is thus free to maintain it.

5. The Use of Technicians and Others

The pharmacists who own their business must run it personally. They may, or may have to in certain cases, employ one or more assistant pharmacists. Pharmacists must do so if their total turnover (Art. L. 579) exceeds a certain amount fixed by the Ministry (the current figure is above FF 2.95 million and by fraction of the same amount). Over 14,500 of these qualified pharmacists, practicing in this way, are members of the Order.

If not, the Public Health Code (Art. L. 581 and following articles) enables the retail pharmacist to be assisted in the preparation and sale of drugs (human and veterinary) by a single category of personnel, i.e., dispensers, who have a professional qualification given by the ministry responsible for technical training. Students taking this qualification must have passed at least their middle school grades.

The course lasts two years and consists of :

− Practical training in retail or hospital pharmacy
− 250 hours/year of teaching provided by an approved establishment.

A professional certificate is awarded to students who have passed the examination.

Under certain conditions, pharmacy students may replace the retail pharmacist during the pharmacist's absence or may be considered as the equivalent of a dispenser.

Pharmacists and dispensers (Art. L. 593-1) must wear an identity badge (for pharmacists and pharmacy students, the pharmaceutical wand, and for technicians, a mortar).

For the sale of parapharmaceutical products (see Section C.4) pharmacists may employ ordinary sales staff. Conversely, if pharmacists wish to open an optical, medical acoustics, or orthopedic counter (depending on the items considered), they will need personnel with the appropriate qualifications if the pharmacists do not possess them.

6. Pharmacy and Primary Care

Pharmacists (Art. R. 5015-4 Public Health Code) must provide assistance to patients in immediate danger if no medical care is available (first aid courses are organized in university pharmacy courses). They act, as already disscussed in Section C.3, as health and social educators to the public.

Pharmacists can perform certain medical tests as a service to customers. This activity is of only marginal importance.

7. Drug Insurance and Third Parties

The retail pharmacists are legally responsible at three levels: penally (with regard to the criminal courts), professionally (with regard to the Order of Pharmacists, which may forbid them to practice their activity), and civilly (with regard to the civil courts). The pharmacists' civil liability means that any victim of damages suffered through the fault of retail chemists or staff for whom they are responsible may obtain as compensation the payment of damages and/or payment of a pension related to the corporal injury received. With regard to the pharmacists' civil liability, they have the right to obtain insurance through any company of their choice. The Mutuelle des Pharmaciens is a special company for pharmacists.

It should be noted that the Securite Sociale, which pays for any medical and pharmaceutical expenses following the injury, may sue pharmacists (in practice, the insurance company).

D. UNIQUE OR INTERESTING FEATURES OR SPECIAL SITUATIONS IN FRANCE

In France, traditional medicine, based on universally recognized scientific evaluation, has been accompanied by the growth in popularity of alternative and natural medicine. Examples of these types of medicine are: homeopathy, phytotherapy, auriculotherapy, iridology, chiropractic, and ethiopathy.[32] Many of them do not use drugs.

Conversely, preparations based on medicinal plants are considered to be fully-fledged in France. Their medicines marketing in the form of drugs have led the minister responsible to design product marketing authorization applications procedures that do not require pharmacotoxicological or clinical testing provided that:

- The medicinal plants belong to the list of 115 herbs that have been specifically tested.
- The laboratory claims only the therapeutic indications officially approved by the Ministry.

Homeopathy has also enjoyed considerable popularity.[33] About 6,000 doctors practice it, one-third exclusively, and 8 million French patients use it to some extent.[31] In 1984, the market was worth FF 800 million plus over FF 1 billion in exports. Its existence is recognized by an official monograph in the tenth edition of the French Pharmacopoeia (January 1983) and by a chapter in the 1985 GMPs.

The laboratories manufacturing such products (five main ones) are experiencing buoyant growth, also abroad, and there are also specialized homeopathic pharmacies. Their nonhomeopathic colleagues send them preparations according to prescription and official preparations they are not able to make. Homeopathy is reimbursed at the rate of 70%. No directive has yet been published which controls homeopathic drugs.

Meanwhile, about 400 drugs are sold under their invented names which already existed in 1941 and received at the time the equivalent of the current product marketing authorization. For other such drugs, which all have common names, the Ministry has authorized their manufacturers as a whole which has enabled them to prepare dilutions down to 30 hundredths.

E. CONCLUSIONS

The French pharmaceutical system depends on a basic network of nearly 22,000 community pharmacies, whose density ensures that drug dispensing is available sufficiently close to patients. This is accompanied by pharmacies in hospitals and clinics. The balance between these two types of dispensing is similar to that between private and hospital medicine.

The system for distributing drugs owes a large amount of its efficiency and flexibility to its distributors, which give rapid service, supplying the customers of retail chemists several times daily.

The French pharmaceutical industry works at international quality standards. Although there are currently fewer new inventions made by its laboratories than previously, this is perhaps because research has not received as much support as it requires.

Although the French Securite Sociale system of health insurance is complicated, it supplies very extensive coverage despite the difficulties encountered. People have the further option of obtaining additional insurance coverage through payment of a fee or a premium. Pharmacists have strived to obtain agreements that exempt insured persons from advancing the cost of medication. Furthermore, self-medication practiced with the assistance of the pharmacist gives the public the chance of obtaining minor treatment under the advice of a trained professional.

REFERENCES

1. Dupeyroux J.J., "Droit de la Securite sociale," Paris, Dalloz ed., 1988: 23, 69.

2. Laroque P., Rivero J., Chesnais J.Cl. et al., "Quarante annees de Securite sociale," Revue francaise des affaires sociales, Paris, Ministere des Affaires sociales et de la Solidarite nationale, Ministere du Travail, de l'Emploi et de la Formation professionnelle, 1985.

3. Tisseyre-Berry M. and Viala G., "Legislation et deontologie de l'industrie pharmaceutique," Paris, Masson ed., 1984: 345, 357.

4. Tisseyre-Berry M. and Viala G., "Legislation et deontologie de l'officine pharmaceutique," Paris, Masson ed., 1983: 322, 324.

5. "REM-France," Paris, Droit et pharmacie ed., 1988: 180-193.

6. Circ. ECO 89-12, Droit et pharmacie ed.

7. Labourdette A., "Economie de la sante," Paris, PUF ed., 1988: 25, 54.

8. Majnoni D'Intignano, "Sante, mon cher souci," Paris, Economica ed., 1987: 23, 52.

9. "L'industrie pharmaceutique et ses realites," 1989, Paris, SNIP ed.

10. Labourdette A., op. cit.: 140, 151.

11. "Le secteur liberal des professions de sante," Paris, CNAMTS, no. 41, 42, 43, 1988.

12. Rousselet F., "Les etudes de pharmacie en France," Lyon pharmaceutique ed., 1988: 353, 357.

13. Ido, "Le 3e cycle d'etudes specialisees de pharmacie et l'internat en pharmacie," Lyon pharmaceutique ed., 1988: 359, 360.

14. Dupeyroux J.J., op. cit.: 611 s.

15. "Memento social Lefebvre," Paris 1989: 724 s.

16. Auby J.M. et al., "Droit medical et hospitalier," Paris, Litec ed.

17. "REM-France," op. cit.: 233.

18. Ido: 238.

19. Circ. ECO 89-21, Droit et pharmacie ed.

20. "Tableau de bord de l'industrie pharmaceutique," 1988, Paris, SNIP ed.

21. Barral P.E., "Douze ans de resultats de la recherche pharmaceutique dans le monde (1975-1976)," Paris, Prospective et sante publique ed.

22. "L'industrie pharmaceutique, ses realites," op. cit.: 17.

23. Braud G., "Made in France," Le Moniteur des pharmacies et des laboratoires ed., Paris 1989, no. 1840: 26 s.

24. Auby J.M. and Coustou F., "Droit pharmaceutique," Paris, Litec ed., fasc. 30, 31.

25. Ido, fasc. 45 s.

26. Ido, fasc. 39.

27. Ido, fasc. 33 s.

28. Ido, fasc. 38-43.

29. Tisseyre-Berry M., "Abrege de legislation et de deontologie pharmaceutiques," 1983, Paris, Masson ed.: 219 s.

30. Huttin Ch., "Le medicament: contraintes et enjeux d'un marche," 1989, La Documentation francaise ed.: 38 s.

31. Tisseyre-Berry M. and Viala G., "Legislation et deontologie de l'officine pharmaceutique," op. cit.: 12 s.

32. Rapports de l'Academie nationale de pharmacie, Informations pharmaceutiques, Paris, Ordre national des pharmaciens ed., 1988: 286 s., 439 s., 889 s.

33. Aulas J.J. et al., "L'homeopathie," Paris, Editions medicales Roland Betex, 1985.

34. Nougerede P., "Approche economique du medicament homeopathique," pharm. thesis Bordeaux II, 1986.

Chapter 9

Federal Republic of Germany

Hannelore Sitzius-Zehender
Bertram Dervenilch
Frank Diener
Gerd M. Foh

A. THE NATIONAL HEALTH CARE SYSTEM

1. History and Background[1]

Health insurance in preindustrial Germany was, like in other nations, organized in the form of the extended family, i.e., the community was jointly responsible for bearing the risk. The churches, and to a lesser extent, the government, assisted in care for the poor and charity cases. During the nineteenth century, industrialization in Germany progressed rapidly. The labor question became virulent; being deprived of the security of the extended family, sickness, age, or inability to work resulted in existence-threatening situations for laborers and their dependents. Single, solely regional and branch-specific supporting funds (Unterstuetzungskassen) were not able to adequately solve the problem, since small enrollment increased the risk of high variance in expenditures. Therefore, rates for contribution payments needed to be extremely high, which prevented members of the low-income working classes from obtaining access to insurance protection. Population explosion through medical progress further aggravated the problem.

The Reich's Chancellor Fuerst Otto von Bismarck, whose term in office extended from 1839 to 1890, perceived the existence of the German *Kaiserreich* as being in jeopardy and thus decided to introduce a modern federal social policy. Because the 1878 law against the publicly dangerous endeavors of social democracy (*Gesetz gegen die gemeingefaehrlichen Bestrebungen der Sozialdemokratie*) failed to stop the more and more powerful rising of party and union movements, Bismarck tried to defeat their appeal by devising a policy to solve the basic economic problem, i.e., the lack of social safeguards for major groups of the population. As a consequence, three laws were approved: On June 15, 1883, the law concerning sickness insurance for workers (*Gesetz, betreffend die Krankenversicherung der Arbeiter*); on July 6, 1884, the accident insurance law (*Unfallversicherungsgesetz*); and on July 22, 1889, the law concerned with safeguarding against disability and old age (*Gesetz betreffend Invaliditaets – und Alterssicherung*). With the political and power-preserving impetus to give political influence to the working classes while at the same time pursuing social security, the three pillars of German social insurance were established.

Up to certain limits of income, all persons receiving wages or salaries in industry, trade, handicraft, inland navigation, and certain service companies were automatically subject to legal sickness insurance. Moreover, in some funds, family members could be coinsured. Contributions were limited to 6% of earnings; one-third had to be paid by the employee and two-thirds by the employer. The insurance protection covered medical treatment and the provision of medications. In addition, starting with the third day of illness, but for a maximum of 13 weeks, sickness money (*Krankengeld*) was paid which equaled at least 50% of earnings. Further, postpartum care for four weeks was covered. The sickness funds that developed subsequently were allowed to administer these benefits as corporate bodies under public law (*Koerperschaften des oeffentlichen Rechtes*), an organizational principle which 100 years later would be labeled as countervailing power between government and the market.

Fundamental and social-political differences between Emperor Wilhelm II and Bismarck led to the dismissal of the Reich's Chancellor in 1890. For the purpose of regaining the population's confidence in the monarchy, Wilhelm II renounced suppression of the working class and instead intensified social security legislation and codetermination. Epoch-making for the German social insurance legislation of the entire twentieth century was the unification of the insurance laws in the federal insurance regulations (*Reichsversicherungsordnung*, RVO) on July 19, 1911. The RVO also codified sickness insurance for the working class. Later, on December 20, 1911, the insurance law for employees (*Versicherungsgesetz fuer Anges-*

tellte) was approved. The law provided social insurance protection for employees with monthly salaries between 2000 and 5000 Deutsch Mark (DM); although being modeled closely after the RVO, the laws were not exactly equivalent.

During the Weimarer Republic (1919-1933), disability, age, and sickness insurance for miners were transferred into the miners society insurance (*Knappschaftsversicherung*) through the Reich's miners' society law (*Reichsknappschaftsgesetz*) on June 23, 1923.

While overcoming the worldwide economic crisis, rapid economic growth brought new financial strength for sickness insurance in the Third Reich (1933-1945). These programs were used by Hitler for realizing his *Weltanschauliche* and political interests. Social insurance was consequently used as an instrument for stabilizing the system, mobilizing worker allegiance, and absorbing the inflation-driving purchasing power of the masses. Expenditures for public health care and accident prevention were drastically increased. Motivated by population politics, a new social-political strategy was introduced: insurance benefits related to family aid were now based upon the number of children. At the same time, the regional principle of social insurance, a relic derived from social classes, was replaced by a centralized administration. Important laws were the law for building up the social insurance (*Gesetz zum Aufbau der Sozialversicherung*), July 5, 1934; and the law for postpartum care and care for the convalescent in sickness insurance (*Gesetz ueber Wochenhilfe und Genesendenfuersorge in der Krankenversicherung*) from June 28, 1935.

Social legislation in the Federal Republic of Germany (FRG) (1949 until today) is largely molded after the legislative situation in the Weimarer Republic. The federal insurance regulations (RVO) continued to be the legal foundation for sickness insurance until 1988, whereby the national-socialistic elements, which pertained to social policy, were eliminated. The growth of the economy in postwar Germany resulted in a dynamically growing financial basis for sickness insurance which in turn produced one of the most powerful and efficient health care systems in the world. Only the first oil crisis in 1973 and the world economy crisis in 1974 brought about financial problems that led to a number of cost-containment laws (*Kostendaempfungsgesetze*) in the health care system.

On January 1, 1989, the so-called health reform law (*Gesundheitsreformgesetz*) was enforced; its main part consists of the fifth book of the social law book (*Sozialgesetzbuch*, SGB V), in which the entire statute concerning sickness insurance is newly codified and reformed.

2. Social and Political Environment

The economic situation in the Federal Republic has been marked by continuous growth since 1982 (Table 1). However, there are special problems in the labor market characterized by a continuously high unemployment rate (on the average, around two million unemployed persons in 1989). Deviation of the labor market from the generally positive economic development causes problems in domestic policy. Explanations range from a demographically unfavorable age structure of the population, to the high number of refugees coming from areas within the former German Reich, to the increasing number of women in the labor market. Thus, the potential number of persons available for employment is growing more rapidly than the number of places for employment. On the other hand, there are also important structural problems of adjustment in areas of primary industry (charcoal, steel, ship building) coupled with a high portion of unemployed persons who are difficult to place. All these factors contribute to the continuously high rate of unemployment.

In comparison with other nations, the political constellation in the Federal Republic may be regarded as stable. Since 1982, the federal government has been led by a conservative-liberal coalition of Christian Democratic/Socialistic Union (CDU/CSU) and Liberal Democratic Party (FDP).

3. Public and Private Financing, Health Costs

In 1986, expenditures for health in the Federal Republic totaled 251 billion DM, which can be separated into 71% provided by public* and 29% by private** carriers. Classifying 1986 health costs into major types of outlays reveals: 59% for treatment, 29% for costs following illness, 6% for prevention and prophylaxis, 2% for education and research, and 4% for miscellaneous expenditures.

The direct health expenditures of the social (legal) and private sickness insurances are usually divided into the sectors: outpatient/ambulant treatment, inpatient/hospital treatment, dental treatment and dentures, and expenditures for medical aids and appliances (Tables 2 and 3). It is instruc-

---

* Public households, 33 billion DM; social sickness insurance, 117 billion DM; social pension insurance, 20 billion DM; social accident insurance, 8 billion DM.
** Private sickness insurance, 13 billion DM; employer, 40 billion DM; private households, 20 billion DM.

tive to note that expenditures under social sickness insurance in 1987 averaged 2,212 DM per insured person compared to only 1,866 DM for those covered by private sickness insurance (Table 4).

4. Facilities and Manpower

Development in various areas of the health delivery sector has differed remarkably. Technologically innovative areas of the health care system increased by more than average rates (i.e., 15% for large medical apparatus) while the drug sector attained — compared to the total market volume of pharmacists — growth rates of 6.5% per anum on average (for further details, see Table 5).

5. The Social Security System

The cornerstones of social security in the Federal Republic are the social unemployment insurance (*Gesetzliche Arbeitslosenversicherung*, ALV), social pension insurance (*Gesetzliche Rentenversicherung*, GRV), and social sickness insurance (*Gesetzliche Krankenversicherung*, GKV). There is no independent insurance for sickness care; instead, it was integrated, together with the health reform law, into Social Law Book V (SGB V). The social accident insurance is financed by collecting contributions from employers; all other branches of social insurance are financed by equal contributions on the part of employers and employees. Presently, these rates amount to 4.3% for social unemployment insurance, around 13% for social sickness insurance, and 18.7% for social pension insurance. There are certain limits for deciding about the level of contributions (*Beitragsbemessungsgrenze*). The limit for social pension and social unemployment insurance was 6,100 DM per month in 1989; that for social sickness insurance is set at 75% of the current social pension insurance limit — thus, it equaled 4,575 DM per month in 1989. If the employee's income exceeds the limit, only the part of the income below the limit is subject to social pension and unemployment insurance. However, membership in social sickness insurance is no longer mandatory.

A fundamental principle of social sickness insurance is the application of coinsurance to the employee's dependents and family members. Employers are obliged to continue paying salaries/wages during the first six weeks of an employee's illness. Therefore, potentially high costs, associated with the risk of becoming sick, are excluded from social sickness insurance operations.

TABLE 1. Selected components of economy, partially estimated for 1988, prognostic for 1989.

COMPONENT	UNIT	ABSOLUTE NUMBER			PERCENT VS. PRECEDING YEAR		
		1987	1988	1989	1987	1988	1989
1. UTILIZATION OF NATIONAL PRODUCT expressed in respective prices private							
consumption	Mrd DM	1,112.0	1,161.0	1,212.0	+4.1	+4 1/2	+4 1/2
government consumption	Mrd DM	397.2	411.5	419.5	+3.8	+3 1/2	+2
investment	Mrd DM	389.3	419.5	448.0	+3.1	+7 1/2	+7
equipment	Mrd DM	168.7	181.5	198.0	+4.5	+7 1/2	+9
building	Mrd DM	220.6	238.0	250.0	+2.1	+8	+5
stock changes	Mrd DM	+10.5	+16.5	+14.5			
contribution from outside	Mrd DM	+111.1	+122.5	+133.5			
export	Mrd DM	638.3	686.5	742.0	+0.0	+7 1/2	+8
import	Mrd DM	527.2	564.0	608.5	+0.7	+7	+8
2. GROSS NATIONAL PRODUCT (GNP)	Mrd DM	2,020.1	2,131.0	2,228.0	+3.9	+5 1/2	+4 1/2
3. IN PRICES FROM 1980:							
private consumption	Mrd DM	924.7	953.0	975.5	+3.5	+3	+2 1/2
government consumption	Mrd DM	328.3	335.0	336.5	+1.6	+2	+1/2
investment	Mrd DM	331.3	351.5	366.5	+1.8	+6	+4 1/2
equipment	Mrd DM	141.3	150.5	160.5	+4.0	+6 1/2	+6 1/2
building	Mrd DM	190.0	201.0	206.0	+0.2	+6	+2 1/2
export	Mrd DM	550.2	579.5	609.0	+0.8	+5 1/2	+5
import	Mrd DM	501.2	533.0	557.0	+4.9	+6 1/2	+4 1/2

4.	GNP	Mrd DM	1,643.2	1,702.0	1,744.0	+1.8	+3 1/2	+2 1/2
5.	REAL VALUE OF GNP	Mrd DM	1,687.0	1,756.5	1,800.0	+2.7	+4	+2 1/2
6.	PRICE DEVELOPMENT							
	private consumption	1980=100	120.3	122.0	124.0	+0.5	+1 1/2	+2
	gnp	1980=100	122.9	125.0	128.0	+2.0	+2	+2
	last domestic utilization	1980=100	119.7	121.5	124.0	+1.2	+1 1/2	+2
7.	ORIGIN OF NATIONAL PRODUCT							
	employed persons (domestic)	1,000	25,891	26,035	26,230	+0.7	+1/2	+1
	working time	hours				+0.7	0	-1
	working volume	Mrd Dm				+0.0	+1/2	0
	productivity	DM per hour				+1.9	+3	+2 1/2
8.	DOMESTIC GNP IN PRICES FROM 1980	Mrd DM	1,634.3	1,693.0	1,736.0	+1.9	+3 1/2	+2 1/2
9.	DISTRIBUTION OF PUBLIC INCOME							
	gross income							
	out of employed work	Mrd DM	1,081.5	1,124.5	1,166.0	+3.9	+4	+3 1/2
	out of enterprises and properties	Mrd DM	486.8	535.0	563.5	+4.0	+4	+3 1/2
10.	PUBLIC INCOME	Mrd DM	1,568.3	1,659.5	1,729.5	+3.9	+6	+4
11.	REAL PUBLIC INCOME	Mrd DM	1,309.7	1,368.0	1,397.0	+2.7	+4 1/2	+2
12.	NET INCOME							
	out of employed work	Mrd DM	580.6	605.0	623.5	+2.8	+4	+3
	out of enterprises and properties	Mrd DM	415.9	459.5	480.5	+5.5	+10 1/2	+4 1/2
13.	AVAILABLE INCOME IN PUBLIC HOUSEHOLDS	Mrd DM	1,269.0	1,322.5	1,374.5	+4.2	+4	+4

TABLE 2

Year	Medical Treatment	Dental Treatment	Dentures	Drug Products From Pharmacies	Medical Aids And Appliances	Hospital Treatment	Sickness Money	Prevention1	Maternity2	Miscellaneous Services3
1970	5.458	1.708	0.828	4.226	0.667	6.009	2.467	248	1.101	1.138
1971	6.809	2.022	1.209	4.971	0.891	7.653	2.958	453	1.286	1.336
1972	7.584	2.250	1.524	5.754	1.179	9.362	3.436	631	1.321	1.516
1973	8.602	2.671	1.860	6.753	1.631	11.700	3.896	744	1.366	1.786
1974	9.930	3.399	2.086	7.883	2.095	15.246	4.263	870	1.604	1.810
1975	11.258	4.129	4.180	8.901	2.582	17.534	4.664	1.010	1.689	2.172
1976	11.923	4.297	5.312	9.642	3.054	19.256	4.733	872	1.774	2.675
1977	12.489	4.607	5.403	9.849	3.339	20.464	4.909	896	1.771	2.831
1978	13.194	4.967	5.754	10.651	3.844	21.865	5.308	780	1.827	3.205
1979	14.122	5.222	6.471	11.372	4.355	23.252	5.941	714	2.146	3.760
1980	15.358	5.517	7.351	12.572	4.881	25.465	6.654	772	3.036	4.246
1981	16.491	5.936	8.110	13.631	5.273	27.321	6.440	906	3.249	4.728
1982	16.917	6.072	6.988	13.776	5.045	29.596	5.896	729	3.068	4.458
1983	17.763	6.280	6.664	14.449	5.234	30.969	5.781	725	2.956	5.074
1984	18.924	6.563	7.338	15.545	6.064	33.215	6.301	875	2.657	6.079
1985	19.660	6.656	7.666	16.603	6.512	35.049	6.378	955	2.736	6.487
1986	20.295	7.164	6.897	17.626	7.221	37.489	6.875	1.036	2.517	6.941
1987	20.966	7.370	6.283	18.889	7.849	39.282	7.391	1.168	2.384	7.419

1. Prevention = measures for early recognition of diseases, and benefits (service and financial) for other preventive measures, particularly for cures.
2. Materity = benefits (service and financial) for motherhood and preventive examinations for pregnant women.
3. Miscellaneous benefits/services = treatment through other health persons, second -opinion physician services, convalescence aid, business and household aid, death grant.

SOURCE: BMA

TABLE 3. Expenditures and refunds of contributions in private insurance (million DM).

Year	Total Insurance Benefits	Hospital	Out-Patient Treatment	Drug Products	Dental Treatment Dentures	Other Services	Refunds of Contributions
1970	2.887	900	503	436	131	635	283
1971	3.223	1.017	551	454	150	712	359
1972	3.567	1.186	578	466	174	798	365
1973	3.833	1.346	620	454	209	905	362
1974	4.201	1.564	717	478	277	1.027	218
1975	4.792	1.793	823	503	352	1.141	179
1976	5.209	1.913	883	517	418	1.261	218
1977	5.655	2.017	890	502	511	1.310	424
1978	6.337	2.162	1.014	518	625	1.420	598
1979	6.900	2.357	1.123	543	738	1.559	580
1980	7.671	2.632	1.272	579	856	1.687	645
1981	8.359	2.980	1.428	667	980	1.780	525
1982	8.895	3.220	1.524	677	1.028	1.825	621
1983	9.033	3.344	1.576	687	1.044	1.701	679
1984	9.915	3.584	1.661	728	1.081	1.827	1.034
1985	10.700	3.751	1.728	808	1.155	1.925	1.334
1986	11.026	3.975	1.770	802	1.216	2.117	1.146
1987	12.211	4.244	1.926	858	1.334	2.278	1.571

1. Hospital day pay, medical aids and appliances, additional benefits in case of death and other services not listed separately. (Source: PKV)

TABLE 4. Expenditures of social and private sickness insurances per insured person in 1987.

	GKV	PKV
HOSPITAL TREATMENT	730,40 DM	744,56 DM
OUT-PATIENT TREATMENT	389,80 DM	468,56 DM
DRUG PRODUCTS	351,20 DM	150,53 DM
DENTAL TREATMENT AND DENTURES	253,80 DM	234,04 DM
OTHER BENEFITS	487,20 DM	399,65 DM
TOTAL	2212,40 DM	1866,67 DM

NOTE: GKV=social sickness insurance
　　　　 PKV=private sickness insurance

SOURCE: Eigene Berechnung Anhand Von SVF KAIG 1989,
Tabellen T 300, T 465 und T 601.

6. The Medical Establishment: Entry, Specialists, Referrals

Physicians (and also dentists) and pharmacists are so-called free professions, which are legally organized in their respective councils (public bodies, *Kammern*). A license for practice is granted through *Approbationsordnungen* and only formal criteria have to be fulfilled. Restrictions on licensure are generally regarded as necessary, but implementing them would be a violation against constitutional regulations. Modifications might become possible when the Common EC-home market will be developed in 1993; by then the principle of professional permissiveness will become generally established.

7. Payment and Reimbursement

Private sickness insurance operates under the principle of reimbursing the patient for previous expenditures, while under social insurance, provider reimbursement is controlled by a complex network of permits, receipts, and prescriptions (Figure 1). It is typical for the underlying service principle that the patient does not pay when service is rendered; instead, health care providers bill the sickness funds directly. The patient pays his/

TABLE 5. Recent change in health care providers and suppliers.

	Time Period	Changes in Percent	Changes Per Year
Physicians in Practice	1980-86	+ 18,3	+ 2,6
Physicians Approved for Rendering			
Sickness Fund Services	1980-87	+ 18,2	- 2,3
District Physicians - Administrative	1980-87	+ 23,9	+ 3,0
Hospital Beds Scheduled	1980-86	- 4,7	- 0,7
Hospital Beds -Attending Physicians	1980-86	- 12,3	- 1,8
Total Hospital Personnel	1980-85	+ 5,5	+ 0,9
- medical personnel	1980-85	+ 8,7	+ 1,4
- physicians	1980-85	+14,6	+2,4
- nursing (students excluded)	1980-85	+ 9,5	+1,6
- medical / technical personnel	1980-85	+ 3,5	+0,6
Medical Bathing Services	1980-86	+ 23,4	+ 3,3
Messeurs	1980-86	+ 23,9	+ 3,4
Physical Therapists	1980-86	+ 37,8	+ 5,4
Other Supporting Personnel	1980-86	+ 66,2	+ 9,5
Optician Services	1980-86	+ 30,4	+ 4,3
Hearing Aid Services	1980-86	+ 16,0	+ 2,3
Dentists	1980-86	+ 14,5	+ 2,1
Employees in Laboratory Enterprises	1980-86	+ 26,2	+ 3,7
Dentists Technicians in Office-Based			
Laboratories	1980-85	+ 32,3	+ 5,4
Medical / Technicians Apparatus - Total	1984-88	+ 73,3	+ 14,9
Apparatus - Ambulatory Services	1984-88	+ 100,0	+ 20,0
Pharmacists	1980-87	+ 13,0	+ 1,6
Total Market Volume of Pharmacies	1980-87	+ 53,0	+ 6,6
Drugs Sold (Single Doses)	1980- 87	+ 5,8	+ 0,7

SOURCE: SVR KAIG 1989, S. 77

FIGURE 1. Administrative flow chart outlining implementation of sickness benefits.

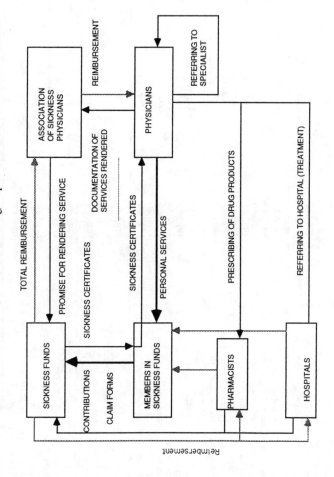

238

her contributions to the sickness fund and receives sickness certificates (*Krankenscheine*) which can be exchanged for medical services. Delivery of services can be initiated only by physicians, who also refer the patients to other health services (hospital treatment, drugs, and medical aids and appliances). Details of these procedures are regulated by specific contracts for delivery and accounting. Using the Social Law Book V (SGB V) as a basis, skeleton treaties are agreed upon between the top associations of sickness funds and health professions. Negotiations can take place on the federal and (in more detail) on the state level. The drug price regulation (*Arzneimittelpreisverordnung*) is of central importance for the medicament sector because it guarantees uniformity of prescription prices for pharmacy-restricted drugs all over the Federal Republic.

8. Coverage

Out of the approximately 61 million inhabitants of the Federal Republic (1987), nearly 54 million are covered through the 1,182 social sickness insurance schemes, while 5.7 million are members of private sickness insurance. Social aid and welfare services provide protection for sickness expenses for 1.5 million persons. However, nearly 200,000 persons lack sickness insurance altogether.

B. PHARMACEUTICAL INDUSTRY AND DRUG DISTRIBUTION

1. Structure of the Industry

The number of drug manufacturers in the Federal Republic is estimated at more than 1,000. Of these, 480 are members of the federal association of the pharmaceutical industry (*Bundesverband der Pharmazeutischen Industrie*, BPI). Drugs produced by these 480 companies represent around 95% of the total value of drug production in West Germany.

The German pharmaceutical industry, as a branch of the industry, grew organically over many decades. But the relatively high share of newly established pharmaceutical companies since 1950 also demonstrates that access to the German drug market can take place rather easily and that potential competition represents an important factor.

The German pharmaceutical industry employs approximately 100,000 persons. This number is nearly equivalent to the number of persons working in the 17,800 community pharmacies in the Federal Republic of Germany. A distribution of pharmacies according to sales shows most of them can be classified as businesses of middle-class size (Table 6).

TABLE 6. Community pharmacy sales classified by volume level and percent of vendors.

Amount of Sales (100,000 DM)	Share of Companies (Percent)
under 7.5	22%
7.5 but under 15	8
15 but under 45	29
45 but under 150	23
150 and over	18

2. Research and Development, Imports, Exports, Patents, Licensing

Research and development costs of a medicament containing a new active ingredient average 250 million DM. From the first experiments for synthesis to the complete drug product, roughly 800 working steps are necessary. For many years, the German pharmaceutical industry has been spending roughly 15% of sales for research and development; expressed in absolute number, the total sum amounted to nearly 4 billion DM in 1988.

In 1987, nearly 47% of pharmaceutical goods produced in the Federal Republic of Germany went into export. Germany has the highest export rate among all drug-exporting countries in the world. On the opposite side of these pharmaceutical exports covering 10 billion DM in 1987 were 5.9 billion DM worth of imports (source: BPI/pharma daten 1988).

Since 1978, patent protection for product innovations has been guaranteed for 20 years. In the case of medicaments, protection by patent can be granted for the active ingredient as well as for the manufacturing method. Time of patent protection starts with the day of application. On the average, though, effective patent life yields only 7.7 years for recovering the cost of research and development and providing profit opportunities. Thus experiments and clinical investigation, the evaluation of these examinations, and the ensuing registration procedures at the federal health office (*Bundesgesundheitsamt*, BGA) take up a considerable amount of patent protection time. The balance for processing license applications in the Federal Republic of Germany, which showed negative numbers in the 1970s, has changed to positive balances in recent years. According to BPI in 1986, licensing yielded 176.8 million DM more than was spent in the same period.

3. Manufacturing

In 1987, the total value of goods produced by the pharmaceutical industry totaled 21.4 billion DM, expressed in manufacturer prices. Of this output, 78.6% (16.8 billion) consisted of human pharmaceuticals, 5.3% was for vitamins and hormones; 5.2% for bandaging materials, cotton, and adhesive plasters; while veterinary products held a share of 1.3%.

In the Federal Republic of Germany, drugs needed for medical treatment are produced predominantly by pharmaceutical companies. Approximately 70,000 preparations are registered with the federal health office (BGA), including some 23,000 homeopathic medicaments.

Drug manufacturers need to have a federal permit for the production of drugs (*Staatliche Herstellungserlaunis*). The manufacturing process has to be performed according to the rules of GMP and GLP.

The ten leading companies together are responsible for around one-third of drug sales. Thirty-five German drug manufacturers conduct their own research.

4. Advertising and Promotion, Price Regulation

On average, pharmaceutical companies spent 12.4% of sales for scientific information and 5.2% for advertisement. In addition, general marketing costs totaled 8.8% of sales. According to the association of the pharmaceutical industry (BPI), roughly 10,000 pharmaceutical salespersons visited physicians for delivering information about possible applications of drugs produced by their companies. In addition, they also gathered the physicians' experiences, whether positive or negative, when using these drugs in patients.

Advertisement for drugs is regulated rather strictly in the remedy advertisement law (*Heilmittelwerbegesetz*). And going beyond, the pharmaceutical industry also follows voluntary restrictions on competition (Codex for Members of the Federal Association of the Pharmaceutical Industry). Advertising for prescription drugs to the lay public is prohibited. The distribution of so-called physician samples, for reasons of becoming acquainted with drugs, is restricted.

In principle, manufacturers are free to decide about the level of their prices. There are no government restrictions on pricing (see Section B.7 under Cost-Containment Activities). However, there are margins for pharmaceutical wholesalers and pharmacies laid down in the drug price regulation (*Arzneimittelpreisverordnung*). For products up to a manufacturer's price of 1.65 DM, pharmaceutical wholesalers are allowed to add a minimum of 21%. With an increasing purchasing price, the margin decreases to 12% for medicaments of 108.71 DM manufacturer's price and above. Pharmacies are also subject to decreasing margin regulations on the wholesale price. For low-priced medicaments (up to 2.40 DM purchasing price), the margin equals 68%. But the margin decreases to 30% for preparations whose purchasing price is 70.30 DM and above. In Germany, a value-added tax has been mandated, which raises all product costs by 14%.

Pharmacies are obliged to grant a discount of 5% to the social sickness funds, which currently insure approximately 90% of the population.

5. Drug Approval and Government Regulation

In the Federal Republic of Germany, registration of pharmaceutical products and market approval are regulated in a drug law (*Arzneimittelgesetz*, AMG). The AMG separates between licensing, where the efficacy of the drug has to be proven, and registering without proof of efficacy. The latter is only possible for homeopathic drugs.

The registration of a new drug has to be applied for to the federal health office (*Bundesgesundheitsamt*, BGA). Scientific documentation, which has to be submitted together with the application, should prove the product's quality, efficacy, and safety. For new drug registration, the documentation must consist of pharmacologic-toxicologic and clinical research data. For drugs already on the market, presenting certain scientific documentation material by referring to the licensing documentation of the original manufacturer is allowed. However, the new AMG contains a restriction on using this information (*Verwertungssperre*) for ten years. During this time, reference to the original documentation is permissible only when consent of the respective manufacturer has been given.

Before deciding about the approval of a drug, expert committees have to be heard. Besides individual licensing procedures, there are also standard licenses. These are possible for groups of chemical agents where quality, efficacy, and safety have already been proven.

There exists a legal claim on licensing a drug, provided the legal requirements are met. However, neither a needs assessment nor an examination of the price-benefit ratio is performed.

6. Competition: Brand and Generic

During recent years, the market share of generics has expanded remarkably. In Germany, the term generics refers to all pharmaceuticals that are copies (second application drugs), even when they are sold under their own brand name. Often the product's name consists of the active ingredient plus the manufacturer's name. Pure generics, where the manufacturer's name does not appear, are seldom found. In 1981, generics accounted for only 6% of pharmacy sales to the social sickness funds, but this share increased to 15.3% by 1988. With generics being, on average, one-third less expensive than the original drugs, the portion of generics associated with the total number of packages sold totaled 20%. Considering that only 40% of the drugs on the market are available from multiple suppliers, this situation demonstrates the importance of generics in the drug market. However, a recent introduction of fixed prices (*Festbetraege*) for pharmaceuticals reimbursable by the social sickness funds (see Section B.7, Cost-Containment Activities), will worsen the competitive position of generic manufacturers in the future.

7. Cost-Containment Activities

The goal of stabilizing the development of expenditures among the social sickness funds led to approving a health reform law (*Gesundheitsreformgesetz*) on January 1, 1989. Cost-containment measures are focusing on the drug sector. Besides intensifying efficiency examination of physician prescribing habits, fixed prices (*Festbetraege*) for drug products were introduced and the so-called negative list was expanded. Constructing the negative list (i.e., list of medicaments for various indications that are not to be covered by sickness funds) has the goal of excluding drugs that are not economical or whose relative therapeutic contributions are heavily disputed from being reimbursed by the social health insurance.

Introducing so-called fixed prices (i.e., price up to which the sickness funds will cover costs for the single agent) is entering a new era. In a first step, fixed prices are to be decided upon for products containing the same active ingredients. Most manufacturers whose prices were higher than the fixed prices lowered their prices so that patients could avoid surcharges and the firm avoided disadvantages in competition. For generic manufacturers, this means that price differences between original drugs and generics are shrinking drastically. Therefore, the pressure on physicians to prescribe generics is declining. Thus, it seems possible that the share of generically written prescriptions may decrease again in the future. The health reform law also provides the option that the physician could allow the pharmacist to select the drug.

Regulations of the health reform law are especially incisive concerning the drug sector. Thus, no increase in drug sales can be expected for the coming years. On the contrary, a reduction of sales in some sectors is not to be excluded. Drug manufacturers whose product lines are strongly affected by the reform measures may find their existence in jeopardy.

C. PHARMACY PRACTICE

1. Dispensing a Prescription: Where, What, Payment, Formulary

Medications ordered on a physician's prescription form, if pharmacy restricted, are to be dispensed exclusively in pharmacies. Thus, pharmacies have a monopoly for dispensing pharmacy-restricted drugs which is regulated in Paragraph 43 of the drug law. With the exception of veterinarians, physicians are not permitted to dispense medications. This dispensing right is explained by the peculiarities of practicing veterinary medicine.

Pharmacists must dispense exactly the drug that was ordered by the physician. This applies equally to strength of the active ingredient, size of package, and dosage form. Pharmacists have the right for drug product selection only when allowed by the physician on the prescription form and then only for generic substitution. Since January 1, 1989 according to Paragraph 73 Abs. 5, SGB V, the prescribing physician should indicate on the prescription form whether the pharmacist is to be allowed to select a cheaper, but chemically equivalent, medicament in place of the one prescribed. Whether establishing a system of fixed prices that limits reimbursement for chemically equivalent drugs to a maximum price, as determined by the sickness funds, will lead to an increased utilization of pharmacists for drug product selection remains to be seen.

Since January 1, 1989, patients who are insured with the social sickness funds and are 18 years of age and older must pay a surcharge of 3 DM per prescribed item for sharing drug costs. As of January 1, 1989, this share of drug costs will be changed to 15%, but at most 15 DM per prescribed medicament. These mandatory surcharges, however, do not apply to medicaments with fixed prices.

Pharmacies use prescription forms for billing the sickness funds. Patients who are insured with private health insurance plans first pay drug charges according to pharmacy sales prices in the pharmacy. After that, they submit receipts to their insurance carrier. Reimbursement is implemented according to the conditions of the respective insurance contracts.

2. Record Keeping

According to pharmacy management regulation (*Apothekenbetriebsordnung*) Paragraphs 6 and 8, if drugs are manufactured in the pharmacy, production and examinations must be documented. These protocols are to be kept in the pharmacy for at least three years.

According to the narcotic drug law (*Betaubungsmittelgesetz*), inventory and dispensing of narcotic drugs has to be documented meticulously. Prescriptions are to be entered in the documentation, and the original record must be stored for at least three years.

Documentation for storage and selling of toxic substances has to be performed according to the dangerous substance regulation (*Gefahrstoffverordnung*).

3. Patient Education

According to the pharmacy management regulation (*Apothekenbetriebsordnung*) Paragraph 20 Abs. 1, pharmacists must render information and counseling to customers (and to persons licensed for practicing medicine, dentistry, and veterinary medicine) as far as necessary for reasons of drug safety. The pharmacist decides whether information or advice is indicated either because of the nature of the submitted prescription or because the patient/customer has asked for it.

Counseling and informing patients/customers is limited through Paragraph 20 Abs. 1, Sentence 2 of the pharmacy management regulation: "The therapy, decided upon by persons licensed to practice medicine, dentistry or veterinary medicine, is not to be affected by informing or counseling the customer." This stipulation prohibits pharmacist comment on drug therapy, which is determined exclusively by physicians, in a way that could affect patient compliance.

The pharmacist's duty for rendering information and counseling applies also to self-medication. Paragraph 20 Abs. 1, Sentence 3 of the pharmacy management regulation says: "In the case of drugs being dispensed without a prescription, it is the pharmacist's duty to render the information necessary for proper administration to the customer." Thus patient counseling is mandated so that information pertaining to drug safety can be presented to the individual.

Patient package inserts (PPI) also serve as devices for patient education and information. Paragraph 11 of the drug law (AMG) states that PPIs must be added to each pharmaceutical speciality (*Fertigarzneimittel*). PPIs must contain documentation about the kind and amount of active ingredients, indications, contraindications, side effects, interactions, and dosage regimen, including references not required elsewhere.

In addition, Paragraph 11a of the drug law (AMG) requires drug manufacturers to provide pharmacists with additional scientific information (*Fachinformation*), which should enable the pharmacist to fulfill his/her duty to inform and counsel.

4. OTCs and Other Classes and Schedules of Drugs, Nonpharmacy Outlets

In principle, drug classes are separated into prescription, pharmacy restricted, and free-to-sell. Prescription, and thus automatically pharmacy-restricted drugs, and also regular pharmacy-restricted drugs, are to be sold exclusively in pharmacies. Corresponding regulations can be found in Paragraphs 43 and 49 of the drug law (AMG).

Nonprescription drugs are separated into pharmacy-restricted drugs, which are only to be sold in pharmacies, and free-to-sell drugs, which can be sold more generally. In 1987, the Supreme Court (*Bundesverfassungsgericht*) decided to approve self-service as a way for offering free-to-sell drugs, inside and outside of pharmacies.

Paragraph 25 of the pharmacy management regulation (*Apothekenbetriebsordnung*) defines which goods besides drugs are allowed to be sold in pharmacies. Among other goods, these are: bandaging materials; products for illness and baby care; medicinal, dental, and veterinarian instruments; products for personal hygiene (as far as these do not serve predominantly decorative purposes); pesticides; and products for animal breeding.

In 1988, the various classes of pharmaceuticals achieved the following portions of pharmacy sales (without value-added tax) totaling 27.3 billion DM: prescription drugs made up 16.6 billion DM (61%), pharmacy-restricted but nonprescription drugs were 8.25 billion DM (30%), while free-to-sell drugs achieved 700 million DM and therefore accounted for 2.5% of total pharmacy sales.

Thus, all classes of drugs together attained 25.55 billion DM or 93.5% of pharmacy sales. An additional 3% was accounted for by drug sundries, whereas nursing care products and other goods account for the remaining 3.5%.

Besides grouping drugs into legislative classes, there are also various lists available that describe drugs. The so-called *Rote Liste* (red list) contains drug products that are being offered exclusively by members of the national association of the pharmaceutical industry (BPI). Eighty-seven main indication areas are covered. The 1989 edition encompasses 8,550 drug products (*Praeparateeintraege*), with 10,832 dosage forms and 21,554 price declarations. At present, the *Rote Liste* is the most important information source for physicians when prescribing drugs.

The big German drug product tax (*Grosse Deutsche Spezialitaeten-Taxe, Lauertaxe*) is the most comprehensive collection of all pharmaceuticals currently available on the German market which are labeled with a pharma-central number. The list renders complete information for pharmacies about all drugs currently on the market, together with up-to-date prices.

The price comparison list (*Preisvergleicheliste*) is part of the medicament guidelines agreed upon between the national boards (*Ausschuesse*) of physicians, dentists, and sickness funds. According to Paragraph 92 Abs. 2, SGB V, these guidelines should guarantee sufficient, expedient, and economical provision of drugs for the insured.

In order that physicians may compare prices and select therapeutically appropriate amounts of drugs, the price comparison list currently classifies products used for the top 14 indications, as follows:

— Medicaments generally suitable for treatment in the respective area of indication
— Medicaments only suitable for some patients or in special cases for treatment in the respective area of indication
— Medicaments with known risks or doubtful therapeutic usefulness and which therefore require special attention when prescribed.

The transparence committee (*Transparenzkommission*) of the federal health office (BGA) publishes transparence lists (*Transparenzlisten*) for parts of the drug market, limited by areas of indications. The goal is to bring about a pharmacologic-therapeutic and price transparence for these groups of medicaments. Paragraph 39b Abs. 1, AMG, states the requirements for these lists: "Transparence lists contain comparisons of therapeutic concepts which explain the therapeutical efficacy of drugs used for the same indication. They show the appropriate amount of drug, therapeutically needed, together with its price per treatment-day or other basis which allows for comparison, and also price per package and per single dose." The transparence committee is comprised of representatives from sickness insurances, physicians, drug manufacturers, consumers, and pharmacy experts.

According to Paragraph 34 Abs. 1, SGB V, a negative list for insured persons age 18 and older designates the following pharmaceuticals as being ineligible for prescribing and reimbursement:

1. Medicaments for use in common colds and minor flus, including catarrh and pain medications, and cough suppressants and expectorants that could be used for these indications
2. Therapeutics for mouth and throat, with the exception of fungus infections
3. Laxatives
4. Medicaments effective against travel nausea.

The Minister for Labor and Social Order, in agreement with the Minister for Youth, Family, Women and Health and with the Minister for Economy, can expand this negative list through exclusion of medicaments that contain unnecessary ingredients or a multitude of ingredients or where the existence of therapeutic benefits is not proven.

In July 1989, a corresponding bill for a legal regulation about uneconomical drugs in the social sickness insurance (*Rechtsverordnung ueber unwirtschaftliche Arzneimittel in der gesetzlichen Krankenversicherung*) was submitted. Also introduced was a bill according to Paragraph 34 Abs. 4, SGB V, for a legal regulation about remedies of inferior therapeutic benefits or with low sales prices in the social sickness insurance (*Rechtsverordnung ueber Hilfsmittel von geringem therapeutischen Nutzen oder geringem Abgabepresis in der gesetzlichen Krankenversicherung*).

In the Federal Republic of Germany, positive lists are not employed to designate drugs eligible for reimbursement.

5. The Use of Pharmacists, Technicians, and Others

According to Paragraph 3 Abs. 1 of the pharmacy management regulation (*Apothekenbetriebsordnung*), persons working in a pharmacy are separated into pharmaceutical and nonpharmaceutical personnel. The latter group can only provide services corresponding to their level of education and knowledge.

Pharmaceutical personnel consist of pharmacists and persons receiving pharmacy education, pharmacy technicians and persons obtaining technician education, and pharmacist assistants. Only these persons are allowed to perform pharmaceutical tasks within the meaning of the pharmacy management regulation. These tasks encompass development, manufacturing, examination, and dispensing of drugs; rendering information and counseling about drugs; and also, checking on drugs stored in hospitals.

The nonpharmaceutical personnel consist mainly of helpers/aids (*Apothekenhelfer*). They support the pharmaceutical personnel in manufacturing and examination of drugs, as well as through operating and taking care of the technical instruments; they also assist in filling and packaging as well as in the preparation of drugs for dispensing. Thus, to support the safe and orderly distribution of drugs, the scope of activities that can be delegated to pharmacy helpers is defined rather narrowly.

6. Pharmacy and Primary Care

Since there is no drug dispensing outside of pharmacies, primary care with the meaning of a basic supply is not known in the system of drug provision in the Federal Republic of Germany; rather, all drugs, with the exception of free-to-sell drugs, are dispensed exclusively through pharmacies. In particular, there is no drug dispensing to the indigent through social welfare or other social institutions. Likewise, sickness funds are not permitted to dispense drugs directly to the insured persons. In addition, hospital pharmacies provide drugs to inpatients only. However, some retail pharmacies sign contracts to provide drugs to patients in hospitals without a pharmacy in the facility.

7. Drug Insurance and Third Parties

The social insurance system in the Federal Republic of Germany does not provide separate insurance coverage for drugs. Rather, all insured persons, according to Paragraph 31 Abs. 1, SGB V, are entitled to drug benefits unless products are excluded from reimbursement by the negative list.

The provision of medicaments, bandages, medical aids, and appliances is, according to Paragraph 27, SGB V, part of the insured person's sickness benefits. This protection also encompasses medicinal and dental treatment, including the provision of dentures. Also covered are bandages, medical aids and appliances, nursing at home and household aid, hospital treatment, and medicinal and complementary services for rehabilitation, as well as stress examination and work therapy.

REFERENCE

1. Cmp. Heinz Lampert. *Lehrbuch der Sozialpolitik*. Springer: Berlin et al. 1985; 1-122.

SELECTED READINGS

Werbe- und Vertriebsgesellschaft Deutscher Apotheker mbH, Apothekenreport Nr. 34, Frankfurt (Main), 1989.

Schiedermair, Rudolf. Gesetzeskunde fuer Apotheker, Frankfurt (Main), 1987.

Buchholz, Edwin H. (publisher). Das Gesundheitswesen in der Bundesrepublik Deutschland, Berlin, Heidelberg, 1988.

Adam, Deiter (publisher). Handbuch fuer den Pharmareferenten, Band 2, Stuttgart, New York, 1985.

Chapter 10

Hungary

Zoltan Vincze

Hungary is a mountain-bordered country with extensive central lowlands. It is a land-locked country bordered by Czechoslovakia, the U.S.S.R., Romania, Yugoslavia, and Austria. The total land area is 93,032 square kilometers.

In 1987, of the 4,885,000 people employed, over 38% worked in industry and construction, 19% in agriculture, and 12.5% in health care and social and cultural activities.

The country has a planned economy, and in 1987 the gross domestic product (GDP) was 1,224 billion forints (HUF), while the net material product was 999 billion forints.[1]

GOVERNMENT

Hungary is a people's republic. The Parliament, the supreme body of state power, composed of 387 deputies, elects the Presidential Council, the Council of Ministers, the President of the Supreme Court, and the Chief Prosecutor. The Presidential Council is collectively the country's titular head of state. Executive and administrative functions of the government are carried out by the Council of Ministers.

For regional government, the country is divided into 19 counties, the capital Budapest, and four other county-ranked towns. Parliament and local government councils are elected by universal suffrage for five-year terms.

DEMOGRAPHY

The population amounts to 10,604,000 (114 per square kilometer), 59.2% of whom live in urban areas (1988 data). About 20% of Hungarians live in Budapest and surrounding suburbs (Table 1), which occupy about 8% of the total area of the country.

The birth rate for the year 1987 was 11.9 per 1,000. The overall death rate was 13.4 per 1,000 population; the infant mortality rate was 17.3 per 1,000 live births (under one year of age). Life expectancy is 65.3 years for men and 73.2 for women. Distribution of the population by age group is as follows: 21.1% under 15; 19.5% aged 15-29; 34.9% aged 30-54; 5.9% aged 55-59; and 18.6% aged 60 and over.[1,2]

A. THE NATIONAL HEALTH CARE SYSTEM

1. Historical Background

Before World War II, only 30% of the population belonged to some health insurance scheme. The accessibility to the health service was rather selective, depending on the individual's socioeconomic situation. After World War II, a new health service strategy was adopted that resulted in a

TABLE 1. Populations of important towns (thousands).

Budapest	2,104
Debrecen	217
Miskolc	210
Szeged	188
Pecs	182
Gyor	131
Nyiregyhaza	119
Szekesfehervar	113
Kecskemet	105

reconstruction of the health care infrastructure. The development of the health service system over the past 40 years can be divided into the following periods:

— A reconstruction period from 1945 to 1950
— The first extensive development phase from 1951 to 1965
— The second extensive phase of accelerated development from 1966 to 1975
— A transitional period between 1976 and 1980 in preparation for the following intensive period
— The introduction of intensive development from 1981, with special emphasis on the strategy of "Health for All by 2000," accompanied by an increasing shortage of the health budget as a consequence of the economic recession.

During the first extensive phase, the moderate development of health care investment was accompanied by a rapid and significant improvement in the health status of the population. However, while between 1950 and 1965 mortality decreased in both sexes and in all age groups, it increased between 1965 and 1985 with regard to both sexes, especially among males in the age group of 35 to 59. It is paradoxical that during the extensive phase of accelerated investment (development) the health status of the population deteriorated, especially as far as parameters of noninfectious diseases are concerned.

The chief political principles of health care and health services were enacted by the Second Health Act of 1972. The main issues were:

— Free-of-charge health service
— Equal accessibility to all forms of adequate health facilities
— Priority for prevention and rehabilitation
— Community participation.

The free-of-charge service promotes higher utilization, especially at the primary level. Accessibility requires a better geographical distribution of the health manpower. Equality in access postulated changes in institutional structure and function. Since the mid-1970s, an integrated inpatient and outpatient hospital system has existed in Hungary at town, county, regional, and national levels. From town to national level, the diagnostic and therapeutic facilities increase progressively, involving a higher degree of manpower specialization.

The priority of prevention (e.g., mass screening, vaccination, etc.) also contributed to an increasing demand for manpower.[3]

2. Social and Political Environment

Hungary operates on a five-year planning cycle, including health services. The share of health and social services from the national budget is determined at the governmental level. The counties then prepare their own plans according to this system. At the same time, the Ministry of Health and Social Affairs determines professional programs (e.g., prevention of cardiovascular and malignant disease, mental hygiene, etc.), which not only have priority in the regional plans, but also receive central financial support from the Ministry of Health and Social Affairs. The plans are brought down to the level of the local integrated hospital polyclinic units, where the managing director is responsible for the economic decisions. This also applies to the primary health care units, which are integrated in the hospital polyclinic unit.

The villages that employ their own primary health care staff are primarily responsible for local financial support and management. For the sake of uniform priorities and supply, there is, in most cases, a contract between the village council and the local hospital polyclinic unit. Thus, the latter takes over the whole of health planning and management, leaving local authorities to ensure only the financial base.

The social services planning and management runs through the county councils, and the system here is the same as above. There are also health-related social problems to which the Ministry of Health and Social Affairs gives high priority and financial support.

The system described above makes the overall planning and management of health and social services possible.

3. Health Costs

The Second Health Act (1972) stipulates, among other things, that each patient should have timely access to adequate, high-level, preventative-curative care required by his or her condition, irrespective of the place of residence. This requirement can be met only if all preventative-curative institutions operating in the country belong to the same homogenous system in which the tasks and the operation of the subsystems are defined, their functions and cooperation are specified, and if the different levels of progressive health care delivery within the system are established.

As a result of this gradual development, health care is available, free-of-charge, as a citizen's right. Thus, the state is responsible for providing full-scale health care for the entire population, and it also provides the corresponding basic conditions. Consequently, the process started earlier has received priority to provide proportionately distributed health care among settlements as well as the same level of care irrespective of location.

Social Insurance

Social insurance covers the whole population of the country. The number of those entitled to social insurance on their own rights was 7 million, and those entitled by being a family member of an insured person numbered 3.6 million, a total of 10.6 million in 1987.

As compared to the year of 1986, the expenditures of social insurance (without the costs of health care, drugs, and medical appliances) have risen from HUF 142.9 billion to HUF 155.5 billion (by 9%). The rate of social insurance costs in the national income has increased from 16.2% to 16.9%. Social insurance costs per worker were 29,368 forints. (This increase is 900% over 1986.) Within the expenditures, the rate of pensions was the highest (70.8%). The rate of maternity and child care as well as family allowances was 20.6%.[2]

In the branch, the sum allocated for development (investments) and overhauls (renovations) was HUF 8.2 billion in 1987, of which the central estimates were HUF 2 billion and council allowance HUF 6.2 billion. The actual consumption was HUF 8.3 billion as a total, of which the central institutions consumed HUF 2.1 billion and the council institutions HUF 6.2 billion.[2,3]

Planning and Budget	*HUF million*
Investments	5,752
Renovations	2,536
Health and social expenditures	45,900

4. The Medical Establishment: Structure and Function of Health Services

The institutions of the preventative-curative services are operated by the Ministry of Health and Social Affairs on the one hand (university clinics, national institutes, sanatoria) and by various levels of councils (village, municipal, and county councils) on the other hand. The preventive-curative institutions run by the councils are also supervised by the Ministry of Health and Social Affairs through its National Institutes.[1,2,3]

Primary Health Care

Primary health care is the basis of the Hungarian health services, which are comprised of the following: family physician (general practitioner), pediatrician, school health service, occupational health service, maternity and infant care (including family planning, genetic counseling, etc.), family dental service, ambulance service, certain outpatient departments accepting patients without referral, basic services (comprising hygiene of settlements, environment, labor, food and nutrition), epidemiology, and health education. Primary health care has been fully organized, including general practitioners, family pediatricians, and occupational services in factories and other places of production. Even the average of one GP/2,300 population does not reflect the actual situation, since a part of this population is treated by physicians in factories and by district pediatricians for children in the towns.[2,3]

Today, every industrial and other establishment employing more than 500 persons has an industrial physician. The school health services, composed of a doctor and an MCH nurse, provide school health care for 2,000-3,000 adolescent school children, 14-18 years, and ensure continuous care on the basis of screening.[3]

Number of GP districts	4,455
Number of pediatric districts	1,378
Number of dental districts	801
Number of school health districts	192

The Ambulance Service

The ambulance service is a national organization under the direct supervision of the Ministry of Health and Social Affairs. There are 159 ambulance stations in the country. The ambulance service has a double task: lifesaving and transportation of patients. Special units start emergency treatment on the spot and in the vehicle to control vital functions.

Inpatient and Outpatient Services

There is a compulsory referral system for inpatient care and most outpatient service. If the primary level is unable or unfit to treat a patient, among other reasons because of the lack of the necessary sophisticated equipment, it is compulsory to refer the patient to the respective outpatient clinics, which are located in towns, cities, or as parts of municipal hospitals. On the average, a municipal hospital is responsible for the outpatient and inpatient care of 110,000-120,000 inhabitants. Cases requiring more complicated diagnostic or therapeutic treatment are taken care of in county hospitals where almost all medical disciplines are represented and sophisticated medical equipment is available. If necessary, municipal hospitals refer their patients to county hospitals. Of course, the county hospitals also handle routine cases requiring no high specialization in their catchment areas. They provide services for an average of 400,000 to 600,000 inhabitants.[3]

In certain specialized areas of medicine where the incidence and prevalence of diseases is rather low (neurosurgery, transplantations, etc.), highly specialized and equipped hospitals with the functions of regional or national centers are operating, which cover several counties or the whole country. Every needy patient can be referred to such institutions. Their responsibility extends over 2 or 3 counties or 1-2 million inhabitants. When necessary, the patient may be moved to this level from the primary sphere, bypassing the intermediate levels. There are seven regional centers working in the country.

At night and during holidays, patients are handled by regular duty services organized all over the country, both at the primary and at the inpatient levels.

Sanitary and epidemiological prevention is supervised by the National Institute of Hygiene. In every county, sanitary and epidemiology stations operate well-equipped laboratories performing microbiological, water, air, and soil tests and investigations, offer laboratory monitoring of production technologies and biological and chemical substances for the safety of food production and marketing, and provide protection against radioactive materials, etc. These institutions conduct the necessary public health investigations and screening and perform some of the compulsory vaccinations.[3]

Special care is provided by tuberculosis, dermatovenereal, oncological, and neuropsychiatric dispensaries. In the 164 tuberculosis dispensaries, 5 million screenings were performed (in 1987). In the 123 dermatovenereal dispensaries, the case load was 2.2 million (2.7% of the total of venereal patients) in 1987. In the 153 oncological dispensaries, 1 million tumor screenings were performed in 1987. In the 126 neuropsychiatric dispensaries, 19,500 new patients and 100,000 long-term care patients were registered in 1987.[2]

The conditions of hospital care were improved by increasing the manpower and the number of technological facilities. In 1965, the number of physicians per 100 beds was 9, and the number of qualified health personnel was 33; these figures increased, respectively, to 12 and 53 in 1987. In

1987, the average number of hospital beds was 98.2 per 10,000 inhabitants.[2]

B. PHARMACEUTICAL INDUSTRY AND DRUG DISTRIBUTION

1. Historical Background

In 1867, the first pharmaceutical enterprise was founded under the name of the Hungarian Pharmaceutical and Technical-Chemical Company. At that time in Europe, only London, Paris, Milan, and Brussels had similar functioning firms. The first Hungarian pharmaceutical company operated for 45 years until it was liquidated in 1912.

The most successful early companies were the firm of Gedeon Richter, incorporated in 1901; the company of Dr. Emil Wolf; the Alka Chemical Works, founded in 1910, which adopted the name Chinoin in 1913; and the subsidiary company of the Swiss firm that Dr. A. Wander founded in 1912 in Budapest, which had the aim of supplying Austria-Hungary, the Balkan states, and Russia with the well-proven preparations of the parent company in Bern. In this field, they could count on domestic raw materials necessary for making preparations of vegetable and animal origin. In 1927, the Alkaloida Chemical Company was founded by pharmacist Janos Kabay, who developed a new industrial process and was granted a patent for the production of morphine alkaloids from poppy heads without the further need of opium as a raw material. By the end of the 1930s, 36 drug manufacturing companies were operating in Hungary, some of which also had subsidiaries abroad.

In 1949, factories, plants, and laboratories were nationalized, and after reasonable amalgamation and reorganizations, the Hungarian drug industry reached its prewar level (i.e., that of 1938). Production was centralized in eight large enterprises whose production range was strictly delineated and the necessary modernization of which was effected by means of a resolute policy of development and investment. Six of these enterprises (Alkaloida, Biogal, Chinoin, EGIS, Richter, and Reanal) continued to function under the direction of the Ministry of Industries. The coordination of their activity was overseen by the Union of the Hungarian Pharmaceutical Industry. The Phylaxis Veterinary Biological Co. came under the auspices of the Ministry of Agriculture and Food, and the Human Institute for Serobacteriological Production and Research came under that of the Ministry of Health and Social Affairs. The right of drug export and import was assigned to one enterprise, Medimpex Hungarian Trading Company for Pharmaceutical Products, which was controlled by the Ministry of Foreign Trade. Domestic distribution was carried out through one wholesaler (Wholesale Company for Pharmaceutical Products) and 20 pharmacy centers. The Union of the Hungarian Pharmaceutical Industry makes the arrangements for efficient strategic decisions by the board of directors and cooperates in their execution. Another body, the board of directors of the respective research institutes, is the controlling and leading organ of all three pharmaceutical research establishments (the Institute for Drug Research, the Research Institute for Medicinal Plants, and the Research and Development Company for the Organic Chemical Industry), each forming a joint enterprise.[4]

2. The Pharmaceutical Industry: Manufacturing

The Chemical Works of Gedeon Richter Ltd., founded in 1901, represents the oldest pharmaceutical industry in Hungary. The Chemical Works of Gedeon Richter is the traditional producer of organotherapeutical and phytochemical preparations. Approximately half of the production value of the factory derives from natural substances and their semisynthetic derivatives. For the most part, the nearly 150 substances are issued in the form of finished products.

Chinoin Pharmaceutical and Chemical Works Ltd. was founded in 1910. Since then, the company has expanded considerably and is now one of the largest pharmaceutical and chemical factories in Hungary. The company manufactures a wide variety of synthetic compounds. The most important are papaverine and its derivatives, of which Chinoin was the first company in the world to produce synthetically on a large scale. The multiple-state, high-volume synthesis of antihypertensives, diuretics, analgesics, chemotherapeutics, and other pharmaceutical fine chemicals is also carried out. The research and production of two important compound groups were recently realized in a relatively short period of time: antitumor sugar derivatives and prostaglandins. Chinoin produces about 100 different active substances used to manufacture most of its pharmaceutical specialties in its high-capacity finishing and packaging plants.

Alkaloida Chemical Company was founded in 1927 by pharmacist Janos Kabay, who developed a new industrial process for making morphine from dry poppy heads. Production started with morphine and was later extended to related alkaloids and different semisynthetic morphine derivatives. Beginning in the 1950s, the company's phytochemical line was enlarged by active ingredients extracted from other plants and then, in the 1960s, by the formulation and packaging of pharmaceutical specialties. The company reached its present form as a result of considerable development in the 1960s and 1970s, manufacturing, in addition to its earlier products, large quantities of synthetic pharmaceutical fine chemicals such as chloroquine, sulfadimidine, phenobarbital, and different forms of pharmaceutical specialties for human use, as well as veterinary products, pesticides, and intermediates.

EGIS Pharmaceuticals was merged after World War II from the subsidiary company of the Swiss firm Wander (Bern), established in Hungary in 1912, and from other smaller pharmaceutical laboratories. The manufacture of bioactive substances, mostly pharmaceutical fine chemical intermediates and finished dosage forms, food products, milk powder, powdered creams, phytobiologicals, and animal feeding preparations, represents the main activity of EGIS. Its work embraces the sale of inventions, licenses, know-how, and associated technical services, as well as the purchase of the same.

Medicines for heart and artery disease are the most important preparations and are sold in great quantities. Psychiatric and neurological products, antibiotics, chemotherapeutics, and antiallergic drugs have a considerable place in the manufacturing program as well. EGIS has built up close processing and sales cooperations with well-known foreign drug manufacturers.

Phylaxia Veterinary Biologicals Company, founded in 1912, produced biologicals for both human and veterinary use until 1954. Since the separation of the Human Institute for Serobacteriological Production and Research, Phylaxia has been an exclusive manufacturer of products serving animal health and livestock farming.

Biogal Pharmaceutical Works, the youngest drug manufacturing company in Hungary, started production in 1952 as the Pharmaceutical Works of Hajdusag. In the beginning, manufacture was confined to penicillins only. Since then, the range of products has gradually been extended and now includes penicillin, oxytetracycline, tobramycin, and erythromycin, as well as several nonantibiotic agents. In addition to human pharmaceutical specialties, Biogal manufacturers veterinary products, feed additives, and cosmetics.

The license and collaboration agreements with foreign partners, such as Zyma, Ciba-Geigy, Pfizer, OM, and Biersdor on the one hand and recent in-house research activity on the other have allowed the company to introduce preparations that meet the requirements of up-to-date therapy.

The Human Institute for Serobacteriological Production and Research started its work as a department of biologicals for human use within the framework of Phylaxia, founded in 1912. This "human department" became independent in 1954 under its present name. The human department of Phylaxia began its activity with the production of antidiptheria and antitetanus equine sera and continued, not much later, with the preparation of vaccines and serological diagnostics.

The product range includes vaccines, stimulants of nonspecific immunity, antisera, blood derivatives, plasma substitutes and perfusion solutions, ophthalmic solutions, diagnostics, culture media and peptones, as well as two skin care products, Irix and Naksol sprays for burn therapy. The number of biologicals and other preparations authorized for sale is 211 and that of diagnostics (for immunology and immunochemistry) and culture media is about 140.

Reanal Factory of Laboratory Chemicals. The reorganization into a factory for laboratory chemicals of the former Magyar Pharma Pharmaceutical Factory, a subsidiary of Bayer before World War II, was begun after nationalization. The right to manufacture medicines was granted to other pharmaceutical enterprises, whereas the production of laboratory chemicals and reagents was transferred from these to this company, which has, since 1957, operated under the name of Reanal. Compared to about 200 products at the beginning, the factory now produces nearly 1,500 different articles and markets about 6,000 substances (fine chemicals) for domestic supply through its network of stores.[4]

Some Data About Production

The growth rate of total production can be seen in Figure 1. The distribution of the Hungarian drug production value according to product groups in 1987 is shown in Table 2.

Foreign and Hungarian Joint Enterprises

One increasingly important form of joint activity involves the Hungarian pharmaceutical industry in foreign markets and participation in foreign producing companies. When establishing joint enterprises, the pharmaceutical industry does not confine itself to only the transfer of finishing and packaging technologies, but cedes the manufacturing processes of active substances to the other party. Some examples of this are the Indian firm Themis Chemicals, Labatec Pharma in Switzerland, the Nigerian firm Imarsel, and Ambee Pharmaceuticals in Bangladesh.

FIGURE 1. The growth of Hungarian drug production since 1960. (1960 = 1).

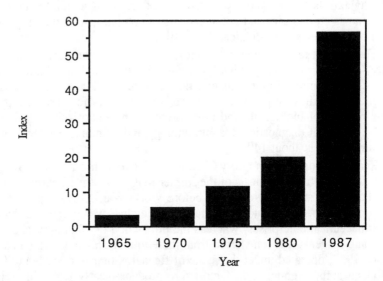

Foreign Cooperations Serving Production

The Hungarian pharmaceutical industry readily takes part in production cooperations where license agreements facilitate exploitation of production capacities. The license relations of the Hungarian pharmaceutical industry with certain multinational concerns (Astra, Bayer, Ciba-Geigy, Eli Lilly, Glaxo, Janssen, Hoechst, Organon, Pfizer, Pharmacia, Sandoz, Schering, Takeda, Zyma, etc.) are well established.

3. Research and Development

Historical Retrospection and Traditions

Pharmaceutical research in the modern sense has been going on in Hungary for about 90 years. Up until the end of World War I, this activity was mostly carried out in the chemical and pharmaceutical departments of universities. Between the two world wars, the larger manufacturing companies organized their own research units in close cooperation with the universities. Since 1945, in addition to the growing activity of the university institutes, the institutions of the Hungarian Academy of Sciences have increasingly taken part in drug research. The research laboratories of the

TABLE 2. Distribution of the Hungarian drug production value according to product groups, 1987.

Preparations for human therapy	55.1%
Active substances for human use	16.4%
Formulated plant-protecting preparations	6.0%
Plant-protecting agents	3.9%
Preparations for veterinary therapy	3.1%
Active substances for veterinary use	1.6%
Laboratory chemicals	1.4%
Food products	0.6%

Source: Medimpex, ed. The Hungarian Pharmaceutical Industry. Budapest: Medimpex, 1989.

pharmaceutical companies have also developed considerably. The Institute for Drug Research was founded. Later, the Research Institute for Medicinal Plants and the Research and Development Company for the Organic Chemical Industry were likewise incorporated into the pharmaceutical industry.

At present, the Board of Science Policy provides overall direction for research work in Hungary. The coordination of research work going on in the pharmaceutical industry is carried out by the Research Council of the Union of the Hungarian Pharmaceutical Industry. Research laboratories of the factories have developed a close relationship with the research institutes of the Hungarian Academy of Sciences, various university institutes, medical research units, and, last, but not least, with the clinical pharmaceutical network, operating under the control of the National Centre for Clinical Pharmacology. Cooperation in a number of scientific subjects is going on exclusively within the industrial branch (e.g., the development of operation techniques, protection of the environment, instrumentation, automation, etc.).

*Organization of the Hungarian Pharmaceutical Research
Institute for Drug Research*

The institute played a leading role in the introduction of nearly all important antibiotics, in connection with both the fermentation and isolation operations (oxytetracycline, streptomycin, chloramphenicol, gentamycin, tobramycin). It participated in the development of production procedures for vitamins (Vitamins B_1, C, and B_{12}), tuberculostatics (PAS, INH), sulfonamides (sulfamethylthiazoel, sulfadimidine, sulfaguanidine), and agents potentiating the activity of these latter (trimethoprim, diaveridine). It played a great part in the establishment of the steroid industry in Hungary by developing processes for the microbial transformation of steroids, resulting in the implementation of manufacturing anti-inflammatory and contraceptive drugs (prednisolone, estrone, norgestrel).[4]

Research Institute for Medicinal Plants

The legal predecessor of the institute was established in 1915 by the Ministry of Agriculture. The institute obtained considerable results in the research of morphine alkaloids, the starting and development of the home volatile oil production, as well as in exploration and enrichment of the home medicinal herb flora (revival of mint cultures, acclimatization of castor-oil plant, the starting of the cultivation of digitalis lanata). The institute, in cooperation with agricultural institutions, has worked out and introduced the system of poppy and ergot cultivation on an industrial level in Hungary.

Research and Development Company
for the Organic Chemical Industry

The basic activity of the company has been research and development work for the organic chemical industries, primarily for the pharmaceutical, plant protection, and plastic industries. The company, in the course of its development work, designs the necessary equipment and processes and provides the specifications of the apparatuses and the technological prescriptions. It also makes smaller equipment. The company provides assistance in construction, erection, and starting of plants and continues to develop the necessary operational control and analytical methods for work.

Drug Research in the Factories[4]

Alkaloids Chemical Company. Research projects are carried out partly by the company's own research unit and partly in cooperation with other research institutes. The main fields of research are substances acting on the central nervous system, nonsteroidal anti-inflammatory drugs, cardiovascular agents, antiparasitics, pesticides, etc.

Biogal Pharmaceutical Works have a growing research base involving the development of several novel drugs during the last ten years, the major pharmacological groups of which are: chemotherapeutics, cardiovascular agents, and immune modulators. The research in the fields of biotechnology, organic synthesis, operation technique, pharmacy, and agriculture is of utmost importance, as it constitutes the guarantee of new, effective compounds for the above pharmacological groups.

EGIS Pharmaceuticals. The most successful original compound of the factory is bencyclane (Halidor®). This vasoactivator has been exported to over 50 countries. Further results of EGIS's research made jointly with the Institute for Drug Research are trimetozine (Trioxazin®), metofenazate (Frenolon®), and tofisopam (Grandaxin®), which have proved very useful in psychotherapy.

Chinoin Pharmaceutical and Chemical Works has attached particular importance to research for as long as it has been in existence. Recent developments include an original new analgesic and anti-inflammatory preparation, rimazolium (Probon®) and selegiline (Jumex®), a novel agent against Parkinson's disease. In addition to synthetic compounds, the company is engaged in research into semisynthetic derivatives of fermentation products and other natural substances. This work has resulted in the implementation of manufacturing several penicillin derivatives (Meticillin®, oxacillin, ampicillin) on an industrial scale and in success in the field of prostaglandins.

The Chemical Works of Gedeon Richter Ltd. The company is a traditional producer of organotherapeutical and phytochemical preparations. Approximately half of the production value of the factory derives from natural substances (partly produced by total synthesis) and their semisynthetic derivatives (steroids, peptides, Vitamin B_{12}, ergot alkaloids, etc.). One of the factory's successes has been the synthesis and development of Cavinton® (vinpocetine), and indol derivative which selectively improves oxygenation and energy metabolism of the brain. Human ACTH was first synthesized in a Richter laboratory.

Hungarian Drug Patents

A good index to the efficiency of drug research is the number of patents granted. In Hungary, according to the patent law for medicines, only the manufacturing process can be protected by a patent not the product itself. Accordingly, Hungarian patent applications refer to manufacturing process, but in foreign countries, if legal prescriptions allow, Hungarian inventors may of course ask for the protection of the product, too. The number of pharmaceutical patents granted or applications under judgment (resulting from Hungarian applicants) during the period between January 1, 1971 and 1987 was:

Item	Patents granted	Applications under judgement
Alkaloida	36	49
Biogal	69	102
Chinoin	567	666
EGIS	251	327
Richter	530	699
Reanal	58	94
Inst. for Drug Research	232	275
R.Inst. for Mod. Plants	33	42
R. and D. Co. for Org. Chem. Ind.	14	24

In addition to these, patents granted abroad ensure legal protection for the export of products made by the Hungarian drug industry.[4]

4. Drug Approval and Government Regulation

Drug Control

In Hungary, the organization of drug registration and control is closely related to the development of the drug industry, drug supply, and drug exports. Thus, the National Institute of Pharmacy today is the national drug control agency. Its main responsibilities are as follows:[4,7]

— Authorization of therapeutic trials on humans with substances previously not used as medicines in Hungary
— Drug registration
— Enforcement and continuous development of drug safety requirements
— Health regulatory supervision of the manufacture of drugs
— Regulatory quality control of drugs
— Withdrawal of pharmaceutical products unsuitable for marketing
— Amending the Hungarian Pharmacopeia and granting occasional exemptions of products from quality or labeling requirements
— Study of drug utilization and promotion of rational drug therapy
— Supervision of drug information

— Monitoring of adverse drug reactions and side effects
— Fulfillment of responsibilities within its international obligations for cooperation.

Introduction of New Drugs

In Hungary, one of the basic principles regarding the introduction of drugs is that an unlimited increase in the number of pharmaceutical preparations does not serve public health. Preclinical requirements are regulated by detailed prescriptions. Expert reports containing the methods and results of experiments are overseen by the National Institute of Pharmacy and evaluated by two expert committees of the Scientific Health Council, the Committee on Drug Administration and Clinical Pharmacology, and the Ethical Committee for Medical Research. The point of the evaluation is that prior to authorization for human trials, evidence should be presented on the efficacy and relative safety of a new drug in animal tests.

Human Phase I and II trials are carried out by the units of the Clinical Pharmacological Network (CPN) referred to above, and the results are assessed by the two aforesaid Committees of the Scientific Health Council. Phase III human trials depend on the judgment of these bodies.

The positive evaluation of the results of therapeutic experiments, as well as the decision of the National Institute of Pharmacy in respect to the product's unsuitability, are both preconditions for the registration of a new drug. The application form containing presentation of the data, proposed control methods, description of stability tests, etc., is prescribed in detail by the institute.[4,7,8]

Quality Control

Modern analytical, biological, microbiological, and other control laboratories were established in the seven large factories. The head of the factory's department for quality control is personally responsible for the quality of the drugs. This responsibility entails not only seeing that the quality of all drugs produced by the factory meets the requirements of the Hungarian Pharmacopeia and the registration prescriptions of the National Institute of Pharmacy, but that each of the substances and auxiliary materials used by the factory, as well as each batch (in-process, semifinished, and finished), complies with the standard.[4,7]

Inspection of Manufacturing

The execution of regular plant inspections is a relatively new activity. The basic inspections, repeated every five years in the seven pharmaceutical factories, constitute the starting point for the regular production control program of the National Institute of Pharmacy. Six different forms of inspections can be distinguished:

— Authorization for drug production in a new plant
— Registration of a new drug
— Suspicion of drug quality defects
— Request of health authorities of a country importing drugs from Hungary (on the basis of bilateral or multilateral agreements or individual requests)
— Follow-up controls
— Random inspections.

International Cooperation

Hungary's international relations are important in every field, and drug control is no exception.

World Health Organization (WHO). Hungarian experts play an active role in the compilation of the International Pharmacopeia. WHO frequently invites Hungarian experts to perform duties as members of professional committees or consultative groups or as WHO consultants in developing countries.[4]

CMEA. Hungary has undertaken a considerable task — in the framework of CMEA's standing Health Commission — in the field of the investigation, evaluation, and standardization of pharmaceutical preparations through international coordination of cooperation, started with the participation of Bulgaria, Cuba, Czechoslovakia, GDR, Hungary, Mongolia, Poland, U.S.S.R., and Vietnam. Hungarian experts take an active part in the elaboration of the methods and prescriptions of the Compendium Medicamentorum.[4,7]

EFTA Pharmaceutical Inspection Convention. Hungary is a member of the international convention elaborated by EFTA states [Pharmaceutical Inspection Convention (PIC)] which over and above the GMP recommendations of WHO, regulates the application of safe conditions in drug manufacture and exercises official control over the observance of the respective prescriptions. Hungary cooperates closely with other member states (Austria, Denmark, Finland, F.R.G., Iceland, Ireland, Liechtenstein, Norway, Portugal, Romania, Sweden, Switzerland, and the United Kingdom) in implementing the provisions of the convention.[7] In Hungary there are no "two different" registration procedures. The stipulations of the Hungarian Pharmacopeia and the quality specifications of the National Institute of Pharmacy refer to the registered drug preparations independently, whether used domestically or abroad.

5. Foreign Trade

The rapid development of Hungarian pharmaceutical factories created an early basis for exports. Prior to the first world war, Chinoin and Gedeon Richter were already selling and forwarding a considerable quantity of drugs to foreign countries.

Medimpex Hungarian Trading Company for Pharmaceutical Products was founded in 1949, and since that time has been representing the foreign trade interests of Hungarian pharmaceutical factories. Although the import and export of pharmaceutical fine chemicals and specialities makes up a decisive part of Medimpex's activity, the company achieves a considerable turnover in other fields, too. Its export range is complemented with feed additives, medicated feeds, laboratory fine chemicals and biochemical products.

The development of the pharmaceutical industry made it necessary to trade in industrial intellectual goods (patents, manufacturing processes, etc.). License and know-how agreements have contributed considerably to the extension of the industry's product range, as well as to the introduction in foreign countries of preparations based on Hungarian research (Figure 2).

The commercial relations of Medimpex extend almost around the world. The company buys and sells in nearly 100 countries. In addition to socialist countries, the most developed nonsocialist countries, as well as developing countries can be found among their partners (Table 3).[4,5]

FIGURE 2. The growth of Hungarian drug export since 1960. (1960 = 1).

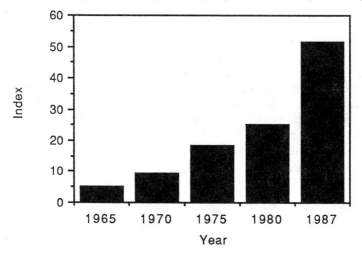

TABLE 3. Distribution of the Hungarian drug export and import according to the main country groups.[4]

Country Groups	1987 Exports Million	Imports Forints
Countries of Rouble settlement	11,770	2,619
CMEA	11,496	2,616
Others	274	3
Countries of non-Rouble settlement	8,305	8,814
Developed	5,531	8,661
EEC	2,141	6,942
Other European	1,089	810
Overseas	2,301	909
Developing	2,774	153
Total	20,075	11,433

Socialist Countries

Hungarian drugs enjoy a good reputation in the socialist countries and are much in demand. The significance of drug consignments to the socialist countries is shown by the fact that their value doubled between 1980 and 1987, representing a growth rate of more than 12% a year. Among the socialist countries, the highest turnover is achieved with the Soviet Union. This is followed by the European CMEA countries. At the same time, Vietnam, Mongolia, the Korean People's Republic, China, Laos, and Cambodia are increasing their trade.

Developed Nonsocialist Countries

Since 1980, developed nonsocialist countries have been the most quickly growing markets of the Hungarian pharmaceutical industry. While the exports to socialist countries increased 40% between 1983 and

1987, exports to nonsocialist countries were up by 48% and those to developed countries by 60% (i.e., 17% per year). The sales effected to overseas nonsocialist countries, first of all to the U.S. and Japan, increased in an especially high degree (three and a half fold) over the same four-year period.

The export turnover (in U.S. millions) in this area over four years was as follows:[4]

Countries	1987	1983-1987 (%)
U.S.A.	21.6	392.7
Japan	25.3	506.0
Switzerland	11.1	114.4
F.R.G.	10.3	72.0
France	8.9	197.8
U.K.	5.7	116.3
Total:	82.9	188.8

The share of the six countries in the total export effected to the nonsocialist area rose from 34% to 46% between 1983 and 1987.

Developing Countries

The predominant markets of Medimpex in the third world — in terms of 1987 turnover (in U.S. millions) — were as follows:[4]

Countries	1987	1983-1987 (%)
Brazil	7.1	169.0
Iran	7.3	104.3
Algeria	3.5	55.6
India	3.5	85.4
Pakistan	3.1	221.4
Bangladesh	1.5	100.0
Hong-Kong	1.2	300.0
Total:	27.2	109.2

The share of these seven countries in the total Hungarian drug export to nonsocialist countries was 15.1% and in the sales directed to developing countries 46% in 1987.

C. PHARMACY PRACTICE:
PUBLIC HEALTH AND ITS ORGANIZATION
IN HUNGARY

1. The Ministry of Health and Social Affairs

Founded in 1950, the leading authority for Hungarian public health is the Ministry of Health and Social Affairs. It is the supreme official body for the superintendence of public health affairs. The advisory and information organ of the Ministry in professional issues concerning public health and medical and pharmaceutical sciences is the Scientific Board for Public Health Affairs. This body forms opinions and makes recommendations on questions like the improvement of public health conditions, the development of health care, and education and further training of medical personnel and experts. It expresses opinions about new medical methods, new diagnostic techniques, the therapeutical value of new drugs, etc. The Scientific Board for Public Health Affairs and the Hungarian Academy of Sciences have several joint committees, the task of which is to form opinions in theoretical and practical questions of medical and pharmaceutical sciences. Such committees are the Committee for Pharmacy and Pharmaceutical Affairs and the Committee for Drug Research and Registration.

The eighth Division of the Ministry of Health and Social Affairs is the Division of Pharmacy. It covers the determination of professional development directions in pharmacy, the management of drug affairs, and the supervision of organs and bodies responsible for drug supply. The Division performs tasks concerning the function of Officinal Centres and institutional drug supply.[5]

Regional Health Organs

Local supervision of health affairs is the task of the public health divisions of county councils. These organs are headed by the Head Physician of the County. In the public health divisions, the functions concerning drug and instrument supply are performed by the head pharmacist of the country. Furthermore, the head pharmacist is responsible for supervision of the Officinal Centre inspection of public and institutional pharmacies.

2. Drug Supply

In Hungary today, 20 Officinal Centres are organized for the management and supervision of the drug supply of the 19 counties and the capital, as well as for the inspection of the professional activity of pharmacies. Decisions about the establishment of new pharmacies are made by the public health divisions of county and municipal councils. In the possession of the preliminary permission, it is the Officinal Centre that has to establish the new pharmacy. This is permitted only in cases where the population in the region exceeds 6,000 people.

The institutional pharmacies cover an important task in the drug supply of the country. Such pharmacies can be found in medical schools (4), in the county hospitals, in the hospitals of the capital, and in the national medical institutions (78 total). The task of institutional pharmacies is to provide drug supply for inpatients, and they generally work in institutes attending a minimum of 450 hospital patients.

3. Dispensing a Prescription: Where, What, Payment, Formulary

In terms of current legal rules, drugs can be distributed only by pharmacies. Furthermore, the pharmacies may distribute medical and health-related articles and instruments (medical hand instruments, dressing materials, diagnostics, chemicals, first-aid sets, medicinal waters, etc.), as well as products for hygienic and health protection purposes (dietetic food, medicinal products not qualified as drugs, paramedicinal products, books for medical education, etc.).

A special legal rule provides for drug prescription and expedition. The physician is entitled to prescribe only Hungarian and foreign registered drug products, medical food preparations, immunobiological products, drugs, galenic preparations, dressing materials, etc., which are listed in the Standard Formulary (Formulae Normales), in the Hungarian Pharmacopeia, in the *Guidebook for the Prescription of Pharmaceutical Products*, or published in the *Gazette of Social and Public Health Affairs*.

The physician may order drugs in prescription formulas. In such formulas, the patient's name, address, and age, as well as the date of prescription are to be indicated. The physician has to sign and stamp the prescription.[1] Drugs also may be prescribed in private practice prescription formulas. These should indicate the name and address of the physician, as well as the number of permission for private practice.[5,9,10]

Drugs that may be prescribed in the scope of outpatient care are arranged into the following categories:

Group 1: Drugs that can be issued without prescription (about 110 products)

Group 2: Drugs that can be issued only by medical prescription (about 690 products)

Group 3: Drugs that can be issued only for medical prescription by the support of social insurance and drugs that can be issued without prescription at consumer's price (about 60 products).

For inpatients in clinics and hospitals, drugs are supplied by the institutional pharmacy. However, in those cases which are not prescription formulas, drug prescription booklets are used. Here the patient's name and age, etc., are indicated. Drugs listed above (Groups 1, 2, 3) may also be prescribed for inpatients, but the number of drugs is supplemented by those distributed exclusively for institutionalized patients, about 195 products. The scope of drugs applied in medical institutional care is supplemented by 250-300 products that are imported for particular patients on the grounds of individual judgment. For the import of such drugs, the permission of the Ministry of Health and Social Affairs as well as the National Institute of Pharmacy is essential.[9,11] These rules apply to products manufactured by pharmaceutical companies. This scope is supplemented by galenic products included in various professional standards (Ph.Hg., Formulae Normales), as well as by magistral products prepared according to the individual prescription of the physician.

Medical prescriptions are valid for one month.

The physician has to specify a prescription in an unambiguous, readable form, and has to use drug names and abbreviations included in the Hungarian Pharmacopeia, in the Formula Normales, and in the orders of the National Institute of Pharmacy.

The pharmacist judges the prescriptions, and is entitled to issue the drug only if the prescription complies with both formal and professional requirements. This rule can be disobeyed only in cases when the word *statim* is indicated in the prescription formula. These cases are comprehensively provided for by current legal rules.

The physician may prescribe the drug either as "free of charge" or as "to be paid for."

4. Drug Insurance and Payment

The pharmacies were nationalized in 1950, following the drug industry and wholesale trade of pharmaceuticals. The price to be paid by the patient for the drugs contained the following: producer's price, price gap between wholesale and retail trades, and the manufacturing price. Also, the sum to be paid for drugs has been changed significantly in the nationalized pharmacies. The pharmacists and the assistant personnel became state employees, for a determined monthly salary, and their income did not depend on the turnover of the pharmacy. At the same time, the charge for the patient was set at 15% of the actual price of the drug. Thus, drug supply was granted a substantial support by the state. This sum reached HUF 19 million in 1988.[2]

In 1977, the compensation fee system was introduced, which meant a rather low sum that was determined for each drug (generally HUF 3.00, or about 15% of the total price). Simultaneously, the drug prescription regulations had undergone certain changes: it was now possible for one prescription to be used repeatedly (one prescription was permitted for three subsequent applications), and thus drug prescribing activity of physicians was significantly reduced. Further simplification was achieved by expanding the number of drugs available without prescription.

In January 1989, the system of drug supply was changed. Under this new system, the producer's price of the drugs, as supplemented by the 3% wholesale trade and 20% retail sale trade price gap, forms the consumer's price. This means that the price of drugs has increased by 80% on the average as compared to the earlier compensation. This remarkably high price is compensated in different degrees (100%, 90%, and 80%) by the Social Insurance Inspectorate. Certain drugs are not supported by social insurance; these are available for patients only at the full price. A new feature of this system is the reintroduction of manufacturing fees.

The Social Insurance Inspectorate provides 100% support for drugs (i.e., the patient receives the drugs free of charge in hospitals and clinics; in trauma care; in cases of chronic disease, e.g., antidiabetics, anti-tuberlotics, antiepileptics, etc.); as well as in the fields of public health and epidemiology; mother, infant, and child protection; and other areas that involve public health policy.

In the determination of the degree of social insurance support, various health political considerations were observed. For example, in Hungary, the rate of patients with cardiovascular diseases is very high. Therefore, the degree of compensation is 90% in this group of diseases. However, to restrain the overconsumption of sedatives, the social insurance provides no support, and the patient has to pay the total price of such drugs.

Eighty percent of magistral drugs (those prepared in pharmacies) is compensated by the social insurance. The degree of the newly introduced manufacturing fee depends on the form of the drug product (e.g., HUF 1.00/tablets, HUF 2.00/suppository, HUF 15.00/5 g of eyedrops, etc.).[9,10,12]

5. Patient Education

Health education plays an organic role in the context of public health in Hungary. The responsibility for the guidance of these tasks belongs to the National Institute of Health Education. The aim of the health education program is to educate the population for a healthy way of life, the prevention of disease, and conscious cooperation in the realization of public health directives. Pharmacists play an important role in the performance of these tasks. They read several hundred lectures in specific topics, such as the importance of regular physical activity, proper nutrition, hygiene behavior, and fight against smoking and participate in the health education of patients. These topics cover all requirements essential to participate in the maintenance of good health. The main supervisor of this activity is the Ministry of Health and Social Affairs and its institute for organization and methodology, the National Institute of Health Education. Operative execution, though, is a task of physicians and pharmacists, who are supported by other medical personnel (nurses, health visitors, assistants) in their work.[5,6]

Health Education Activity of Pharmacists

The regional situation of pharmacies, as well as the drug supply work, provides comprehensive responsibilities for health education activity. This sphere of work is twofold: it covers tasks to be performed within pharmacies and activities outside the pharmacy. As to tasks within the pharmacy, an ethical duty of the pharmacist is to inform the patients about proper dosage of drugs and all knowledge necessary in the given case. The mental hygienic role of health education may be reflected in activities within the pharmacy. The strengthening of the patient's belief in the physician, in recovery, and in the drug is very important.

A form of health education outside the pharmacy is stimulated by the reading of lectures. Lectures about the proper storage of drugs; about expedient, well-maintained domestic drugstores; and about toxicological problems are useful and necessary. For fifth-year students of the Hungarian faculties of pharmaceutical sciences, it is obligatory to read two lectures about health education. In this way, they are also prepared for the practice of their future profession. The topics to be selected are rather multifaceted; some major ones are, for example, information about drugs available without prescription, the prevention of mushroom poisoning, proper nutrition of infants, consultation about the use of nutritive foods, health education of schoolchildren, and the harmful aftermaths of alcohol and smoking.[5]

Methods of Health Education

In the health education work, we always consider the social features and distribution of the population (industrial workers, intellectual workers, etc.), the cultural level of the region, and the physiological facts (health status, sex, age, way of life, etc.). We also distinguish between problems to be solved within the pharmacy and those to be discussed in the scope of lectures or communications outside the pharmacy. Also, written materials are applied in the form of informative leaflets for the people.

The most common communication methods in Hungary are: roundtable conferences; lectures supported by films; exhibitions (connected with lectures); and meetings of physicians, pharmacists, and patients. In addition to these possibilities, pharmacists are encouraged to employ all occasions for health education and have to select their methods in the most advantageous way. These methods may form an organic part of the home health care trend, which has just started in Hungary, and includes important roles for pharmacists.[10]

6. Manpower

The Use of Technicians and Others

The functioning of pharmacies requires personnel and material conditions. The personnel conditions include experts of higher graduation (pharmacists) for certain tasks, persons of secondary education (assistants) for tasks not requiring specific knowledge, and auxiliary personnel for other work (packaging, cleaning).

Education and Further Training of the Health Professional

Medical higher education is available in five medical universities. General practitioners and dentists are educated in four universities, and pharmacists in two universities (in Budapest and Szeged). The Postgraduate Medical School is responsible for the further training of graduated experts. This training is performed in two faculties: the Faculty of Postgraduate Medical Training is for graduated physicians, and the Faculty of Medical Colleges is for graduated health workers.

In the medical universities, there are 12 faculties, 82 institutes for theoretical research, 80 clinics, and 28 departments. The number of teachers and other workers with higher education was 3,825 in the medical universities in 1987; 38.3% of them are women. Among these people, there are 2,361 specialists (physicians or pharmacists), and 930 persons hold scientific degrees.[2]

In the medical universities, the total number of students was 7,505 in the 1987-88 academic year; 742 of them were foreign students (the number of pharmacy students was 1,483). Women accounted for 54.4% of the students. The total number of first-year students in the 1987-88 academic year was 1,483. Of these 1,013 were medical students, 192 dental, and 278 pharmacy.[2]

The number of specialist physicians has increased: it was 1,483 in 1986 and 1,516 in 1987. The number of specialist pharmacists has also increased. In 1986, 155 pharmacists acquired the qualification of specialist, and in 1987, 191, in nine pharmaceutical branches.

The number of pharmacy assistants was 7,069 in 1987, while 494 assistants were graduated in 1987. The pharmacy assistants participate in a two-year education after the secondary school, then they have the possibility of obtaining a qualification for specialist assistant work in three subjects. In 1987, 224 specialist assistants in drug issue, 26 specialist assistants in analytics, and 32 specialist assistants in drug management obtained their degrees.

The Tasks of Pharmacists

The head of pharmacy and the deputy head of pharmacy are responsible for the operation of the pharmacy, and they are obliged to organize, manage, and supervise the work of the pharmacy employees. If the number of workers in the pharmacy reaches 14, a deputy head of pharmacy can be appointed. The head of pharmacy is appointed by the director of the pharmacy.[5,6]

The assigned pharmacist takes part in all work, the responsibility for which is defined by the head of the pharmacy. His or her most important tasks include the preparation, issue, and testing of drugs; the supervision of the assistants' work; and health education.

The pharmacy assistant is a direct coworker of the pharmacist. The most important tasks here are the quantitative reception of drugs and medicinal items and storage of these products. The assistant is also responsible for the technical tasks of drug preparation, for certain parts of pharmacy office work, and the issue of drugs that may be purchased without medical prescription.

Auxiliary pharmacy personnel consist of workers with no special qualification (e.g., cashiers, packagers, and cleaners). Their number was 4,466 in 1987.

Statistical Data[2]

Turnover of public pharmacies
in million Ft   4,477

Drug turnover per 1 inhabitant in Ft   422

Population per pharmacist   3,115

Population per pharmacy   7,126

Number of pharmacists   4,506

    in community pharmacies   3,404
    in hospital pharmacies   430
    in the industry and other 672

Number of assistants   7,069

Number of other workers   4,466

D. CONCLUSIONS

Recently, Hungary has undergone substantial changes both in political and economic respects. The elaboration of the new constitution, as well as the order for the regulation of the establishment of new parties, is in process. In the field of economy, the most important task involves change-over to a market system.

As in political and economic life, important changes can be expected in the field of health care. The Chamber of Medicine has been established; it intends to deal with safeguarding professional interests. The foundation of the Chamber of Pharmacists is also in process.

A considerable reprivatization process has been started, this also concerns health institutions. For example, pharmacies can be leased, and possibly, later, pharmacies and private sanatoriums can be opened.

REFERENCES

1. EISZI Statisztikai Irodaja, ed.: Magyarorszag 1988. Statisztikai adatok. Budapest: Kozponti Statisztikai Hivatal, 1988.

2. Statisztikai adatok Magyarorszag 1987. Evi egeszesegugyi helyzeterol. Nepegeszsegugy. 1988: LXIX/1/:193-270.

3. Forgacs I.: Structure and Function of the Health Services. In: Kokeny M., ed.: Promoting Health in Hungary. Budapest: Central Statistical Office, 1987: 25-32.

4. Medimpex, ed.: The Hungarian Pharmaceutical Industry. Budapest: Medimpex, 1989.

5. Zalai K., ed.: Gyogyszerugyi szervezes. I. Egyetemi jegyzet. Budapest, 1984.

6. Zalai K., ed.: Gyogyszerugyi svervezes. II. Egyetemi jegyzet. Budapest, 1984.

7. Bayer I.: Drug registration and drug control in Hungary. OGYI. Budapest, 1983.

8. Szekely G., Vincze Z. et al.: A human IV fazisu gyogyszervizsgalatok. Gyogyszereszet (in press).

9. Szocialis esegeszsegugyi kozloni. XXXIX/1/1-131.

10. Vincze Z. and Nikolics K.: A Magyar Gyogyszereszeti Tarsag szerepe es jelentosege a gyogyszereszet fejleszteseben. Gyogyszereszet, 32. 453-456 (1988).

11. Vincze Z. and Zalai K.: Az allatgyogyaszati gyogyszerellatas jelentosege es fejlesztesenek lehetosegei. Gyogyszereszet, 31. 293-296 (1987).

12. Vincze Z. and Nikolics K.: A Magyar Gyogyszereszeti Tarasag tevekenysege. Gyogyszereszet, 32. 627-629 (1988)

Chapter 11

Italy

Marcello Marchetti
Paola Minghetti

The population of Italy in 1988 was 57,399,108. The land area is 301,277 square kilometers, and it is divided into 19 regions and 2 autonomous provinces. The total provinces instituted correspond to 8,097 municipalities (City Council).[1] There are 191 inhabitants per square kilometer. Nearly 14% of the population is older than 64, and 17.8% are under 14.[2] The index of aging (population 65 and over/population 0-14) is 77.2%. The total male demographic concentration is equal to 48.6%. Men make up 40.1% of the population aged 64 and over, while male concentration in infancy (from 0-14 years) is 51.3%.[2] Gross national product in 1988 was L. 1,078,863 billion. Total public health care expenditure in 1988 was L. 55,887 billion (5.2% of GNP). Total drug expenditure in the same year was L. 14,464 billion (1.3% of GNP), of which L. 10,019 billion was paid by the state. Approximately 18% of state health expenditure was dedicated to drugs.[3] Per capita public health costs are L. 973,600, and L. 174,500 is spent on medicines.[3] Each family generally spends L. 39,079 on health costs and services each month.[2]

A. THE NATIONAL HEALTH CARE SYSTEM

1. Social Security System

The National Health Service (NHS) [Servizio Sanitario Nazionale (SSN)] was set up by law No. 833 of December 23, 1978. Before then, pharmaceutical and medical aid was available on the basis of insurance schemes which were obligatory. The numerous bodies that concerned themselves with this type of insurance were successively abolished. This system was in force in the years preceding the second world war and was founded on the principle that it would be available for both present and previous employees.

With the advent of the new constitution, the principle of the social state was established. It does not seem likely that it will disappear, even if the realization of the social program is due to undergo modifications determined by management intervention in the private sector.

The objectives of the NHS are the promotion, maintenance, and recovery of the mental and physical health of the population, as well as the provision of health education for each citizen, training of health workers, food and drink hygiene, safety at work, and the identification and elimination of environmental pollutants. The health service, which includes the pharmaceutical service, is therefore guaranteed by the state for all citizens, even foreigners as long as they are resident in Italy. The operative structures of the NHS are the local health units [Unita Sanitaria Locale (USL)], each unit being responsible for groups of from 50,000 to 200,000 inhabitants and referrable to a single city council, a part of that city council, or to more than one city council. Each individual region has the task of coordinating the activities of the USLs in its area, since they generally enjoy almost complete autonomy. There are 698 USLs, each being responsible for an average of 82,234 inhabitants.[4]

Hospitals generally depend on the local health unit in charge of the area, and each citizen has the right to free hospital treatment (room and board, medicines, doctors' fees, etc.). This treatment can be taken either in public or private hospitals, the latter being integrated through national agreement with the NHS. Primary medical care is also free to each patient by means of general practitioners. In 1987, primary health care included 56,023 general doctors and 3,835 pediatricians.[4]

The citizen can have access to specialists and have laboratory analyses carried out either totally free or by contributing to the cost by means of a quota (ticket), as long as such a request comes from his or her own doctor. Everyone can freely choose a general doctor from the NHS list. This doctor is then the only one able to prescribe visits, hospital treatment, examinations, and drugs.

Primary care doctors are paid according to the number of patients that they have been assigned, which is never more than 1,500. It is in this way that the previous system of payment for services has been replaced. The state also controls pharmaceutical assistance outside hospitals, and it provides this service by means of pharmacies which operate throughout the country, both public and private, all under national agreement with the NHS. Drugs that can be prescribed within the NHS are listed in the National Therapeutic List (NTL) [Prontuario Terapeutico Nazionale (PTN)].

This is the operative instrument, approved and continuously updated by the Ministry of Health, in which, according to specific criteria, those proprietary medicinal products and galenics that can be prescribed at the expense of the state are listed. Each citizen, to a greater or lesser degree, contributes toward the cost of the medicine (ticket), according to his or her fiscal position, social and health conditions, and the treatment that he or she will receive. Such a contribution is formed by a fixed quota for each prescription (each prescription form can contain up to two predetermined packages of medicaments or up to six if they are single-dose injectable antibiotics) and by a percentage of the retail price of the medicines. The size of both these contributions and the categories of citizens who are exempted from any payment are fixed from time to time by the National Parliament when the state budget is being approved.

2. Pharmacies and Hospitals

Pharmaceutical assistance can be obtained free of charge in 15,591 pharmacies in Italy.[5] Of these, 1,164 are run by city councils and 14,427 are privately managed. There is a total of 1,752 treatment centers in Italy: 1,112 (63.5%) are run by public organizations (hospitals), and 640 (36.5%) are private (clinics). These provide a total of 450,377 beds.[1] Of these, 1,291 are general treatment centers, 291 are specialized, and 170 are psychiatric.[1] In 1986, there were 9,532,094 patients in various hospitals: 8,288,100 were in public institutions and 1,243,994 in private ones.

The average hospital stay was calculated at 12 days per patient per year, and 70.2% of available beds were occupied each day.[1] There is a total of 75,367 doctors operating in public treatment centers, and 60.7% (45,746) of these work full-time.[1] Auxiliary health staff employed in the same hospitals total 223,037, among which 130,126 are professional or head nurses and 66,203 are general nurses.

There are about 1,500 pharmacists working in around 700 hospital pharmacies.[6] The hospital pharmacy is dedicated exclusively to the needs of the hospital without any possibility of dispensing directly to the public.

About 300 pharmacists operate within the NHS, their only purpose being the monitoring of primary care prescriptions, the controlling of pharmaceutical costs, and, more generally, the activities of the community pharmacies open in the pertinent area of the local health unit. In most cases, the head of a hospital pharmacy covers all tasks and is in charge of all activities relating to the public center and the community.

B. PHARMACEUTICAL INDUSTRY
AND DRUG DISTRIBUTION

1. The Pharmaceutical Industry

Of the 310 industrial firms operating in Italy (1988) (3.1% fewer than in 1987), many are partly or totally financed by foreign capital. In 1988, 58.68% of the market share, based on retail price, was controlled by foreign capital, and it is a phenomenon that is growing with the years. The greatest slice of this percentage is represented by the U.S.A. (22.44% of the total), followed by the Federal Republic of Germany (11.72%), Switzerland (9.01%), the U.K. (7.89%), and France (5.25%).[3] The leading 200 firms share 98.87% of the market; the first 100, 88.70%; the first 25, 49.13%; and the first 10, 28.92%, with a general trend towards a greater concentration over the years.[3]

There are 65,673 personnel employed in the industry (1988), a 2% increase over the previous year, while research workers number 6,747, an increase of 5%. Research operators are classified as follows: 3.6% in pure research, 58.9% in applied research, and 37.43% in development research. Research costs have also increased over the previous year, from L. 853.3 billion to L. 979.9 billion, which indicates a total increase of 11%.

In 1988, export of proprietary medicinal products and raw materials amounted to L. 1,863.7 billion, with an increase of 12.14% over the preceding year, due exclusively to the greater quantities involved. Pharmaceutical imports reached a figure of L. 2,878.6 billion, an increase of 22.5% due to an increment in price rather than quantity, which has led to a major deficit in the balance of trade figures for pharmaceuticals. Italy exports to 175 countries and imports from 66. Imports from European countries play the largest part (83.19%); 59.44% of this total comes from members of the EEC, while North America contributes 10.67%. Over 78% of Italian pharmaceutical exports go to industrialized nations.

The total proceeds of the national market have increased by 14.9%, despite the fact that prices for proprietary medicinal products have remained blocked and sales have increased by only 2.33% in volume. The greater value of the pharmaceutical market is due to a shift in consumption toward drugs that have been placed on the market more recently and have a higher retail price as a result. These drugs, newer but not necessarily innovative, are much more widely prescribed and as a result, drugs that are still efficient but no longer remunerative due to their low cost are discarded by the companies that produce them. Highest sales are for those products that have been on the market for four years, while in the other countries, both European and non-European, the greatest part of sales concerns older products that were put on the market six, seven, or even nine years before.[3] In general, though, the price of proprietary medicinal products is low (L. 9,205 compared to an EEC average of 10,037).

There are 5,309 proprietary medicinal products on the market corresponding to 10,748 confections (predetermined packages), both of which are experiencing a continual decrease and notable shift toward single component drugs. In Italy, a proprietary medicament is available only in packages that have already been made up by the firm according to registration specifications and that cannot be unpacked by the pharmacist. In this way, sale to the public does not correspond to the quantity actually needed by patients but to the content of the package. The authorizing system therefore approves not only the other characteristics of the product but also the final packaging.

As far as therapeutic categories are concerned, consumption has undergone some modifications in the last two years as a result of which drugs for the digestive system and metabolism, cholagogics and hepatoprotectors, vitamins, antianemics, cough medicines, and anti-influenza remedies have lost out on the market. On the other hand, medicaments for the cardiovascular system, dermatologics, systemic antibiotics, psycholeptics, antacids, antiulcerants and ophthalmics have seen an increase in their share of the market.

Pharmaceutical consumption steadily diminished in the period 1960-1988. An average annual rate of market development of 5.7% in the period 1960-1970 fell to 3.3% in the period 1970-1979 (in 1979 the highest level of pharmaceutical consumption was recorded). In the period 1979-1988, the average annual rate became negative, equal to -0.5%, despite the fact that some slight recovery in consumption has been evident in recent years.[3]

In Italy, patent protection was unheard of before 1978 and the ensuing legislative modification, which was passed in 1979. According to the ruling, no distinction, not even on duration, can be made between a patent for medicines and a patent for other commercial products. Thus the problem of the interaction between the economic effects of industrial protection and the period of effective commercialization of medicaments arises.

2. Drug Approval and Administrative Regulation

Before a proprietary medicinal product may be marketed, it must be granted two different authorizations by the competent authority. The first concerns production and is issued to the manufacturer for the preparation of a given pharmaceutical dosage form after an inspection, which ascertains the suitability of the premises, apparatus, manufacturing processes, and staff. The second, however, concerns the specific product and is granted after an evaluation by the United Commission for Drugs [Commissione Unica per il Farmaco (CUF)] of the corresponding protocols (pharmaceutical, technological, toxicological, pharmacological, and clinical tests).

The passage from the preclinical to the clinical phase is granted on the basis of animal data and requires a particular formal authorization awarded by the Central Health Authority.

In addition to reports from experts in various sectors, other relevant information is required regarding the complete sales presentation of the products (labeling, leaflet, packaging, etc). Therefore, the label that is to appear on containers and outer packages and the leaflet that is to be enclosed with each medicine must be subjected to approval. On the label and on the outside of the package, the name of the product and of the producer, the qualitative and quantitative composition, the preparation and expiration date, the manufacturer's batch number, the method of administration, the special storage precautions, and the number of the authorization to place on the market must be shown. Recently, it has been stipulated that a self-adhesive label be added to the outer package carrying the product name and the authorization number given by the authorities at the time of registration, the corresponding bar code of suitable proportions for automatic reading, and the eventual contribution to the cost on the part of the patient. According to NHS regulations, the label must be detached from the product and applied to the prescription form by the pharmacist at the time of dispensing. Thus both epidemiological and administrative monitoring are permitted in a general and easy way.

The use of leaflets is compulsory for all proprietary medicinal products, both ethical and OTC, and must be approved beforehand by the Ministry of Health, as must any successive change to this same leaflet. The leaflet reports on the composition, characteristics, therapeutic indications, interaction with other medicaments, dosage form, directions for use, contraindications, warnings, and side effects.

Registration can be carried out by the multistate procedure according to particular conditions and in all cases when the medicinal products are developed by high technology or derived by biotechnology. Multistate registration, which provides reciprocal recognition in at least two other member states of the EEC, authorizes the placing on the market of the medicinal product in all countries of the Community. The procedure presumes that each member state recognizes the validity of protocols for marketing authorization presented by another country. This creates a difficulty that is not easy to overcome, thus limiting the number of registrations issued up until now.

For medicaments that are replicas of others already on the market, whether exactly the same or essentially similar, authorization can be obtained by means of a simpler procedure that does not require the presentation of the results of pharmacological and toxicological tests or of clinical trials, but provision of technological and analytical protocols only. In this case, the medicament is identified by the international nonproprietary name recommended by the World Health Organization, where such a name exists, or adopted by the National Pharmacopoeia, followed by the name of the producer.[7] The replica of a drug is permitted on the market ten years from when the original product was registered, and it is given a lower price than the original, since there are no costs intended to remunerate scientific information and research.[8]

3. Generics

The situation for generic products is in a state of continuous evolution, thus being still rather vague. At present, pharmacies and firms authorized to produce drugs can prepare generics. A list of products exists whose monographs are included in the National Pharmacopoeia. Although not representing a closed number of preparations — in fact it cannot limit industrial production — the Pharmacopoeia is an authoritative, scientific validation of quality, safety, and efficacy of this class of medicaments. Reference to the Pharmacopoeia of a member state has been formally made by the EEC in a recent directive which extends to all industrial medicaments the provisions laid down up to now for proprietary medicines only. When a formulation is not listed in the National Pharmacopoeia, the manufacturer will need to substantiate efficacy and safety requirements of the product in a different way, which always presumes simplified experimental tests and clinical trials.

The generic market is currently almost nonexistent because until April 1989, generics were not available under the NHS program. There are few generics (only 49) that can be prescribed, and they belong exclusively to the systemic antibiotic and chemotherapeutic categories. No incentive is provided for their prescription as an alternative to the proprietary medicinal products, nor can the pharmacist exchange the prescribed proprietary product for the corresponding generic, since substitution can be made only with the equivalent proprietary product. The development of the generics market depends mainly on its position in the pharmaceutical services within the NHS as an alternative, either obligatory or discretional, to the equivalent proprietary medicinal product and on its economic incentive.

4. Wholesale Distribution

This activity is the least regulated in the pharmaceutical field, since the law provides only for the part-time presence of a technical director with a suitable academic qualification (pharmacy or chemistry) who is responsible for the health and hygiene aspects. This situation implies that in the setting up of a European market there must be almost immediate harmonization, since elements of pressure are principally economic. In recent years, a cooperative movement has been developing as a result of which intermediate distribution can now be carried out by pharmacists and not necessarily solely by independent firms created for this purpose. The form of cooperation between pharmacies involves not only the pooling of goods, like medicines, but also services that become centralized (data banks, quality control, specific and permanent training activities, etc.). There are 261 firms and 334 wholesalers, the latter figure representing different forms in which the intermediate distribution takes place and including cooperation among pharmacists.

5. Administrative Classification of Drugs, Price Regulation, Advertising and Promotion

Among the various duties of the United Commission for Drugs is that of assigning the proprietary medicinal product to one of four administrative classes indicated by law in 1988. These are:

— Drugs that can be prescribed within the NHS and are included in the National Therapeutic List (NTL)
— Drugs that are to be used only in hospitals or in other centers by specialists
— Self-medication drugs that cannot be included in the NTL
— Other drugs that cannot be prescribed by the NHS, for example, benzodiazepines and other minor tranquilizers.

The Commission also provides for periodic modifications in these classes when new scientific discoveries make it necessary. Prior to 1988, it was the manufacturer who proposed the administrative class of the drug that was to be placed on the market and its inclusion in the NTL, while the public administration had only the right to refuse it.

Classification is not only significant from the NHS point of view, but the ruling that is applied is different for the marketing of the various categories of medicaments. It is absolutely forbidden to advertise those drugs that require a medical prescription, or are dispensed under the NHS program, or that contain narcotics and psychotropics. For the OTC medicaments advertising is permitted, but the Ministry of Health checks in advance the contents and pictures relating to the message and issues a specific authorization. Any type of incentive to buy by means of discounts, presents, and free samples is forbidden.

Scientific information to health operators, as distinct from advertising, is allowed and does not require a previous authorization, provided that it is limited to the contents shown on the technical report of the product, approved by the Ministry of Health.[4] Medical samples can be distributed to prescribers only and not to pharmacists, and they can be freely distributed only in the two years immediately following the product's placement on the market. After that period, medical samples are given subject to a formal written request on the part of the prescriber.

Retail price of the proprietary medicinal product is the same throughout the country and is fixed by the state [Interministerial Price Commission (Commissione Interministeriale Prezzi)] for both proprietary medicinal products and for those generics that appear in the NTL.

The price of OTC products is no longer imposed by the authority of the state but is fixed by the producer and reported to the Ministry, which ensures that the price is the same throughout the country. Price is also authoritatively administered for those medicines prepared from time to time by the pharmacist on the personal prescription of the doctor. For these, the National Medicine Tariff (Tariffa Nazionale dei Medicinali), approved by the Ministry of Health, establishes the price of the various ingredients that make up the prescribed formula and professional fees in accordance with the dosage form and the eventual additional duties for toxic, strongly colored or corrosive substances, poisons, and narcotics. Service fees outside normal opening hours are also remunerated with a separate fee provided by the National Tariff and added to the final price of medicines. Even the remuneration of wholesalers and pharmacists is fixed by law, respectively in the measure of 8% and 25% calculated on the final price of the medicine, for which no form of competition is possible.

6. Cost-Containment Activities

The problem of national pharmaceutical costs is becoming increasingly more evident, with unavoidable effects on the national budget. Thus various measures have been taken to contain economic pressure, which in the outlined scheme cannot occur without provisions adopted by the National Parliament or the Central Health Authority. One emergency measure, which has recently been recognized as legitimate by the EEC, provides that there can be a block on the price of medicines for a defined length of time if it is justified by proper motivations. Up to now, price block has not given the hoped-for results, since in actual fact, the effect has been a shift in prescriptions toward newly registered and therefore more expensive medicines.

The citizen contribution has also been increased. This year it rose from 20% to 30% of the retail price and stands at 40% for some categories of medicine that are held to be less important from both a pathological and a social point of view. In addition, the fixed quota for every prescription has been increased a number of times in recent years. One group of products, the lifesavers, does not call for any contribution.

The unification in the United Commission for Drugs in 1988 of every competence relative to medicines, with the exception of price and to their prescription within the NHS, also allows for the systematic review of medicines that have been on the market for more than three years. The dynamism of the administrative classes makes it possible to modulate the effect of the pharmaceutical market on the public economy.

The NHS is progressively reduced by the elimination of proprietary medicinal products used for the treatment of minor ailments. The introduction of generics, even if formally drawn up, has not yet been put into practice, and its effect on the economy as an alternative and cheaper way of prescribing has been nonexistent until now because these medicines have lacked any form of incentives. Finally, attempts have been made, using various methods, to modify the prescribing habits of the doctor, thanks mainly to the introduction of the bar code for all proprietary medicinal products that can be prescribed within the NHS and the continuous monitoring of medicinal prescriptions. This has prepared the way for a computerized system of control of medical prescriptions through which the Public Administration knows the exact qualitative and quantitative prescribing habits of each doctor within the service. Therefore, the local health unit can act in various ways against those who are markedly exceeding the local, regional, and national average in quantity or in value of drugs prescribed.

C. PHARMACY PRACTICE

1. The Pharmacist

The Faculty of Pharmacy awards two degrees, one in pharmacy and the other in chemistry and pharmaceutical technology. Both degrees permit entry to the same area of activities, in a way that is either (1) exclusive, as in preparation, detention, and distribution of medicines, in hospitals and to the public in community pharmacies, or (2) shared with graduates in other disciplines, as in production and control of medicines, intermediate distribution of the finished products and the raw materials employed in medicines for use in humans or animals, management of production and wholesale trade of medical aids and devices, health devices, cosmetics, dietetics, and baby foods.

There are at present 24 Faculties of Pharmacy (soon to be 27) spread throughout the state universities of Italy, with an estimated student population of about 24,000.

The two academic qualifications, although allowing entry to the same area of activities, differ from a formative point of view, one being oriented more specifically to distribution management and the other to production and quality control of medicines. Present reforms in the course of study stipulate that, starting from the academic year 1990-1991, both will last for five years and will include a six-month practice period to be spent in a community or hospital pharmacy or, partly, in a pharmaceutical company.

Both degrees fulfill the general criteria laid down by the EEC, so they are mutually recognized and give the Italian pharmacist the right to be employed in one of the member states of the Community. The opening and managing of a pharmacy still follow regulations issued by each country.

Graduates of the Faculty of Pharmacy are subject to the control of the Public Health Administration, initially through State examination and mandatory membership of the professional body and then through regular surveillance of the related activities.

Professional bodies are set up provincially (Provincial Order of Pharmacists) and in turn form a national association (FOFI). They depend on the Central Public Administration and exercise disciplinary power over their members. Enrollment in the professional body confers the right to practice the profession in a subordinate form throughout Italy, while the owner of a pharmacy must be a member of the professional body whose authority coincides with the provincial area in which his or her pharmacy operates. The system involves a subordinate relationship with local or Central Public Health Administration that is direct in cases where the pharmacist operates in a public organization and indirect or special in all other situations. Penal and administrative responsibility derives from this, as does the recently added (1988) civil liability for damages caused by medicines or, more generally, by defective products.

2. Pharmacy As a System

The pharmaceutical service is assigned by the constitution to the state, which puts it into effect in both a direct form through its own organizations or indirectly by giving the right to private individuals to perform this professional activity. Distribution of medicines to the public is allowed only through pharmacists operating in a pharmacy, while the health service, both public (hospitals) and private (clinics), can use medicaments in its own operations. An essential distinction therefore exists between use and distribution of medicines, and the pharmaceutical service leads on from these two assumptions. As a consequence, while the first situation permits the service to be administered at home should circumstances require (type of medication, for example permanent oxygen therapy, or the complexity of the therapeutic operation, as in total parenteral nutrition), in all other cases any form of dispensing that does not take place in the pharmacy is forbidden (home delivery, mailing, etc.).

Distribution of medicines is assured through a network of community pharmacies whose location is established according to criteria laid down by law. These are in keeping both with population size and distribution and refer to each municipal area. The regional reference is thus the local council (municipality), and the location map (pianta organica) of the pharmacies is the technical and administrative tool of planning, reviewed every two years and approved by each individual region. Council areas are subdivided into topographical zones inside which the pharmacy operates administratively, according to the population size of the entire location and to the demands on the pharmaceutical service.

The opening of a new pharmacy does not take place on the basis of free enterprise but by permission of the regional authority and by the consequent authorization on the part of the Local Health Unit, and by means of a public competition in which all pharmacists who are members of the professional body and possess the requisite qualifications can take part. The ratio of pharmacies to number of inhabitants is fixed by law at 1 in 4,000 in communities of more than 25,000 inhabitants and at 1 in 5,000 in the others. At the same time, there is a distinction between rural and urban practices, the former operating in communities of up to 5,000 inhabitants and the latter in larger locations. In places where the population is less than 3,000, the pharmacy is due an annual contribution from the region, the council, and urban pharmacies. Furthermore, practicing in a rural zone is associated with incentives affecting the public competition list through which pharmacies are assigned.

The branch pharmacy takes into account the possible temporary fluctuations in the number of residents following particular local situations, as in the case with resorts. This is the only situation in which a pharmacist can be the owner of two premises, the main one and the branch, the latter, however, operating only at limited times of the year.

The pharmaceutical dispensary, run by the nearest pharmacy, ensures that the pharmaceutical service operates even in those areas whose size would not justify the opening of a pharmacy.

To sum up, the main instruments for regionalization of the pharmaceutical service are the *pianta organica* of pharmacies and the public competitions for assigning them, together with the branch pharmacy and the pharmaceutical dispensary.

Since 1968, law requires that the same person must both own the pharmacy and run the activity professionally. In this way, company management, thereby implying chains of pharmacies, and the ownership of more than one shop on the part of a single pharmacist, are forbidden. In addition, a strict ruling of incompatibility has been established whereby the owner of a pharmacy, even if in possession of a pharmacy degree, is barred from any other activity that calls for this qualification (teaching, intermediate distribution, manufacturing, etc.).

The only exception is that of the council, which can own and manage more pharmacies. This originates from the right of preemption introduced in 1968 and applicable to 50% of the number of pharmacies to be instituted through a public competition. This is a unique example among the member states of the community; therefore, privately run pharmacies that have received special permission to operate exist, as do pharmacies whose owners are not private individuals but are, in fact, the council. The latter, even though they are public, operate as if they were private, especially in their relations with the NHS; that is, they put the pharmaceutical service into effect in the name and on behalf of the state like the private ones and are ruled by the same regulations. The distributive network is therefore affected by both private and public pharmacies, in respect to the exclusive right of the pharmacist and the pharmacy to distribute medicaments and to the criteria laid down by the regionalization of the pharmacy system, with an individuality that reflects the political choices of each council.

Public pharmacies, after more than 20 years of the preemption law being applied, form a limited percentage of the distributive system (8.6%) and are concentrated by preference in the northern regions of the country. They are an expression of political trends rather than a result of competitive practices because their location, although favored by the mechanism of preemption, follows general regulations. A common characteristic of all pharmacies, be they public or private, is the lack of competition, since the four basic factors (place, price, product, promotion) are established beforehand by state regulations and cannot be put to use for profit ends.

The continuity of the pharmaceutical service is ensured by the institution of temporary management, by which the owner is substituted for by a pharmacist who manages in his or her place. The substitution can be determined either by public demands (these cases are decided by the Health Authority), or communicated formally by the owner in well-defined circumstances (illness, for a maximum of 5 consecutive years or 6 years in a decade, call to arms, serious family reasons for a maximum of 30 days per year, for annual holidays depending on the regional law, election to public office, pregnancy, union duties, etc.). In these latter situations the pharmacist can choose his or her own substitute, who must have the formal requirements to professionally manage the activity, while private relations between the substituted and the substituter are regulated according to the civil code.

Continuity is also guaranteed by the health authority through a system of mandatory duty times regulated by regional law. In this way, one or more pharmacies are open even during the daily and nightly closing times of the rest and also during holiday periods.

Since 1968, as opposed to the previous system, it has been possible for pharmacies to be sold, even if the conditions imposed for their transfer are such as to make this situation exceptional. On the basis of this regulation, pharmacies can be transferred only once in the life of the owner and only five years after their acquisition. If the owner should die, the heirs can arrange for the sale within three years, extendable to seven if the spouse or child enrolls in the Faculty of Pharmacy. The proprietor of the pharmacy who has sold his or her practice can buy another after two years. The purchaser, apart from being in possession of the requisite academic and moral qualifications, must have practiced this profession as an employee for at least two years or have been declared suitable for ownership in a previous public competition. The competition therefore serves both the purpose of assigning new pharmacies and of declaring as suitable for ownership those who, although having taken part, do not win pharmacies to be set up. The transferral of a pharmacy must cover both professional and ownership aspects and cannot be applied to public pharmacies.

3. Pharmacy Practice

There are two distinct areas of activity in a pharmaceutical practice, one of which is more strictly professional and which refers to the particular regulations governing the sector of medicines. The other, although still concerned with health, is not exclusively and absolutely reserved to pharmacists and pharmacies only. Other persons who do not possess specific professional qualifications can operate in the latter field and are subject to the general laws on trade (dietetics, baby foods, cosmetics, products for personal hygiene, herbal remedies, etc.). Therefore, two types of activity coexist within the same practice, both regulated by legislative acts but only one being the preserve of the pharmacist. The legislation that safeguards public health stipulates that in the area of medicines there can be no form of competition, while it is tolerated to a certain extent in the commercial sector. In a similar way, the public duty of the pharmacist and the pharmacy is modulated, although some areas of therapy support (prosthetics and dietetics) are still activities that operate within the domain of the National Health Service.

The medicinal sector has as its object medicines in their various administrative forms of marketing; that is, the proprietary medicinal products and galenics. The former are divided into ethicals, copies, medicaments without prescription, and those intended for self-medication; galenics are those medicines prescribed personally by the doctor or official preparations whose formulae are included in the National Pharmacopoeia or, according to the ruling recently adopted by the EEC, in the pharmacopoeia of a member state.

The absolute right of medicament dispensing by the pharmacist and pharmacy is likely to continue for some time after 1993, even if in some countries of the EEC, namely the U.K., Denmark, and the Netherlands, medicines are present in other distributive channels. Public access to medicines can take place after the doctor prescribes a drug either in a repeatable or nonrepeatable prescription, to be dealt with or kept by the pharmacy, or in a special form set up by the Health Authority. The latter is reserved for controlled preparations that appear in the first three lists of the special law for narcotic and psychotropic substances. The type of prescription required for drug dispensing is prechosen and made obligatory at the moment of registration if it concerns relative proprietary medicinal products or by the specific provisions dictated by the National Pharmacopoeia, as in the case of galenics.

The repeatable prescription is the mildest form of public protection provided for by the rulings. It is automatically repeatable five times during the validity period of three months, if not otherwise indicated by the prescriber. The pharmacist is obliged to apply the date and the price as well as the pharmacy's stamp at the bottom of each prescription form. Outside NHS programs, this kind of prescription is usually given back to the purchaser.

The nonrepeatable prescription can be used only once, is retained by the pharmacist, and is valid for three months from the date of compilation. It must be drawn up according to special procedures that require the name, surname, and address of the patient in addition to a detailed description of the quantity, dosage, and posology to be administered and must be signed in full (and not just initialed) by the doctor. The nonrepeatable prescription involves two different actions on the part of the pharmacist, according to the regulations. The first case is when the prescription must only be retained; that is, not given back to the customer. The second case is when the original of the prescription must be kept for five years for documentation purposes.

The nonrepeatable prescription to be retained for five years is obligatory for galenics that contain one or more of the substances indicated as poisonous by the Pharmacopoeia. For proprietary medicinal products that contain these substances, such a prescriptive protection is imposed at the time of marketing authorization.

The repeatable and nonrepeatable prescription is also used for those substances subject to control listed in Tables IV, V, and VI of the law on narcotics and psychotropics, while the special ministerial form is confined to the proprietary medicinal products and galenics listed in the first three tables (schedules).

The prescription intended for controlled medicaments has the same form throughout the country and is given out to doctors by the respective professional bodies. It has a reduced validity of ten days, and can contain only one preparation and dosage for a request of eight days of therapy. It must be drawn up according to particular provisions, and the original prescription is kept for five years by the pharmacist. It is made up of three sections, the first of which remains with the doctor for the files and is retained for two years. The other two are given to the pharmacy, which keeps one for the files and sends the other to the Local Health Unit for refund if the service was carried out within the NHS.

Within the NHS, the validity of the prescription is fixed at ten days, whatever treatment is requested and therefore whatever type of prescription may be required. In any case, the prescription is always retained by the pharmacist, who then sends it to the Local Health Unit for refund after having applied the relative self-adhesive tag attached to the packaging of each dispensed medicine.

The public can have access to some medicines without any recourse to a doctor by a simple verbal request or on the advice of the pharmacist. The first case deals exclusively with self-medication medicines, which the individual has heard about through some form of advertising. In the second case, the pharmacist suggests the use of those medicines that do not call for a medical prescription, following the registration if they are proprietary medicinal products or following the National Pharmacopoeia if they are galenics.

Besides a totally autonomous self-medication, there is, therefore, a therapeutic treatment guided by the pharmacist that refers on the one hand to the proprietary medicinal products included in this category and on the other to the proprietary medicinal products and galenics that do not require the intervention of the doctor. In both cases, the ruling stipulates that the pharmacist is the protective figure whose role is intended to guarantee the patient in absence of medical intervention and to provide him or her with sufficient information for proper use, which cannot rely on medicament presentation only.

The action of the pharmacist consequently differs in its various assumptions, as a result of which he or she represents an integrative support to the doctor when the medicine can only be dispensed on a prescription, while the pharmacist's role is much more incisive, particularly on the correct and appropriate use of a product, when he or she deals with a guided medicament or one for self-medication.

Therefore, the fact that a medicine is assigned to one administrative class rather than another has a decisive effect on the professional practice, given that the classification is not static but dynamic. As a result, the same medicine over time can be put in a category different from the one in which it was originally placed. This is, in part, the phenomenon of re-regulation (formally deregulation) that in the field of medicines acts through the administrative classification, the process of demedicalization, and the substituting program.

Substituting a prescribed medicine with a pharmaceutical equivalent was introduced in 1979 and renewed in 1989, but it is still limited to proprietary medicinal products containing the same active ingredient and to services within the NHS. It can be performed by the pharmacist only in exceptional cases, as when the pharmacy does not have the prescribed product in stock or when the proprietary medicament cannot be temporarily supplied by the intermediate distribution.

The right of the pharmacist to substitute medicines that have been prescribed by the doctor can have concrete developments if it is extended to all pharmaceutical services, both within and outside NHS programs, if the pharmacist can resort to the equivalent generics and not only to proprietary medicinal products, and if there is sufficient incentive for generic prescription within the NHS.

Classification, partial demedicalization, and substituting power are the critical points of the profession and need a cultural background that cannot be limited to academic training, but this presumes a permanent program of education linked to an easily accessible data bank, which must be single and based on official sources. In a country where prescriptions within the NHS make up 82% of the value of the pharmaceutical market, the need for documentation is accomplished mainly by the Local Health Unit through analyzing and keeping the prescriptions sent for accountancy and administrative purposes. Despite this, the pharmacist is still obliged to document the services that have been effected outside the NHS program and in particular those pertaining to the preparation of medicines with particular reference to raw materials, compounding processes, and the apparatus used. The National Pharmacopoeia Commission is moving in this direction and preparing itself to draw up good preparation practices that, in a way, are similar to the GMP that define a regulation which is both applicable to the pharmacy and to the operations that are carried out there.

Within the pharmacy there are two figures who, although possessing the same academic qualifications, have different responsibilities in respect to the Public Health Administration: the pharmacist and/or director, who is by law also the owner in private management practice, and the colleague, who is an employee of the former. Both have personal professional responsibilities, those related to dispensing medicines, while only the pharmacist owner and/or director has the task of ensuring the proper running of the pharmacy and the duties imposed by the Public Health Administration to this end. For example, the obligations relating to documentation, to stocking of given medicines, to the application of good storage practice to raw materials and finished products, to the legitimation of sales staff, to the observance of opening times and adherence to the system of duty time imposed by the Local Health Unit, to services within the NHS and account reports that derive from them, and to suitable hygiene of the premises and equipment are all duties for which the pharmacist owner and/or director is responsible.

In the retail price, fixed and unchangeable for all medicines imposed by the State or suggested by the producing company on various assumptions, there is a fee of 25% fixed by regulations. This quota is intended to remunerate intellectual services and management activities of the owner of the pharmacy, whether this be an actual person, as is the case with private management, or an authorized person, as in public activities owned by the council. In this case, as is the general rule for all employees of a practice, the equivalent of these intellectual services is stated by the salary awarded according to national agreement. The same regulatory fixed remuneration is applied to services within the NHS, minus the contribution to the cost on the part of the patient when it is provided for.

The pharmacy, whether public or private, anticipates the value of the medicines supplied to the NHS, and it is successively reimbursed each month by the local or regional health authority on the presentation of prescriptions filled in the corresponding period. The citizen therefore does not pay at the moment in which he or she receives the medicines from the pharmacist or, when provided for, limits himself or herself to returning one quota for the prescription and one as a percentage of the fixed retail price of each medicine. Exemption from contribution to pharmaceutical costs can be determined on the basis of income or on the social nature of the illness from which the patient is suffering. Thus the economy of the pharmacy depends on the exemption system, too, which reflects the demographic composition of the community.

The frequency and regularity of refund by the Public Health Administration is another critical factor in the pharmacy's economy. Even if they are established by the National Triennal Agreement (Accordo Nazionale Triennale) that regulates relations between the NHS and both public and private pharmacies, refunds do not arrive at regular intervals in all parts of the country and because of this, even the value of a pharmacy suffers according to its location.

In determining the value of a pharmacy, in addition to the regularity of refund by the Public Health Administration, the gross business turnover and the percentage of that due to the NHS is taken into account, with the value of stock, furnishings, and equipment remaining invariable.

The presence of the NHS excludes the possibility of refunds for private assistance.

4. Health Products

Besides the medicinal field, the pharmacy practices in a nonexclusive way, such as the marketing of products of varying health content, where other distribution points can intervene according to the general law for retail sale. This law stipulates that pharmacies are authorized to sell a limited range of products that are cataloged in a national list by the Public Health Administration. In spite of this, due to the prechosen system of listing, which refers in many cases to the marketing category in general and not to its specific contents, the situation is not the same in all parts of the country, resulting in a great variety of nonmedicinal products put on sale by pharmacies.

Among the member states of the EEC, the situation differs markedly, so it will be necessary to bring the already existing provisions into harmony, with a trend towards deregulation that could transform the role of the state from one of prevention to merely repression. This is a particularly critical sector since it largely identifies the image of the pharmacist and the pharmacy, which by now are believed by the community to be legitimizing the products they sell. The distinction between the areas of medicines and health products forms the basis of every evaluation of this sector, which reflects in a very obvious way on the presentation of the products (labeling, brochures, advertising, etc.) and therefore on the content of the market that it is following.

In a well-developed nation, the fragmentation of community needs is shown by an ever-increasing demand for products that do not have the requisites of medicaments. These are products for the restoring and correcting of the organic functions of humans. The requisites of health products are only those of fostering the physiological activities of the organism. Consumer protection is carried out differently in this sector, moving from state intervention of a preventive type through a system of authorizations, twofold for medical aids and medical devices, single for dietetics and baby foods, to intervention of a repressive form, as in the case of cosmetics and personal hygiene products.

Dietetics, medical aids, and medical devices can form part of the services offered by the NHS through medical prescription. The relevant products are distributed directly by the Local Health Unit or given to the patient either by public or private pharmacy or by other points of sale.

Some emergent areas, in particular dietary supplements and herbal and homeopathic products, still lack specific regulation, in particular dietary supplements despite their widespread use. The National Parliament, in approving the state budget for 1989, expressly allocated financial resources for the drawing up of specific regulations for herbalism and homeopathy, while there is still no provision for dietary supplements. In the absence of specific norms, these products can be used therapeutically only if they are put on the market as galenics; otherwise, they must be considered as foodstuffs.

Herbal remedies and dietary supplements are undergoing a significant development, and there is an ever-increasing demand for these products, satisfied in part by pharmacies and in part by other points of sale, whether belonging to small- (reform houses) or large-scale distribution. These can be connected, to a greater or lesser degree, to the presence of the pharmacist. There is, therefore, in the health product area, a dichotomy between pharmacist and pharmacy due to which consumer protection is entrusted to product presentation and in some cases to the specific intervention of the pharmacist, who, however, carries out this activity off pharmacy premises.

Homeopathic products, although not formally belonging to the medicinal area, represent a particular exception because they are prescribed exclusively by doctors and are considered remedies by the community. In the absence of specific regulations, these products are tolerated by the Public Health Administration, but they can be distributed only on the issue of a medical prescription and dispensed only in the pharmacy, while their production and import is subject to health authorization. Homeopathic products are still excluded from the NHS program, and their refund is not even authorized in an indirect form (1989).

D. CONCLUSIONS

The outlined system is characterized by a marked lack of the competition that exists in various forms in other countries. The intention of the Italian legislator is to free the whole area from the common system of rules, to subject it to particular provisions aimed primarily at safeguarding public health, and at the same time, to provide economic protection for every activity that concerns the field of medication. The situation could undergo some modifications with the advent of the Common Market in 1993, but these will come about at varying speeds and will presumably be more marked in the health product field than in the medicinal area.

From this outlook, the social choice appears irreversible. Even though we can predict modifications in its realization, it will involve the management of the social program more than the political attitude, and will remain public.

REFERENCES

1. Istituto centrale di Statistica. Le regioni in cifre. Roma, 1989.
2. Istituto Centrale di Statistica. Annuario Statistico Italiano. Roma, 1988.
3. Farmindustria. Indicatori Farmaceutici. Roma, 1989.
4. F. Angeli. Censis. XXI rapporto/1987 sulla situazione sociale del paese. Roma, 1987.
5. Federfarma. Personal Communication. Roma, 1989.
6. Societa Italiana di Farmacia Ospedaliera. Statistiche SIFO. Milano, 1989.
7. A. Sordi Sabatini. Denominazioni Comuni Italiane dei Principi Attivi Contenuti nei Medicamenti. Milano: OEMF, 1988.
8. G. Battaglino. EEC Directives and Recommendations in the Pharmaceutical Field. Milano: OEMF, 1987.

SELECTED READINGS

M. Marchetti, B. R. Nicoloso. Servizio Farmaceutico nella U.S.L. La vigilanza sulle farmacie. II edizione. Milano: OEMF, 1987.

M. Marchetti. Esami di farmacia. Vol. II. Legislazione farmaceutica. Milano: Guadagni Ed., 1987.

Chapter 12

Japan

Minori Tatsuno
Kiyoshi Muraoka
Tsuneji Nagai

A. THE NATIONAL HEALTH CARE SYSTEM

1. Historical Background

Medical systems, like social systems in general, are embedded in a cultural matrix from which is derived the coherent body of ideas of which the system is composed. The practice of all medicine, therefore, including that of industrialized societies, has evolved as a result of its setting in a unique cultural context. Medical and pharmaceutical techniques and therapies, including instrument-aided therapies, are all modified by cultural values. Medical and pharmaceutical therapies, even those that are scientifically established, are culture bound in the sense that the questions raised by theoreticians and the methods used to answer them are products of a particular historical period.[1]

Ancient medical practices of Japan combined Japanese indigenous folk medicine with medicinal herbs and indigenous faith healer religions characterized by a special value on cleanliness and a fear of contamination. The impetus for medical technological change has mainly been introduced from other societies, first from China and later, Europe and America. At the end of the fourth century, highly developed forms of Chinese politics, science, technology, medicine, and religion (Buddhism) greatly spurred emulation. The system of social welfare was also influenced by Buddhism. In 593 A.D., the first national social institution for the benefit and welfare of mankind (hospital or asylum for patients, children, and the poor) was organized at the Buddhist temple in Osaka.[2] In the seventh century, the Japanese central government was established to enact the law of the land, which included decrees concerning medicine. The emperor of the era introduced a Chinese taxation system that preempted the sick from taxes and their families from military and public service, but he offered no remedy for the destitute. In the Tempyo Era, Buddhist priests began not only to cure the sick but also to improve their living conditions. The government, at first having oppressed the priests, made the best use of them by legalizing their activities and succeeded in pacifying the people, who suffered from smallpox spreading all over Japan. In 730, Empress Komyo set up a social settlement and made a medicinal herb garden from which products were to be given to the sick and the poor.

In the twelfth century, the medical activities of the Zen Buddhist priest doctors, for example Ninsho, contributed to building the Gokuraku-ji Temple, where a sanatorium, or leper house, as well as a room for administering medicinal herbs, was established.

In the sixteenth century, Christianity and Western medicine were introduced by the Jesuits and then by the Franciscans from Portugal and Spain. The first European hospital and leper house was established in 1557 by the Portuguese monastic, Luis de Almeida, a surgical expert, in southern Japan. Almeida introduced surgical treatment that incorporated a method of psychological suggestion facilitated by his belief in religion. In the same hospital, Japanese doctors treated patients by Japanese folk medicines and Chinese medicine (Kampo medicine or Japanized traditional Chinese medicine) with herbs and drugs. At this time, the Japanese government, which was chiefly interested in European military technology and the other new advances in scientific technology, readily encouraged the practice of Western medicine for its strategic advantage in warfare. However, it began to witness integrated European colonization in the South Pacific, and fearing its domination, the government outlawed the propagation of Western medicine along with Christianity at the end of the sixteenth century. With the exception of China and Holland, all contacts with the outside world were forbidden in 1639. Thus traditional Chinese medicine became dominant, and the vestiges of Dutch medicine remained only to some extent in Nagasaki. In a closed Japan, Chinese-style medicine developed in two ways; one was a reevaluation of the very traditional Chinese medicine in light of the original tests, and the other was the newly developed Japanized Kampo medical treatment. At this time, the Japanese used many herbs and drugs that were made with new instruments, which were used to cut the herbs and to mix various powders.

At the end of the eighteenth century, the new technology of Western medicine was reintroduced from Holland to Japan. Most authors believe that the translation of a Dutch anatomy book *Tafel Anatomia* was the turning point in the Japanese history of medicine.[2] With the translation of this book, Western medicine (Dutch medicine, called Rampo medicine in Japan) was presented as an alternative to the Chinese (Kampo medicine*) interpretations and not merely as a supplement. Younger medical practitioners of Kampo and Rampo medicines stood against such a cooperative or conciliatory attitude. P. von Siebold, a physician from a prominent German medical family, established in Nagasaki a private school for Dutch language technology and medicine. Dutch medicine began to flourish when J. L. C. Pompe van Meedervort was allowed to establish a school with a full Dutch medical curriculum, including instruction in the basic sciences and a hospital of Western medicine.

In the nineteenth century, a reopened Japan again reconstructed the health care system with a Western-style or cosmopolitan medicine** to prevent acute infectious diseases. This change was a part of Westernization promoted by the new government of the Meiji era (1868 to 1912) that also Westernized industry, political institutions, social patterns, and so on. Medicine was a small but highly visible part of their effort to quickly modernize and so avoid colonization.[2] German medicine was introduced officially in place of the traditional Chinese medicine. Health legislation was focused on economic development and sustaining the military (i.e., national prosperity and strength). A European-style pharmaceutical industry was encouraged, while Kampo medicine was discouraged by the new government. A thorough transformation based upon Western medical de-

---

*Kampo medicine: otherwise known as traditional Chinese medicine, Oriental medicine or East Asian medicine by M. M. Lock. Kampo medicine is Japanized Chinese medicine developed under the Japanese cultural conditions for 1300 years, while retaining much of its original Chinese character.[5]

Kampo medical system differs from Western medical system in the basic premise and perspective. The former emphasizes that the imbalance of the bodily or mental functions of the patient is the fundamental cause of the illness, while the latter focuses mainly on pathological changes of affected organs. The goal of Kampo diagnostics is to determine a *sho* or a Kampo-type syndrome for each patient according to both the conditions of the patient and the environmental factors around him or her. Once the *sho* is determined, it immediately indicates the need for appropriate Kampo therapy such as medication or acupuncture. In Kampo medicine, therefore, diagnosis goes directly to the therapy, which is not always the case in Western medicine.

**Cosmopolitan medicine[6] is otherwise termed Western medicine, modern medicine, biomedicine, or scientific medicine. Since it has become the dominant method of healing in Japan and is supported by the state, it is simply called medicine without any adjective.

livery holds the key to the successful penetration and growth of cosmopolitan or Western medicine in the pluralistic system of Japanese medicine.[3]

After World War II, the public health care administrative system was patterned after the American style. A delegation of the American Pharmaceutical Association visited Japan in 1949 to advise the Japanese government that it should separate the system for pharmaceutical dispensing from medical practice. The movement for the separation, which had been disbanded in prewar days, was encouraged by this advice. Thus during the first decade after the war, the nation's health administration was significantly improved; life expectancy increased, infant mortality decreased by more than half, and the tuberculosis death rate decreased by more than three-quarters.

The 1960s were years of drastic social change and technological progress accompanied by rapid growth of the gross national product (GNP), the establishment of an extensive chemical industry, and the fostering of mass media. With an aging population, chronic and degenerative diseases, such as cardiovascular disease and cancer, became increasingly prevalent. At the same time, growing environmental pollution caused grave episodes, most of which were associated with water polluted by toxic substances discharged from chemical plants.[4]

2. Social and Political Environment

Japan has an area of approximately 377,400 square kilometers (145,700 square miles) and a population of 122.3 million based on 1987 data. Japan's real gross national product of ¥364,200 billion and 3.8% real growth in 1988 is now the world's second highest after the U.S. The political and economic situations are stable, and the social standard has become higher every year since World War II.

After the 1960s, the interest was to supply fine-grained and kindly health care services and education so that people will have an affluent society in Japan. Most Japanese parents are enthusiastically educating their own children, when they are youthful, while the middle-aged or elderly are concentrating on maintaining good health.

Japanese society faces a change in population structure. The proportion of elderly people in Japan (65 years of age and over) in 1987 was about 10% but is expected to increase by the year 2020 to 23% of the total population. Thereby, Japan will be rapidly becoming one of the leading old-age societies in the world. Furthermore, it is predicted that among those 65 and over, the proportion over 80 years old is on the increase (16.4% in 1982; 19.1% in 2000; 26.8% in 2025). In addition, the image of the family is now changing in Japan. The average number of members per household was 3.25 (4.55 in 1955). From the point of view of family structure, more than 60% of the total were nuclear households. Moreover, 60% of all women have a job. A greater proportion of these families will increase the proportion of elderly couples and of elderly persons living alone, 80% of whom will be women. The rate of elderly with chronic diseases is also growing. Of all people over 70 years of age, about 5% are permanently bedridden. However, the proportion of the elderly in institutions for the aged (other than hospitals) is now about 1% in Japan, compared with 5% in Sweden. Most of the Japanese elderly want to live in their own home in the community, if possible with their son's or daughter's family and with their grandsons or granddaughters (three-generation households), and want to die in their own homes, not in hospitals or other institutions for the aged. If they get sick, they want to be cared for by family (e.g., their spouse, parents, and children) in their own home or even in a hospital. As a result, a social welfare system with public community care of the aged is still lacking in Japan. If hospital care is indicated, not only hospital nurses but also the members of the patient's family, such as wife and mother, take care of the patient in the ward.

Environmental pollution has been an extremely serious problem in Japan since the 1960s; various national and voluntary countermeasures have been developed. However, prediction of environmental pollution in the coming years is extremely difficult, and in view of the ever-developing technology, particularly in industry and transportation within this densely populated, small land area, no optimistic forecast is warranted.

3. Health and Medical Environment

Japan had the lowest infant mortality rate in the world — 5 per 1,000 live births in 1987. Life expectancy was 75.61 years for males and 81.37 years for females in 1987. It is generally thought that the Japanese standard for health and hygiene is very high.

However, there are many problems in medical care and the pharmaceutical environment in Japan. First, the principal causes of death (Table 1) and disease structure (i.e., the kind of diseases prevalent and of greatest concern to the health care in a region) have changed remarkably (as a typical pattern seen in the industrialized countries) over the past 50 years. The former medical supply system is now old-fashioned. In the last 30 years, the number of patients with infectious diseases has decreased to one-third of the former. The dominant diseases have become the noninfectious, chronic, and degenerative disorders. The number of patients with

TABLE 1. Death rates by five principal causes[7] (1950 to 1987, rate per 100,000 population).

year	cause 1	cause 2	cause 3	cause 4	cause 5
1950	Tuberculosis (all forms)	Cerebro-Vascular Disease	Pneumonia and Bronchitis	Gastritis and Enteritis	Malignat Neoplasms
	146.4	127.1	93.2	82.4	77.4
1960	Cerebro-Vascular Disease	Malignat Neoplasms	Pneumonia and Bronchitis	Heart-Disease	Senility
	160.7	100.4	93.2	73.2	58.0
1970	Cerebro-Vascular Disease	Malignat Neoplasms	Heart-disease	Accidental Death	Senility
	176.5	116.8	87.0	42.6	38.2
1980	Cerebro-Vascular Disease	Malignat Neoplasms	Heart-Disease	Pneumonia and Bronchitis	Senility*
	139.5	139.1	106.2	33.7	27.6
1987	Malignat Neoplasms	Heart-Disease	Cerebro-Vascular Disease	Pneumonia and Bronchitis	Accidental Death
	164.1	118.3	101.7	44.9	23.0

*Senility without mention of psychosis

heart disease has increased eight times, while patients with malignant neoplasms have increased fivefold. Defined causes of death have also changed (Table 1). In view of the rapid increase in the number of elderly, the trend in the change in disease structure will likely be accelerated in the coming 20 years.

Secondly, while many patients in Japan find that cosmopolitan medical therapy furnishes them with temporary relief, they also try an alternative medical system in their search for a complete cure when faced with a debilitating chronic condition.[10] The increase in the number of chronic diseases raises their concern about the side effects of frequent ingestion of synthetic medicine; for example, subacute myelo-optic neuropathy (SMON) disease caused by Quino Form or 5,7-diiodo-8-hydroxyquinoline. This concern has been actively promoted by the mass media, and after the 1970s, it has made many patients prefer traditional, nonsynthetic pharmaceuticals to synthetic ones, although only a few nonsynthetic pharmaceuticals are available.

Most Kampo drugs are made from dried medicinal herbs. When prescribing a Kampo medication, doctors choose and mix several kinds of medicinal herbs according to the patients' *sho* or Kampo-type diagnosis. This Kampo medication is usually not covered by insurance, resulting in quite high costs. In 1954, however, a pharmaceutical company succeeded in extracting essences of herbs and converting them into fine granules, and in 1975, insurance covered some herbal prescriptions and about 120 kinds of extracts from herbs. Thus, many Western physicians have been able to prescribe a new Kampo-type drug in hospitals or clinics without precise knowledge of Kampo medicine. Traditional medicine for rehabilitation of chronic diseases has entered the world of big business and become an enormous component of practice.[11] It is, therefore, important to understand that Japan has had a variety of medical experiences composed of Western medicine, Kampo medicine, and other folk and popular medicines (see Section D).

Finally, the problem is seen in the increasing trend in medical care expenditure. The national medical expenditure has increased by 40 times (or two times, if corrected according to national income) during the past 25 years. In recent years, the national medical expenditure has increased by ¥ 100 million every year. It reached ¥ 1,800 billion in 1988. This point relates to shortcomings in the method of payment. Under the present point system, the amount of payment is computed on the basis of number of items of medical service rendered, with the result that the more medicine dispensed, injections given, and examinations done, the larger the income of the hospitals and clinics. As a result, the use of drugs tended to be great and the per capita cost of medicine dispensed per day has increased year after year. In addition, the ratio of pharmaceutical costs to the total national medical care expenditure increased 30% in 1987. Moreover, the Japanese people pay an exorbitant amount for OTC drugs in pharmacies and for Kampo drugs in clinics or pharmacies. In general, pharmaceuticals are the favorite with Japanese. Estimating from surveys performed by the Central Social Insurance Medical Council and other sources, the situation of cost sharing and distribution of medical care expenditures in Japan is as shown in Figure 1.

4. Facilities (Hospitals, Clinics, etc.)

According to a medical facilities' survey by the Statistics Department of the Ministry of Health and Welfare, at the end of 1986, the country had 9,699 hospitals with a total of 1,533,887 beds, 79,369 general clinics with a total of 282,046 beds, 47,174 dental clinics, and 35,783 pharmacies (29 per 100,000 population). The composition of hospitals, clinics, and hospital beds by category of patients is shown in Table 2.

FIGURE 1. The flow of national medical care expenditure.

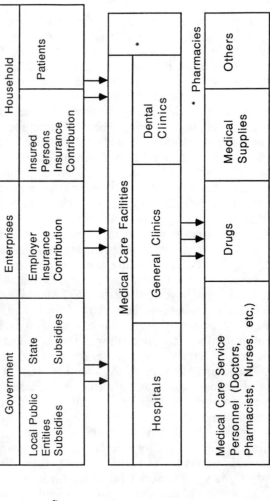

TABLE 2. Number of hospitals, clinics, and beds in hospitals (end of year).

	1965	1975	1985
Hospitals Total	7.047	8.294	9.699
Mental	725	929	1.035
Tuberculosis	340	87	22
Leprosy	14	16	16
Infectious Disease	46	27	13
General	5.922	7.235	8.613
Clinics Total	64.524	73.114	93.416
Clinics with Beds	27.332	39.104	25.740
Clinics without Beds	37.192	44.010	67.676
Dental Clinics	28.602	32.565	47.174
Beds in Hospital Total	837.652	1.164.098	1.533.887
Mental	172.950	279.123	340.506
Tuberculosis	220.757	129.055	51.367
Leprosy	13.230	14.020	10.205
Infectious Disease	24.042	21.042	14.109
General	442.536	721.858	1.117.700
Beds per 100.000 person Total	889	1.040	1.260.7
Mental	176	249	279.9
Tuberculosis	225	115	42.2
Leprosy	13	13	8.4
Infectious Disease	25	20	11.6
General	450	645	918.6

Source: "Survey of Medical Institution," Ministry of Health and Welfare.

The Medical Service Law of 1948 provides detailed licensing regulations for hospitals that have 20 or more beds. The same law does not provide such detailed requirements for the number of nurses per patient or safety standards for clinics. The latter can have up to 19 beds. Article 13 of the law does state a restriction for clinics, saying that the managers of clinics shall endeavor not to accommodate a patient for more than 48 hours, except in a case where there is an unavoidable reason in medical consultation. There are always exceptions. So there is almost no difference between clinics and hospitals (especially small hospitals) in the care they provide.

5. Health Manpower

The number and ratio of major categories of medical and other health personnel at the end of 1986 are shown in Table 3.

The educational system of Japan has five stages: kindergarten (one to four years), elementary school (six years), junior high school (three years), senior high school (three years), and college or university (six years for medical and dental schools and four years for others). In addition, there are junior colleges (two to three years), and in many universities, graduate courses are offered. After World War II, compulsory education was extended from six years to nine years, up to junior high school.

With more than 90% of junior high school graduates advancing to senior high schools, general education in Japan actually runs for 12 years. Each health professional, except an assistant nurse, is required to have finished a general education before moving on to health professional education. Those trained in vocational schools, usually attached to hospitals, do not enjoy the benefits of the School Educational Law. The educational requirements for various health professionals are summarized in Table 4.[8]

6. Social Security System

In the field of social security, the following ministries or agencies have responsibilities: the Ministry of Health and Welfare, the Ministry of Labor, and the Environmental Agency. The national government supplied ¥ 38,700 billion for Social Security pensions, medical care, labor insurance, and welfare in 1986. A 33.6% social insurance tax on the people

TABLE 3. Health personnel (the end of 1986).

Category	Total active number	Ratio per 100.000 population
Physician	191.346	157.3
Dentist	66.796	54.9
Pharmacist	135.990	111.8
Public-health nurse	22.050	18.1
Midwife	24.058	19.8
Clinical nurse	639.936	526.0
Dental hygienist	32.666	26.8
Dental technician	31.139	25.6
Acupuncturist	108.782	89.4

TABLE 4. Educational qualification requirement for health personnel.

Qualification	Professional Education		
Physician	University	pre-medical medical	(2 yrs.) (4 yrs.)
Dentist	University	pre-dental dental	(2 yrs.) (4 yrs.)
Pharmacist	University	pre-pharm pharm	(2 yrs.) (2 yrs.)
Radiology technician	Training school		(3 yrs.)
X-ray technician	Training school		(3 yrs.)
Health laboratory technician	Training school		(2 yrs.)
Clinical laboratory technician	Training school		(3 yrs.)
Physical therapist	Training school		(3 yrs.)
Occupational therapist	Training school		(3 yrs.)
Dental hygienist	Training school		(1-2 yrs.)
Dental technician	Training school		(2 yrs.)
Midwife	Training school for Registered Nurse		(1 yrs.)
Acupuncturist	Training school		(3 yrs.)
Nutritionist	University, or Junior college, or Training school		(4 yrs.) (3 yrs.) (2 yrs.)
Registered nurse	University, or Junior college, or Training school		(4 yrs.) (3 yrs.) (3 yrs.)
Assistant nurse*	Training school		(2 yrs.)

*Assistant nurse is able to enter training school after junior high school.

against the national income is 15% to 20% less than the levy by European countries. However, this rate will be increasing by over 40% against national income in the twenty-first century.

Insurance coverage in Japan is based on benefits-in-kind and fee-for-service and is implemented for the whole population. Thus, the people can receive the necessary medical care relatively easily. On the other hand, certain problems have arisen, and a means of solving them must be found.

The first health insurance program was established in 1938, and this has gradually expanded so that nearly 100% of the total population of Japan uses this coverage. The health insurance can be categorized into three patterns. First, there is the medical insurance system for employees and their dependents. The employees' health insurance covers all employed persons, embracing nearly 60% of the population, and consists of the state-administered insurance carried by the government, the association-administered health insurance carried by large enterprises, and the dairy workers' health insurance carried by the government.

Employees are 90% covered for medical and dental care, within a prescribed range which sets the rate of contribution. The white-collar workers pay about 10% of their salary per month for social insurance fees. The second component is various mutual aid associations. The insured are national and local government officials, public corporation employees, private school teachers, and clerks. Seamen's health insurance is carried by the government. The third level of protection involves a national health insurance system for the self-employed and their dependents, the unemployed, or the retired in which municipal governments act as the insurer. The other special insurance system is actually a supplementary insurance for those over age 70. The cost of medication for the poorest and/or handicapped and the cost of public health care for people, such as vaccination or checkup, is borne by the National Treasury.

All drugs eligible for insurance reimbursement are designated by the Ministry of Health and Welfare and the National Health Insurance (NHI). The rate of reimbursement is published in an advisory booklet available to doctors and hospitals. In most cases, the NHI price is by brand, although in some cases, such as oxytetracycline, where the drug is frequently prescribed, reimbursement is quoted by generic name. The price at which the doctor is reimbursed is unrelated to the actual price he or she has paid for the product.

B. PHARMACEUTICAL INDUSTRY AND DRUG DISTRIBUTION

1. Historical Aspect of Pharmaceutical Companies

Japanese pharmaceutical companies may be classified into four groups based on origins: (1) traditional wholesalers who founded pharmaceutical companies during the Meiji era (the latter half of the nineteenth century), (2) drug importers trading with European countries who came to manage industrialized pharmaceutical companies, (3) many pharmaceutical companies having factories founded during World War I, and (4) others that were established to manufacture new synthetic medicines. Most of the big companies, such as Takeda, Shinogi, Fugisawa, and Tanabe, were established by the traditional wholesalers.

The Japanese pharmaceutical industry, therefore, exhibits a commercial background where firms have been stronger in sales promotion than in academic research. However, activities in research and development are being strengthened rapidly. For example, the export of royalty technologies is increasing, as it overcame the import in 1983 to 1984, although the ratio of import of the finished products to exports is still 2:5 (see Table 5).

2. General Views of the Pharmaceutical Industry

Japan is the second largest drug market in the world following the U.S. However, high technological advancement in general, stimulated by industrial efforts to recover from the disastrous impact of the oil crisis in 1973-1974, has spread to the pharmaceutical industry. This has resulted in an enhancement in productivity in the manufacturing process and also in research and development activities.

For a long time, the Ministry of Health and Welfare has been concerned mainly with regulating pharmaceutical manufacturing, but it has recently changed to take a lead role in the promotion of research and development. For example, the Adverse Drug Reaction Injury Relief and Research Promoting Fund, which was established by the Ministry in 1979, provides research and development-oriented companies with capital in the form of investments or loans and with other promoting services.

3. Production of Drugs

The production of drugs amounted to ¥ 4,280.7 billion in 1986, an increase of 7% from 1985. These patterns continued from 1952 to today. As a result of a statistical analysis, most years reflected increased drug production, except for 1984 and 1985 (Table 6).

TABLE 5. Percentage of foreign market of pharmaceutical industry in Japan, U.S.A., and Europe.

Japan	5.7%*	(the average of 12 pharmaceutical companies)
U.S.A.	33.6%	(the average of 12 pharmaceutical companies)
Europe	78.4%	(the average of 8 pharmaceutical companies)

*The percentage of foreign market of Japan is the smallest of the three.
Source: leaflet of JAMA (Japan Pharmaceutical Industries Association), p. 13

TABLE 6. Trend of medicinal and pharmaceutical production.[8]

*** Classified by principal pharmaceutical categories

(in million yen)

	1984	1985	1986	1986	
	Amount	Amount	Amount	Increase over the Preceding Year	Composition Ratio
Medicines	4,026,985	4,001,807	4,280,732	7.0%	100.0%
Antibiotics	742,496	690,505	683,361	-1.0%	16.0%
Miscellaneous agents affecting metabolism	330,522	314,132	328,629	4.6%	7.7%
Cardiovascular agents	531,299	519,683	563,301	8.4%	13.2%
Agents affecting central nervous system	396,492	383,855	430,516	12.2%	10.1%
Agents affecting digestive organs	344,273	353,799	372,044	5.2%	8.7%
Vitamin preparations	245,514	238,541	234,712	2.2%	5.7%
Agents for epidermis	245,530	239,679	261,030	8.9%	6.1%
Others	1,190,895	1,261,613	1,398,139	10.8%	32.5%
Quasi-medicines	364,680	370,441	389,944	5.3%	-
Medical Devices	932,753	968,187	979,817	1.2%	-

*** Product Amount (At the end of 1986)

(in million yen)

	Production Amount		Increase over the Preceding Year		Composition Ratio	
	1986	1985	Increase	Ratio	1986	1985
Total (Medicinal Products)	4,280,732	4,001,807	278,925	7.0%	100.0%	100.0%
Medical Services	3,649,842	3,383,710	266,132	7.9%	85.3%	84.6%
(Domestic Production)	2,547,249	2,386,022	161,227	6.8%	59.5%	59.6%
(Imports)	1,102,593	997,688	104,905	10.5%	25.8%	24.9%
Other	630,890	618,096	12,794	2.1%	14.7%	15.4%
(General Uses)	582,685	572,802	9,883	1.7%	13.6%	14.3%
Household Medicines	48,205	45,294	2,911	6.4%	1.1%	1.1%

*Source: "Pharmaceutical Industry Production Trend Statistics," Pharmaceutical Affairs Bureau, MHW.

4. General Situation for the Export
and Import of Pharmaceuticals

The total value of pharmaceutical exports during 1986 totaled ¥ 123.3 billion, a decrease of ¥ 8.7 billion, or 6.6% from the previous year, which registered a total of ¥ 132 billion. The percentage of exports to total pharmaceutical output (¥ 4,280.7 billion) was as low as 2.9%. When classified by destination country, exports to the United States accounted for the biggest portion, at ¥ 33.5 billion (27.2%), followed by the Federal Republic of Germany, Italy, Taiwan, and France. Exports to these five countries accounted for 52% of total exports. When classified by kind of pharmaceutical, the export of antibiotics accounted for the largest portion, at ¥ 36.3 billion, or 29.5% of the total, followed by vitamins and nutritive and restorative alternatives.

Regarding pharmaceutical imports, on the other hand, the total for 1986 was ¥ 310 billion, a decline of ¥ 21.1 billion, or 6.4% from the previous year, which registered a total of ¥ 331.1 billion. When classified by supplier countries, imports from the United States represented the largest proportion, at ¥ 98.3 billion (31.7%), followed by the Federal Republic of Germany, Switzerland, England, and France. Imports from these countries accounted for 73.5% of total imports. When classified by kind of pharmaceutical, antibiotic imports accounted for the largest portion, at ¥ 82.6 billion (26.6%), followed by import of biological preparations and hormones.

Economic internationalization and expansion of foreign trade in pharmaceuticals should be welcomed and promoted, as far as they serve to improve the health of people around the world. If there are still many who die (or suffer) due to lack of medicine, then we are faced with an important task in which human beings should successfully deal with the problem of health inequality. There is, however, always the danger that trade in pharmaceuticals will turn into trade in poisonous substances; therefore, constant attention should be paid so that import-export trade in pharmaceuticals will not turn into import and export of hazardous pharmaceuticals.

The Japanese people concerned with pharmaceuticals and/or engaged in medical service are requested not only to exercise surveillance over the effectiveness and safety of pharmaceuticals imported into Japan for application to Japanese patients, but also to take extreme care in the exportation of pharmaceuticals to Southeast Asia and Africa to avoid exporting hazardous pharmaceuticals.[17]

5. Research and Development of Pharmaceuticals

Development of new effective elements to supply pharmaceuticals that may meet the requirements of present-day medical treatment, including improvement of existing pharmaceuticals, are tasks essential for the survival of pharmaceutical manufacturers. Some people say that the development of new medicines in Japan has often been conducted with dependence upon licenses introduced from or bulk medicines imported from Western countries to move ahead of domestic competition. That position probably was excusable during the period of confusion immediately after the war and the following period that required expeditious rehabilitation. But Japan, which has now come to occupy an important position as a major economic world power, is required to make an adequate contribution as an advanced country in the field of new medicine development as well. Efforts have been made in Japan, both on corporate and industrial levels, for the promotion of research and development of pharmaceuticals. Japanese technological know-how of pharmaceutical manufacturing and improvement is said to have reached a world-class level. It is currently being pointed out, however, that future development of original medicines will require greater fundamental research than ever before. As such, there is an increasing tendency to secure laboratory facilities and personnel for fundamental research; consideration is also being given to better resource allocation.

The Health and Welfare Ministry has been taking various measures to support development of new medicines based on new technologies. For the purpose of examining new biotechnologies, the Central Pharmaceutical Affairs Council has set up its Biotechnology Special Committee which is expected to aid in the development of new anticancer drugs, antisenility dementia drugs, and anti-AIDS drugs.

Clinical tests on patients are indispensable from the viewpoint of securing safety and efficacy in the development of new pharmaceuticals. Proper clinical tests seem necessary not only to examine pharmaceuticals for official approval but also to offer new safe and efficacious pharmaceuticals for practical use as promptly as possible. It was decided, therefore, to set up within the said section in October 1988 a special pharmaceutical examination system to provide administrative guidance for the appropriate planning of clinical tests of pharmaceuticals made from new ingredients that are expected to yield gains in efficacy. For the purpose of promoting the development of medicines against intractable or rare diseases which private enterprises find difficult to develop due to the problem of profitability, the New Medicine Development Promotion conference was set up in 1979, and efforts have been being made for these purposes by supporting researchers with research expenses and other necessities. As a result, for example, the Adverse Drug Reaction Injury Relief and Research Promoting Fund was organized by the Ministry of Health and Welfare (see Section B. 2).

6. Pharmaceutical Approval System

Under the Pharmaceutical Affairs Law (hereinafter referred to as the Law) necessary regulation is being formed concerning pharmaceuticals, quasi-drugs, cosmetics, and medical devices (hereinafter referred to as pharmaceuticals, etc.) to ensure their quality, effectiveness, and safety. Manufacture (importation) of pharmaceuticals, etc., requires approval and/or permit as provided in the Law. The approval is given to pharmaceuticals, etc. which may be manufactured or imported to be sold to the public for medical treatment or preservation of health, while the permit is given to factories (sales offices) with qualified personnel and appropriate materials, equipment, etc., for commercial manufacture (importation) of pharmaceuticals, etc.

Manufacture (importation) of the following items requires both the approval and the permit of the Welfare Minister as stated above, but other items require only the permit as stated above:

1. Pharmaceuticals (contained in Japanese Pharmacopoeia, except for those designated as items requiring no approval)
2. Quasi-drugs
3. Cosmetics containing hormone(s)
4. Medical devices other than those conforming to Japan Industrial Standards (JIS).

Approval of pharmaceuticals, etc., is given by the Welfare Minister through examination of each item seeking approval for manufacturing or importation, including name, components, quantity, use, dosage, efficacy, effect, performance, adverse reactions, etc., in the context of the current scientific level of the Welfare Minister, but approval will not be given in any of the following cases:

1. If the efficacy, effect, or performance as stated in the application is not obtained
2. If there is no value in the item's use because it is expected that unfavorable reaction will exceed favorable efficacy, effect, or performance

3. If the properties or quality of the item is not appropriate for the preservation of health.

Thus, careful examination is essential to ensure the safety of new pharmaceuticals. The application is submitted for deliberation to the Central Pharmaceutical Affairs Council, which has several special committees and which conducts evaluation and review of the effectiveness, safety, etc., of the new pharmaceuticals. Furthermore, to make application data more reliable, standards to be observed when safety tests are conducted on animals (commonly called good laboratory practices, or GLP) have been in force since April 1, 1983, and facility inspections are also conducted. With regard to clinical tests to obtain reliable data and to pay due ethical consideration to the rights of persons who undergo clinical tests, standards for clinical tests of pharmaceuticals (commonly called good clinical practices, or GCP) were drafted and published in December 1985.

Pharmaceuticals containing newly-developed active ingredients must be reexamined six years after approval, and pharmaceuticals made of established active ingredients but administered through a new administration route (for example) must be reexamined after four years.

On the other hand, a permit to a manufacturer (importer) of pharmaceuticals, etc., will be given when the manufacturer's factory (importer's sales office) conforms with standards prescribed by the Welfare Ministry, and when said manufacturer (importer) is not subject to any disqualifications as provided in the Law. Manufacturers of pharmaceuticals are especially required to observe regulations for the manufacturing control and quality control of pharmaceuticals (commonly called good manufacturing practices) to assure the quality of the pharmaceuticals they manufacture.

7. Administrative Regulation of Pharmaceutical Affairs

The health of the public will be greatly damaged if inferior or inappropriate pharmaceuticals, quasi-drugs, cosmetics, or medical devices are manufactured or imported and sold. Therefore, state and prefectural governments have pharmaceutical inspectors stationed throughout the country to find inferior pharmaceuticals, etc., to conduct the supervision, regulation, and guidance for the observance of the Pharmaceutical Affairs Law and related regulations, and to prevent damage arising from inferior pharmaceuticals, etc. The inspectors enter pharmaceutical factories and stores and conduct inspections to see that pharmaceuticals, etc., are manufactured and sold under the conditions as provided in the Pharmaceuticals Affairs Law. As of April 1, 1987, the total number of pharmaceutical inspectors stationed in Japan amounted to 2,751 (including 31 inspectors appointed the state government).

8. Advertising of Pharmaceuticals

Advertising of pharmaceuticals could exert a seriously harmful influence upon the health of the people, if it was false, extravagant, or otherwise inappropriate. It is, therefore, regulated by the Pharmaceutical Affairs Law (Articles 66 to 68) and the standards for adequate advertisement of pharmaceutical products in Japan. Advertisement against the regulation is handled by the state or perfectural inspectors. Each industrial association monitors the advertising of pharmaceuticals in accordance with its own self-control guidelines. For example, the Japan Popularly-Used Medicines Industrial Association examines its own advertising of pharmaceuticals by conducting periodic review of advertisements broadcast through the media of television, radio, newspapers, and magazines.

Drug information services furnished by D-men (detail men) are one of the interesting promotional systems of a Japanese pharmaceutical company. D-men number more than 40,000. Further, salesmen for drug wholesalers number more than 35,000. The proportion of sales promoters to medical doctors, who number approximately 170,000, thus seems unnaturally high.

According to a survey, factors that motivated Japanese hospital doctors to use some new drugs were as follows: information from D-men, 42.1%; medical journals, 21.1%; lecture meetings, 21.1%; and direct mailings, 15.8%. However, in the case of British general practitioners, these were: medical journals, 31%; recommendation by medical specialist, 25%; advice by peer doctors, 13%; information from D-men, 12%; and lecture meetings, 7%. In Japan, therefore, information from D-men constitutes a major source of pharmaceutical information provided to medical doctors.

C. PHARMACY PRACTICE

1. Pharmacist

At the end of 1986, the total number of pharmacists stood at 135,990; this means there are 111.8 pharmacists for every 100,000 people, or 895 people per pharmacist. In recent years, more female pharmacists have come into the profession, and at the end of 1986, they accounted for 56.5% of the total practitioners (see Table 7). In terms of services, 78,578 pharmacists were engaged in pharmacies and medical clinics, while

36,132 pharmacists were engaged in nonmedical settings, of which 14,147 worked in a pharmaceutical company (see Figure 2).

2. New Developments in Pharmacists' Profession in Japan

Since pharmaceutical technology has advanced markedly in developing new dosage forms, dispensing by pharmacists has changed considerably. Counting tablets or capsules recently became the major task in dispensing, but it is too simple for pharmacists educated in colleges of pharmacy or pharmaceutical sciences. Such handwork is now being replaced by automatic dispensing machines. So the pharmacist's role has been extended to newly developed scientific areas, which are as follows:

1. Inspecting prescriptions from community clinics
   a. About 300 errors in 30,000 prescriptions from community clinics
   b. About 1%-2% errors in 60,000 prescriptions from a university hospital
   Important errors are found in drug names, dosage, duplication of same or similar drugs, and drug interaction. Of course these errors are comparatively few, but they are often very dangerous for patients.
2. Finding and correcting dispensing errors
   a. About 2% error in an average of 950 prescriptions (1976)
   b. About 0.65% error in 160,000 prescriptions (1986)
   These errors were found and corrected by dispensing inspection.
3. Direction and explanation for drug use
   Drug therapy will be ineffective and even dangerous if patients do not administer drug preparations as directed. Pharmacists should explain drug administration to patients in an understandable way.
4. Collecting and supply drug information
   Now drug information centers have been established in 37 out of 47 prefectures.
5. Drug history of each patient
   Recently, recording of drug history for each patient has been encouraged by medical care insurance. Ideally, all prescribed drugs, including OTC drugs used by the patient, and patient experiences of hypersensitivity to special drugs should be recorded. Drug interactions and duplicated administration of the same drug or similar drugs thus may be avoided.
6. Participation in drug regimen by monitoring blood concentration of the administered drugs

7. Participation in clinical drug evaluation before approval for manufacturing drugs
8. Quantity and quality control of drug preparations stored in the pharmacy and minimization of dead stock by computer control
9. Participation in monitoring side effects of drugs for medical care and OTC
10. Patient consultation on OTC drugs

    Pharmacists should counsel patients on efficacy and safety of OTC drugs and on appropriateness of the drug for symptoms after some information is collected from the patient. When pharmacists find that the patient should consult a doctor, they advise the patient to visit a doctor at once. By this process, OTC drug selling is connected with medical care. Of course the drug history is prepared in the pharmacy and sent to a doctor.
11. Participation in community medical care and health programs concerned with individual health

    Pharmacists consult on public or environmental health and student health in schools.

3. Pharmacies

There were 35,783 retail drugstores in 1986, 18,935 general pharmacies (pharmacist controlled), 21,917 specially licensed dealers (limited drugs only), and 16,246 door-to-door peddlers, all registered to sell drugs. A pharmacy is defined as an establishment where drugs are prepared and dispensed. The floor area must exceed 19.8 square meters and include a dispensary of not less than 6.6 square meters. The drugstore must have a floor area of not less than 13.2 square meters. It may sell all types of drugs except designated products. A specially licensed dealer can be located in areas where there is an insufficient number of pharmacies. The prefectural governor designates such dealers and licenses them to sell drugs. A home deposit drug dealer is one who deposits drugs with consumers and at a later date collects revenues for what has been used.

4. Pharmaceutical Education

As stipulated in the Pharmacists Law, the graduates of a pharmaceutical faculty of a university or a pharmaceutical college must pass the national examination on pharmaceutical theory and practice to obtain a pharmacist's license.

The following postgraduate courses are available to these graduates: Master's course in pharmacy, two years (40 pharmaceutical faculties of

TABLE 7. Number of pharmacists in 1986.

Pharmacists(total)	135,990
1) Engaged in pharmacy and in medical clinic (total)	78,548
Owners of pharmacy (employers)	17,379
Employees of community pharmacy	26,370
Employees of hospital pharmacy	33,947
Employees as laboratory technician in medical hospital and clinic	852
2) Engaged in non-medical facilities	36,132
Pharmaceutical education and institution in pharmaceutical school	3,082
Administrative of health care	5,007
Pharmacists in pharmaceutical industry	14,147
Sellers of pharmaceuticals and chemicals	12,646
Pharmacists in toxicological industry	240
Pharmacists in chemical industry	1,010
3) Engaged in others	21,310
4) Non profession	17,579

Source: "Survey of physicians, dentists and pharmacists," Ministry of Health and Welfare.

FIGURE 2

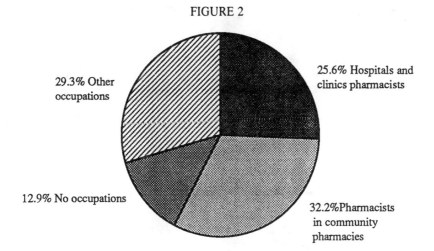

29.3% Other occupations

25.6% Hospitals and clinics pharmacists

12.9% No occupations

32.2%Pharmacists in community pharmacies

university or pharmaceutical colleges); Master's course in pharmacy, three years (31 pharmaceutical faculties of university or pharmaceutical colleges).

There are 46 pharmaceutical faculties of universities or pharmaceutical colleges consisting of 14 national, 3 public, and 29 private institutions. The total student enrollment of these faculties of universities or pharmaceutical colleges is about 33,000, and every year about 8,000 graduates are turned out. The ratio of male students to female students is about 3:7.

A pharmaceutical faculty of a university or a pharmaceutical college usually has, in addition to the department of pharmacy, one or two of the following departments: pharmaceutical manufacturing, hygienic pharmacy, and biological pharmacy.

Employment opportunities for graduates of pharmaceutical faculties of universities, and pharmaceutical colleges, classified by the type of service are shown in Figure 3.

5. Continuing Education for Pharmacists

The Japanese Ministry of Health and Welfare started a Study Committee for Pharmacists Training in 1987 for internal discussion of continuing education of pharmacists. In the fall of 1988, the committee published an intermediate report suggesting a need for a system under which the continuing education of pharmacists can be promoted. The Japan Pharmaceutical Association is endeavoring to establish the Foundation for Promotion

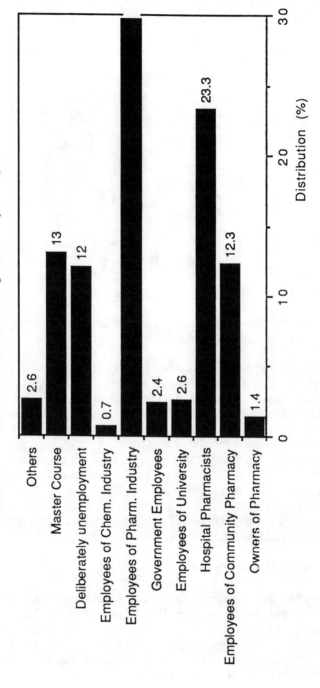

FIGURE 3. Professional distribution of new graduates of pharmacy.

of Continuing Education for Pharmacists to implement the spirit of the report.

Pharmacists' activities are not limited to the field directly related to medical care, such as pharmacies, hospitals, and clinics; they also encompass such diversified areas as the pharmaceutical industry, distribution of medicines, and public health; education and training have been given in these areas. Currently, however, the need for reinforcing scientific and technical expertise in administrating medicines to patients has surfaced as a common problem for all.

The Japan Pharmaceutical Association hopes that by integrating continuing education of pharmacists in these areas under the programming by the Foundation with a particular emphasis on the above issue, to achieve more effective results.

6. Pharmaceutical Research

As a result of radical social changes, an adjustment in health research efforts, as well as in education and training, is urgently required. In April 1971, the Science and Technology Council of the National Government stressed environmental science, soft science, and life science as new fields to be developed. In relation to health, the need for an interdisciplinary comprehensive health science has been recognized increasingly in recent years. Accordingly, increased research efforts should be directed to applied soft science. Laboratory research, which gave brilliant results in infectious and other acute diseases, offers small hope of gain against the prevalent chronic, noninfectious diseases, which call for well-designed, large-scale, long-range, interdisciplinary approaches based on people's daily lives. The interdisciplinary approach has already been developed substantially in response to serious environmental pollution episodes.

Prior to the establishment of the Comprehensive 10-Year Strategy for Cancer Control (see Table 8), as a result of talks held between the Minister for Health and Welfare in Japan and the U.S. Secretary of Health and Human Services in May 1983, it was agreed to promote joint research activities in cancer between the two countries. It is proposed that these activities be centered in the National Cancer Center of Japan and the Nation Cancer Institute of the United States and that there be progressive interchange among researchers.

Biotechnology represents a wide sphere of application and can contribute in a very significant way to health and medical care. Progress is particularly apparent in research concerning pharmaceutical products, some of which are now reaching the application stage.

Recent striking advances in medical care technology have brought med-

TABLE 8. Main research topics of the "Comprehensive 10-Year Strategy for Cancer Control."

Topic	Contents
1. Research related to Human Oncogenes	In addition to investigating the functions of human oncogenes, pursuing research into 20 forms of oncogenes including those for lung cancer and cancer of the urinary bladder which have to some extent been clarified.
2. Research into Human Viral Carcinogenesis	It is thought that in some cases of leukemia, Liver cancer, etc. a certain type of virus enters the cells, thereby giving rise to cancer. It is intended to investigate the process of carcinogenesis brought about by infection.
3. Research on Tumor Promotion and Control	Looking into carcinogenic acceleration materials (salt, hormones, etc.), determining their operating functions and researching into methods of control.
4. Research Concerning the Development of New Technologies for Early Diagnosis	The development of early diagnosis technology through the application of laser and other high technology and the analysis of protein produced by oncogenes.
5. Research Concerning The Development of Treatment Methods in accordance with New Theories	Research into the cancer cell membrane structure and functions and, by controlling the operation whereby protein is produced by oncogenes, developing an effective new treatment with few side-effects.
6. Research into Mechanisms for Immunomodulation and Immunomodulation	Research into not cancer per se, but into the defense responses of the body which harbors cancer, and countering cancer by using the body's power of immunity through strengthening the defensive mechanisms.

328

ical care face-to-face with new situations. Even for sicknesses from which it has not been possible to recover, new horizons for survival have been opened by the extensive use of this technology. On the other hand, this evolution in medical care has induced conflicting problems concerning the thoughts and values about life and death that are deeply rooted in society and attitudes toward life. The effect of this process is not merely restricted to our health problems, but extends to social norms and how human beings should live.

D. UNIQUE OR INTERESTING FEATURES OR SPECIAL SITUATIONS IN JAPAN

In Japan, as in other societies with complex economies, there is a wide range of medical and pharmaceutical options available for treatment of physical discomfort and emotional distress. In looking at Japan, one can identify three overlapping sectors of health care, which have been called by Kleinman the popular sector, the folk sector, and the professional sector.[11] This brief section focuses mainly on some interesting features of the popular and professional sectors of medical pluralism in Japan, which have yet to be studied in detail by medical or pharmaceutical sociologists.

1. Popular Sector

This sector includes the lay, nonprofessional, nonspecialist domain of society where illness is first recognized and health care activities are initiated. It embraces all medical and pharmaceutical options that people utilize without consulting medical practitioners. Among these options are self-treatment or self-medication, and advice or treatment given by a relative, friend, neighbor, or workmate. In the popular sector, the main arena of health care is the family.[12] As in both Western and non-Western societies, most health care in Japan takes place within this sector.

One of the most important components of this sector is self-medication. Preparations in this area are taken most commonly for fever, headache, stomachache, indigestion, sore throats, etc. It is the usual practice for people to use self-prescribed patent medicines when they have drunk too much, caught cold, or gotten tired at work. To supply the great demand, a wide variety of drugs are sold at *yakkyoku*, or a pharmacy, all over the country. When people buy a drug, they often choose the well-known brand name promoted by mass media advertising of drugs directed at the public. People usually name a patent medicine by its effect, e.g., they take "stomach medicine" against stomachache, "headache medicine" against headache, "cold medicine" against a common cold, and "medicine to remove itch" against itch. Many workaholics take vitamin tablets against fatigue instead of resting.

The ordinary user, however, has few concerns about the pharmaceutical ingredients. Individuals are concerned about whether the drug is effective against the illness or the symptom from which they are suffering. They take the same attitude toward the medicine prescribed by doctors in hospitals or clinics. Most of the patients never ask their doctors about the name of the medicine, and most doctors never tell their patients. Individuals take the doctor-prescribed medicine upon trust, in part, because the dominant doctor-patient relationship in Japan is paternalistic.

In recent years, several pharmaceutical or other companies have supplied many kinds of drug-like articles for people, especially for those who desire to get health easily. Among them are bottled or canned drinks containing small amounts of nutrients such as fructose, glucose, amino acids, and dietary fibers. These items are sold at a station stall, at a drugstore, and at a vending machine. Although the nutritive value of a bottle of the drug-like article is less than that of an egg or a glass of milk, the former costs several times as much as the latter, if one consumes the equivalent amount of nutrition. Many people, nevertheless, prefer these drinks to natural foods, mainly because they can feel refreshed after drinking them. Thus they simply come to believe in the efficacy of drug-like articles, as is the case with medicine.

Today few people practice home treatment by using medicinal herbs. Most of the Japanese people have forgotten the virtue of the herbs and how to distinguish them from other nonmedicinal herbs, and it seems to many people that herb treatment is old-fashioned. They prefer more sophisticated forms, such as tablets, capsules, and ampules of Western medicine.

Those who have a stiff neck or lumbago often apply a compress to the affected part. For that purpose, many kinds of cold or hot compresses are sold at every pharmacy. Recently, magnetic compresses have become popular among those who believe that the magnetic force is every effective.

2. Folk Sector

In Japan, as in Western countries, the folk sector is small. The healers in this sector are not part of the official medical system. There is a variation in their types, from technical experts like bonesetters, midwives, or herbalists to spiritual healers and shamans. The field also includes Yoga groups and health food movement groups.

But when people become ill and the illness is not or seems not to be cured in the popular sector, most believe that it is best to consult Western-trained physicians. As Namihira pointed out, 88.2% of the Japanese people who sought treatment in 1985 visited Western-style doctors practicing in hospitals or in clinics.[13] There were few who had chosen and consulted alternative healers in the folk sector. This is mainly because virtually all Japanese are covered by compulsory health insurance that is valid only for Western medical practices and several traditional medical practices, like acupuncture and massage. Therefore, as long as patients consult Western-style doctors, they only have to pay within the limit of ¥ 50,000 or so a month.

To practice acupuncture, massage, or bonesetting, which is called quasi-medical practice officially, practitioners have to be licensed. Western-style medical doctors are allowed to practice it, although neither acupuncture nor massage is included in the curricula of medical schools in Japan, and medical students are hardly interested in or study non-Western medicines of the folk sector. Therefore, healers in this sector are trying not to break a law. Yoga is not regarded as a medical practice but as an exercise for the health of body and mind. The practice of laying-on of hands in spiritual healing is not considered to be a medical practice, but a religious one.

A few shamanistic medicine women are working in Japan (e.g., *itakos* in Tohoku and *yutas* in Okinawa). According to Namihira, a new type of *yutas* appeared in recent years, and their practice consists chiefly of treating ill people, massaging them, taking history and explaining their illness in Western medical terms, and advising them on lifestyle, without using possession, which is their traditional way.[14] An investigator concluded that the traditional magical or shamanistic curing system is changing under the influence of modern medicine.

3. Professional Sector

This sector consists of the organized, legally-sanctioned healing professions, such as modern Western scientific medicine and, to some extent, some traditional medical systems like acupuncture and bonesetting. In this country, scientific medicine is the basis of the professional sector. It includes the wide range of medical and paramedical professionals found in Western societies.

Ikegami pointed out the basic characteristics of the Japanese health care system.[15] Firstly, it is market oriented. Nearly all hospitals maintain a large outpatient department from which they obtain virtually most of their inpatients. The bulk of health care is provided by the private sector, which constitutes 80% of the hospitals and 93% of the clinics. Hospitals are defined in Japan as medical facilities having more than 20 beds, while clinics have no beds or 19 at the most. Second, it is hierarchically structured. A great deal of the power and the prestige tends to be concentrated in the professors of the premier medical schools, and their decisions are important in the diffusion of technology. The most prestigious hospitals, such as university hospitals, can more easily afford the high-technology equipment, like a CT scanner, due to the subsidies they receive. Most hospitals in Japan recruit their doctors through the informal patronage and recommendations of the professors of the medical schools. Some of the doctors are part-timers in the hospitals, while they belong to their *ikyoku*, which is situated in the medical school they graduated from, where the professor rules his "family" of doctors. Third, hospitals have become primary institutions for the indigent and sick lacking social support. While this is quite obvious in psychiatric and geriatric care, it is also apparent in hospital care in general. Japan has one of the highest number of beds per population (1,461 beds per 100,000 population) and one of the longest average lengths of hospital stay (55 days for all hospitals). Thus for most elderly persons who are sick and not ambulant because of senility, the hospitals in Japan also take on the role of asylum, like an old people's home, and treat the old people with lots of drugs, like treating other inpatients regularly.

In Japan, physicians can practice Kampo medicine only after they graduate from medical school and obtain a license to practice Western medicine, as noted above in the case of acupuncture. Today, as Ohnuki pointed out, there are three kinds of licensed doctors in Japan: those who practice only biomedicine or Western medicine, those who practice Kampo medicine exclusively, and those who use both.[16] The great majority belong to the first group. Although precise statistics are unavailable, she estimated that among the 100,000 doctors in Japan, only 100 to 150 practice Kampo medicine exclusively. Recently, the doctors who use both Western medicine and Kampo medicine have become more numerous. It is very difficult for only biomedically trained doctors to master prescription of herbal medications in the traditional way of Kampo medicine. Nowadays, more than 20 pharmaceutical companies in Japan sell 163 kinds of new Kampo-type drugs (or mixed herbal prescription) so that Western doctors can easily prescribe medication using only their knowledge of Western medicine.[18] Strictly speaking, drugs prescribed in this way are not exactly Kampo drugs. In Kampo medicine, physicians have to prescribe medication not only according to signs and symptoms of the patient, but also to the *sho*, or the Kampo diagnosis about the condition of the patient based on the Kampo medical theory or its cosmology. Even if, for example, there are two patients suffering from the same disease, such as hypertension in Western medical terminology, Kampo medical doctors may diagnose them as two different *shos* and then prescribe two or more different Kampo medications due to the difference between the *shos*.

The advent of new Kampo-type drugs in Japan, therefore, does not mean that most of the Western medical doctors who prescribe them can realize or have adopted the cosmology of Kampo medicine. They believe that Kampo herbal drugs are harmless because both effect and side effect of Kampo herbal drugs are milder than those of Western synthesized drugs (although this lay concept is not always true), and that new Kampo-type drugs are scientifically sophisticated, with the effect of traditional Kampo herbal drugs preserved. Thus they use new Kampo-type drugs mainly against common or chronic diseases such as colds, asthma, bronchitis, vertigo, hepatitis, cancer, etc. Most ill people tend to be satisfied with prescribed Kampo-type drugs because they believe that Kampo drugs are effective and harmless.

REFERENCES

1. Lock MM. East Asian Medicine in Urban Japan. Berkeley: University of California Press, 1980: p. 11.

2. Long SO. Health Care Providers, Technology, Policy, and Professional Dominance. In: Nobeck E, Lock MM, eds. Health, Illness and Medical Care in Japan. Honolulu: University of Hawaii Press, 1987: p. 69.

3. Ohnuki-Tierney E. Illness and Culture in Contemporary Japan. Cambridge: Cambridge University Press, 1984: p. 221.

4. Hashimoto M. Japan. In: Douglas-Wilson I, McLachlan, eds. Health Service Prospects, an International Survey. London: The Lancet Ltd. and the Nuffield Provincial Hospital Trust, 1973: p. 219.

5. Lock MM. op. cit.: p. 2, p. 14.

6. Dunn FL. Traditional Asian Medicine and Cosmopolitan Medicine as Adaptive Systems. In: Leslie C., ed. Asian Medical Systems: A Comparative Study. Berkeley: University of California Press, 1976: p. 135.

7. Ministry of Health and Welfare. *Kokumin Eisei no Dokoh* (Japan's National Health Yearbook). Tokyo: Health and Welfare Statistics Association, 1988.

8. Hashimoto S. Personal communication, 1977.

9. Japan Pharmaceutical Association. Pharmacists in Japan. Tokyo: 1988.

10. Lock MM. op. cit.: p. 253.

11. Kleinman A. Patients and Healers in the Contest of Culture. Berkeley: University of California Press, 1980: pp. 49-70.

12. Helman C. Culture Health and Illness. Bristol: Wright, 1984: p. 43.

13. Namihira E. *Byoin-iryo no Airo* (The Bottleneck of Hospital Care). Life Science Vol. 15, 7, 0. 14, 1988.

14. Namihira E. *Dentohteki Chiryoh Kodoh to Kindai Igaku no Setten* (Relationship Between Traditional Curing and Modern Medicine). The Japanese Journal of Health Behavioral Science. 1987;2:150-163.

15. Ikegami N. Health Technology Development in Japan. International Journal of Technology Assessment in Health Care. 1988;4:239-254.

16. Ohnuki-Tierney E. op. cit.: pp. 107-108.

17. Katahira, K. A Lecture of Social Pharmaceutical Conference in 1987 in Tokyo, Japanese Association of Social Pharmacy.

18. Japan Pharmaceutical Information Center. *Nihon Iyakuhin-Shu 1988* (Yearbook: Drugs in Japan, Ethical Drugs in 1988). Tokyo: Yakugyo-no-Hihon-Sha, 1988.

SELECTED READINGS

Tatsuno M. History of Social Pharmacy. In: Noguchi M, ed. *Shakai Yakugaku Nyumon* (An Introduction to Social Pharmacy). Tokyo: Nanzando Press, 1987: pp. 1-26.

Ministry of Health and Welfare. Annual Report on Health and Welfare for 1983. Tokyo: Japan International Corporation of Welfare Services, 1984.

Hashimoto M. Japan. In: Douglas-Wilson I, McLachlan, eds. Health Service Prospects, an International Survey. London: The Lancet Ltd. and the Nuffield Provincial Hospital Trust, 1973: pp. 211-30.

Japan Pharmaceutical Association. The Foundation of Continuing Education for Pharmacists. Tokyo: 1989.

Chapter 13

Mexico

Fela Viso Gurovich
Carmen Giral Barnes

A. THE NATIONAL HEALTH CARE SYSTEM

1. History and Background

The first public health regulatory body in Mexico was born in 1527 and called Protomedicato. Its objective was to regulate all aspects of public health and hygiene. After the Independence War in 1831, a law abolished the Protomedicato, and a board called the Medical Board of the Federal District was created. The first Mexican sanitary code was published in 1891 and the second in 1904. It was not until 1917, when the new Mexican Constitution was established, that health matters in Mexico were considered properly. The Health Council was autonomous and dependent, directly, from the President of Mexico. By presidential decree in 1943, the Health Department (Secretaria de Salud y Asistencia, or SSA) was created, with two departments: Public Assistance and Public Health.

General Health Law

The General Health Law was published in 1984. It is the legal document that governs all public health matters in Mexico. This law established the constitutional right of all individuals to health (Article 4 of the Political Constitution of Mexico), and established the following sanitary authorities: the President of Mexico, the General Health Council, the Public Health and Assistance Secretary (SSA), and the state governments, including the Federal District. The General Health Law was modified in 1987.

The National Health System is composed of federal and local public agencies and organizations and individuals and organizations from public and private sectors, which render health services. Its objectives are, among others, to provide health services to all the Mexican population, with special emphasis on preventive actions; to contribute to the harmonic demographic development of the country; to develop social assistance services for young children, aged and crippled people; to aid in the development of the family and the community; and to organize a rational administration and development of manpower to improve health. The Health Secretary (SSA) or Ministry of Health is responsible for coordinating the National Health System (see Figure 1).

2. Social and Political Environment

Mexico in the 1980s is different from the country that is still remembered by our parents or grandparents. In the last part of the twentieth century, Mexico has been a country characterized by a complex social, political, economic, and cultural dynamism. The changes have occurred very rapidly, and the challenges produced are more difficult considering the situation and the bonds that exist with the economies of other countries.

The social dynamic has not been caused only by the celerity of the changes, but also in response to a number of situations, cultural patterns, lifestyles, and economic capacities. It is the result of: history and circumstances; independent life and interaction with the rest of the world; modernity and tradition; prosperity and inequality; justice and equity; and security and uncertainty. All these factors provoked, at the beginning of the 1980s, one of the most significant crises during the twentieth century.

The crisis is evidenced by a high external debt, an unbalanced productive and distributive system, insufficient internal savings, shortage of foreign currency, and unequal distribution of the benefits of development. In this economic crisis, a National Plan of Development (1983-1988) was created. Among its objectives was to start the qualitative changes that the country needs to become an equal society. Therefore, the structural change is oriented to social development, with special emphasis on food problems, popular supply, education, housing, health, social security, and ecology. Among the strategies of renovation and consolidation of the National Health System that were developed from 1983 to 1988, were sectorization, decentralization, administrative modernization, community participation, operative efficiency, and fortification of services.

FIGURE 1. Health Secretary Organizational Chart (1989).

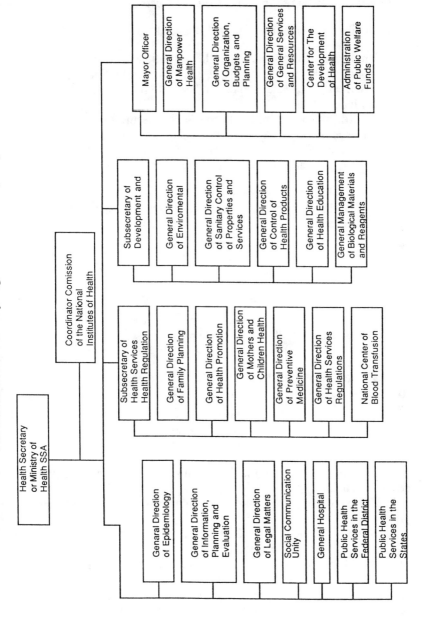

3. Public and Private Financing, Health Costs

In Mexico, there are three internal financial resources:

1. Unipartite — Corresponding to the institutions that cover open population (all the population that it is not covered by Social Security or has access to private medicine). The state finances this service with the federal budget.
2. Bipartite — Corresponding to the Social Security of state workers, financed by the workers and the federal government
3. Tripartite — Corresponding to the Social Security of all the workers. The incomes are paid by the employer, the employee, and the state.

The external financial resources are formed by the international organizations of credit and by private institutions or international official institutions. These resources are usually used for new buildings and to buy equipment for the hospitals and clinics.

From the total public expenditure, 8% was used for health in 1986, compared with 10.1% in 1978.

Cost in Health and Social Security

1970	15,530 million pesos
1980	162,490 million pesos
1986	1.9 billion pesos

Percentage of Health Expenditure, Considering GNP

1978	2.23
1980	2.18
1984	1.80
1986	1.60

The National Formulary allows the health system to strike in the type and cost of medicines.

In Mexico two important markets of medicines exist, the public and the private.

1. In the public market, 95% of the medicines belong to the National Formulary and are obtained through consolidated contests, which represent 50% of the units of the whole market and 28% of the value of this whole market. These contests are held once a year, but the next one will be organized in such a way that they will be every 3.5 years. The price for the consolidated purchase is through an offer of the supplier around a limit value given by the government. The assignment of the suppliers depends not only on the price, but also on: the specific model in which the quality

of the product is evaluated (at present and in a retrospective situation) by the national program of sample analysis; the degree of integration of the process of manufacture of the pharmochemicals (if the suppliers are producers as well); and whether the industry is national or transnational.

The medicines of the health system have the same label on which the generic is printed in big letters (75%) and the trade name in small ones (25%). This is opposite to the situation in the private market. Therefore, the private market can be considered the only market for generics.

There are some monopolistic medicines which are not offered by the industry, and therefore the health system has to subrogate the purchase through the private pharmacies. This price is very high for the government to update; it represents 35% of the medicines of the National Formulary. The government pays the industry 30 days after delivery, and if they do not deliver on time, they will suffer fines and the details will be written in the respective dossier for the next contest.

2. The private market works through seven distributors, which monopolize the distribution by 90% and supply 20,000 private pharmacies around the country. The distributors earn 25% of the total cost of the medicine, and the pharmacies earn 12%. In general, the payment system of the distributors to the industry is 45 days.

It is important to indicate that the distributors are the owners of a great number of pharmacies and pharmaceutical industries.

4. Facilities

Including private medicine, in 1983 there existed 10,233 medical units in the country:

9,294 (91%) Outpatient
  828 (8.1%) General Hospitals
  111 (1.1%) Specialty Hospitals

Number of beds/1000 people = 0.78

It is important to indicate that in 1985, before the earthquake, there existed:

For open population:	6,742 beds
Social Security	11,178 beds
(IMSS and ISSSTE)*	
Total	17,920 beds

---

* IMSS—Social Security for all the workers. ISSSTE—Social Security for government workers.

After the earthquake, 5,625 beds were lost, but there is currently a recovery of 65%.

In 1987, there were 11,300 medical health units in existence.

5. Manpower

Considering official data, there are 130,000 physicians in Mexico. (In 1983, there was 1 physician/1,095 people.)

Physicians and Nurses

(1973 - 1983)

Institutions	Physicians/100,000 people			Nurses /100,000 people		
	1973	1980	1983	1973	1980	1983
Country	4.1	4.3	4.8	8.1	8.8	11.6
Open Population	1.3	3.9	3.6	3.7	3.2	10.3
Social Security	10.7	6.5	8.4	17.9	17.3	18.8
IMSS	9.3	5.3	5.9	17.2	17.1	18.0
ISSSTE	17.0	8.8	17.8	22.0	16.1	19.7
PEMEX[2]	13.7	17.5	18.6	16.5	20.9	23.7

Nurses/Physicians

1973	1980	1983
1.9	2.0	2.4
2.9	0.8	2.9
1.7	2.7	2.2
1.8	3.2	3.1
1.3	1.8	1.1
1 2	1.2	1.2

(Agenda Estadistica DGE - SIC 1975 - 1985)
(Boletin Informativo DGEI, SSA)

---

2. Mexican Oil Company

6. The Social Security System

The Health Ministry is responsible for the health system. In the public sector, there are two different areas of development: the Social Welfare and the National Health Institutes. The Social Welfare includes the National System for the Integral Development of the Family (DIF), the Youth Integration Centers, and the National Institute for the Elderly. There are ten National Health Institutes: Cardiology, Cancerology, Respiratory Diseases, Pediatry, Nutrition, Neurology and Neurosurgery, Perinatology, Psychiatry, Public Health, and Children's Hospital of Mexico. Besides these areas, entities of the federal public administration are included: the Institute of Social Security (IMSS) and the Institute of Social Security and Services for Government Workers (ISSSTE). Both are organized according to each defined sector for which they are responsible. There are other institutions in this sector, such as the Mexican National Oil Corporation (PEMEX) and the military structure.

Since 1975 health service institutions have been required to use a National Formulary (as published in the official government journal). Since then, the Formulary has undergone different modifications; in 1983, an Interinstitutional Commission of the National Formulary was created to bring the Formulary up to date. The National Formulary includes only generics, and whenever possible, combination of drugs has been avoided. In 1989, it included 307 generics and 452 dosage forms.

7. The Medical Establishment: Entry, Specialists, Referrals

There are 15,164 general practitioners, divided among 32 states, who work only at the Health Ministry:

General Practitioners

State	Number	Percentage (%)
Federal District or Mexico City	3,048	20
Aguascalientes		
Baja California	297	2
Baja California Sur	145	1
Campeche	97	1
Coahuila	324	2
Colima	164	1
Chiapas	334	2
Chihuahua	320	2
Durango	154	1
Guanajuato	674	4

Guerrero	375	2
Hidalgo	488	3
Jalisco	768	5
Mexico	1,843	12
Michoacan	593	4
Morelos	170	1
Nayarit	91	1
Nuevo Leon	561	4
Oaxaca	414	3
Puebla	616	4
Queretaro	202	1
Quintana Roo	176	1
San Luis Potosi	314	2
Sinaloa	174	1
Sonora	359	2
Tabasco	649	4
Tamaulipas	297	2
Tlaxcala	323	2
Veracruz	572	4
Yucatan	233	2
Zacatecas	200	1

There are 5,028 specialists, in 49 specialties, who work in second- and third-level hospitals all around the country. Of these, 38% are in Mexico City and 62% are in the rest of the 31 states.

8. Coverage

Although coverage has increased to 87% of the total population, 10.3 million people do not yet have easy access to health services. In comparison, 14 million did not have health services in 1982.

National Coverage in 1984

	Population	Percentage
IMSS-ISSSTE	36.6 million	47%
Open population (SSA)	26.9 million	36%
Private medicine	3.8 million	4.8%
Total	67.3 million	87.8%

National Coverage in 1987

Within the population of 81.2 million people, a coverage of 91% was achieved. Therefore, 73.9 million Mexicans had access to permanent health services.

B. PHARMACEUTICAL INDUSTRY AND DRUG DISTRIBUTION

1. Structure of the Industry

At present, there are 288 enterprises that manufacture medicines in the country. Two hundred and seventeen are national, and twenty-nine produce pharmachemicals for the pharmaceutical industry.

Geographic Distribution

Concerning distribution, 74% of the companies are located in the Federal District, 8% in Mexico State, 7% in Jalisco, and 3.6% in Puebla.

Jobs

Internal employees number 46,000; of these, 22,000 are laborers. The remainder work in technical, administrative, or research activities.

Supply

Of all the medicines consumed in Mexico, 99.6% are produced locally. Sales in 1986 were 720,000 million pesos.

2. Research and Development, Imports, Exports, Patents, Licensing

Research and Development

The first step in planning the basic points concerning research and development of drugs is to establish national drug policies. At present, Mexico is just beginning to coordinate the activities concerning this issue.

Costs of Health Care and Drug Consumption

	GNP Health Costs	Drug Costs
F.R. Germany	8.7	18.0
U.S.	7.7	9.3

Sweden	7.7	9.9
France	7.5	13.0
Italy	6.5	13.1
Japan	6.0	14.0
Switzerland	5.9	17.7
Spain	5.7	28.3
England	4.7	8.3
Mexico	1.8	38.0

(Source: Consideraciones Generales para la Formulacion y Establecimientos de Politicas Nacionales en Materia de Medicamentos PNS P/8520 WHO (OMS) Ginebra 1985).

Victor L. Urquidi, in the newspaper *Excelsior* (October 1, 1988), indicates that the national expenditure in R&D in relation to the GNP in 1985 was 0.53%, and in 1988, 0.3%. In the industrialized countries of the United States, Japan, and Germany, the relationship is 27%, 28%, and 28%, respectively.

In Mexico, there are few answers for the many questions concerning R&D, such as: Research and development of what? Research and development, how and why?

Imports and Exports

Concerning the commercial balance, the exportation of medicines increased in 1986 ($77 million) and importation decreased ($5 million). The exportation of pharmochemcials increased during 1984-1988 by 60%, representing 30% of national production. This effort allowed the country to equalize its foreign currency balance during 1986 and 1987.

Patents

In January 1987, a decree concerning reforms, the Law of Inventions and Trademarks, was published. According to the law of 1976, the companies established in Mexico were allowed to produce and sell medicines that in other countries needed the benefits of a patent. Therefore the country had all the new drugs, since the salts were obtained outside of the country or were produced locally if the technology was bought. The law was modified in 1978, 1981, and 1987. According to Article 10, the following are not patentable:

VIII — The biotechnological processing of pharmaceuticals, drugs in general, beverages and foods for animal use

XI — Chemical pharmaceutical products, drugs in general, beverages and foods for animal use

3. Manufacturing

Of the total dosage forms 98.2% are manufactured in the country. The remaining 1.8% is attributed to new drugs, oncologic agents, and blood products of low consumption.

4. Advertising and Promotion, Price Regulation

The General Health Law regulation concerning the sanitary control of advertising (1986) implies that advertising should promote behaviors, practices, and habits encouraging the physical and mental health of the individual. It includes specific rules related to advertising of health care systems, food, nonalcoholic beverages, alcoholic beverages, tobacco, drugs, medicinal plants, medical equipment, diagnostic agents, surgical materials, etc. Finally, it explains the mechanisms of authorizations, regulatory surveillance, sanctions, and safety measures.

Price Policy

Public

The selling prices will be established by the Ministry or Secretary of Commerce and Industrial Foment (SECOFI) based on studies of cost and price.

Private

During 1987, SECOFI defined a more efficient mechanism to review prices. The products were divided into three categories:

a. Products equivalent to those in the National Formulary
b. Basic products of popular consumption
c. Complementary products.

For Groups A and B, the review includes a determination based on the system of cost-price-utility. It is compulsory to register the selling prices. Group C is ruled by the system of automatic registration of prices. A statistical measure is undertaken, and there is a six-month review of the costs.

It is necessary to indicate that there are products with exorbitant utilities and there are others in which the operation range is very small and sometimes they are not produced anymore. It is considered that in this chain, the distributors earn too much money, which is reflected in the final price of the products.

5. Drug Approval and Government Regulations

One way to exert sanitary control on medicines and similar products is through the registration granted by the SSA. Based on technical and scientific developments, the registration system was clarified and updated. To avoid fraudulent use of the product, it is necessary that the registration dossier contain all the data needed to provide clear information for the pharmaceutical industry and the product. Therefore, a registration certificate should be granted containing all the specifications and conditions of the product.

There are two basic factors that are considered when a dossier is going to be prepared: is the active substance new to the Mexican market or has the active substance already been registered in the Mexican market. One important fact is that the SSA's criteria is to avoid registration of products containing two or more active substances unless their clinical advantage is fully demonstrated.

According to the Regulation for Registration and Review of Pharmaceutical Specialties, the dossier should contain official and legal documents, as well as pharmaceutical and clinical information. Drug registration requirements and procedures for active substances are different depending on whether the substance is new or already registered in the Mexican market, is a licensed or imported product, or is a product that is going to be exported.

6. Competition: Brand and Generic

In Mexico, there is no real generic market, although the physicians within the National Health System do prescribe the generics contained in the National Formulary.

7. Cost Containment

Drug Classification

Group One — Includes all the medicines that have the same pharmo-chemical and dosage forms of the National Formulary

Group Two — Includes all the medicines of the National Formulary and some that are sold over the counter

Group Three — All the medicines that are not included in Groups One and Two.

Control System

The medicines included in Groups One and Two will be judged by the system of cost-price-utility. The ones included in Group Three will be reviewed by the system of cost-price every six months.

Factors

For the medicines of Groups One and Two, a factor of the operative expense (80% on average), will be applied in addition to a utility of 18% and the percentage accepted for the distributors and the pharmacies. Concerning the medicines of Group Three, there are two options: to apply the same factors used in Groups One and Two or to keep the same utility factors and the percentages accepted for the distributors and pharmacies, adding the operative expenses not considered in the medicines of Groups One and Two.

C. PHARMACY PRACTICE

1. Dispensing a Prescription: Where, What, Payment, Formulary

El Reglamento de la Ley General de Salud en Materia de Control Sanitario de Actividades, Establecimientos, Productos y Servicios (General Health Law), published in January 1988, indicates that the only personnel authorized to prescribe are physicians, dentists, homeopathic physicians, veterinary physicians, health professionals in social service, nurses, and midwives. All authorized personnel should be registered and authorized by the relevant educational authorities.

The National Formulary is compulsory for all health professionals who work in hospitals and clinics of the National Health Care System. In this system, patients receive medicines without charge.

2. Record Keeping

Private pharmacies sell drugs with or without prescription. They do not keep records of the prescriptions, with the exception of controlled drugs.

By law: (1) medicines containing psychotropic drugs similar to narcotic drugs require for their sale to the public a special prescription notebook edited by and given by the Health Ministry to the health professionals; (2) medicines containing psychotropic drugs, which will be supplied only once require a prescription, which should be retained by the pharmacist; and (3) medicines containing psychotropic drugs, which will be supplied three times only require a prescription, which should be sealed each time with the date of purchase. A list of medicines is given for each of these types of controlled drugs.

In the public pharmacies, a copy of the prescription is kept, forming a record of prescriptions of the medicines included in the National Formulary.

3. Patient Education

In Mexico, the pharmacist is not considered a member of the health team, and therefore does not function as a patient educator. Clinical pharmacy is not yet developed in the country, although isolated efforts have been made to organize it.

One of the projects that is gaining force in Mexico is the creation of an Adverse Drug Reaction Information Center sponsored by the national University of Mexico (UNAM), the Institute of Social Security (IMSS), and the Health Ministry (SSA). This project was begun after nine years of planning, with patient education as one of its services. To initiate the project, necessary information for 11 drugs (antibiotics and analgesics) were compiled.

4. OTCs and Other Classes and Schedules of Drugs, Other Nonpharmacy Outlets

At present, the SSA has reviewed and registered 93 OTC products, with others in the process of being updated. The registration dossier of OTC drugs should contain official and legal documents, as well as pharmaceutical and clinical information, and should follow the Regulation for Registration and Review of Pharmaceutical Specialties.

Other classes of scheduled drugs are homeopathic drugs, which are reviewed and registered at the SSA. Usually these are very old products, and few innovations are presented. Homeopathy was recognized officially in Mexico at the beginning of the twentieth century.

5. The Use of Technicians and Others

One of the main problems concerning individual pharmacy practice in Mexico is that although by law each pharmacy or drugstore should have a responsible pharmacist, it is not necessary for the owner to be a professional of the pharmaceutical sciences.

There are 22,000 pharmacies, 20,000 of which are private; the rest are owned by the government and give service at the Social Security Hospitals (IMSS). All pharmacies are responsible for the identity, distribution, and conservation of the drugs sold, as well as the correct preparation of magistral prescriptions and activities defined in La Ley General de Salud 1984 (General Health Law) and in the Reglamento para Droquerias, Farmacias y Demas Establecimientos Similares (Pharmacies Regulation). In all these activities, the pharmacist requires the cooperation of auxiliary personnel who assist with the technical direction of the pharmacy and during temporary absences.

It is important to indicate that a shortage of pharmacists exists, since there are only 35,000 in the entire country. Consequently, it has been very difficult to follow the law, and the majority of pharmacies have been attended by personnel of lesser capabilities. This situation has provoked the payment of fines and has established a vicious pattern of lack of prepared personnel, the rapid appearance of new pharmacies, and the weak application of the law by the sanitary authorities.

Considering this situation, the following are possible solutions:

1. Increase substantially the registered list of pharmacy careers. It is important to remember that the substructure needed is very expensive, that for almost 40 years it has remained static, and that most of the pharmacists are absorbed by other areas where they have better potential for high salaries and professional identification.

2. Relax the sanitary regulations for pharmacies. However, this is inadvisable from the ethical point of view, since it would increase the misuse, abuse, and traffic of drugs.

3. Professionalize the sanitary control by introducing a technical assistant for pharmacy operations. Without interfering with the commercial aspects of the dispensation system, the technical assistant, under the tutelage of the pharmacist, could not only solve problems but could improve the sanitary guarantee of drugs for patients. The primary importance of this paraprofessional would be as a health educator in nonurban areas,

which have limited access to health professionals and sanitary authorities. The technical assistant program would be of low cost to the government, and considering the special situation of the Mexican pharmacies, would fill a national need by assisting the government and the public in drug quality assurance and in the prevention of drug misuse and abuse.

At present, there are 979 students in 5 different schools of CONALEP (the National College of Professional Technical Education) around the country studying this technical career. In 1990, the first professional technicians are expected to be incorporated into some of the pharmacies of the country. By law, the pharmacies must hire them to help the responsible pharmacist in each pharmacy.

6. Pharmacy and Primary Care

Since the pharmacist is not considered a member of the health team, there is no relationship between pharmacy and primary care in Mexico.

D. UNIQUE OR INTERESTING FEATURES
OR SPECIAL SITUATIONS IN MEXICO

Besides Western medicine, Mexican culture and education also includes traditional remedies. Considering geographic characteristics of the country, as well as traditional and cultural aspects of each group of people, it is found, especially in small towns, that people do not have access to health clinics or to physicians, and therefore rely on the witch, or the grassman or grasswoman (*hierbero* or *hierbera*). Even in those places where there is a physician doing Social Service (which is compulsory for all professionals in the country), traditional medicine is preferred, or sometimes a combination of medicines and teas, with some magic practices as well.

It is important to indicate that there is laxity in the sale of medicines without prescriptions, even though they are required for the sale of, for example, antibiotics. This is due to the fact that in 1942, pharmacies were included in the Commerce Chambers, and therefore the sale of medicines (except the ones controlled by the Health Ministry) is conducted in the same way as the sale of any other merchandise.

E. CONCLUSIONS

The National Health System with the Social Security Service is one of the most congruent plans in structure, considering the objectives desired in the political and social environment of the country. The pharmaceutical policies in Mexico in the last five years have been accelerated after many years of neglect. Examples of these actions are, among others, the General Health Law actualization, the fifth edition of the Mexican Pharmacopeia (1988), the Guide of Good Manufacturing Practice (1985), and the Guide of Validation (1988). These will guide Mexico in establishing a healthy pharmaceutical industry manufacturing high quality medicines and self-sufficient in the production of dosage forms, and with the interest and necessity of supporting bioavailability studies parallel to clinical studies. Mexico is currently not self-sufficient in the manufacturing of pharmochemicals, and does not have manpower for research in clinical studies, biopharmacy, or the development of technology and equipment.

Concerning regulatory affairs, the Health Ministry is finishing its evaluation of drug registration. In 1980, there were 20,000 medicines registered; currently, there are only 6,000, and the aim is to lower the number to only 2,000 medicines.

Even though the International Marketing Service reports that the average price of medicines in Mexico was one of the lowest in the world in 1987 ($1.96) there is chaos in the assignment of prices considering cost structure. It is necessary to review the price of medicines. Many efforts have been made, but since there is no coordination between regulatory actions and the pharmaceutical industry, there has been no real impact and progress has been slow. Some recommended actions are:

1. The national manufacture of pharmochemicals and containers
2. Fitness of the process of good manufacturing practices and validation in the manufacturing of dosage forms
3. Support for the exportation of pharmochemicals and medicines
4. Since there is not good control in the process of commercialization of medicines, it is necessary to support and consolidate the modernization of the legislation to mend the mistakes and contradictions of the law and to create coordination among sectors.
5. It is absolutely necessary to incorporate the pharmacies into the sanitary regulation system.
6. Since the pharmaceutical service does not exist within the National Health System, it is very important to support the incorporation of the pharmacist into the health team.
7. To create schools or colleges of pharmacy within the universities,

where pharmacists, identified as professionals, are restored to their positions in health service to the country

8. To support, within the universities, postgraduate studies to train researchers in biopharmacy, pharmacology, and biotechnology
9. To support, within the universities, postgraduate studies to develop technology and equipment for the manufacturing of pharmochemcials and dosage forms
10. To support within the universities, studies in traditional medicine, to rationalize it and to obtain advantages of its practice in the country
11. To create a postmarketing surveillance of drug service around the country.

SELECTED READINGS

Censo de Recursos Humanos. Mexico. Oficialia Mayor. Secretaria de Salud; 1988.

Frenk, J., Robledo-Vera, C., Nigenda-Lopez, G. et al. Subempleo y Desempleo entre los Medicos de las Areas Urbanas de Mexico. Salud Publica de Mexico, 1988; 30(5): 691-699.

Giral-Barnes, C. Investigacion y Desarrollo de Medicamentos en America Latina. In: Memorias de la I Conferencia Latinoamericana sobre Politicas Farmaceuticas y Medicamentos Esneciales. Mexico; 1988.

Gonzalez-Block, M.A. El Traslape de la Demanda en el Sistema Nacional de Salud de Mexico: Limitaciones en la Integridad Sectorial. Salud Publica de Mexico. 1988; 30(6): 804-814.

Gonzalez Martinez, R. Situacion Actual de la Industria Farmaceutica. In: Memoria de la Cuarta Convencion Nacional de la Industria Farmaceutica. CANIFARMA. Mexico. San Juan del Rio, Queretaro; 1988: 43-66.

Informe Anual de la Camara de la Industria Farmaceutica. Mexico. CANIFARMA; 1988.

La Industria Farmaceutica en Difras; 1978-1987. Mexico. CANIFARMA; 1988.

Ley General de Salud y modificaciones. Mexico. Secretaria de Salud; 1984, 1987.

Ley de Invenciones y Marcas. Mexico. Secretaria de Comercio y Fomento Industrial; 1987.

Reglamento de la Ley General de Salud en Materia de Control Sanitario de la Publicidad. Mexico. Secretaria de Salud; 1986.

Salazar, G., Giral, C., and Viso, F. Medicines Registration Review Process in Mexico. J. Clin. Pharmacol. 1988; 28: 577-583.

Salud y Seguridad Social; Mexico. Secretaria de Salud. Fondo de Cultura Economica; 1988: 83-84, 92-94, 103-104.

Segunda Revision de Fomento de la Industria Farmaceutica. Mexico. CANIFARMA; 1987.

Soberon Acevedo, G. El Derecho a la Salud. Mexico. Partido Revolucionario Institucional. Secretaria de Divulgacion Ideologica; 1987.

Soberon, G., Kumate, J., y Laguna, J. La Salud en Mexico: Testimonios 1988. Vol. I. Fundamentos del Cambio Estructural. Mexico. Secretaria de Salud. Instituto Nacional de Salud Publica. El Colegio Nacional. Fondo de Cultura Economica; 1988: 64-70, 359-365.

Soberon, G., Kumate, J., y Laguna, J. La Salud en Mexico: Testimonios 1988. Vol. II. Problemas y Programas de Salud. Mexico. Secretaria de Salud, Instituto Nacional de Salud Publica. El Colegio Nacional. Fondo de Cultura Economico; 1988: 37-39.

Valdes-Olmedo, C. Apuntes sobre el Financiamiento de la Salud en Mexico. Salud Publica de Mexico; 1988; *30*(6): 815-826.

Chapter 14

Netherlands

Hubert G. Leufkens
Albert Bakker

A. THE NATIONAL HEALTH CARE SYSTEM

1. History and Background

In the Dutch health care system, charity and private initiatives have always played a significant role. In the period before 1970, the relationship between the state and the health care system was based on principles of providing a maximum of freedom to private institutes and foundations, which often had a religious affiliation, for funding and owning health service facilities. Health care legislation was very limited, and the most important government regulations were concerned with entry requirements for the medical professions (physician, pharmacist, dentist). These regulations were established in the mid-1880s. Since then, only graduates from the medical schools of a university have been allowed to practice medicine, and pharmacists, for instance, must have a university degree in pharmacy.

Government policies on health care before 1970 were influenced by the political attitude that autonomy and self-regulation are the best ingredients for achieving high quality in health care. The major responsibility of the state was to ensure enough financial resources for an autonomous process of developments in the health care system. This has resulted in a public health insurance scheme that is compulsory for all individuals who earn less than a certain threshold income. The level of threshold income leads to an enrollment of about 60% of the total population into the public health insurance scheme. Individuals with a higher income are almost always privately insured. Only 5% of all health care expenses are being paid out-of-pocket.

Thus, the history of Dutch health care shows on one hand a system with features of an independent and autonomous growth, stimulated by professional dominance of the providers of care and continuous increase in the demand for care, and on the other hand the absence of economic barriers due to a comprehensive (for the majority of the population, compulsory) public health insurance scheme. Prof. Ruud Lapre has compared Dutch health care before 1970 with "a highly-developed automobile with a powerful medical-professional motor, a government and advisory bodies which wish to make continuous improvements in its quality, insurers who charge almost nothing for the petrol and an admiring public who watch the car race by; it is however, a car without any steering."[1]

2. Social and Political Environment

In the mid-1970s, the state government started developing a system of legislation and policies with features of more government involvement, planning, and cost control. A major impetus for this development was a shift in the political arena by which state involvement was stimulated and enhanced. In that period, the social-democratic party was gaining influence. Since then, several committees have advised the Ministry of Health to restrict the growth of the Dutch health care system, and a number of laws have been proposed in the Parliament. However, the providers of care offered opposition and were successful in mobilizing strong political power. As a result, the approved laws in the Parliament were compromises, and in those fields where no legislation could be established due to the absence of sufficient political support, ad hoc decision making took place. Especially, these short-term decisions have caused disturbance and frustrations in the health care sector. The ultimate improvements in the quality of care and reductions in costs were very limited.

The period of more government involvement ended in the mid-1980s, and this switch has been highlighted by the publication of the report of the Dekker Committee in 1987.[2] Dekker (former president of the Philips Company) and his fellow members of the committee proposed a compulsory basic (limited coverage) insurance scheme for the total population, introduction of more market elements in health care (competition among health insurers and among providers of care) and improvement of quality of care by developing utilization and quality review schemes as part of the contract conditions between insurers and the providers of care. The Dutch health care system is moving into the Dekker era, which will probably bring a total new environment for health care providers, patients, and insurers. How this radical change to a market-oriented health care system will succeed remains unclear. There is, however, strong and stable political support for continuing this development.

3. Public and Private Financing, Health Costs

The major source of finance for the Dutch health care system is a public health insurance scheme, the Sickfund. This scheme contributed 34.4% of the total in 1987. All individuals who earn yearly less than about Dfl. 48,000 (approximately two-thirds of the population) are covered under the Sickfund scheme. The insurance premiums are collected centrally and administratively handled through the salary paying system (reduction of 5.05% + 5.05% contribution of the employer in 1987).

Another important public source is a scheme (AWBAZ) that originated from an insurance for exceptional medical expenditures. During the past decades, this scheme has been expanded more and more for funding general health services (i.e., public health nurses). In 1987, this part of the public insurance system contributed 23.7% of the total. Private contributions, which can be insurance based or out-of-pocket, accounted for 27% in 1987. Government contributions are limited in the Dutch health care financing system (14.9% in 1987). In conclusion, funding of health services is based on two major sources, about two-thirds from public health schemes and one-quarter from private sources. Government policies in the Dekker era are aimed at establishing a basic (limited coverage) public insurance scheme for everybody, with supplementary coverage of health care services provided by a competitive private insurance market.

As in most Western countries, health care costs have increased continuously in the last decades. In 1987, health care expenditures were 10.1%, expressed as a proportion of GNP. In Table 1, the costs of the various services over several years are summarized. During the period 1982-1987, there was an average annual growth of 2.8%. This is much less than in the period 1975-1982, which had an average annual growth of 9%.

The hospital sector accounts for about one-third of total health care expenditure. This sector has been a major target for cost-saving measures for many years. The most significant measure was the hospital budgeting system, introduced in 1984. Since then, hospital funds are being fixed to a certain yearly budget by a central government body. Before 1984, there was an open-ended financial system with absolutely no incentive for cost control. Extramural health services (i.e., general practice) account for 17% of the costs. Since 1974, there is a strong political pressure to stimulate extramural services, mainly to reduce health expenditures. The intention of substitution of expensive (hospital) services with cheap (primary care) services has proved a difficult task to accomplish due to strong opposition from hospitals and specialists.

The figures for drug costs are concerned only with expenditures for

TABLE 1. Health care cost 1982-1987.

	1982 10^6 Dfl	1983	1984	1985	1986	1987	mean annual growth
1. Hospitals and medical specialists	12750	13087	13250	13402	13617	13634	1.3
2. Mental care services	2509	2594	2656	2739	2809	2847	2.6
3. Services for the disabled	2882	3045	3234	3285	3399	3465	3.8
4. Services for the elderly	7246	7578	7626	7891	8111	8186	2.5
5. services for youngsters	689	701	710	716	723	747	1.6
6. Extramural services	6437	6644	6711	6924	7167	7463	3.0
7. Drugs, aids and appliances	2673	2796	3034	3302	3641	3973	8.2
8. Public preventive care	771	763	761	773	772	786	0.4
9. Management, admin. and misc.	2108	2289	2434	2472	2611	2643	4.6
Total 1 through 9	38065	39497	40426	41504	42850	43743	2.8

pharmaceutical care outside hospitals. Funds for drug expenditures inside hospitals are part of the hospital budget. In the sector of extramural drug use, an average annual growth of 8.2% has resulted in a proportion of total health costs of 9.1% in 1987. Pharmaceutical care shows the strongest growth in costs among all services. In Section B, drug costs will be discussed in more detail.

4. Facilities

In the Dutch health care system, two major sectors (echelons) can be distinguished:

First Echelon: Primary care, including general practice (GP), public health nurses, midwives, dentistry, and pharmacy.

Second Echelon: Intramural care including hospitals, medical specialists, psychiatric clinics, mental retardation institutes, and nursing homes.

In the last decades, several health centers for primary care have been established, where GPs, public health nurses, dentists, pharmacists, etc., cooperate closely. This development has been funded in a number of cases by the state and the Sickfunds under the premise that these facilities provide cost-effective services.

In 1987, there were about 200 general hospitals (eight university hospitals included) with a total number of 67,530 beds, which means a bed/person ratio of 4.5 per 1,000. The volume of the hospital sector is decreasing gradually during the last decade due to a government policy of limited licensing in terms of number of beds and number of specialists and strict budgeting schemes.

In Table 2, the developments in available hospital beds, occupancy rates, and average care period over the past period are shown. The introduction of day care and home health care, efficient planning, and the introduction of new medical technologies are factors that have stimulated these developments. The formal policy of the Ministry of Health is to reduce the volume of hospital services to a target bed/person ratio of 3.8 per 1,000.

The number of available beds in psychiatric hospitals and mental retardation institutes is 24,599 and 30,542, respectively (1987). Both sectors show very high occupancy rates of 95%-99% and face significant capacity shortage. In 1987, there were 49,974 beds in nursing homes. The sector of geriatric care needs significant expansion in the near future to meet all the needs of the growing elderly population.

TABLE 2. Occupancy rate and days of care.

	1982	1983	1984	1985	1986	1987
No. of hospital beds	69612	69602	68990	68461	67970	67530
Occupancy rate (%)	85.8	83.1	79.9	78.4	76.5	73.9
Av. no. days of care	12.7	12.4	12.1	11.9	11.7	11.5

5. Manpower

Manpower in health care is an important determinant for access to and quality of health services. In Table 3, a quantitative overview is given of manpower in some major fields in 1987. For a number of years, state policies aimed at limiting training facilities for health professionals to control available manpower in health care. However, as a result, forecasts predict a shortage in manpower in certain fields in the near future (e.g., nursing).

6. The Social Security System

The Netherlands possess a comprehensive and well-developed social security system, which originated in the early 1960s and expanded strongly in the 1970s during a period of social-democratic administration. The state is responsible for providing every citizen with a minimum financial budget. As a result, numerous regulations for financial support of those who are not able to have a paid job have been introduced: financial support schemes for the elderly, unemployment insurance, public schemes for mentally and physically disabled persons, welfare schemes for those who are unemployed for longer than two years, welfare for divorced persons, etc. The Dutch social security system shows features of both wealth and a strong social consciousness, but also of complexity, bureaucracy, high costs, and inefficiency.

Recently, a tendency has emerged in Dutch society to reflect upon the social security system. A major impetus here is the changing socioeco-

TABLE 3. Number of workers in health care.

	1983	1987
GP's	5634	6243
Medical specialists	9937	11612
Dentists	6271	7405
Pharmacists	1672	1991
Midwives	974	1014
Pharmacy technicians	6645	7766

nomic environment in Europe looking forward to the year 1992 when many different systems within the EEC are to be harmonized and unified. There is a great variety in social security systems in Europe, and the Dutch system is probably one of the most comprehensive. A certain degree of equalization has to be expected.

7. The Medical Establishment: Entry, Specialties, Referrals

The GP plays a key role as gatekeeper in the health care system. It is common practice that a patient presents his or her complaint first to the GP, who does a primary examination and makes an estimation of the severeness of the complaint. The GP may decide that the complaint needs further specialist care. Referral to a medical specialist is only possible (reimbursable) after consultation with the general practitioner.

8. Payment and Reimbursement

The structure of financing and reimbursement in health care is in a process of continuous change. In Section A.3, the major financial resources have been discussed. The funds collected through the Sickfund system, the scheme for exceptional expenses, private contributions, and the state are used to pay for health services. Basically, there are three major ways to pay for health services in the Netherlands:

- Fee-for-service: Medical specialists, GP services for privately insured patients, pharmacy services (fee per prescription), dental services for privately insured patients, hospital care (fee per day).
- Capitation fee: GP services for Sickfund patients, dental services for Sickfund patients, services provided by HMO-like organizations.
- Facility subsidy: Many facilities (hospitals, public health nurse, institutions for preventive medicine) provide care by salaried professionals; most of these facilities follow a mixed system of funding composed of fee-for-service as well as subsidy systems based, for instance, on number of salaried employees or number of beds.

Until recently, community pharmacy services had a mixed system for Sickfund patients: a capitation fee for each registered patient and a fee-for-service per dispensed prescription. For privately insured patients, there was a selling price, representing the purchasing price plus margin (30%-35%). Since 1988, the reimbursement system has been changed both for Sickfund patients and for privately insured patients. The pharmacist now receives a fixed fee (in 1988, Dfl. 11.50) per prescription. This change has caused a severe disturbance among pharmacists.

Major changes are to be expected for the reimbursement of services due to the process of reorganizing the health care system. Capitation fees will probably play an important role in HMO-like health service organizations. The fee-for-service will become a major paying system for health professionals. However, it remains unclear how the developments in reimbursement and payment for health services will continue. Competition among insurers will probably lead to strong differentiation in reimbursement systems.

9. Coverage

The same uncertainty concerns the extent of coverage. Today, coverage by the Sickfund system knows almost no limitations (a few exceptions: Dfl. 2.50 copayment per prescription with a maximum of Dfl. 125 per year; OTC drugs are excluded from reimbursement; coverage of advanced dental care is poor; and physiotherapy is limited to a certain number of treatments). Private insurers provide a broad variety of (very competitive) packages of coverage with several levels of copayment. The future will probably bring a basic (minimum) insurance for everybody, with limited coverage. For all additional services, one has to contract a private insurance company. The content of the basic insurance is the subject of hot political debate.

B. PHARMACEUTICAL INDUSTRY AND DRUG DISTRIBUTION

1. Structure of the Industry

There are about 100 pharmaceutical companies which are mostly subsidiaries of large multinational firms. Domestic firms include three innovative companies (Organon, Duphar, and Gist-Brocades), wholesalers with manufacturing divisions, and a growing number of generic companies. The pharmaceutical industry employs about 12,600 persons, including approximately 200 pharmacists.

Almost all major companies are members of the Dutch Association of Pharmaceutical Industry (NEFARMA). This association represents the pharmaceutical industry in all political, economic, and social segments of Dutch society. NEFARMA publishes each year a repertory on all specialities on the market, maintains a computerized drug database which is accessible to health professionals, and has a proactive policy on communicating the societal value of pharmaceutical research.

2. Research and Development, Imports, Exports, Patents, Licensing

Research and development (R & D) is concentrated mainly in the three innovative firms, Organon, Duphar, and Gist-Brocades. R&D at Organon (a subsidiary of AKZO) has always had a strong tradition in sex hormones (oral contraceptives). Over the past few years, R&D efforts have been expanded into new fields of interest (psychotropic drugs, cardiovasculars, immunomodulators). Duphar is a subsidiary of the Belgium firm Slavay, and R&D covers fields of interest such as gastrointestinal and neurological diseases. Gist-Brocades has concentrated most R&D efforts on biotechnology. Biotechnology will become more important in the near future in Dutch pharmaceutical R&D. Another subsidiary of AKZO (Intervet) was the first company in the world that marketed veterinary vaccines manufactured by recombinant DNA technology.

Most of the drugs on the Dutch market are imported (about 90%). Domestic firms do not have special privileges and have to compete normally with foreign firms. Export of drug products (90% of all production) is mainly directed to the member states of the EEC. Expansion to non-EEC countries is expected.

With relatively little domestic R&D operations, the issue of patents is considered more in the general context of how to guarantee the progress of pharmaceutical research. Political support for more patent protection has grown in the past decade. NEFARMA advocates strongly the increase of the effective patent protection period. Decision making depends, however, on the developments within the EEC. Licensing of drug production and drug marketing is regulated by the medicines Supply Act of 1963. This act covers the whole spectrum of manufacturing, approval, marketing, and distribution. Once a drug is approved by the registration authorities, the manufacturer is free to launch the drug on the market. Special licenses are possible for investigational new drugs. These drugs are allowed to be distributed only after full written consent of the physician.

3. Manufacturing

Although the domestic pharmaceutical industry in the Netherlands is small in terms of domestic R&D and manufacturing, there is a positive trade balance. The industry belongs to the top ten drug manufacturers in the global pharmaceutical industry. In 1987, sales accounted for Dfl.

2,376 million. Major areas of production are sex hormones, antibiotics, and raw products for the fermentation industry. Merck Sharpe & Dohme has a significant manufacturing plant in Haarlem. Most other foreign firms consist solely of sales and marketing departments.

Dutch pharmacists manufacture about 15%-20% of all drug products dispensed in pharmacies. This drug compounding is still a major task of pharmacists, although most products are being provided by the pharmaceutical industry. The Royal Dutch Pharmaceutical Society (KNMP) supports drug compounding by pharmacists by preparing protocols, quality control protocols, and continuing education. Recently, a network of regional quality control laboratories has been developed to meet full industrial quality standards. This network is a joint operation between the KNMP and the major wholesalers. Despite continuous opposition of the pharmaceutical industry to compounding in pharmacies and some public debate about the quality of compounded products, pharmacists are resolved to continue compounding.

4. Advertising and Promotion, Price Regulation

Promotional activities on pharmaceuticals are mainly directed to prescribers (GPs, medical specialists). Recently, promotion focused on pharmacists has increased due to the growing impact of pharmacy on prescriber decisions. A special issue here is generic substitution, where pharmacists are playing an important role. Moreover, it is becoming more common that GPs are not willing to welcome medical representatives. They send them to the pharmacists so that they can assess the value of, for instance, a new introduction, a new drug application, etc. These developments have led to greater industry interest in the pharmaceutical profession. Especially in hospitals with formulary committees (pharmacists are frequently secretaries of these committees), pharmacists have become a major target for promotional activities.

Advertising for prescription-only and pharmacy-only drugs to the general public is prohibited. Since 1981, sampling is restricted to two units per physician. Promotion is regulated by legal criteria and by code of good practice. The legal requirements are concerned with the indications for which promotion is allowed, minimum information to be presented about the qualitative and quantitative composition of the product, etc. Monitoring of promotional activities is done by the government (inspector for health care) and by an independent review committee representing industry, medical professions, and independent experts. There is no formal price regulation yet, although several proposals of the Ministry of Health to reduce costs are composed of price control measures (i.e., reference price lists). Still, the mechanism of what the market will bear significantly determines pricing of a new product. For a new product in a market segment with competitive products, this price is mostly a little higher than that of existing products. There is no control on pricing of products for new markets (e.g., the recently introduced cholesterol synthesis inhibitors).

Dutch drug prices are among the highest in Western Europe. In 1983, drug prices were 62% higher than the average level in the EEC countries. Looking forward to the free exchange of services and products within the EEC after 1992, it is expected that Dutch drug prices will become lower.

5. Drug Approval and Government Regulation

Legislation for drug approval was introduced in 1963. This legislation provides for a Committee for the Evaluation of Medicines, whose mandate is to determine the efficacy and safety of drugs. The Committee, which consists of 18 experts on pharmacy, clinical pharmacology, toxicology, and related scientific fields, is independent in its decision making about the approval. The requirements for drug approval in the Netherlands are among the strictest in Western Europe. Applicants for drug approval provide the financial resources (annual fee, application fee, modification fee) for the work of the Committee, which is supported with a total staff of about 60 (mostly scientific experts). Applicants for drug approval have to submit full documentation on the efficacy, safety, and pharmaceutical quality of the product. Within 30 days of receiving the registration documents, the authorities may request additional information. After this period, the Committee must come to a decision within 120 days. Marketing approvals are valid for five years. The approval decision also concerns the schedule of marketing (prescription-only, pharmacy-only, OTC) status.

To comply with European standards, the Committee has participated since 1978 in the EEC Committee for Proprietary Medical Products (CMPC). Cooperation on the level of the EEC is becoming more important, when in 1992 a free exchange of products and services between the member states of the EEC should be accomplished.

The drug approval process started in 1963, when there were already about 5,000 products on the market. These products were registered provisionally then. Yet not all those products have been officially approved. By May 20, 1990, all these drugs were expected to meet the same standards as other drugs. The Committee is now in the process of negotiations with the manufacturers on how to handle their EEC requirement efficiently. Since 1978, a company seeking approval of a generic product has been allowed to use information of already approved (speciality) products (open file). According to EEC directives, there is no difference now between the requirements for approval for generic products and specialities. The Committee is also responsible for the licensing of parallel-import products. All of these responsibilities have increased tremendously the working burden of the registration authorities, consequently causing delays in decision making. This has led to severe tension between the pharmaceutical industry and the registration authorities. In 1987, a reorganization of the Committee's staff was carried out to speed up the approval process. In Table 4, a number of key figures of the drug approval process of 1982 and 1986 are shown. There are new developments in the field of postmarketing surveillance (PMS). The (existing) voluntary ADR reporting system will remain important, but new, additional schemes are under way. Restricted release, release under intense monitoring, and record linkage between drug use and morbidity automated databases are options for future PMS approaches.

6. Competition: Brand and Generic

In 1987, there were 6,416 drug products on the Dutch pharmaceutical market, 4,297 specialities and 2,119 generic products. The total value of the market was Dfl. 1,946 million (excluding distribution and pharmacy costs). The market shows an annual growth of 6%-10%, with an increase of 8.2% for 1986-1987. In Table 5, a breakdown of pharmaceutical sales by ATC main group category is given.

The main categories in terms of sales are the cardiovascular drugs (22.2%), gastrointestinals (17.8%), psychotropic drugs (11.9%), respiratory drugs (11.2%), and the systemic anti-infectives (9.4%). The major growth categories are the anticancer drugs, gynecological/urological drugs (including sex hormones), respiratory drugs, and gastrointestinals. The growth in sales of gastrointestinals has been stable during the past few years due to the contribution of sales of H_2-blockers, which are leading products in the market.

TABLE 4. Number of applications, registrations, available specialities and generics, parallel licenses, 1982 and 1986.

year	applic	regist	special	gener	parall
1982	426	389	3857	2033	275
1986	634	359	5354	2145	282

TABLE 5. Breakdown of pharmaceutical sales in 1986 and 1987 (in Dfl million).

ATC group		1986	1987	% change
A	aliment tract & metab	310	346	11.6
B	blood and blood	44	48	9.1
C	cardiovascular syst.	404	433	7.2
D	dermatologicals	· 84	91	8.3
G	genito urinary syst. incl. sex horm	120	135	12.5
H	syst. hormones excl. sex horm	19	19	0
J	general antiinfectives	176	183	4.0
L	cytotoxic &immunosuper	34	39	14.7
M	musculo-skeletal syst.	116	124	6.9
N	central nervous syst.	225	232	3.1
P	antiparasitic syst.	2	2	0
R	respiratory syst.	195	219	12.3
S	sensory organs	32	34	6.3
V	various	37	41	10.8
Total		1798	1946	8.2

Drug utilization is lower than in most other countries (Dfl. 251 per capita in 1987 for Sickfund patients, Dfl. 187 for privately insured patients). In only 54% (1986) of the consultations between patient and physician is a prescription for a drug written. In the Netherlands, the number of prescriptions per capita per year is 3.3, whereas it is 11.6 in West Germany. There exists a general attitude of reluctance among patients regarding the use of drugs. This attitude is similar to that seen in the Scandinavian countries.

The market has become disturbed in the past few years due to growing political pressure to control costs. The Ministry of Health has taken quite a number of ad hoc restrictive measures since 1981, with very limited coherence. Panic reactions have followed, and the pharmaceutical branch has entered an arena full of threats, restrictive policies, and societal pressure.

In 1987, the University of Rotterdam published a survey (funded by the Ministry of Health) on the Dutch pharmaceutical market. A major conclusion of the report was that the pharmaceutical market shows a clear lack of competition. No price control (in 1985 a price index of 199 versus 104 for Spain or 100 for France), high wholesaler margin (about 20%), high pharmacy costs, and the absence of incentives in the pharmaceutical branch for generic substitution were identified by the investigators as significant factors in a constant rise of costs. They recommended more price competition between providers of pharmaceutical products and giving pharmacists financial incentives to substitute.

It is estimated that generic sales account for 13% of the market and parallel-imported drugs for 7%. Of the remaining 80%, three-fourths consists of specialities for which the patent has already expired. This means that only 20% of prescribed drugs can be considered as real innovative products.

The government, as well as the health insurers, endorses generic substitution. It is expected that the share of generic sales will grow in the near future. The innovative industry has been concerned about this growth for many years and has developed several strategies to counter this development. Several multinational pharmaceutical firms now produce specialities as well as generics containing the same active ingredient. Within the pharmaceutical industry, this has led to a debate, and a number of German and Swiss firms refuse to move in that direction. Wholesalers are especially competitive in the generic drug market. All major wholesalers have generic production divisions, and expansion of these activities is to be expected. The generic market is attractive due to relatively high prices for generics (generic/brand price ratio 65%).

7. Cost-Containment Activities

As already discussed earlier in this chapter, there is strong political pressure to control drug costs. For a number of years, the Ministry of Health has taken many cost control measures (negative lists, copayment of Dfl. 2.50 per prescription, limitation of number of items per prescription, stimulation of generic substitution, reduction of pharmacy fees). However, drug costs continue to grow, with an annual increase of about 8%. The shift in prescribing to newly introduced (more expensive) drug products has contributed significantly to this increase. Analyses by the staff of the Sickfund Council have shown evidence that this phenomenon is especially apparent in the field of cardiovascular therapy. The introduction of modern ACE-inhibitors and calcium-entry blockers has led to substitution in those market segments, where cheaper diuretics and beta-blockers were leaders in the past.

Probably the major reason for the failure to control drug costs is the short-term character and incoherence of the measures taken by the government. Therefore, physicians, pharmacists, etc., became frustrated and showed limited willingness to cooperate. The economic-political culture in Dutch health care shows a high preference for self-regulation. Any measure without support of the health professionals has a great chance of failing. In 1988, most participants (industry, physicians, pharmacists, wholesalers, Sickfunds) in the pharmaceutical arena have decided to counter the Ministry of Health and to undertake the initiative for cost-cutting measures themselves. They have reached an omni parties agreement (OPA) including the following major elements: voluntary reduction (6%) of the drug prices by the industry, reduction (1%) of the wholesaler margin, a small reduction (2%) of pharmacy fees, and a limitation of the financial incentive for generic substitution by the pharmacists. This last element shows a contradiction because it will probably limit generic substitution. This compromise has been adopted by the Ministry of Health, mainly because other earlier attempts to control costs have failed, and this agreement originated with the interested parties themselves.

C. PHARMACY PRACTICE

1. Dispensing a Prescription: Where, What, Payment, Formulary

In the community, prescriptions for a medication can be filled in two different ways: in a pharmacy or by a dispensing physician. There are no outpatient pharmacies in hospitals, at least not yet. Hospital pharmacists are advocating outpatient pharmacies in certain circumstances, but community pharmacists oppose this. Other outlets for prescription drugs are not allowed yet, but this may change in the future. The introduction of more market elements in health care may lead to abolishing the monopoly the pharmacists and dispensing physicians have on dispensing prescription drugs. All hospitals with more than 300 beds do have a pharmacy for drug dispensing and clinical pharmacy services. Approximately 18% of all drugs are being used in hospitals.

In 1987, there were 1,320 pharmacies and 826 dispensing physicians. Community pharmacy has always advocated a ratio of 1 pharmacy per 8,000-10,000 patients. Although the entry of young pharmacists has decreased the ratio somewhat in the last decade, the mean ratio is still approximately 1 pharmacy per 8,000. Before 1987, the pharmacy association (KNMP) controlled the establishment of new pharmacies. This system had no legal justification, but in practice it was fairly difficult to start a pharmacy without approval of the KNMP. Although every pharmacist is now free to open a new practice, there are certain limitations, mainly the economic feasibility of expansion. Reduction of remuneration and expected additional cuts on pharmacy fees have caused significant reluctance to invest in new pharmacies. The market for community pharmacies is becoming tight. Moreover, there is now a certain balance between available pharmaceutical manpower and market opportunities for pharmacist positions.

The number of dispensing physicians has decreased gradually in the last decade due to a strong increase of new pharmacies in rural areas. The Medicines Supply Act says that when a patient has to travel more than five kilometers before he or she can fill a prescription, a GP is allowed to operate a dispensing unit. For pharmacy and physician organizations, these regulations have been for many years an issue of hot debate and cumbersome legal affairs. It has to be expected that the number of dispensing physicians will decrease somewhat further but will then remain stable. In rural areas, it will always be difficult to run a pharmacy in a cost-effective manner.

2. Record Keeping

When a patient brings a prescription to a pharmacy, in most cases an automated prescription filling process starts. Approximately 85% of all community pharmacies are computerized. There exists a strong liaison between patient and pharmacy due to the Sickfund system, which allows patients to frequent only a single pharmacy. This important feature of Dutch pharmacy makes it possible to keep fairly complete patient drug use profiles. These profiles are applied to medication surveillance, to drug utilization review, to consulting with physicians, and to provide patients with talor-made drug information. Moreover, these profiles are being collected and used in joint pharmacoepidemiologic research projects of scientific researchers and pharmacists. Pharmacoepidemiology is becoming a major field of interest in Dutch pharmacy. The access to a complete drug use profile is one of the major factors in the trend in Dutch health care, in which pharmacists are becoming a keystone in optimizing drug utilization.

There are, however, threats to this developing trend. A sanctioned patient-pharmacy liaison conflicts with the proposals of the government to stimulate competition in health care. A competitive environment, according to the Ministry of Health, should give patients maximum freedom to select among the available pharmacies. It is clear that if this proposal is adopted, a unique feature of Dutch pharmacy will diminish.

To be prepared for circumstances where patients can go to different pharmacies, pharmacists are developing new computer technologies: smart card, computer networks, electronic data exchange, etc. All these endeavors aim to save the complete patient drug use profile, the keystone for professional pharmacy.

3. Patient Education

Patient education and counseling were adopted by the pharmacy profession in the early 1980s as important tasks to accomplish. At that time, there was wide support growing for health education and, in particular, information on drugs. Oral as well as written information is important. Several continuing education programs offer courses in communication skills and counseling methods. To improve written drug information, patient information leaflets, patient compendia on drugs, and modern applications of audiovisual methods have been introduced.

Computer technology has become an important tool in patient information (i.e., CD-ROM). A few pharmacy computer systems provide options for promptly printing a tailor-made patient leaflet, which can be given to the patient when the drug is dispensed. Although computer technology has facilitated the development of pharmacists as health educators, oral communication and personal empathy remain necessary in patient education.

4. OTCs and Other Classes and Schedules of Drugs, Nonpharmacy Outlets

Over-the-counter (OTC) drugs are available at pharmacies as well as at druggist shops. There are about 3,000 druggist shops, with a very good regional dispersion. A better access to these outlets is probably a major reason for the fact that druggists cover approximately 65% of all OTC sales. Pharmacists are developing competitive strategies to improve their market position on OTC sales. A major wholesaler now offers a line of branded OTCs which are only marketed through pharmacies. OTC sales are generally lower than in most other countries. It is a formal policy of the Ministry of Health to stimulate self-care and self-medication. In the last few years, a number of previously prescription-only drugs have received OTC status (e.g., ibuprofen, hydrocortisone, ointments).

Moreover, the government expects that limited insurance coverage will stimulate patients to handle more health problems themselves. Contradictory to this policy is the recent measure of excluding many OTC drugs from the drug coverage scheme. It has to be expected that this will cause substitution of expensive (reimbursable) prescription drugs.

5. The Use of Technicians and Others

Technicians play an important role in Dutch pharmacy practice. In 1987, there were 7,766 technicians with a position in either community or hospital pharmacy. Technicians receive three years of training on all aspects of pharmacy practice, and there are continuing education courses (specializations) on analytical chemistry, pharmacotherapy, compounding, and hospital services. Although the pharmacist has the ultimate responsibility, pharmacy technicians have full responsibility in pharmacy practice regarding their tasks of compounding and dispensing. Duties (and responsibilities) of pharmacy technicians have become heavier in the last decade due to developments such as medication surveillance, patient education, etc. There are negotiations between the KNMP and the unions of pharmacy technicians about a raise in salary and improvement of working conditions.

6. Pharmacy in Primary Care

The distribution of drugs in Dutch health care can be broken down into three channels: community pharmacies (70%), dispensing physicians (12%), and hospital pharmacies (18%). As most drugs are being used in primary care, pharmaceutical services should provide an environment for both optimal distribution of drugs as well as for supporting rational decision making concerning the prescribing and use of drugs. In Dutch pharmacy, a third part of pharmaceutical services is concerned with compounding medicines to guarantee tailor-made pharmacotherapy. As in most countries, pharmacy in primary care has become a key discipline in assuring the rational use of drugs. Computerized medication surveillance, patient education, and consulting with physicians are three major tasks that have been adopted by the pharmaceutical profession. The expansion of extramural clinical care (home health care) needs a close cooperation between the two echelons of Dutch health care. Community as well as hospital pharmacists are cooperating closely to guarantee optimal pharmaceutical services to patients leaving the hospital and requiring additional care at home.

Threats for pharmacy in primary care are related to two basic issues: economic pressure due to a sequence of government measures for controlling drug costs, and professional dominance of physicians. The first issue has been discussed in earlier sections of this chapter. With respect to the relationship between pharmacy and medicine, it has to be noted that although more and more community pharmacists are entering the field of clinical pharmacy, physicians (mainly GPs) are continuously alert for possible turf battles.

7. Drug Insurance and Third Parties

For Sickfund patients, drug insurance covers almost all drugs on the market. Excluded are a great number of OTC products. Reimbursement is *in natura* with a copayment of Dfl. 2.50 per prescription, with a maximum of Dfl. 125 per year. The coverage of drug insurance under the proposed basic health insurance scheme is unclear.

Private insurance schemes have almost the same coverage as the Sickfunds, but they have a restitution system. Patients first have to pay out-of-pocket and then apply for restitution.

D. FEATURES OF DUTCH HEALTH CARE AND CONCLUSIONS

The Dutch health care system can be characterized by a number of features:

— There exists a primary echelon with the GP as key professional and gatekeeper for the secondary echelon of care
— In primary care, cooperation between GP and other health professionals (public health nurses, pharmacists, midwives) is being considered as a major determinant of the quality of care.
— The costs of health care account for 10% of the GNP and increase annually at 2%-3%.
— Drug utilization in the Netherlands is relatively low. Drug costs represent about 9% of all health expenditures, with an annual growth of more than 8%. There is high political pressure to reduce costs.

A strong pharmacy-patient liaison offers an optimal environment for patient education and keeping drug use records. Drug use records are important for medication surveillance, drug utilization review, and pharmacoepidemiology.

Although threatened by factors such as economics and professional turfs, the future of Dutch pharmacy looks positive. There is a growing societal acceptance of the pharmacists' tasks in supporting the rational use of drugs.

Developments at the EEC level will be of importance in shaping the future position of pharmacy in Dutch health care.

REFERENCES

1. Lapre RM. A change of direction in the Dutch health care system? Health Policy 1988;10:21-32.

2. Committee on the structure and financing of the health care system (Dekker Committee). Readiness for change. The Hague, 1987.

SELECTED READINGS

Financial Survey on Health Care 1989. Ministry of Health, The Hague, 1988.

Rutten FFH, Freens RJM. Health care financing in the Netherlands: recent changes and future options. Health Policy 1986;6:313-20.

Drost RA, Reijnders PJM. The registration of medicines in the Netherlands. J Clin Pharmacol 1987;27:937-44.

The pharmaceutical market in the Netherlands. Scrip World Pharmaceutical News, 1983.

Mantel AF, Wierenga B et al., eds. The Dutch pharmaceutical market in observation. Univ. of Rotterdam, 1987.

FIP special: pharmacy in the Netherlands. Pharm Weekbl 1987;122:737-92.

Chapter 15

New Zealand

Jenny Cade

A. THE NATIONAL HEALTH CARE SYSTEM

1. History and Background

New Zealand has a land area of 268,675 square kilometers (103,736 square miles), being approximately equal in size to Japan and the British Isles. It is composed of a number of islands; the major ones are the North Island, where two-thirds of the population resides, and the South Island. The climate is mild and temperate, being ideal for agriculture.

The land was settled initially by the Maoris (Polynesians probably from the Cook or Marquesas Islands) around 1,000 A.D. European migration took place from the mid-nineteenth century, principally from Britain. The population was 3.35 million in 1987, including 295,000 Maoris (8.8%) and 95,000 Pacific Islanders (2.8%). About one-third of the people live in the Auckland area of the North Island, and only 16% now live in rural areas. Of the ten most populous cities (see Table 1), only two, Christchurch and Dunedin, are in the South Island. Greater Auckland has one million people.

The annual average population growth from 1981-1986 was 0.8% with a birth rate of 16.7 per 1,000 per annum. Life expectancy was 71 for males and 77 for females in 1984-1986. While males outnumber females at birth, in the over-80 group there are more than twice as many females as males. The years 1977-1987 saw a net migratory loss of 167,000, which compares with a net gain of 180,000 in the previous decade. Controls reduced the flow of immigrants from Europe, offsetting an increased flow from the Pacific Islands, and in this period many Kiwis (as New Zealand-

---

The assistance of two members of the Council of the Pharmaceutical Society of New Zealand, Glen Caves and Owen Diggelmann, is appreciated greatly.

TABLE 1. Largest cities in New Zealand.

Greater Auckland*	1,000,000
Christchurch	167,700
Wellington	136,000
Hamilton	96,200
Dunedin	76,800
Lower Hutt	63,800
Palmerston North	60,700

* Includes: Manukau, Waitamata, Takapuna
and other communities.

ers are known) sought the apparently greener pastures of Australia. (Some 200,000 New Zealanders currently reside in Australia.)

2. Social and Political Environment

New Zealand is an independent member of the Commonwealth, and as such acknowledges the Queen of England as its head of state. She is represented by a governor-general. In fact, political power and patronage resides with the prime minister, as head of the political party in power. The supreme legislative body is the General Assembly, which consists of the governor-general and the House of Representatives. This has 95 elected members (68 North Island seats and 27 South Island). Four seats are elected by the Maoris. Democratic elections are held every three years.

Farming still contributes 60% of New Zealand's exports, although it is a less dominant influence in the economy than 10 years ago, when meat, wool, and dairy products contributed 70%. Major trading partners are the U.S., Japan, and Australia, a far cry from the 1940s, when Britain took 88% of exports. Along with many other countries, New Zealand has seen great change in the last five years. The Labor government elected in 1974 has proved to be the most radical in New Zealand's history. Controls were abolished, the dollar floated, and the economy opened to competition as never before. A major shake-up of the state sector has seen some government departments corporatized and others called upon to operate competitively and profitably. Farm subsidies have been abolished, causing a great deal of hardship to the rural sector, and unemployment has reached an all-time high, exceeding 10% of the workforce. The winds of change that have been sweeping through the entire fabric of New Zealand society have not left the health area unaffected.

The New Zealand health scheme is administered by the Department of Health. It involves a partnership between central and local government, self-employed doctors, and various voluntary organizations. The Department of Health, in a mission statement in 1988, said that its goal is "to enable all New Zealanders to have access to as comprehensive, equitable, effective and efficient a health system as can be provided within the resources available from public funding." The department has itself been restructured, and the government is reforming the health sector by amalgamating hospital boards and district offices of the Department of Health to form Area Health Boards. This process will be accompanied by decentralization, with local areas being responsible and accountable for the administration of health services.

3. Financing of Health Costs, and Facilities

Vote Health amounted to $NZ 2.9 billion in 1987, with hospital boards and Area Health Boards taking 69% of the vote. In 1986, the 186 New Zealand hospitals employed 48,000 staff to deal with 428,000 admissions and more than 4.2 million outpatients. Even with an increased turnover per bed of 17.4, surgical waiting lists increased to 50,454, 10,000 more than in 1982. Many choose to take out private health insurance to ensure speedy treatment and to subsidize fees paid for health services. The government pays a General Medical Services Benefit to medical practitioners. From January 1, 1989, the rates were set at $NZ 16 per visit for children, $NZ 12 for beneficiaries and the chronically ill, and $NZ 4 for all other adults. These fees form only part of the total charges made by practitioners, and the balance is payable by the patient. Patients in public hospitals receive free treatment, while those in private hospitals pay for most of their treatment but receive a contribution from the government.

In 1987, 6,390 doctors were in practice in New Zealand, of whom approximately half were general practitioners and half were specialists. There were also 1,238 dentists in private practice. The doctors are self-employed but receive substantial subsidies from the government by way of the GMS allowances. They administer and organize the practice and pay ancillary staff. Many work in group practices, which may include associated health professionals such as physiotherapists.

4. Pharmacy Manpower

The Pharmaceutical Society of New Zealand acts as the registering body for pharmacists and pharmacies. Any person wishing to practice pharmacy in New Zealand must pay an annual practicing fee for his or her name to remain on the Register of Pharmacists. Qualification within New Zealand is by way of a degree course at the University of Otago or a diploma course at the Central Institute of Technology, followed in each case by a preregistration year of practical training (internship). Reciprocal registration agreements exist between the Pharmaceutical Society of New Zealand, the Royal Pharmaceutical Society of Great Britain, the Pharmaceutical Societies of Eire and Northern Ireland, and the pharmacist registering bodies in each of the states of Australia. These relationships provide for the mutual recognition of pharmaceutical qualifications based upon registration in one of these countries.

Pharmacists from other countries wishing to register in New Zealand must have their qualifications assessed to determine their equivalence, with respect to content and standard, to current New Zealand qualifications. Those whose primary qualification is not considered equivalent may take a screening examination offered by the Australian Pharmacy Examining Council (but taken in New Zealand), followed by a law examination and a period of practical supervised experience.

At the end of 1988, there were 3,444 pharmacists on the New Zealand register. About 10% of pharmacists are employed in hospitals, and less than 1% in industry and government departments. The vast majority work in the 1,116 retail pharmacies spread throughout the country. Current New Zealand law prevents multiple ownership (except for Boots the Chemists, who had several branches established when the law was introduced). While there are a few dispensing-only pharmacies, most both dispense and offer for sale a comprehensive range of traditional, mainly health-related, products. Average turnover in 1988 was NZ$ 794,700, with 57% being derived from Social Security dispensing.

5. The Social Security System

Pharmaceuticals available under Social Security are defined in the Drug Tariff compiled by the Health Department. Drugs may be "free" (SS), partly subsidized (i.e., part charged to the patient, usually where a "free" cheaper part-charge generic equivalent exists), or non-SS (where the patient pays for the entire cost). The list of products with part charges is used by the government to keep prices down, as patients and doctors prefer to use "free" drugs. In general, the usual period of supply is five days (four days in the case of antibiotics), with or without a repeat supply of a similar quantity. For long-term therapy, up to three months' supply may be prescribed, or six months in the case of oral contraceptives. Some drugs are categorized as "Retail Pharmacy—Specialist" and may only be prescribed on the recommendation of a specialist. Others are classed as "Hospital Only" and, although available on the prescription of a general practitioner, may only be dispensed by hospital pharmacies.

Although the Department of Health has always closely supervised drug costs, escalation of prices saw the Pharmaceutical Benefits bill rise beyond $450 million in 1987. The government then introduced a system of patient charges of $NZ 1 per prescription item. Beneficiaries, children, the chronically ill, and accident cases (ACC) were exempted. This scheme had little impact on the drug bill, as over 60% of patients were exempt. The drug bill reached an all-time high in 1988, when it topped the $NZ 500 million mark. The government then, in February 1989, introduced a system of flat charges payable by all. Those previously exempt paid $NZ 2 per item and all other adults $NZ 5 per item. To assist those who were disadvantaged, a safety-net limits the amount payable. Individuals reaching 25 items or family groups reaching 40 items within the year are exempted from further charges for that year. At the same time, GMS benefits have increased, the object being to enable easier access to the general practitioner. It is too early to assess the value and impact of the scheme.

6. The Medical Establishment: Entry, Specialists, Referrals

The point of entry into the health system is the general practitioner (GP). The patient is free to choose a GP but may find that busy practices have a full book, thus limiting choice. X rays, laboratory tests, physiotherapy, audiometric testing, and the like may be ordered by the GP at little or no cost to the patient. Access to specialists is by way of referral by GPs. The referral may be to a specialist in private practice or to a hospital clinic where there is no charge to the patient.

7. Coverage

Any person normally resident in New Zealand is eligible for health benefits. There is also a reciprocal arrangement with Australia. While even in isolated districts people have access to a GP (the government sponsors special area doctors in remote areas), many people must travel for hospital and specialist services.

B. PHARMACEUTICAL INDUSTRY
AND DRUG DISTRIBUTION

1. Structure of the Industry

Many of the world's larger multinational pharmaceutical companies are represented in New Zealand either by subsidiaries or agencies. The companies represented are mainly from the U.S., U.K., Scandinavia, West Germany, and Switzerland. Their New Zealand subsidiaries report either directly to their head office or through Australian or Asia/Pacific area offices.

Most companies import virtually all of their requirements. There is some third-party manufacturing in New Zealand, but the number of companies that have their own manufacturing plants, never great at any time, is steadily diminishing, and by the end of 1988, there were only a handful (6) remaining.

There are two main generic companies in New Zealand, although one of these also represents overseas research-based organizations. Several of the larger companies have full-time medical directors on staff, others have part-time medical directors, while the remainder are generally covered from Australia.

The trend toward multiple representative teams within the same company is gaining momentum, certainly within the larger companies. Two companies at the end of 1988 had four separate teams of medical representatives, each promoting a small and generally specialized group of products.

The research-based companies are represented by the Researched Medicines Industry Association of New Zealand Inc. (RMI), formerly the Pharmaceutical Manufacturers Association (New Zealand) Inc., and there are some 40 members representing approximately 50 major multinational pharmaceutical houses. The RMI fosters the New Zealand Institute of Medical Representatives, which has as its aim the improvement in standards and product knowledge of members. This is brought about by three study courses and examinations over a two-year period.

2. Research and Development, Imports, Exports, Patents, Licensing

Research

No basic research on new pharmaceuticals is done in New Zealand. It is generally only at the Phase II or Phase III trial stage of development that local investigators become involved. These trials are regulated by the Medical Research Council, a government-appointed body, through the Standing Committee on Therapeutic Trials. Approval by this body and the ethics committee of the hospital concerned are required before any trial of a nonregistered medicine can proceed.

Importers of pharmaceuticals must keep product monographs and certificates of analysis for each batch of each product brought into the country.

Import Licensing

Pharmaceuticals for the last year have been exempt from the need for import licenses in line with the government policy of phasing out all licensing. However, there is currently a move to change the Concessionary Entry Policy, which removes any duty on pharmaceuticals that are on the Pharmaceutical Benefits Scheme. About 23,000 individual concessions are likely to be revoked, and there will be a policy of making a total item code free of duty where there is no objection from the local manufacturers. If there is no agreement, it is likely that all individual products imported fully finished will carry duty at the current rate of 22.5%. It is the intention of the government, however, to phase down the duty rates in the next few years so that most items will carry a duty rate of about 15%. In addition, it is likely that the percentage of New Zealand content required to qualify for local manufacture will increase from the current 25% to 50% in the next 3-4 years.

The above policies are in line with the government's intention to open up the market and make local manufacture competitive on a worldwide scale. However, there has been considerable lobbying from the local manufacturers to slow down the process, in view of the significant number of factory closures and the high level of unemployment.

Exporting

There is a limited amount of export of pharmaceuticals from New Zealand. This may be the total production of a specific line for a geographic area (e.g., Southeast Asia), or it may be the manufacture of a product that for a larger country would constitute a comparatively uneconomical run in its production plant. The smaller New Zealand plants are better able to handle such manufacture.

Manufacturing premises of companies which produce for export are inspected by regulatory authority personnel of the countries concerned.

Patents

Pharmaceutical patents may cover the product itself or the process of manufacture. Patents are not granted for methods of administration to human beings, but they are granted for methods of administering certain compounds to animals.

The term of a patent is 16 years. Under Section 31 of the Act, there is provision to extend the original patent by a maximum of ten years for exceptional circumstances where there is inadequate remuneration. There have already been a number of extensions granted by the High Court in New Zealand under this section.

Under Sections 46 to 50 of the Act, at least three years after the grant of a patent any interested party can apply for a compulsory license or for an endorsement of the patent with the words "licenses-of-right." Such applications are made on the grounds that the patent is not being commercially worked in New Zealand, or on other specified grounds which may be broadly classified as abuses of monopoly. Such a compulsory license or license-of-right endorsement may also be granted on the application of a government department, after at least three years from the date of grant of the patent, on any of the grounds relating to abuses of monopoly (Section 49). The commissioner is required to consider a number of factors in exercising discretion, including the fact that the patent holder should receive a reasonable remuneration.

These regulations covering abuses of monopoly have a general application. There are also compulsory license regulations that apply solely to inventions relating to food and medicines. According to these, the Commissioner should grant a compulsory license on application by an interested party at any time after the grant of a patent unless it is believed that "there are good reasons for refusing the application." In setting the royalty and other terms of a compulsory license, the Commissioner is required to help maintain food and medicines at the lowest prices to the public, consistent with a reasonable return to the patentees. Only one compulsory license has been awarded by the Commissioner (in 1974), but the license was never operated by the licensee. There was another case pending as of December 1988.

The Industrial Property Advisory Committee has made a majority recommendation that the normal patent term be extended by four years to a maximum of 20 years to take account of regulatory delay. However, to date, there has been no move in New Zealand to revise the Patents Act owing to other legislative priorities.

The Pharmaceutical Manufacturers Association of New Zealand has made unsuccessful representations against the power of the Commissioner to grant compulsory licenses.

In addition to the compulsory license provisions, the New Zealand Patents Act contains general provisions (Sections 55 to 58) enabling any government department and any person authorized in writing by a government department to "make use and exercise any patented invention for the services of the Crown." Such use is made upon such terms as may be agreed, either before or after use, by the government concerned and by the patentee with the approval of the Ministry of Finance. The provisions include, but are not restricted to, periods of emergency. They are quite general and could be invoked to authorize the manufacture in New Zealand or the importation into New Zealand of a patented compound and/or pharmaceutical composition, regardless of the patent rights or normal license agreements. The provisions are limited by the requirement that the use must be "for the services of the Crown." But an amendment to the Hospitals Act provides that the services of a hospital board should be deemed to be the services of the Crown. The effect is that the boards are empowered to purchase drugs from nonpatented sources for dispensing solely by the boards.

3. Manufacturing

Pharmaceutical manufacturing in New Zealand has always been wholly concerned with the production of the product in its final form. Manufacture of pharmaceutical substances has never been undertaken. Besides the six RMI members still manufacturing in New Zealand, two generic manufacturers and three contract manufacturers make up the local scene. Several multinational pharmaceutical companies have in recent years closed their New Zealand plants for a variety of reasons:

a. Greater ease of importation (fewer trade barriers)
b. Cost of upgrade
c. Rationalization of plants worldwide
d. Generic competition for older products as newer products are being manufactured in one plant for the company's total world requirements.

Plant inspections are carried out by the Health Department inspectors on an annual basis. The Health Department has its own Code of Good Manufacturing Practice, which parallels the multinationals' own codes.

Qualified personnel working in pharmaceutical manufacturing are science graduates and pharmacists.

4. Drug Approval and Government Regulation

Product Registration Procedure

New Medicines

The drug registration procedure is organized in two parts. First the company makes an Application for Consent to Distribute a New Therapeutic Medicine. The second stage is to make an Application for Inclusion of Medicines in the Drug Tariff. If the first application is successful, the drug is registered and the company is allowed to market the product.

Application is made to the Ministry of Health, which may then refer it to the Medicines Assessment Advisory Committee for detailed consideration. The information required by the Ministry includes:

a. All names of the drug (official, approved, chemical, and proprietary). Reference should be made to the publication in which these names appear. Registered owners of proprietary names should also be listed. The structural formula, where appropriate, should be indicated.
b. Name and address of manufacturer
c. Form of drug (e.g., pure substance, tablet, mixture with other drugs). Strength and identifying marks should be specified, as well as all active and inert ingredients and their quantities.
d. Method of distribution
e. Status in other countries
f. Size of container as well as details on labeling, package inserts, and all descriptive material that is to be attached
g. Dosage and method of administration
h. Purposes for which the drug is recommended
i. Evidence of clinical efficacy
j. Adverse medicine reaction data.

Finally, a full report must be supplied of manufacturing methods and quality control. This must include details of stability testing, predicted shelf life, procedures for determining the drug in biological fluids, pharmacological/toxicological reports and reprints of supporting papers, and photocopies of certificates of medicine approval in other countries.

In vitro and animal studies can be carried out in countries outside New Zealand provided they comply with conditions laid down for this work by the official administrations in the United Kingdom, United States, Canada, and Australia or by the World Health Organization. Details on the bioavailability of the drug are also required; this should be carried out in vivo by an independent investigator. This in vivo work can either be based on pharmacological responses or chemical data relating to blood levels or on urinary excretion of drug or metabolites. Clinical trials may also be conducted.

Bioavailability studies should be conducted in comparison to an existing preparation, an appropriately formulated solution of the active ingredient, or a placebo if appropriate. In the case of combinations of drugs, the bioavailability of each active ingredient must be established where possible.

If in vivo studies are not practical, then in vitro evidence will be accepted. The New Zealand Medicines Regulations note that the following characteristics require special scrutiny:

a. Relatively low solubility and tendency to polymorphism
b. Narrow range of therapeutic index (critical concentration drug): e.g., digoxin, levadopa
c. Microdose formulations; i.e, less than 5 mg
d. Substances with suspect availability, including drugs known to have a tendency to combine with excipients
e. Substances where absorption is undesirable
f. Substances whose too-rapid release is undesirable

The Medicines Assessment Advisory Committee meets three times a year. Material for its consideration should be submitted not later than two months before the designated meeting.

Changed Medicines

Notification of change to a registered medicine is made under Section 24 of the Act and is referred to as a Changed Medicine Notification.

Under Section 24 of the Medicines Act of 1981, a changed medicine may not be distributed until 90 days after notification of the change has been given to the Director-General of Health. The Director-General may modify this either by giving consent to earlier distribution if there is satisfaction of all material points, or by referring the notification of change to the Minister, in which case the medicine may not be distributed until the consent of the Minister has been notified in the *Gazette*.

The changes for which notification must be made are:

a. The purpose for which the medicine is represented to be used, the recommended dosage, or the recommended manner of administration
b. The labeling of the medicine, or of any container or package in which the medicine is packed, or any descriptive matter accompanying or enclosed in any such medicine, container, or package
c. The strength, quality, or purity of the medicine
d. The methods of manufacture of the medicine or the facilities for testing its strength, quality, purity, or safety
e. The location of the premises in which the medicine is manufactured.

Reports of tests made to establish the safety and/or efficacy of the changed medicine should be submitted in appropriate cases. Where relevant safety data have already been lodged with the department, this may be referred to.

Within 45 days of receipt of notification, the Director-General of Health may require additional information on any matter relating to the medicine. The further information, or an interim reply, is to be supplied within 90 days of receipt of the notification by the department. Failure to do so may result in the notification being referred to the Minister as described above.

If a drug is changed, it may not be distributed until 90 days after notification of the change to the Director-General of Health.

The Department of Health notes that a large majority of new and changed drug applications fail to include complete information. The most common omissions that cause delay in application processing are lack of a local agent, which is essential; proof of approval to distribute in countries other than New Zealand; labeling errors, which arise through not conforming to the requirements of the specific legislation; and incomplete quality control data and omission of information on inactive ingredients and stability studies. An expiration date is not required by law but is generally encouraged.

Registration: Drug Tariff

Not until marketing approval has been gained can an application be made for inclusion in the Drug Tariff. In practice, only very rarely will it be included without experience of its use in New Zealand. Applications for Drug Tariff listing are considered first by the Pharmacology and Therapeutics Advisory Committee, which is composed of members of the medical profession. Its recommendations are submitted to the Ministry of Health, and a Drug Tariff Amendment is made if the application is approved. Amendments are usually made in April, August, and December.

Applications for listing close on February 15, June 15, and October 15. The applications must contain the following information:

a. Official or approved name
b. Form, strength, and recommended dosage
c. Recommended indications
d. Prices to retail pharmacists in New Zealand
e. Prices to retail pharmacists in the United Kingdom, Australia, and the country from which the preparation is exported
f. Availability of the preparation in the country's four main centers
g. Names of doctors who have used the preparation in New Zealand for at least four months
h. Nine copies of published papers from recognized Western medical publications.

In general, drugs are not included in the Drug Tariff unless they are developments sought by the medical profession itself. Combination drugs are not usually accepted, and there are always newer preparations awaiting consideration by the Tariff Committee, as well as older drugs that are considered to have been superseded.

It is useful for inquiring committees to know the status of the drug in other countries, in particular, whether it has been approved by the Food and Drug Administration in the United States, the Medicines Commission in the United Kingdom, or the Australian Department of Health.

5. Advertising and Promotion, Price Regulation

The Researched Medicines Industry Association Code of Practice sets out the following general provisions.

Provisions of the Code

Marketing of Products

At all times, marketing activities must comply with requirements laid down by the following current legislation and with any subsequent legislation enacted:

Misuse of Drugs Act	1975
Toxic Substances Act	1979
Medicines Act	1981
Medicines Regulations	1984

General Provisions Applicable to All Marketing Activitists

a. Methods of marketing must never be such as to incite unfavorable comment and bring discredit upon the pharmaceutical industry.
b. Information furnished to the medical profession about a medical specialty product must be accurate and balanced and must not be misleading, either directly or by implication.
c. Claims for the usefulness of a product must be based on current knowledge and reflect this evidence accurately and clearly.
d. There shall be a clear differentiation between statements based on clinical and pharmacological evidence and any theoretical projection of that evidence. Such evidence shall be referenced.
e. Communications on medical specialties must reflect an attitude of caution: superlatives are to be avoided. Words such as "safe" must not be used without qualification, and it must not be stated categorically that a product has no side effects, toxic hazards, or risk of dependence.
f. Where reference is made to the prescribing of a preparation in terms of the Drug Tariff, the phrase "freely" prescribed on the Drug Tariff and similar misleading phrases must not be used.
g. Disparaging references to other products or manufacturers, either directly or by implication, must be avoided. Authoritative clinical trials, however, may be quoted provided that the full text of these published trials is readily accessible in New Zealand for verification and adequate reference is given. Competitive products shall not be identified by brand name. Comparisons to other manufacturers' products shall be on a factual and fair basis. Where unpublished references are used to support comparisons with competing products, the source material (in the English language) must be made available on request. Care should be exercised to ensure that the above provisions, where relevant, are applied when advertising restricted and pharmacy-only medicines to the public.
h. The chief executive of a member company is responsible for observance of and adherence to the Code.

Price Regulation

Prices of pharmaceuticals are regulated to a large extent by the Drug Tariff. If a product is included in the Drug Tariff (free of charge to the patient), some measure of negotiation between the manufacturer and the customer (the government through the Health Department) will have taken place. Prices of the product in the United Kingdom, Australia, and the country of origin are included in the submission for both listing on the Drug Tariff and any subsequent applications for price adjustments.

A national tender system for the supply of pharmaceuticals to the hospital system provides keen competition within the industry.

6. Competition

The Market

Ideally, products in a market should be substitutes for each other, and there should be minimal substitutability between markets. In practice, one must choose boundaries for markets somewhat arbitrarily, as there tend not to be sharp divisions in substitutability.[1]

The Firms

For each medicine (counting different brands separately from generic equivalents), IMS data gives 137 firms producing medicines in the 77 therapeutic groups in 1986-1987. Not all of these firms, however, can be regarded as independent. A number of overseas companies use the same distributor in New Zealand. Some firms are branches of others, and some share the same address. In each of these cases, firms with different names are unlikely to be competitors. This reduces the number of separate firms in 1986-1987 from 137 to 105.

The ethicals market, taken as a whole, is moderately concentrated. The aggregate shares held by the leading firms (1986-1987) are:

1 firm	13%
2 firms	22%
3 firms	32%
4 firms	48%
5 firms	71%

To examine entry in the pharmaceutical industry in New Zealand, Sutton first looked at individual products.[1] She took as her sample the top 50 medicines (by wholesale value) in the 1981-1982 and 1986-1987 financial years. Different brands of the same generic substance were counted as separate medicines. Figure 1 shows the changes in rank. Five of the products in the top ten in 1981-1982 were still there five years later. About half (23) of the 50 top products in the earlier year had slipped to below fiftieth place 5 years later. Thus entry of new products appears relatively easy.

FIGURE 1. Changes in rank (by sales) of the top 50 preparations in 1981/82.

Rank in	Rank in 1986/87			
1981/82	1-10	11-30	31-50	>50
1-10	5	3	0	2
11-30	1	5	5	9
31-50	1	3	4	12
				23

This squares with a slowing rate of innovation if the new products are either only minor advances or simply changes in form, such as new compositions of two or more existing medicines previously sold separately.

Sutton next looked at firms to see how their rankings have changed in recent years. Of the 11 leading therapeutic classes (defined above), the market leader changed in four cases over the three-year period of 1983-1984 to 1986-1987. In each case, this was due to a change in the leading product rather than a change in distributing firms in New Zealand. These results demonstrate that, as with the earlier U.K. studies, entry of new products is relatively easy. And through such entry, new firms gain leadership.

There are a large number of firms in the pharmaceutical industry in New Zealand, and the level of overall concentration (degree of monopoly) is moderate. Concentration is high in the submarkets defined by therapeutic groups, but leadership in therapeutic classes changes fairly rapidly. Entry of new products is not unduly difficult.[1]

Generic Competition

There are two major generic houses in New Zealand, although several multinational pharmaceutical companies also market small ranges. Generics have between 7% and 10% of the current ethical market, but with the coming legalization of substitution by the pharmacist, albeit with some conditions, this percentage seems set to rise.

7. Cost-Containment Activities

Within the pharmaceutical industry, rising costs are being felt as in most other industries. Companies are moving to counter this trend in a number of ways, most of which reflect the international nature of the business:

    a. Greater movement toward the total world production of the final dosage form of a particular medicine in just one or two plants, usually in countries that offer tax incentives to new industries

    b. Reduction of the number of products or packs marketed

    c. Closure of local production facilities with a move to total importation of stock.

The Health Department, as the monopoly purchaser of the vast majority of ethical pharmaceuticals, is also looking hard at containing the cost of prescribed medicines. Recent moves include the introduction of patient charges (see Section A.6 — The Social Security System) and a much tougher stance on negotiating price increases. Various other moves are being studied and could be introduced depending on the success of these measures.

C. PHARMACY PRACTICE

1. Dispensing a Prescription

Pharmacy practice in New Zealand reflects the British heritage of the country. Laws are based, in many instances, on British equivalents, the British Pharmacopoeia, or the British Pharmaceutical Codex set standards, and Martindale is present in almost every pharmacy. Again, the British National Formulary contains most commonly compounded preparations.

The average retail pharmacy dispenses annually around 24,000 prescriptions. Twice a month, batches of prescriptions are submitted to pricing offices for pricing and payment made by the Treasury, usually by direct credit into the pharmacy's bank account. A system whereby prescriptions are sent via modem to a large central computer and priced electronically has been successfully tested in 1988-1989.

2. Record Keeping

Prescriptions are required to be recorded in numerical order, and often the records are kept on computers. New Zealand pharmacy has a high degree of computerization, with some two-thirds of retail pharmacies operating computer systems. Other pharmacies maintain records on microfilm or use manual systems. Narcotics records are kept in a bound register which must be balanced every six months. Advisory pharmacists employed by the Department of Health carry out inspections on behalf of the Pharmaceutical Society. However, with the advent of area health boards, their inspecting roles are less well-defined.

3. Patient Education

A system of cautionary and advisory labels is used to assist patients in the use of their medication (see Table 2). This is used as an adjunct to counseling of the patient when the prescription is handed out. In addition "How to Use" leaflets are made available by the Pharmaceutical Society and distributed by the Pharmacy Guild. Pharmacies also distribute a variety of informational leaflets and magazines on various aspects of health, supplied by the Department of Health, manufacturers, and others. The Society and the Guild have been to the fore in the promotion of national campaigns to inform the public about skin cancers, poisoning, etc.

Campaigns run by the Heart Foundation, Asthma Society, Neurological Foundation, and others are also supported actively by pharmacies, which are acknowledged as appropriate centers for the dissemination of health information.

4. OTCs and Other Classes of Drugs, Other Nonpharmacy Outlets

Over-the-counter medicines for trivial complaints (e.g., aspirin, paracetamol, cough drops) may be sold by any retail outlet in New Zealand, but most pharmaceuticals for OTC sale are classified by Medicines Regulations. Restricted Medicines may only be sold from a pharmacy, and the details of the sale must be recorded (e.g., mefenamic acid, idoxuridine 0.1% cream). Pharmacy-only medicines may only be purchased from a pharmacy (e.g., combination analgesics containing codeine, antihistamines, cough and cold remedies containing sympathetic amines and/or dextromethorphan).

In remote areas where the nearest pharmacy is ten or more miles away, other retail outlets are occasionally given permission by the Pharmaceuti-

TABLE 2. Cautionary and advisory labeling scheme – Summary. This chart to be used in conjunction with the explanatory booklet.

LABEL 1	LABEL 2	LABEL 3	LABEL4	LABEL 5	LABEL6	LABELS 7 & 7A		LABEL 8	LABEL 9
This medicine may cause drowsiness and affect your ability to drive or to operate machinery. AVOID ALCOHOL.	Avoid taking alcohol while you are having treatment with this medicine.	TAKE EACH DOSE ON AN EMPTY STOMACH one hour before OR two hours after food.	Do not take this medicine with milk, antacids or preparations containing iron or calcium.	Alcohol Certain foods and other medicine should not be taken while you are having treatment with this medicine. SEE SEPARATE CARD.	REFRIGERATE - DO NOT FREEZE	DISCARD CONTENTS AFTER	DISCARD DAYS AFTER OPENING	Avoid excessive exposure to direct sunlight while you are undergoing treatment with this medicine.	Do not stop taking this medicine without your doctor's permission. ALWAYS KEEP A SUPPLY.
Antihistamines (daytime use) Antidepressants Buprenorphine Clonidine Pethgidine Pentazocine Sedatives (daytime use) Tranquillisers	Disulfiram Hypnotics Oral - Hypoglycaemics Sedatives (nighttime use) Trichomonacides	Many Oral antibiotics Chloramphenicol Erythromycin Stearate caps ERA caps Flucloxacillin Linconmycin Phenoxymethyl Penicillins Tetracyclines and Derivatives	Tetracyclines and Derivatives Delete milk with Doxycycline Minocycline Penicillamine	Monoamine-oxidase inhititors Isocarboxazid Phenelzine Tranyl-cypromine	Reconstituted antibiotic syrups and suspensions Insulins	Some reconstituted antibiotic syrups and suspensions.......7 days Aspirin Suspension...........7 days Soluble aspirin mixtures............3 days Diluted or admixed liquid medicines and external Preparations......14 days Eye Drops/ Ointments............30 days Glyceryl Trinitrate in original pack.......12 days Glyceryl Trinitrate not in original pack........8 days Sodium Cromoglycate solutions............30 days		Topical Tar preparations Nalidixic Acid Griseofulvin Demeclocycline	Beta Blockers Antihypertensives Steroids for long term oral use Medicines used in the treatment and prophylaxis of asthma.

395

TABLE 2 (continued)

REFERENCE A	REFERENCE B	REFERENCE C	REFERENCE D	REFERENCE E
The words "whole" to be inserted on main label immediately after "table, capsules" etc	The words "immediately after food" to be inserted on main label in an appropriate position	The words " half an hour before food" to be inserted on main label in an appropriate position	The words "until finished" to be inserted on the main label in an appropriate position	The words "with a glass of water" to be inserted on main label in an appropriate position
Enteric coated tablets Repeat action tablets Sustained action tablets and capsules	Medicines likely to cause gastro - insertinal irritation Antirheumatic agents Cortisone and derivatives Potassium proparations Theophylline and derivatives Medicines whose maximum therapeutic effect is enhanced by food or where food is used to aid compliance.	Oral anti - cholinergic preparations Atropine preparations Belladona Hyoscine preparations Rifampicin	Antibiotics Antifungal agents Sulphonamides and other preparations when completion of the course is important	Medicines likely to cause oesophageal irritation Antirheumatic agents potassium products Tetracyclines Theophylline and derivatives

cal Society to sell a limited range of specified pharmacy-only medicines. These situations are monitored closely.

5. The Use of Technicians and Others

Technicians, known in New Zealand as dispensary assistants, are widely used in both hospital and community pharmacy. They undertake a year-long theoretical course in conjunction with 18 months of practical experience before certification. Under supervision, dispensary assistants perform the recording, labeling, and a variety of simple dispensing tasks (e.g., counting tablets, pouring simple mixtures). Each pharmacist may supervise only one dispensary assistant. A submission has been made by the Pharmaceutical Society to the Minister of Health to widen the range of duties that may be undertaken by dispensary assistants under the supervision of a pharmacist and thus free the pharmacists to spend more time advising patients on their medication.

6. Pharmacy and Primary Care

Pharmacy has always been regarded by the New Zealand public as a focus of primary health care. Because of the ready availability of the pharmacist and the absence of a consultation fee, the public freely seeks advice on health matters from pharmacists. Minor ailments are handled and referrals made for more serious conditions. In line with the rest of the Western world, there has been, over the past few years, an upsurge in interest in alternative health treatments, and some pharmacies stock and sell homeopathic and herbal remedies. Liaison with other primary care providers has not been particularly close in the past, but one of the stated aims of the Pharmaceutical Society is to encourage close working relationships between pharmacists and other health professionals.

7. Drug Insurance and Third Parties

As has been previously mentioned, although the New Zealand government takes prime responsibility for the health of New Zealanders, many take out health insurance with private companies to ensure prompt surgical treatment, should it be required. The insurance companies also provide subsidies on fees incurred for doctor visits, prescriptions, etc. Any accidental injuries and consequent medical conditions, regardless of blame, are handled by the Accident Compensation Corporation, which is financed by levies on employers. The corporation provides compensation for permanent physical disability and for loss of earnings on an income-related basis.

D. SPECIAL FEATURES
IN NEW ZEALAND

The Maori population of New Zealand has for centuries used various native plant remedies for a variety of ailments. These have not been thoroughly investigated, so the evidence concerning their efficacy is anecdotal. The Maori perspective is seen as increasingly important by the Department of Health. The issue of biculturalism is one that is being addressed on many fronts in New Zealand society, not the least of which is health. Undoubtedly, the *pakeha* (European) section of the population can learn much from the traditions of the Maori.

E. CONCLUSIONS

New Zealand has a well-established pharmaceutical network that provides convenient and efficient service. Approximately 1,100 community pharmacies are distributed throughout the country and sell a diverse range of OTC health-related commodities in addition to their dispensing practices. There is a high degree of computerization.

The costs of the Pharmaceutical Benefits Scheme are high, and although the government is making strenuous efforts to contain costs, it seems likely that the contribution payable by the patient will increase. The main problem is that there is no control over the point of the pen where prescriptions are generated. The computerized pricing system that has been successfully tested will yield valuable information on drug usage which has been formerly lacking. The Drug Tariff, which lists pharmaceutical benefits, is in need of reform. While it has served the country well for 40 years, it has, like Topsy, just grown. Various aspects of the Tariff are in need of simplification and would not stand up under close scrutiny. Government attempts at cost containment are having a severe impact on community pharmacy, and the country may well see a decline in numbers on economic grounds. Some of the diminution is likely to be by amalgamation, which could well be a good thing for pharmacy practice, but smaller pharmacies in rural areas could be the first to go, to the detriment of the community.

A change in emphasis is occurring as the New Zealand public shows more interest in folk medicines and alternative health approaches. This is likely to alter the somewhat conservative face of New Zealand pharmacy as it adapts to changes in society. The future role of pharmacy technicians will widen in scope, freeing the pharmacist from the technical role to spend more time giving advice to patients. This will call for a greater emphasis on continuing education and, in particular, the cultivation of better communications skills by the pharmacist.

REFERENCE

1. Sutton, F. Trends in Pharmaceutical Benefits Expenditures over the last decade. New Zealand Year Book 1988.

SELECTED READINGS

Air New Zealand, Almanac 1989.
Department of Health. Clinical Services Newsletter No. 254 — Pharmaceutical and GMS Benefit Changes.
Department of Health. Report for the year ended March 31, 1988.
Pharmacy Guild of New Zealand Annual Report 1988.
Pharmaceutical Society of New Zealand Annual Report 1988.

Chapter 16

Nigeria

Olanrewaju Ogunlana

A. THE NATIONAL HEALTH CARE SYSTEM

1. History and Background

Nigeria is a federation of 21 semiautonomous states with a capital territory named Abuja. Nigeria is located on the Gulf of Guinea and has an area of about 932,773 square kilometers. It is a tropical country that lies between latitudes 4° and 14°N and experiences two main seasons, dry and rainy. The climate changes from the coastline to the north as it becomes progressively dry. A span of swampy forests in the south to the grasslands in the north covers the almost 1,050 kilometers of land from the coast.

The population of Nigeria was estimated in 1979 by the United Nations to be 74.6 million, while the 1984 estimate was 93.7 million. It has recently been estimated to be 120 million, as annual population growth was put at 3%.

There are five major ethnic groups in Nigeria: Hausa/Fulani, Ibo, Ibibio, Kanuri, and Yoruba. It should be mentioned, however, that there are over 260 ethnic groups in the country. Consequently, one is not surprised to note the extent that the culture, tradition, socioeconomic development, and religions of these have affected the history of geographic exploration, trade, and medical care.

The history of medical care in Nigeria has been recorded back to 600 B.C., when the Phoenicians and the Carthaginians came across the Sahara to northern Nigeria. Traditional medicine was practiced; the *wombai* and the *gozau* of the Hausas and Nupes in the north and the *adahunse* and *dibia* of the Yoruba and Ibos of the south constituted the known primitive forms of medical care.

By the thirteenth century A.D., Islamic and Arabic culture moved toward northern Nigeria and influenced the medical care. Furthermore, it is recognized that one of the earliest contacts of the Western world with Nigeria was with the Portuguese toward the end of the first half of the fifteenth century. This was followed by the transatlantic slave trade with Western Europe and the Americas, which started in the sixteenth century. Records show that physicians accompanied the slave ships, but their services were mainly for their staff, with occasional services to the slaves.

The development of a health care system in Nigeria can therefore be linked with the early traders, explorers, and religious groups. These Europeans and Arabs, who came primarily in the last 500 years with their specific objectives, had to practice various health care delivery systems. Observations show that the types of health care that they practiced at the time were far from being scientific, as there were clear indications that the medical care practiced was mainly empirical, similar in character to the traditional medical care. It is believed that the health services in Nigeria, therefore, essentially developed with traditional services and in parallel with the political administration of the country. This is noteworthy because the medical services were initially centralized before decentralizing into regions, states, and local governments. Such a developmental process provides some explanation for the urban areas receiving better attention than the rural areas where, in effect, traditional medical care maintained services with greater zeal and attention.

The history of medical planning in Nigeria can be said to have commenced with the Ten-Year Development Plan of 1946. The plans were not comprehensive but showed genuine desire on the part of the British government to assist the development efforts of the government of Nigeria.

The advent of regionalization in 1952 caused the first policy paper on public health for the Western Region:

> The Health policy of the present Government is based on three fundamentals. Firstly, the provision of adequate pure water; secondly, the progressive building up of environmental hygiene; and, thirdly, the expansion of hospitals, maternity and child welfare and dispensary services coupled with a vigorous campaign on preventive medicine in the field.

To have a clearer approach to the broadly stated policy, one has to refer to the involvement of voluntary agencies in the earlier supply of health care. Toward the end of the nineteenth century, in 1895, the Sacred Heart Hospital was established by the Roman Catholic Mission at Abeokuta, and the pattern was for the missionary religious bodies to provide modern medical care to the people. The missionary organizations were the Anglicans, Presbyterians, Methodists, Baptists, and Roman Catholics. They have played more important and significant roles than government in the provision of modern medical care to the inhabitants, up to the middle of this century. Considering such situations, the northern areas of the country had to rely on government medical services, and the military clinics and hospitals formed the beginning of these services. The traditional health care practitioners were, however, functioning in all areas of the country. Some historians have commented that the rivalry that developed between the traditional and Western medical practitioners has persisted until today.

At the onset of regionalization and self-government in the country (Table 1 shows the statistics of medical institutions), most of Nigeria's manpower was engaged in the government and voluntary services and also in commercial and industrial houses. A very small number had private practices. One commentor noted "the wholly independent general practitioner was rare . . . the average Nigerian could not afford to pay economic fees for private medical care." However, in Lagos, private/general practitioners existed quite early and in higher numbers because many of the populace earned enough to afford such services. The situation has not changed. Figure 1 shows the growth of Nigerian hospitals 1900-1960 (Schram, 1971).

To effectively develop the various arms of the health care delivery services in Nigeria, it was necessary that the states of the federation join forces in effecting specific policies, which have been carefully stated. The Federal Ministry of Health then acted as the coordinating body for planning and coordination of programs, etc.

The federal government of Nigeria has adopted Health for All by the Year 2000 as the fundamental objective of national policy. Furthermore, primary health care (PHC) is being pursued as the strategy to make health care accessible to the majority, if not all of the population, by the year 2000. The national health policy formulated was based on a policy resting on the national philosophy of social justice and equity. The emphasis is being shifted from curative to preventive services with particular reference to primary health care. The health or disease profile of Nigerians presents the typical picture of a poor developing society where infectious and parasitic diseases remain the most common causes of sickness and death. Even now, there is an increasing frequency of chronic diseases like hypertension, diabetes, and cancer as causes of morbidity and mortality.

2. Financing

During the early period of growth in the health care delivery system within the regions and states, most of the voluntary agency hospitals and clinics were gradually taken over by the government. This process was

TABLE 1. Statistics of medical institutions as of 1952 at the beginning of regionalization and self-government in Nigeria.

	Northern Region	Eastern Region & Southern Cameroons	Western Region	Lagos	Total
GOVERNMENT AND NATIVE ADMINISTRATION					
General Hospitals & Nursing Homes	28	24	16	2	70
Maternity Hospitals And Homes	10	104	133	1	248
Special Hospital	-	1	2	1	1
Beds	30663	2366	1691	566	7686
Dispensaries	385	273	228	8	894
Doctors	77	68	44	59	248
MISSION, COMMERCIAL AND PRIVATE					
General Hospital And Nursing Homes	11	30	20	11	72
Maternity Hospitals And Nursing Homes	10	101	56	4	171
Special Hospitals	-	-	-	-	-
Beds	677	2728	1138	84	4627
Dispensaries	148	50	20	8	226
Doctors	<-------------- about 150 -------------->				

FIGURE 1. The growth of Nigerian hospitals (Schram, 1971).

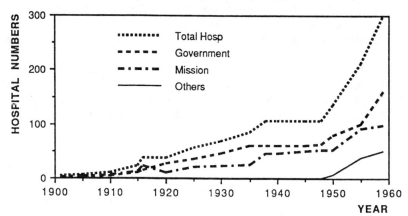

rationalized on the basis that each of the regions/states has responsibility for health care services and that local health authorities should also provide child care and environmental sanitation services. The federal government then had to maintain certain teaching hospitals, port health, and health education services. In spite of these, it should be noted that drugs and matters relating to drugs are still reserved for federal list in the constitution. It is commonly said that drugs have no barrier as regards their movement within a country or even a subregion.

The financing of the health care system in Nigeria, therefore, rests on the federal government and each of the state governments, as well as the federal capital territory, Abuja. The commercial and industrial businesses, however, provide medical services for their employees. Some of these businesses have hospitals, clinics, and health centers. These facilities have caused the development of industrial health care delivery centers.

To fully assess the financing of the national health care delivery system, the combination of federal allocations, state allocations, private concerns provisions, and individual entrepreneurs' contributions to staff health care have to be collated. Under these circumstances, the actual and capital expenditure profile of the federal government can be used as an indicator of the percentage of expenditure generally budgeted for health care delivery, while it should be recognized that the private sector caters mainly to the higher socioeconomic classes, invariably located in large cities.

At all levels, there have been no medical insurance schemes. Some states have provided for some period of time for free medical service, but have since modified the system. The modification may rest on the classification of patients with commensurate deductions acting as a subsidy. Sometimes drug costs are subsidized, but in general, payments for services are required at all levels of service. At this time of economic crisis, the populace has very great difficulty in meeting the cost of health care services.

Due to cuts in allocation of funds as well as the increasing high cost of health care delivery, the lower percentage of federal government recurrent and capital expenditure, which decreased in 1989, creates difficulty in the financing of health care. It has to be checked, and necessary recovery has to be made in subsequent allocations.

3. Facilities

Having considered the historical development of health services in Nigeria, it should be noted that government hospitals, clinics, health centers, and outpatient services run by private corporations (industries) and teaching and specialist hospitals are still operating. It is also noteworthy that some of the hospitals previously developed and manned by various religious organizations, but earlier taken over by government, are in some states being returned to the religious groups. With the activation and development of the local government systems, there are efforts being made to work out fully the involvement of the local government in the Primary Health Care program. There are increased numbers of health centers and clinics with coordination from the local government. Even though this is not fully done by all local governments in the states, there are plans for effecting these in all the states.

In reviewing the facilities shown in Figure 1, since such facilities were mainly, at that time, within the hospital setting, it can be seen that tremendous growth has been achieved. The relevance of the figures under PHC, as shown in Table 2, is that while plan allocations for the various items were not actually met, in the case of the Primary Health Care Scheme, 90.5% of plan allocation was met. In total, only 39% of health allocations were actually met. The capital commitment on the PHC Scheme has caused better input to rural environment to enhance the health care delivery system during the past five years.

In some of the states of the federation, careful studies have been made which show that the health needs of the people far outstrip the expansion rate of health facilities and health manpower development. However, there are differences in the level of health consciousness of the people in the various divisions of the states.

The number of teaching hospitals financed by the Federal Ministry of Health increased from 1 to 2 in 1962, to 6 in 1975, to 11 in 1979, to 12 in

TABLE 2. Registered and licensed pharmacists and registered pharmaceutical premises in Nigeria by state as of December 31, 1987.

STATE OF NIGERIA	1987 MID YEAR POPULATION ESTIMATE	1987 REGISTERED AND LICENCED PHARMACISTS		1987 POPULATION PER PHARMACIST	1987 NUMBER OF REGISTERED PHARMACEUTICAL PREMISES		
		NUMBER	% TOTAL		TOTAL	CAPITAL CITY	OTHER LOCATIONS
ANAMBRA	6,944,800	282	8.2	24,630	140	50	* 90
BAUCHI	4,694,700	37	1.1	126,880	21	12	9
BENDEL	4,751,900	231	6.7	20,570	123	88	35
BENUE	4,686,400	35	1.0	133,900	16	7	9
BORNO	5,787,900	53	1.5	109,210	49	40	9
CROSS RIVER	1,824,100						
AKWA IBOM	4,891,900	52	1.5	129,150	23	9	14
GONGOLA	5,030,600	23	0.7	218,720	21	12	9
IMO	7,091,600	270	7.8	26,670	157	31	** 126
KADUNA	3,192,400						
KATSINA	4,721,100	197	5.7	40,170	124	96	28
KANO	11,100,800	144	4.2	77,090	96	****95	1
KWARA	3,295,100	89	2.6	37,020	54	34	20
LAGOS	3,955,500	1,276	37.0	3,100	814	103	***711
NIGER	2,084,900	42	1.2	39,640	31	14	17
OGUN	2,994,800	80	2.3	37,440	71	26	45
ONDO	270,800	105	3.0	50,200	60	24	36
OYO	10,058,000	313	9.1	32,130	158	112	46
PLATEAU	3,887,900	67	1.9	58,030	49	42	7
RIVERS	3,321,000	109	3.2	30,470	61	****59	2
SOKOTO	8,764,100	28	0.8	313,000	21	14	7
ALL STATES	108,350,300	3,433	99.7	31,560	2,089	868	1,221
FCT, ABUJA	262,500	12	0.3	21,880	3	3	-
TOTAL	108,612,800	3,445	100.0	31,530	2,092	871	1,221

KEY: * 66 at Onitsha, ** 101 at Aba, *** 177 in Lagos Island and 110 at Surulere, **** Almost total neglect of locations outside the state capital.

1980, and to 13 in 1988 as a result of establishing new ones and converting and modernizing some state hospitals.

4. Manpower

The development of health manpower has taken a different turn since 1979, when basic health care delivery was introduced. Both at the federal and state levels, emphasis has been placed on the training of community health workers in established schools of health technology. These community workers have been posted at the rural health centers.

Similarly, the training of doctors, dentists, nurses, midwives, pharmacists, and veterinary officers has been enhanced by the inclusion of the Primary Health Care delivery system in the curricula. However, the difficulty, which being continually met, is to get this cadre of health workers to work in rural areas.

The profile on overall distribution of registered and licensed pharmacists in the federation as of December 31, 1987 is shown in Table 2.

5. The Medical Establishment

There are such places as dispensaries, rural health centers, comprehensive health centers, clinics, maternity centers, general hospitals, specialist hospitals, and teaching hospitals. There are other establishments, as shown in the example of Lagos State (Table 3). During the years 1984-1988 there were more health centers built and put in direct relationship to the local governments.

It is expected that more patients will be cared for following the growth of the Primary Health Care system. The community services provide the first level of care for patients. Patients then go to the general hospitals before referrals to the specialist hospitals. The progression has not been clearly demonstrated, mainly because these centers are currently being developed, and patients arriving in either a teaching or specialist hospital cannot be ignored. There is, therefore, a great need for educating the people on the value of first obtaining treatment at the level of community health centers before receiving treatment at the specialist and teaching hospitals.

There are a growing number of private medical practices in the country. Some of these are group practices, and provisions are made for clinics and annexes in the nonurban areas. They, therefore, provide training for young general medical practitioners in the community.

It is continually being emphasized that no single unit of the government health care delivery system should operate in isolation. General services,

TABLE 3. Type of medical establishments in Lagos City.

Medical Establishment	Quantity	Medical Establishment	Quantity
General Hospital	18	Health Centre	5
Pediatric Hospital	1	Leprosorium Clinics	1
Maternity	1	Dental Clinics	4
MaternityHomes (Registered)	38	Other Clinics & Dispensaries	21
Maternal / Child Health Centre	8	Prison Hospital	1
Infectious Disease Hospital	1	Teaching Hospital	1
Psychiatric Hospital	1	Teaching Hospital Annexe	2
Orthopaedic Hospital	1	Armed Forces Hospital	1
Tuberculosis Hospital	1	Armed Forces Dispensary	8
Opthalmic Unit	1		

X-ray, physiotherapy, laboratories, pharmaceutical services, and special clinics are available in each of these establishments. The X-ray services have improved considerably over the years, and mobile units are available to move into nonurban areas. These services are usually paid for by the individuals. There are some provisions for mass screening of the community for diseases (e.g., tuberculosis). The laboratory services have continued to be in considerable need for the effective diagnosis and control of diseases. However, the funding of such services has rested on the individual, except when a specific campaign is launched in certain areas. The procurement or availability of drugs and medicaments to the community has been mainly borne by individuals. Over a period of five years (1978-1983), there were, in some states, attempts to effectively pursue a policy of free health care. The overall effect of such a policy on the resources has caused many states to review and accept a low level of subsidy for their health care. There is at present subsidized medical care. Even though the level of subsidy is low, there are some people who manage to receive care within the system.

B. PHARMACEUTICAL INDUSTRY AND DRUG DISTRIBUTION

1. Structure of the Industry

The pharmaceutical industry and drug distribution business have been linked appropriately in the Nigerian setting. Manufacturing businesses have developed from the drug wholesale dealers or distributors. Consequently, most of the manufacturing done within the industry has been of a secondary nature and has had to be linked with the main moving items of the pharmaceutical industry (e.g., cough mixtures, antimalarial tablets, etc.).

Upon independence in 1960, nearly all the pharmaceuticals used in the country were imported, but by 1979, following a report by the pharmaceutical trade group, about 23% of the total market was produced locally. By 1986, it was estimated to have risen to about 35%. It was, however, observed in 1982 that the performance level of the local pharmaceutical industry was only at about 60%. This showed that the industry was not operating optimally. There were needs to expand the market for effective distribution and some incentives and protection by government, as well as the commitments of the industry by increased investments in equipment and machinery. Figure 2 illustrates the basic sections of the industry and distributive trade, up to the level of the community pharmacy. This struc-

FIGURE 2

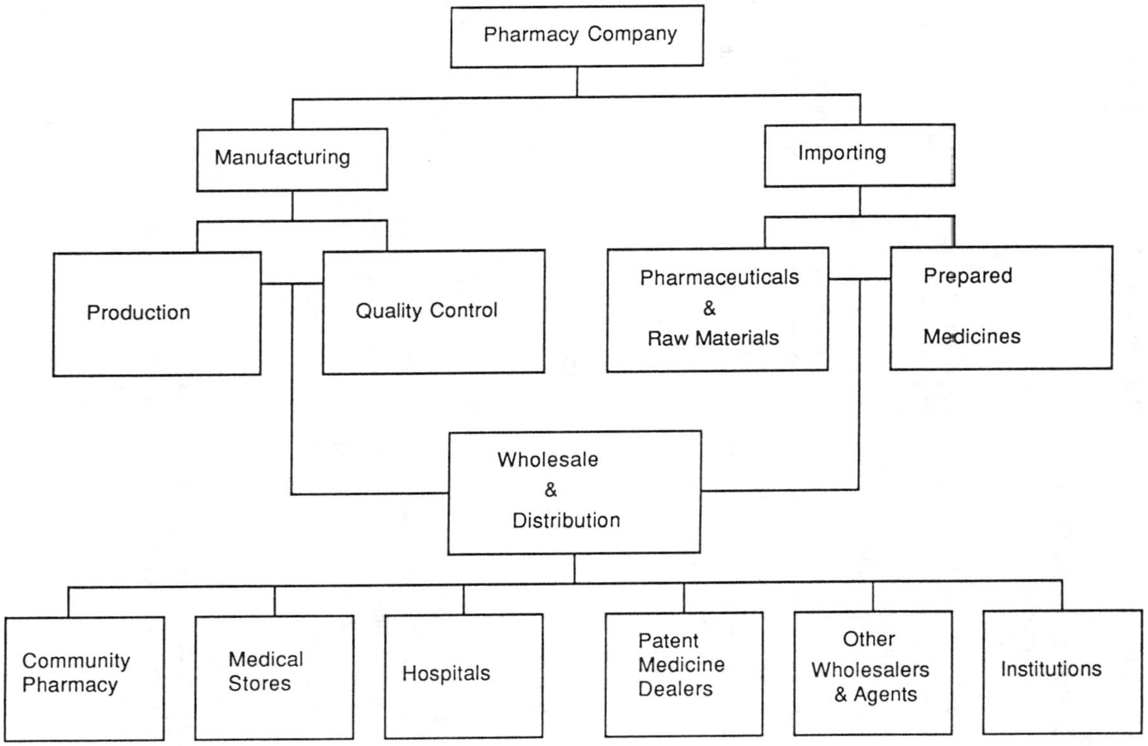
Pharmacy Company
Manufacturing
Importing
Production
Quality Control
Pharmaceuticals & Raw Materials
Prepared Medicines
Wholesale & Distribution
Community Pharmacy
Medical Stores
Hospitals
Patent Medicine Dealers
Other Wholesalers & Agents
Institutions

ture has been maintained over the years; the difference would have been in the proportion of each of the manufacturing and importing activities.

2. Research and Development
and Imports

It has been the responsibility of universities to organize research and develop drugs and medicine. The industries in Nigeria have relied exclusively on the research and development of their overseas-based companies, while the indigenous companies have shown, over the years, that they cannot, at the stages of their development, support meaningful research and development. It must be stated, however, that some of these industries have contributed some seed money to research in institutions, as well as provided some support for attendance and participation in international conferences. The institutions concerned are schools and faculties of pharmacy and some colleges of medicine. The Pharmaceutical Society, in collaboration with its technical groups, presented working papers and seed money for the establishment of a national institute, and it has since been established as the National Institute for Pharmaceutical Research and Development at Abuja, Federal Capital Territory. It is under the supervision of the Federal Ministry of Science and Technology.

In recent times, it has been stated that appropriate research and development must be undertaken by various industries. This has become mandatory mainly because of the policy of utilizing local raw materials for production. The pharmaceutical industries have been involved in seminars and workshops on obtaining raw materials from local sources. In effect, better collaboration has developed between the industries and the academic institutions, which have indicated some local materials that have to be developed for such a move towards self-reliance in the pharmaceutical industry. The overall position is that the restriction and cost of raw materials are continually putting the focus on local sources.

There are increasing raw materials import costs, but there were observable decreases in the numbers of items, even though higher costs were obtained. There should be a reduction in various imports, already affected by foreign exchange availability, and some patents are expected to evolve to the extent that local sources would give better industrial growth and stability. There are provisions for the licensing and registration of products in the country.

3. Manufacturing

Official government policy gives pharmaceutical production a high priority. The policy stresses value-added processes and discourages, if not totally refuses, tertiary production (i.e., exclusively repackaging activities). With regard to the present economic difficulties, the higher the value added in the local industry, the greater the savings of foreign exchange, and the better the balance of payments and the economy.

Up to about 1983, most of the pharmaceutical production in Nigeria was basic and at an average of 15%-20% value added. This falls short of the expectations of the national industrial policy. If the expectations of the Ministry of Industries are not met by the manufacturing houses, it cannot be fully the fault of such industries. The government has been reminded continually of the desired intersectoral collaboration and the necessary integration required for effective implementation. The pharmaceutical houses and individual local manufacturers are unable to set up their own plants for basic materials. Furthermore, adequate planning and protection, which were lacking, are now being tackled. Limiting imports has to be carefully pursued because substantiation has to be emphasized, as it has happened before in cases where government has restricted or banned the importation of products whose local supply is grossly inadequate (e.g., pharmaceutical bottles, caps, and other containers and labels).

As of December 31, 1987, there were 56 registered establishments undertaking local drug production, with the breakdown as shown in Table 4.

The breakdown has shown a large and considerable proportion of the establishments in Lagos State, primarily because of the infrastructural facilities being in the main commercial center. The adjoining or nearest states to Lagos are Ogun and Oyo. There are provisions in many of the industries for unit-dosage packaging, infusion fluids manufacture, tablet

TABLE 4

STATE	NUMBER	STATE	NUMBER
Lagos	33	Kware	2
Ogun	7	Anambra	1
Bendel	4	Imo	1
Oyo	4	Kaduna	1
Kano	2	Rivers	1

and capsule production, fluids and other preparations for oral administration, creams, and semisolids for topical use. In general, Nigeria has an industry that is trying all possibilities to meet the needs of the community.

The dilemma of the industry is that of deciding what is best: importing a product, or importing raw materials and using local packaging materials or systems to produce the final product. The manufacturers have an active group within the Manufacturers Association of Nigeria, apart from the industrial groups of the Chamber of Commerce and the Pharmaceutical Society of Nigeria.

Considerable effort has been made to ensure that registered manufacturers maintain a high level of quality production. It is believed that most, if not all, of the fake, substandard drug products found in the market were either illegally manufactured or imported. Good manufacturing practice has been the posture of the manufacturing houses. The Orange Guide of the British industry, as well as the WHO guidelines, have been found appropriate in developing a necessary environment for production.

4. Advertising and Promotion and Drug Approval Regulations

The Poisons and Pharmacy Act, Chapter 152 of Nigeria Laws, Sections 56 and 57, provides guidelines. Similarly, there are provisions in the Food and Drugs Act regarding advertising. In part, these read that "No person shall label, package, treat, process, sell or advertise any food, drug, cosmetic or device in a manner that is false or misleading or is likely to create a wrong impression as to its quality, character, value, composition, merit or safety."

Section 56(1) of the Poisons and Pharmacy act deals in particular with advertisements relating to certain diseases (e.g., cure of venereal diseases), care of any habit associated with sexual indulgence, or stimulation of the mental facilities. Similarly, there are provisions on the prohibition of advertisements relating to abortion. There have been procedures set out over many years, in the Federal Ministry of Health, for the approval of any advertisement on drug products to be made in the mass media, in leaflets, and to nonprofessional groups of people.

The pharmaceutical industries do use various "catchy" phrases to introduce some of the drug products in their advertisement through mass media and billboards. Promotion of drugs is usually handled by experts after due studies on the effectiveness and trials in local hospitals and environment. The methodology for the promotion of medicines and drug products involves mainly medical practitioners, such as clinicians, pharmacists, and other biomedical sciences personnel. It is necessary to add that most, if not all, of the drugs subject to promotion have been primarily developed by parent pharmaceutical companies, with local input only to justify further their effectiveness.

Manufacturers of drug products have to apply for registration of drugs. The application has to be made to the Federal Ministry of Health Drug Regulation and Information Unit. The information required includes the following:

a. Name and address of the manufacturer of the product
b. Name of product (proprietary and generic)
c. Promotional category (e.g., OTC [nonprescription] or prescription only)
d. Indications
e. Form of presentation in packages — type of package and package sizes
f. Name and quantity of each ingredient, indicating each generic name and company's code name
g. Chemical name and the structural formula of each active ingredient
h. Method of manufacture
i. Route and condition of administration
j. Dosage form
k. Side effects and interaction with other substances
l. Contraindications, indicated for the dosage form
m. Adverse reactions, as seen in trials, and symptoms of overdosage
n. Antidote in the event of overdosage — in terms of the specific drugs symptoms
o. Teratogenicity — results of trials studying the effect during long-term administration to mothers and fetuses, and fertility studies on both males and females
p. Analytical method of each ingredient, chemical or microbiological — raw materials specifications and test methods to be produced along with compendial methodology
q. Shelf life, stability data, release assays, and limits to be provided, and well as the physical and chemical characteristics of the product during storage and identification methodology for degradable products
r. Toxicological data for short-and long-term studies — full laboratory findings to be shown with a rational for tests
s. Clinical data — details on clinical pharmacology, close ranging and clinical trials in specific indices, including all side effects seen, both adverse and otherwise, plus statistical indication of efficacy.

Under the Food and Drugs Act, there are drugs that cannot be manufactured without first having a certificate from the Minister of Health. This includes such drugs as liver extract in all forms, insulin in all forms, living vaccines for oral and parenteral use, drugs prepared from microorganisms or viruses for parenteral use, antibiotics for parenteral use, radioactive isotopes, and sera and drug analogues for parenteral use. The certificate from the Minister of Health ensures that the premises of manufacture and the process and conditions by and under which the manufacture is to be carried on are, in the opinion of the Minister, suitable for ensuring that the drug will be safe for use. Regulations are also available on the prohibition of various misleading practices as they relate to label, package, advertisement, or sale. There are provisions in the Food and Drug Act which state that no person shall sell arsphenamine, dichlorophenarsine hydrochloride, neoarsphenamine, oxophenarsine hydrochloride, or sulfarsphenamine without first obtaining, in accordance with the regulations, certification from the Minister that the batch from which the drug was taken is safe for use.

The provisions of the Poisons and Pharmacy Act, Dangerous Drugs Act, Food and Drugs Act, and various decrees, including the most recent on Prohibition of Drug Sales in markets, buses, etc., have relevance to the effective distribution of drugs in the country. There are continual reviews of some of the regulations, in line with changing attitudes and community responses.

5. Brands and Generics

The availability of drugs to the populace depends on importation of drug products and some local manufacture. Under these circumstances, it is unavoidable for a collection of branded drugs to be imported or produced in the country while various generics enter the market through those sources out to compete with the producer of the brand drugs. The manufacturers of generics have the advantage of low price. When the economy is rather tight, as it has been in Nigeria, there is no doubt that the generics are favored over branded drugs. The essential drugs list scheme, which is being vigorously pursued by governments of the federation has, therefore, provided a firm area for the competition between branded drugs and generics.

In speculating on the level of competition to be reached, there is no doubt that, except for the development of branded generics, there will be a strong move toward generics because of the changing face of the economy.

The free and easy movement of drugs in the subregion has already caused concern. The free flow of drugs has facilitated the manufacture of generics, coupled with the development of many indigenous drug manufacturing houses that would be able to produce generics conveniently. It can be seen, therefore, that the competition between brand and generic drugs would definitely swing in favor of generics. It is also to the advantage of the community that the generics, which generally come at a lower cost, put drugs within easy reach of many people.

C. PHARMACY PRACTICE

1. Dispensing a Prescription and Record Keeping

There are guidelines on what constitutes a prescription in the 1986 maiden edition of the Nigerian National Drug Formulary and Essential Drugs List. Included in these guidelines is the need to provide essential information on a prescription (e.g., name and address of patient; age of patient; name and dosage form of drug; the dose; frequency and duration of administration; the date of prescription; name, signature, and address of the prescriber). Furthermore, it is recommended that names of drugs are best written out in full and that quantities should be stated in the metric system; hence, accepted abbreviations are to be used. Guidelines on prescribing for children and the elderly have also been adequately covered in the booklet, while the requirements of the law for prescribing dangerous drugs and other controlled substances are restated.

Prescriptions are dispensed from registered pharmaceutical chemist premises which, according to the law, should have superintending pharmacists to oversee the disposal of drugs. Similarly, medicines can be dispensed in hospitals and dispensaries by pharmaceutical chemists or dispensary attendants in a government hospital. There are various limitations regarding the overall act of dispensing, such as whether the drugs or medicines being dispensed are in the categories of drugs in Parts I to III of the Poisons and Pharmacy Act and Dangerous Drugs Act; the point or place of dispensing, whether in hospitals or in local pharmacies; and whether the drugs are being dispensed or disposed of by a body corporate.

The Poisons and Pharmacy Act, Chapter 152, Sections 30 and 32, states in detail the procedure for dispensing some medicines, while Section 36 of the Act refers to "Refusal to dispense, or sell or negligently dispensing drugs and poisons." Section 37 of the Poisons and Pharmacy Act makes the provisions of the law earlier indicated applicable to any corporate body carrying on a business in the dispensing or selling of drugs and poisons within registered premises.

Record keeping has been a major part of the law, and formats of record books are prescribed, such as the "Disposal of Poisons Book," each of which must be available for inspection at any time.

Rules on the labeling of prescribed drugs and medicines are also fully stated. A National Drug Formulary and Essential Drugs List has been made available since 1987, and some of the teaching and specialist hospitals have also developed their own formularies. Invariably, these are still being operated in institutions with little or nor impact at all on the general public. In the urban areas, the problems of prescription writing and prescriptions still exist within the system, in that physicians and other medical practitioners in private practice write prescriptions which they then dispense to their own patients. The majority of the public does not have access to most physicians as they cannot afford consultation fees; hence, some noncompliance has crept into the drug disposal system. The prescription-only drugs are invariably being sold without the necessary prescriptions. It is a precarious situation, but with better health education and necessary reorientation on prescription and medical care, this noncompliance can be eliminated. In these circumstances, all record keeping is considerably affected.

Payment for dispensed drugs is made in full by the patient, as there are no insurance schemes or services to take up part of the cost of health care.

2. OTCs and Other Classes of Drugs

Over-the-counter (OTC) drugs are a variety of proprietary drugs, and in view of the noneffective prescription activity within the health system, one is continually concerned about the number of drugs that can be obtained in this way. The problem of rural and urban facilities and accessibility to drugs has resulted in the provision for the sale of Patent and Proprietary Medicines. Such provisions are in Sections 50 to 55. At present, sale of medicines through patent medicine dealers who have a limited list of medicines has become a concern, as there are a considerable number of such patent medicine stores in the various towns and villages in spite of the increasing number of pharmacies. These stores are continually being found to carry lines of drugs and medicines that are outside their permitted lists. Every effort is being made to advise and educate people to obtain drugs from the right places.

In effect, a clearly defined OTC or nonprescription sale of drugs is not being realized in the current practice of pharmacy. This can be effectively challenged through patient education and training of more pharmacists.

3. Patient Education

Many citizens of Nigeria, particularly those in the rural areas, have their first contact in health institutions, like the health care centers, with primary health care workers. These health care workers have the responsibility for treating a variety of endemic diseases. The health education of such patients also rests with the primary health worker. The education of patients on drugs, their uses, storage, and disposal has remained the responsibility of the pharmacists who have stayed mainly in the urban areas.

Patient education has been of primary relevance to effective drug use; hence, a clearly formulated approach to providing patient education has been the subject of discussion and adaptation by local community pharmacies. It should be mentioned that hospital pharmacies are actively involved in such education. The understanding of simple regimens has caused much concern, to the extent that symbols and diagrams have to be developed for better patient understanding. It has been observed that the advice given to patients at the time of delivery of drugs has actually provided better education and approach to their use of drugs.

Many pharmacies have now started to provide adequate space for counseling to supplement the usual drug delivery contact. Other forms of patient education being pursued are handbills, leaflets, and brochures made available for public collection in pharmacies, as well as car stickers and television discussions. The majority of patients appreciate these efforts and ask for more; a few radio or television programs are organized regularly. The problems of self-medication that have reached a high level have also caused a great demand for patient education. There are a few drug information centers already in existence.

4. Use of Technicians and Others in Pharmacy Practice and Within Health Care

Earlier in the development of pharmaceutical services in Nigeria, there were dispensers, dispensing assistants, and dispensing attendants. Then changes in nomenclature and tasks caused a renaming to "chemist" and "druggist," "dispensers," and "dispensing assistants." No doubt metamorphoses have since occurred, and there are now pharmacists, pharmacy technicians, and dispensing assistants/attendants. It was considered that another subskilled group be called "pharmacy technologists," but so far, the expected skilled input of that group has not been fully delineated. Consequent to the development of the Primary Health Care program and the Essential Drugs List scheme, these groups of workers have been further extended to include primary health workers.

The pharmacy technicians are trained in the basic principles of dispensing production, packaging, and stock control of drugs. Their practice and courses at institutions of training give the necessary theoretical and practical insight into drug preparation and disposal.

It has not been possible to cover a great proportion of the country; hence, lower cadres of staff who remain responsible to the pharmacists have evolved. The primary health workers have, by design, been authorized to dispense a limited range of drugs at their discretion for common endemic diseases, but under a control system with a doctor. The pharmacist, therefore, has the responsibility of training these cadres of staff who are subprofessional.

5. Drug Insurance

There are discussions progressing on health insurance schemes; hence, drug insurance has to await the full formulation of a health insurance scheme in Nigeria. It is, however, believed that the availability of drugs to patients will be enhanced if a drug insurance policy is developed, especially as the high cost of drugs and medicines has continued to cause many patients to avoid procuring necessary drugs for treatment and medical care.

When nationally organized health care programs are available, the situation experienced during the Expanded Program on Immunizations (EPI), Family Planning, etc., where the selection of drugs and the criteria for their administration is centrally directed, the issue of insurance will not be relevant. In general, it is desirable that adequate provisions for drug insurance be developed in the face of the increasing costs of all drugs, even the essential ones.

D. UNIQUE OR INTERESTING FEATURES OR SPECIAL SITUATIONS IN NIGERIA

In the African setting and in the Nigerian environment, the protection of health has been a major preoccupation. It has actually occupied a special place in the way of life. This fact has caused the Nigerians, indeed the Africans, to see the role of traditional medicine as being important in their life.

It was pointed out when the historical development of health care was being discussed, that traditional health care has been practiced side by side with other forms of medical care and that most of the care in the rural areas is associated with traditional medicine. Since pharmaceutical services have not spread fully into all rural and urban areas of Nigeria, the effective use of traditional medicines cannot be ruled out. Herbal medicines have, therefore, continued to thrive, while research into them has continually challenged the scientists.

There is a general belief that there are great prospects in obtaining efficacious and useful pharmaceuticals from plants and all natural products. The government of Nigeria has indicated its desire to effectively utilize the skill and natural resources of traditional healers to effect health care along with the Western method. There is no doubt that the developed health policy provides for a possible integration of the traditional with the Western mode of medicine. There are moves to set up recognized institutions for the training of herbalists and the standardization of their various recipes.

Pharmacists are now generally trained to recognize some of the traditional medicines and herbal preparations, as well as the environment in which such practices are effected. Consequently, research is being carried out on herbs and actual resources. It is believed that some of these recipes and preparations can substitute for some Western medicines. The raw materials are usually available from the immediate environment. In spite of the expansion of modern health care, there are indications that traditional systems are not losing their influence and ability to adapt to the environment.

The sociocultural influence on total health care delivery has also provided for other forms of healing systems (e.g., faith healing, homeopathy). There seems to be little interaction between these practitioners and the Western medicine and traditional medicine practitioners; in effect, there is development of a dual health care system from which patients can select. When faced with the task of advising on drug use, the pharmacist must first determine the extent of the patient's use of traditional medicine for any particular complaint.

Herbalists and traditional medicine practitioners have indicated their willingness to submit their drugs/medicines for the necessary analysis. A move in such a direction would benefit all, as the complementary relationship between Western/modern medicine and traditional medicine needs full exploration.

E. CONCLUSIONS

The pharmaceutical services provided in Nigeria are continually changing. The development of a health care delivery system in the country has constantly affected the services and, as expected, adequate responses have to be made to the various needs that may evolve. Since the steady implementation of the Primary Health Scheme, it has become necessary to be specific on a National Health Policy and hence, a National Drug Policy.

There are already clear indications that manpower development of different cadres (i.e., pharmacists and supportive staff) has to be fully pursued.

There is a need for better patient drug use monitoring, especially when considering the parallel and complementary traditional and Western/modern medicine use. To meet this challenge, the training institutions for pharmacy have introduced clinical pharmacy programs which provide adequate approaches for the correlation of drug use with various disease state management.

The need to change the focus from a curative to a preventive medical system within the health policy must be freely brought out within the drug laws. It is generally believed, just as it was shown in India, that a large number of common illnesses are self-curing and/or self-limiting and need only symptomatic treatment with simple remedies, whether herbal or allopathic. It is conceivable that the common illnesses that are infectious can be controlled with simple, economical preventive measures or be treated with cheap, safe, and effective drugs.

The research activities on the materials of traditional medicine are being pursued vigorously, and it is hoped that concerted efforts would help toward some level of integration. However, there is already an unequivocal recognition that obtaining raw materials for any indigenous pharmaceutical industry must be guided by the traditional herbal recipe after appropriate research. Already the results of researchers are showing a positive approach in that direction.

With regard to the increasing population and drug needs, the production of drugs should be related to real health needs. Basic drugs as already presented in the Essential Drugs List and Formulary must be backed up with cheap and easy-to-use preparations. There should be no mistake in evaluating drug costs and satisfying health needs.

The system now appears to present an almost exclusively curative urban sector, while the rural sector is trying to do some promotive and preventive work. Consequently, the level of health education must be raised in all sectors, and pharmacy services need to be fully involved in designing and projecting this communication task.

Trends in the pharmaceutical services include a strong patient-pharmacist interaction. This would facilitate better communication, counseling, and prevention of errors in drug utilization and achieve rational use of drugs either as preventives or curatives. With an increased focus on primary health care, the number of patients and the number of visits per patient will increase. Consequently, with the tendency for drug treatment to be demanded at each visit, the progressive extension of primary health care entails a large increase in the provision of drugs. It is this type of challenge that calls for responsible logistics for drug supply and drug production, as well as rational prescribing and utilization. Drug quality control can therefore be linked, since scarcity of funds would compel governments to purchase drugs at low prices. Effective quality control can best be achieved through quality assurance procedures in local pharmaceutical industries. An improved pharmaceutical industrial system should be encouraged.

While pharmaceutical services have to operate within a health policy with a strong primary health care system, the health authorities in charge of the procurement of drugs and pharmaceuticals must give due consideration to product cost and product quality, along with the development of national and local production, as part of the overall development of the country.

Chapter 17

Norway

Bjorn Joldal
Bjorg Stromncs

A. THE NATIONAL HEALTH CARE SYSTEM

1. History and Background

Norway is situated on the Scandinavian peninsula with half of the country north of the Arctic Circle. To the east, Norway shares borders with Sweden, Finland, and the Soviet Union. To the west, it faces the North Sea and the Atlantic Ocean. Much of Norway's history has been shaped by the sea. In the ninth century, the Norwegian Vikings sailed across the Atlantic to what is now known as the North American continent. Their efforts to explore unknown lands opened commercial and cultural opportunities. Many Norwegians have earned their living from the ocean, from traditional fishing to modern fish farming, through shipping, and through the discovery of oil in the North Sea.

Before 1900, most people lived directly off the land's natural resources through hunting, fishing, farming, and forestry. Industrialization came late. In 1800, merely 5% of the population held industrial jobs.

The Norwegian kingdom, established in the ninth century, maintained a separate monarchy until the fourteenth century, when it united with Denmark. The union with Denmark lasted 400 years. It ended in 1814, when Norway was virtually given to Sweden by the European powers after the Danish king sided with the French in the Napoleonic wars. At this critical turning point in Norwegian history, its people insisted on their right to self-determination. Delegates were elected to a constituent assembly. On May 17, 1814, the assembly adopted a Constitution for a free, independent, and democratically ruled Norway. The union established between Sweden and Norway in 1814 was dissolved by referendum in 1905.

Norway had its first pharmacy in Bergen about 1590. Bergen, lying on the west coast, was at that time the largest city and the center of trade and shipping.

2. Social and Political Environment

Norway is a constitutional monarchy. The legislative power is vested in Parliament (the Storting). Its members are elected every four years as representatives of political parties; the right to vote is universal. For administrative purposes, the country is divided into 19 counties, which are subdivided in 444 municipalities. The municipalities are gaining increasing control over local affairs.

With an average of nearly 24 acres per capita, Norway is one of the most sparsely settled countries in Europe. The population is slightly over 4 million (see Tables 1-3). About 40% of Norwegians are living within 50 miles of the capital, Oslo. About 45% of all Norwegians live in rural communities.

3. Public and Private Financing, Health Costs

All persons resident in Norway are, with a few exceptions, protected against the costs of ill health under a compulsory scheme supervised by the National Insurance Institute. National Insurance provides both medical benefits and income-related daily cash allowances in cases of illness, disability, pregnancy, and childbirth.

Medical treatment in hospitals is given free, but outpatients are required to bear the full cost of drugs, except in connection with serious and chronic diseases. According to special rules and rates, a refund is made for certain expenses (e.g., in connection with treatment prescribed by a doctor, certain medicaments as mentioned above, home nursing, and travelling expenses in connection with treatment). The health figures are general and include hospitals, medical and dental centers, national insurance schemes, etc., plus expenditure on administration and regulation of relevant government departments, in addition to pharmaceuticals (see Table 4).

4. Facilities (Hospitals, Clinics, etc.)

There are four main types of hospitals: regional; district; large; and smaller, local hospitals. Most hospital beds are publicly owned; a few are owned by voluntary organizations. Beds in hospitals and nursing homes totaled 67,000 in 1986, of which 21,700 were in general hospitals (see Table 5).

The Hospital Act of 1969 gives the counties full responsibility for the planning, construction, and running of hospitals, except two hospitals owned and run by the state. According to the Hospital Act, ambulatory care is part of hospital activities.

In recent years, privately owned outpatient clinics have become more common, especially in the large cities. They perform minor surgical operations and other specialist treatment.

TABLE 1. Demography, economy, health expenditure, health care availability, health factors.

Demography

Population (millions)	4,2	1987
Annual average population growth rate (%)	0,3	1980-85
Population aged under 15 (%)	19,4	1986
Urban population (% of total)	60	1980
Birth rate (per 1000 per annum)	13,0	1987
Death rate (per 1000 per annum)	10,7	1987

Economy

Gross Domestic Product (billion NOK)	516.022	1986
GDP growth rate (%)	4,2	1985-86
Imports (billion US$)	31.424	1987
Exports (billion US$)	29.634	1987
Rate of inflation (CPI, %)	6,7	1987-88
Agriculture, forestry, fishing as % of GDP	6	1987
US$ conversion factor US$ = NOK	7,40	1986
NOK	6.74	1987

Health Expenditure

Public health expenditure as %of GDP	6,6	1986
% of private incomes spent on health	3,8	1987
Pharmaceutical market valuation (US$ million)	458	1986
Average annual increase in pharm market (%)	14,9	1981-86

Health Care Availability

Number of doctors (economically active)	9,440	1986
Population per doctor	442	1986
Number of hospitals	112	1986
Population per hospital bed	173	1986
Number of pharmacies	315	1988

TABLE 1 (continued)

Health Factors

Life expectancy, male (years)	73	1986
Life expectancy, female (years)	80	1986
Infant mortality per 1000 live births	8,1	1981-85
Defined causes of death		
(% of recorded cases):		1986
Heart and hypertensive disease	34	
Neoplasms	22	
Cerebrovascular disease	12	
Pneumonia	6	
Accidents, poisoning and violence	5	
Notifiable communicable diseases		
(000's of recorded cases):		1986
Influenza	192	
Mumps	3	
Gonococcal infections	6	
Measles	1	
Whooping cough	1	

Source: Central Bureau of Statistics,Oslo. The Norwegian Association of Proprietor Pharmacists

TABLE 2. Population by sex and broad age groups.

Range	Number	Percentage population	Percentage male
0-14	812,000	19.4	51
15-34	1,283,000	30.7	51
35-64	1,410,000	33.8	50
65-79	524,000	12.6	44
80+	146,000	3.5	34

Source: Central Bureau of Statistics, Oslo, 1986

5. Manpower

General medical care is provided either by local public health officers or by private practitioners.

Public contacts with the health service due to disease were surveyed in 1985. Approximately 345 persons per 1,000 had such contact during a 14-day period; 186 consultations were with a physician, 50 with a physiotherapist, and 30 with a nurse (see Table 6).

Increasing attention is paid to developing the primary health care sec-

TABLE 3. Population of important cities.

	Population
Oslo	451,300
Bergen	208,900
Trondheim	134,500
Stavanger	95,500
Kristiansand	63,300

Source: Central Bureau of Statistics, Oslo, 1987

TABLE 4. Health expenditure and central revenue, 1986.

CENTRAL GOVERNMENT		HOUSEHOLD
Total current revenue (% of GNP)	%of total expenditure Health	%of total consumption* Medical care
48,4	10,5	10

* 1980-85
Source: World Development Report,1988

tor. The municipal health authorities are given the responsibility for planning and running of this sector, including the general somatic nursing homes; general practicing physicians; physiotherapy; and nursing, including public health nurses and home nursing (see Table 7).

6. The Social Security System

As a basic premise, Norwegians in difficult social and economic situations are legally entitled to assistance. The welfare system attempts to give people equal access to such basic necessities as a place to live and work, education, and health care. All the major political parties in Norway support the concepts behind the welfare system. But there is ongoing discussion of how best to organize the system. This debate reflects the fact that a declining work force must pay for a growing number of retired and young people, increasing the per capita cost.

All persons resident in Norway, irrespective of nationality, are (with a few exceptions) protected against the cost of ill health under a compulsory scheme supervised by the National Insurance Institute. National Insurance Scheme funding is calculated on the basis of taxable income, shared by

TABLE 5. Hospital beds.

	Number of beds	Beds per 1000 population
General hospital	21,700	5.2
Nursing homes	30,000	7.2
Mental hospitals	4,400	1.1
Mental nursing homes	4,500	1.1

Source: Central Bureau of Statistics, Oslo 1987

TABLE 6. Health care personnel (economically and not economically active).

		Inhabitants per unit of personnel
Doctors	10,110	410
Dentists	4,400	943
Nurses	44,350	94
Assistant nurses	47,720	87

Source:Central Bureau of Statistics,Oslo, 1986

employers and employed members. National Insurance provides both medical benefits and income-related daily cash allowances in case of illness, disability, pregnancy, and childbirth.

7. The Medical Establishment:
Entry, Specialists, Referrals

General medical care is provided either by local public health officers or by private general practitioners. There is a free choice of doctor, but in the more isolated regions, especially in the north and west, there may be only one doctor who can easily be reached for treatment.

There are about 9,400 economically active physicians in Norway, of whom 8,450 are in active patient care (see Table 8). The rest are primarily in public administration, defense, education, and research.

Data for 1988 on medical specialists in hospitals are given in Table 9. An additional 580 specialists are established in private practice. Increasing specialist care is being given by hospital outpatient departments. Patients are referred to medical specialists by a general practitioner.

TABLE 7. Personnel in muncipal health service, 1986.

Personnel,	Number
Physicians	
total	2,716
inhabitants Per physician	1,530
Physiotherapists	2,426
Nurses	4,142
Auxiliary nurses	1,589
total	13,380

Source: Central Bureau of Statistics, Oslo, 1988

TABLE 8. Economically active physicians, 1986.

Total	Non- institutional health service			Institutional health service	
	Total	Gen. pract.	Spec. Pract.	Total	Gen. hosp.
9,443	3,682	2,768	583	4,768	3,994

Source: NOS Health Personnel Statistics

8. Payment and Reimbursement

Medical treatment in hospitals is given free. Outside hospitals, a cost-sharing system is established. The patient pays a specified fee per consultation with a physician, and the remaining sum is refunded from the National Insurance.

Outpatients are required to bear the full cost of drugs, except in connection with serious and chronic diseases, as will described in Section B.7. When yearly costs exceed a given limit (NOK 950 in 1989), the excess cost will be covered by the insurance. Children and retirement pensioners pay a smaller amount.

TABLE 9. Specialist medical practitioners in hospitals, 1988.

Practitioners	Number
Total	2,760
General medicine	427
General surgery	429
Psychiatry	370
Radiology	225
Gynecology / obstetrics	149
Pediatrics	126

Source: The Norwegian Medical Association

B. PHARMACEUTICAL INDUSTRY AND DRUG DISTRIBUTION

1. Structure in the Industry

The modern Norwegian pharmaceutical industry is composed of five major companies: Nycomed AS, Apothekernes Laboratorium A.S., Hydro Pharma a.s., Weiders FarmasOytiske A/S, and Collett-Marwell Hauge A/S. Two companies in particular have carved out a substantial niche market for themselves and have managed to gain international acclaim in their chosen fields. The first was Nycomed, which, with its expertise and investment in the contrast diagnosis sector, has a leading position within this sector. Its internationalization has progressed from small beginnings, to reaching exports in 1987, amounting to 77% of the total sales. The company has had a policy of investing in R&D for many years, and it collaborates with various establishments, including hospitals, universities, and other pharmaceutical firms, in sharing and developing knowledge.

The second firm is Apothekernes Laboratorium, which is part of the AL group. This group has managed to become the leading world producer of Bacitracin®, which it sells mostly in bulk for veterinary purposes. The company's main expertise is biotechnical production with special application to antibiotic fermentation processes. Both firms are highly international within their niche markets. Each is planning for the next generation of products, which for Nycomed is further development of diagnostic preparations as new biotechniques present new challenges.

AL continues with biotechnology in bulk production but has also branched out into fish vaccines, which, when established, could well be a new international success due to the rapid growth of the fish farming industry and the lack of appropriate products to date. The characteristics of the Norwegian market and industry generally are that the five pharmaceutical companies are not able to cover the whole range of products demanded by the market. Those companies with economies of scale produce niche products and export 70% to 80%. Of the three remaining Norwegian producers, one, Weider, synthesizes and exports bulk pharmaceuticals to a limited international market. The other two produce entirely for the home market. This results in the country importing about 75% of all products.

2. Research and Development, Imports, Exports
Patents, Licensing

Research and Development

For the two major pharmaceutical companies, 10-15% of turnover was invested in R&D. For Nycomed, 50% of the funds invested in R&D in 1987 were spent on research on third-generation X-ray contrast preparations based upon the established market leaders previously developed by Nycomed.

AL, as Nycomed, is aware of the need to support R&D to maintain a leading position. AL invested NOK 37.3 million, or 13% of its revenue, in R&D, and there are 77 of a total of 376 employees engaged in R&D. One of their brightest hopes is the fish vaccine, which is being developed with the University in Tromso. AL is also investigating other veterinary medicines specially for fish farms. The other is the use of minute magnetized monodispersed particles for separation of cells and subcellular fragments and viruses.

Both firms have large numbers of skilled personnel and a large percentage of university-educated personnel. They work in close cooperation with doctors and research institutions, often sponsoring positions both in Norway and abroad.

Imports, Exports

Of the total sale of pharmaceutical specialties in 1986, 75% was imported products, while the remaining 25% was covered by domestic products. The imported products were mainly from Sweden (25%), Denmark (14%), the U.K. (13%), the Federal Republic of Germany (12%), and

Switzerland (9%). Norwegian pharmaceutical firms exported their products to the Federal Republic of Germany (25%), Sweden (19%), Denmark (12%), and the U.K. (10%). The total import of pharmaceuticals amounted to NOK 1400 million and the corresponding export was NOK 490 million. The percentages of export for the different firms are as follows:

	Export	Sales in Norway
Nycomed	79.2%	20.8%
Weifa	16.2%	83.8%
AL	88.5%	11.5%
Hydro-Pharma	7.3%	92.7%
(the former NAF-LAB)		

Patents

Pharmaceutical products themselves are not patentable, but their manufacturing processes are. The patent's term is 17 years, dating from the day of filing. The invention may be compulsory licensed three years after patent grant or the filing date if it is not fully exploited. Licensing may be enforced when considered in the public interest.

At present, the Norwegian government is considering the introduction of product patents, provided that the rules on compulsory licenses are changed to modify the possible adverse effects of a product patent system.

3. Manufacturing

Safeguards regarding good manufacture are based on a manufacturing license issued according to the conditions of the plant, premises, and staff and the presence of a qualified person responsible for manufacture and quality assurance. The Norwegian pharmaceutical manufacturers operate in accordance with international standards for GMP.

The supervision of the manufacture is carried out through plant inspections by medicines inspectors of the Directorate of Health. An inspection may be assigned routinely or for a specific purpose and might also be ordered to follow up on complaints. Pharmaceutical inspections form an important means of ascertaining the quality of drugs brought on the market. As a member of the Pharmaceutical Inspection Convention, Norway exchanges inspection reports with other participating states.

4. Advertising and Promotion, Price Regulation

Advertising and Promotion

In Norway, all advertising, price lists, catalogs, etc., must be approved before use. This provision applies to advertising both to the public and to physicians. Advertising must be moderate and objective, not give a misleading or exaggerated impression of the product's medical value, and not be so formulated as to encourage unnecessary or nonmedical use of the product. Any advertising of nonregistered specialties or of drugs included in the Pharmacopeia or approved formularies is prohibited.

Advertising of drugs to the public is permitted only for nonprescription drugs and on certain conditions. Drugs may not be advertised on radio or television or in cinemas, public premises, or streets or roads. Advertisements to physicians, dentists, and veterinarians must contain only generally approved indications. Quotations, curves, etc., from medical literature should be adequately reproduced, with a complete indication of the source. The advertising must state the composition of the product, the contraindications, and the most important side effects. If a generic name exists, it must be clearly indicated. There are also strict rules on the distribution of samples.

Price Regulation

Norwegian price control covers all categories of pharmaceutical preparations, both prescription and nonprescription drugs. Price control seems rather comprehensive compared with the situation in other European countries. According to the Norwegian legislation, the price of a pharmaceutical specialty shall not be "in disproportion to its value." The cost of a drug should be set against its direct or indirect benefits compared with alternatives. Data on these matters are scarcely available in optimal form, most countries appearing to adopt an arbitrary approach. In Norway, price consideration is an integral part of the registration procedure. Negotiations are conducted with the manufacturer to agree upon an acceptable price. Prices of new products are compared with the prices of similar products on the market and with the prices charged in other European countries, particularly in the country of manufacture.

It is less difficult to judge whether a price increase is reasonable. In many countries, the authorities have concentrated on this kind of control. In recent years, the Norwegian public health authorities and the pharmaceutical industry have developed models for price adjustments. In these formulas, inflation, changes in exchange rates, etc., are taken into account.

The price to the public is calculated by adding 30% plus a NOK 3.50 levy to the wholesaler's selling price. In addition, a value-added tax of 20% is imposed. For drugs on prescription, the pharmacist charges a fee of NOK 1.40. Hospitals may claim a 2% reduction on the price to the public.

5. Drug Approval and Government Regulation

Selection of Drugs

Since 1928, quality, safety, efficacy, and cost requirements have formed the basis for drug evaluation and registration in Norway. Some ten years later, the concept of need was included. The following are the present criteria for selection of drugs:

- Selection should be based on scientific documentation.
- The efficacy/toxicity ratio must be weighed against the severity of the disease.
- New drugs should represent better therapeutic alternatives than those already on the market.
- Drug combinations should be avoided unless the combination shows a clear advantage over that of each ingredient.
- There should be a clear-cut medical need for any new product.
- The number of drugs should be limited.
- Approval should be given for a limited period (five years).
- The drug may be restricted to use by hospitals or specialists.

Additional criteria are price, local therapy traditions, etc.

The Need Clause

According to the Norwegian regulations, a pharmaceutical specialty must be medically justified and be considered needed. As the term "need" has not been defined more precisely, the registration board has had to establish its own practice. Through use of the need clause, the number of similar preparations and synonyms (i.e., generics) has been limited. By allowing some generics, price competition, as well as the supply of drugs has been maintained. Medical need has been used to avoid the registration of too many combinations.

A study of decisions made by the Specialties Board during the years 1981-1983 shows that approximately 40% of applications are rejected.

Need considerations are involved in more than 60% of the rejections. One striking effect of the assessment of need is the limited number of drugs on the market: about 1,100 different drugs (2,100, including different dosage forms and strengths of drugs) are registered in Norway, compared with ten times that number in some other European countries. The number of drugs is probably also influenced by the small size of the Norwegian drug market.

Fixed Combinations

The Norwegian policy has been based on some essential requirements:

—Each component should make a contribution to the claimed effect.
—A component may be added to enhance the effectiveness or safety of the active ingredient or to minimize the potential abuse of the ingredient.
—The components should have approximately the same half-life and duration of action.

In addition, a patient population of reasonable size should benefit from the combination. The limitation of the number of fixed combinations has been made possible by the need clause.

6. Competition: Brand and Generic

The question of generics is complex. There seems to be confusion as to what constitutes a generic product. Generics are simply nonpatented medicines that are supplied under a generic or brand name. The majority of the products made by Norwegian manufacturers have for many years been branded generics. The products have been prescribed by their brand names and priced as other pharmaceuticals.

It is only recently that some foreign manufacturers have started to introduce products by their generic names and by competitive prices—as an average, 30% under the brand prices. So far, only a limited range of generic products has been registered. Since safety and efficacy data for such products are well known, the documentation for quality and bioavailability have been the main requirements for registration. Some manufacturers have argued that generics act differently from the original products or are inferior in quality, but significant differences in potency have not been proved. In addition, well-known manufacturers of original products try to get their share of the generic market by launching generic alternatives.

The government has tried to encourage the prescribing of cheap synonym (generic) products, but doctors have not been receptive to the idea and show great loyalty to brand names. As of 1989, however, doctors prescribing products that are reimbursed by the national health insurance are obliged to prescribe the cheapest of equivalent products unless there are important (significant) medical reasons for not doing so.

7. Cost-Containment Activities

Various approaches have been made to the containment of drug costs. When planning for drug supply, one of the first steps must be to secure a rational selection of drugs. An insufficient selection may result in the purchase of too many products and duplicate items, consuming limited inventory capital and complicating distribution. In many countries, limited funds are used on new drugs whose clinical efficacy has not been established and on doubtful combination products. The criteria for registration and the Norwegian need clause have been described. Need considerations are also applied by hospital drug committees in their selection of drugs. Between 200 and 300 preparations are considered sufficient to cover the need for routine drugs at a Norwegian county hospital with approximately 300-400 beds.

The Norwegian price control covers all categories of pharmaceutical preparations, both prescription and nonprescription drugs. The price control seems rather comprehensive compared to the situation in other European countries. The price of synonym products will usually have to be 20%-30% lower than the price of the original product.

While containing the cost of drugs should not be a goal in itself, most governments or health insurance schemes have introduced measures to limit expenditure on drugs. The bone of the Norwegian drug reimbursement program is that it is applying to the chronic population. From 1960 up to the end of 1980, medications for all persons, irrespective of age, suffering from prolonged or chronic ailments were completely covered by the health insurance. In 1981, the regulations were changed, and cost-sharing was introduced. Up to 1988, the patient share was a fixed amount per prescription. From 1989, the patient share will be 20% (10% for retired people and children) of the total cost per prescription, not exceeding NOK 150 and with a limit of NOK 950 per year. Excess costs will be covered by the insurance. The list of diseases eligible for reimbursement is specified in the regulations. The patients are characterized as having a chronic disease, and the physician must be convinced that long-term medication is necessary. The formulary now includes 36 chronic diagnoses.

The reimbursement scheme includes other measures to limit expenditure. The prescribing of certain categories of drugs is limited to specialists only, and the amounts are limited to a three-month supply.

Generic prescribing has been encouraged for many years, but recently a new requirement has been introduced. Doctors shall prescribe the cheapest generic product unless there are overwhelming medical reasons why another product should be provided.

There has been a steep increase in the total expenditures of the reimbursement scheme. The upward movement of drug reimbursement expenditures in Norway is dependent on both economic and noneconomic forces, namely, the price increases, growth in the number of persons qualifying for the program, introduction of new and more costly drugs that have greater efficacy in healing serious diseases, and reforms within the reimbursement system itself. Because the scope of the drug care program is limited mostly to the chronic population, reimbursement expenditures by the National Insurance contribute a minor part to the total expenditures on health services.

C. PHARMACY PRACTICE

1. Dispensing a Prescription: Where, What, Payment, Formulary

Where

The right to import and distribute medicines to Norwegian pharmacies is the monopoly of the state-owned Norsk Medisinaldepot (NMD). Drugs are sold only in pharmacies, the only exception being parenteral infusions, which may be sold directly from NMD to the hospitals. Drug dispensing by doctors is unusual; the exception is drugs for initial acute treatment.

Norway has 299 privately owned pharmacies, 34 of which are branch pharmacies. In addition, there are 16 hospital pharmacies owned by the counties or the state.

Norway has a rather high population/pharmacy ratio. Within the 19 counties of Norway, the population per pharmacy varies from 20,000 in certain rural areas to 10,000 in the Oslo area. This means a very long distance between pharmacies in some areas, with considerable transportation and communication problems. The number of customers per year in Norwegian pharmacies is 24 million. About 10 million of these want a prescription filled. Because of long distances between pharmacies, as mentioned, about 1.8 million packages are sent to patients from the pharmacies.

The Directorate of Health decides whether pharmacies shall be established or closed. The decision to open a pharmacy is based on such considerations as population per pharmacy, distance between pharmacies, and transportation facilities. The Board of Health or the local authorities are responsible for raising the question of establishing a new pharmacy. They are established when it is desirable or necessary from the point of view of the public. In 1984, a working group under the Directorate of Health presented a plan for the geographical distribution of pharmacies. This plan is now under revision.

To maintain pharmacies that are uneconomical as business ventures but essential as a public service, the government has set up a tax and subsidies system that reduces some of the inequalities of income. If the cost of running the pharmacy is within reasonable limits, all proprietor pharmacists may rely on making a reasonable income.

What

Pharmaceutical products must be approved and registered before marketing (see Section B.5).

There are a limited number of drugs registered in Norway: 2,062 preparations, including different dosage forms and strengths, as of January 1, 1986 (1,009 different trademarks).

The compounding of drugs in pharmacies is now reduced to a minimum, about 3% of the total sales in the pharmacies. The health authorities have issued pharmacy-GMP directives. The directives are based on the PIC guidelines, adjusted to fit the small and varied production in pharmacies.

Veterinary preparations and homeopathic products are sold solely from pharmacies and regarded as drugs according to the law.

Payment

Of the total sale of drugs in 1986, which was NOK 3,389 million, some NOK 1,144 million (about 34%) was reimbursed by the National Insurance Institution. The hospitals account for about 20% of the total sales. Most of the reimbursed drugs are prescription drugs. Of the total sale of prescription drugs, some 75% is reimbursed.

The bone of the Norwegian drug reimbursement program is that it is applying to the chronic population. A cost sharing between the patient and the Health Insurance was introduced in 1981. Up to then, the medicines were delivered free to the patient. The pharmacies get paid partly from the patient and partly from the Health Insurance and administer this part of the reimbursement scheme (see also Section B.7).

Formulary

A Drug and Therapeutic Formulary based on a joint initiative by the Norwegian Association of Proprietor Pharmacists, the Norwegian Medical Association, the Norwegian Medicinal Depot, and the Norwegian Medicines Control Authority was introduced in 1984. It was distributed to relevant categories of health personnel and is sold in bookshops. The book provides qualified and updated advice and recommendations on drug therapy, and it is revised at regular intervals (1986, 1988).

The drug manufacturers and importers issue a catalog annually of registered pharmaceutical specialties (*Felleskatalogen*). The product monographs are presented in alphabetical order according to brand name. The monographs are approved by the health authorities. The thirty-first edition was published in 1989. *Felleskatalogen* is distributed to physicians and pharmacists and is sold in bookshops.

Drug and therapeutic committees in hospitals usually prepare a local formulary. Between 200 and 300 preparations are considered sufficient to cover the routine use of drugs in a county hospital. Selection criteria are scientific documentation, available dosage forms and strengths, and price considerations.

2. Record Keeping

In Norway, the systematic handling of drug data at the prescription level is limited to narcotic drugs. A computer monitoring system based on data from all the prescriptions for narcotic drugs collected in the pharmacies was introduced in 1970. The purpose of the system was to give the health authorities an opportunity to follow and control the use of narcotics and to take action when consumption or prescribing patterns seemed unjustified. For that purpose, reports giving a rapid, detailed, and complete picture of prescribing patterns were needed. Such information has routinely been taken out of the computer system over successive quarterly periods.

Prescriptions for psychotropic drugs are kept in the pharmacies for one year. The purpose of this system is, as for narcotic drugs, to give the health authorities an opportunity to follow and control the use of psychotropic drugs.

About 65% of the pharmacies now have the capability of recording the prescriptions under the reimbursement scheme in a computer system. This is valuable data on drug utilization, but so far it is not used for routine monitoring. Special studies, mainly of a descriptive nature, have been performed and are useful in evaluation of drug use patterns, including patient compliance studies, and in measuring effects of educational, informative, and regulatory efforts.

3. Patient Education

Patients get their information on illness and treatment from several sources: doctors, nurses, pharmacists, family, neighbors. Health personnel are becoming increasingly aware of the importance of patient education. In some communities, groups of health personnel, including pharmacists, work together on questions of drug use in local drug and therapeutic committees.

As all drugs, both OTC and prescription, are sold only through pharmacies, the pharmacists and technical assistants have a central position as to patient education. Information is a natural part of the distribution of drugs from pharmacies. As mentioned earlier, about two million packages with prescription drugs are sent yearly to patients because of long distances. This represents a special challenge concerning information to the patient. Various written information is of importance and in frequent use. Telephone calls could be used more frequently.

The Norwegian Association of Proprietor Pharmacists has initiated public campaigns on different themes related to drugs, such as:

— Letting the pharmacy take care of unwanted leftover drugs in the homes
— Correct intake of oral dosage forms
— Optimal use of laxatives
— Changing to insulin 100 I.U.

A few hospitals have started patient education programs for groups of patients. The hospital pharmacists are involved in education about drug use.

To support and supplement information and advice provided verbally by doctors and by pharmacy personnel, various written information is available from Norwegian Medicines Control Authorities and other sources independent of drug producers.

Brochures are provided for all common self-care drugs and are available from all pharmacies. Drug information leaflets for prescription drugs have been compiled for different pharmacological groups of drugs. As of spring 1989, 15 groups were covered. The leaflets are available to the public through pharmacies, although doctors may obtain supplies.

The question of patient package inserts (PPI) is under discussion in Norway, in light of the EEC Commission's proposal to harmonize patient information.

Basic knowledge of drugs and drug therapy is not yet common in schools. In 1988, suitable teaching material was presented for the first time. This is an important basis for further patient education.

During 1989, a Drug and Therapeutic Formulary for the public was published, based on the information, advice, and recommendations on drug therapy given in the Drug and Therapeutic Formulary previously mentioned (see Section C.1). The Norwegian Association of Proprietor Pharmacists, the Norwegian Medical Association, the Norwegian Medicinal Depot, and the Norwegian Medicines Control Authority have initiated the publication.

4. OTCs and Other Classes and Schedules of Drugs, Other Nonpharmacy Outlets

The sale of all categories of pharmaceutical preparations, both prescription and nonprescription, is generally restricted to pharmacies only. To meet the needs of the public in rural areas, about 1,300 nominated outlets are permitted. These outlets are allowed to sell only nonprescription drugs, which they get from a regular pharmacy. They are run under the responsibility and inspection of the proprietor of the regular pharmacy.

During the last few years, more drugs have been classified as OTC preparations, sometimes in amounts limited to a few days' treatment. This makes it easier for the public to take care of minor ailments without seeing a doctor. It represents a challenge to the pharmacists to give professional advice in these situations.

The Director General of health decides if a drug is on prescription and to which prescription group it belongs.

About 20% of the total sales from pharmacies are OTC drugs or surgical dressings, appliances, or other health/hygiene products. This amounts to about 50% of the total number of packets sold.

5. The Use of Technicians and Others

Only pharmacists are allowed to dispense a prescription, but technicians do a variety of other work in the pharmacies. About 45% of the technicians have secondary education at the advanced level, which takes one year (see Table 10).

There are two categories of pharmacists in Norway, educated at the Institute of Pharmacy, the University of Oslo, or the National College of Dispensers. The university curriculum takes five years, while the National College of Dispensers takes two and a half years and is mainly directed toward the work in a pharmacy. Only pharmacists graduated from the

TABLE 10. Pharmacy personnel, 1988.

	Total	Per pharmacy
Pharmacists, total	1.335	4.2
Owners / managers	315	(1)
Pharmacists, Univ. grad.	407	(1.3)
Pharmacists, dispensers	609	(1.9)
Technicians	2,959	9.4

Source: The Norwegian Association of Proprietor Pharmacists

university may obtain a license to own and manage a pharmacy. Further education has a high priority for all personnel categories.

6. Pharmacy in Primary Care

The municipal health authorities are responsible for the planning and running of primary health care, including general somatic nursing homes. Drug distribution in the general somatic nursing homes is regulated by law and includes pharmaceutical inspection. Home nursing was included in this legislation in 1988.

The fact that about 80% of the drugs sold in Norway are used in primary care presents a challenge for the pharmacists to take responsibility for rational drug use, together with local health personnel. Some local drug and therapeutic committees/groups were established over the last few years; most groups include physicians, pharmacists, and nurses. A few of these groups have a formal agreement with the municipal health authorities.

Collaboration between primary and secondary health care is of great importance. This concerns pharmacists in hospital and primary care.

7. Drug Insurance and Third Parties

Coverage of drug costs has been previously discussed.

A compulsory no-fault pharmaceutical compensation scheme coupled with a patient insurance scheme was introduced by legislation in Norway during 1989. All manufacturers and importers of drugs have a strict product liability, with claimants needing to prove only causality, not negligence. Clinical trials are covered.

The legislation requires that manufacturers join in a drug liability organization that is responsible for a joint insurance. The insurance premiums are covered by an increase in the price of drugs and will be stipulated on the basis of the total sales per manufacturer.

The drug injury compensation scheme does not cover injuries caused by treatment errors alone. A claim for compensation in such cases should be directed towards the doctor, hospital, or pharmacy involved.

D. CONCLUSIONS

The basic aim of a drug policy is to ensure that effective and safe drugs of good quality and reasonable cost are available to meet the health needs of the country. When looking at the solutions found in various countries, consideration must be given to the historical development of health care, to geographical conditions, and other factors that may have considerable influence.

Norway is a large country with a small population. Distances and climate are unusual. The health service is a public service, and it should be available and efficient everywhere in the country. Paradoxically, economy did not seem so important until Norway became rich. The richer the country becomes, the poorer it seems to be, simply because possibilities and expectations grow faster than income. Norwegian drug expenditure is low compared with other European countries (8% of total health care expenditure). This percentage has been decreasing.

Norwegian drug expense reimbursement is a selective system, meaning that reimbursement is confined to selected groups of the population. The list of diseases now includes 36 chronic diagnoses. For each diagnosis there is a specified list of drugs and preparations. Prescribing of certain categories of drugs is limited to specialists only. The amounts are limited to a three-month supply. OTC drugs are excluded from the scheme. Generic prescribing is encouraged.

In Norway, price considerations are an integral part of the registration procedure. Both prescription and nonprescription drugs are subject to control. Prices of new products are compared with the prices of similar products on the market and with the price charged in other European countries, particularly in the country of manufacture. Models for price adjustments have been developed. In these formulas, inflation changes in exchange rates, etc., are reflected.

The wholesale distribution of drugs is carried on by a state monopoly (NMD), which operates with fixed wholesale margins. Part of the net income is used to support drug information and clinical pharmacology research.

At the retail level, the establishment of new pharmacies is controlled. There is a strict governmental control of both professional and economic matters. The Norwegian pharmaceutical service is a self-financing system. The consumer prices are fixed. The system of tax and subsidies is important. The Norwegian pharmaceutical service is characterized by its exclusively professional nature. Drug prices are moderate and the retail margins among the lowest in Europe.

The Norwegian criteria for registration of drugs have been much discussed, especially the requirement for medical need. Inclusion of the need clause in the Norwegian legislation some 40 years ago introduced a social dimension into the drug policies at a very early stage. Drugs were not only assessed purely from a scientific or technical point of view, but in the light of health care for the population as a whole. The WHO approach on essential drugs corresponds very well with this way of thinking. The report on the selection of essential drugs is of importance not only to developing countries but also to developed countries. The Norwegian registration policy for several decades illustrates that it is possible to limit the number of drugs on the market significantly without detrimental effect to the patients.

In small countries, only limited resources are available and extensive programs for the continuous evaluation of all kinds of drug therapy problems are beyond reach. This situation calls for international cooperation. Within the Nordic area, the control authorities have for many years had close cooperation on the evaluation, standardization, and postmarketing control of drugs, including statistics on medicines.

During recent years, this cooperation has been further developed to include harmonizing of requirements for clinical trials, common application forms, labeling, etc. One of the present challenges is the establishment of a unified European Community market for pharmaceutical products by 1992. This will have a major influence on the evolution of medicines regulation in the 1990s.

Drug problems are international by nature. Norway contributes to several activities within the drug field, such as the WHO drug action program, including the development of drug policies, essential drugs programs, training of health personnel, actively supporting the international control of dependence-producing drugs, and on a regional level support and participation in drug utilization studies, studies of drug regulation, etc.

Through bilateral assistance to and cooperation with other countries, Norway has gained experience and developed new approaches to drug problems which could serve as useful models for future work.

Chapter 18

Panama

Jeronimo Averza C.

A. THE NATIONAL HEALTH CARE SYSTEM

1. History and Background

The Republic of Panama, an S-shaped country with an area of 78,000 square kilometers and a population estimated in 1989 at 2,369,850, occupies the narrowest section of the Central American isthmus in the northwesternmost part of South America. It is bounded by Columbia on the east, Costa Rica on the west, the Pacific Ocean on the south, and the Caribbean Sea on the north. Although Panama is geographically a part of Central America, it was historically attached to Columbia until 1903 and has remained aloof from the political affairs of the Central American republics.

The Caribbean coast has a tropical rainy climate. The average annual rainfall is 128 inches, and monthly temperatures average 80° Fahrenheit with little seasonal variation. The Pacific Coast has a tropical wet and dry climate: a five-month dry season from December to May, and a rainy season from May to November. The profile of the population is young: 42% of the population is under 18 years of age, and 5.1% is 65 or over. The drop in mortality and fertility rates in the last decade is 2.19%, compared to 3.1% for the 1970-1975 period.

This growth in population has been accompanied by rural-urban migration, except in 1987-1988 when migration stopped because of the economic crisis. More than 52% of the population lives in urban areas, and approximately 64% of it is located in the capital city of Panama and surrounding areas (San Miguelito and Arraijan). There are various American groups (Cuna, Guaimies, Teribes, Chipchas, and Chocoes) totaling about 150,000 persons.

Spanish is the official language of the country, and English is widely spoken. The principal religion is Roman Catholicism. Panama is divided into nine administrative provinces. Currency unit: 1 Balboa = 100¢ = $1.

2. Social and Political Environment

The health integration system (Ministry of Health and the Social Security system) in 1988 covered 2,148,000 persons. According to the National Health Program, priority has been given to remedying the situation with the help of the Pan American Health Organization (PAHO), regional strategies, and a plan of action for health for all by the year 2000.

During the last decade, educational facilities increased more than 45%. Primary and secondary education enrollment levels (for the population between five and 19 years) were 638,500 students in 1988; the university enrollment rose from 24,000 to 54,557 in the five universities (two are state universities and the other three are private). The level of illiteracy about 11%. Costs for education in the public sector were 11% of the national budget.

Over the last decade, major housing projects have been implemented, particularly in the capital city, surrounding areas, and other major cities in the provinces. Based on data from census sampling, the housing shortage is estimated at 5.3% (1987) of the per capita gross national product (GNP).

Over the last two decades, favorable changes have been registered in health indicators for Panama's population. Life expectancy at birth rose from 67.8 years in 1973 to 72.08 in 1987.

Infant mortality dropped from 33.3 deaths per 1,000 in 1973 to 19.1 in 1987; the neonatal mortality rate dropped from 16.1 in 1973 to 10.6 in 1987. Water-borne diseases, which in 1960 were the most common cause of death, dropped to become the sixteenth most common cause in 1987.

The intensification of immunization activities at all health establishments has resulted in no cases of poliomylitis being recorded since 1972 and only one case of diphtheria since 1975. There have been no reported cases of smallpox, plague, yellow fever, or cholera, and malaria has been limited to very small areas, with less than 100 cases reported in 1984. On the other hand, other chronic diseases are at an all-time high. These include cancer, heart disease, and accidents, the three main causes of death.

In the same period, there has been an increase in the road networks stemming off of the Pan American Highway that runs across the country. The total length of the road network, in 1986, was 9,720 kilometers (810 kms of concrete, 2,355 kms of asphalt paving, 4,079 of asphalt coating, and 2,476 kms of rural earth road).

3. Public and Private Financing, Health Costs

The health sector is financed mainly from the Social Security fund and through contributions from the Ministry of Health, the National Institute for Water Supply and Sewer Systems, and the Metropolitan Sanitation Bureau. During the last decade (1975-1985), the country applied more than 10% of its gross domestic product to the health sector; the contribution from the Social Security fund during 1988 was a little more than U.S.$183 million. In this same period, operating costs for these services increased from U.S.$84.3 million in 1975 to U.S.$265 million in 1985. There are also some additional expenditures based on funds provided for the payment of services to the uninsured population. This source was not significant in 1975 when the integration started with the majority of the provinces (six out of nine). But in 1985, when the integration almost covered the nine provinces, this source was included officially in the national health budget. No official information is available on health costs in the private sector; however, studies done by a private company indicated that these costs probably represented about 40% of total health costs of the country.

4. Facilities

HEALTH INSTITUTIONS

	Hospitals	Policlinic and Health Center	Sub-center and Health Stations
Total	58	178	435
Public	38	178	435
Private	20	---	---

CAPACITY

	Beds	Cradles	Occupancy percentage	Daily Average of Patients	Average of occupancy in Days
Total	7.088	710	65.1	5.074	7.9
Public Inst.	6.072	592	68.9	4.593	8.6
Private Inst.	1.016	118	42.3	481	4.4

5. Manpower (1)

In 1987 the health care team was composed accordingly to the following figures.

a.   Physician: 2,722   with specialty 1442
                            without specialty................................. 491
                            general practitionary......................... 789

                            Total   2,722

b.   Dentist: 519   with specialty.......................................75
                            general dentist................................444

                            Total   519

c.   Nurses: 2,456   with specialty...650
                            general nurses...................................1.806

                            Total   2,456

d.   Pharmacist: 805   with specialty....................................74
                            general pharmacist.................................731

                            Total   805

e.   Medical Techologist: 773

f.   X Ray Technician: 309

g.   Auxiliary Nurses:3724
----
(1) Including all health care professionals registered in the Ministry of Health.

Personnel in Health Institutions

	Physicians	Dentist	Nurses	Pharmacists	Auxiliary Nurses	X Ray Technicians	Medical Technologist
Total	2.722	519	2.456	805	3724	309	773
Public Ins.	2.452	84	2.184	543	3254	240	682
Private Ins.	0.270	435	272	451	451	69	91

6. The Social Security System

The Social Security system was created by Law 23 in 1941 and has developed to an obligatory system in the country, including public and private workers. It has a monthly fixed share of 7.25% of the employee's gross salary, paid by the employee; 10.75% of the employee's gross salary, paid by the employer; and 6.75% paid by the retired (pensionary).

The Social Security system has three large programs:

a. Maternity and illness
b. Invalid, old age, and death
c. Professional hazard.

It is the strongest health institution in the country and covers more than 1.6 million people. It has around 15,000 employees, including physicians, dentists, nurses, pharmacists, and paramedics, and is the first employer of the health care team in the country.

The Social Security system has a drugs committee which prepared a formulary with 500 different pharmaceutical products, including drugs used only in hospital services and in the three different health care stages: primary, secondary, and tertiary. Over the last two decades, favorable changes have been registered in the Social Security health facilities: from one hospital and 280 beds in 1968 to nine hospitals and 2,106 beds; from ten polyclinics in 1968 to 24; from two polyclinics in the capital city in 1968 to seven in 1988. Total expenditures for health services in 1988 were around U.S.$183 million. The Social Security fund paid to its 72,065 pensioners U.S.$251.4 million in 1988, with a minimum pension of $145 and a maximum pension of $1,500.

7. The Medical Establishment

Based on government regulation, the health care providers must have a registration number given by the Ministry of Health to accomplish their professional functions in the country. Physicians, dentists, and nurses, after graduation from a university, have to spend a two-year apprenticeship in a state hospital before their registration.

To become a specialist, a physician has to spend two to five years in residency working in a state hospital. Dentists, nurses, and pharmacists have to study in a university two or more years or complete a fellowship in a hospital.

8. Payment and Reimbursement

The cost of a visit to a physician is from $10 for a general practitioner to $60 for a specialist. There are three companies with illness insurance programs: Mutual of Omaha, Blue Shield and Blue Cross, and Pansanitas, with partial reimbursement according to the coverage acquired by the patient.

9. Coverage

The Social Security system has complete coverage, and the private insurance programs have partial coverage.

B. PHARMACEUTICAL INDUSTRY AND DRUG DISTRIBUTION

1. Structure of the Industry

The country does not have good development in the pharmaceutical industry. There are seven pharmaceutical laboratories and among them, two international branches: Schering Corporation Lab and Winthrop Sterling Products Lab. There are five national laboratories. Two of the pharmaceutical laboratories are located in the second city of the county Colon and five in the capital city. Production of the seven laboratories equalled 10% of the national consumption of pharmaceutical products, plus $4 million that was exported to Central American countries and the Caribbean.

2. Research and Development, Imports, Exports, Patents, Licensing

Research studies are made only at the University of Panama and are very limited; 90% of the drugs consumed are imported from the U.S.A., Central America (Costa Rica, El Salvador, and Guatemala — from laboratory branches), Switzerland, and Germany. The importation is regulated by the government, and there is a small protection in the exportation. Patents are active in the country, having a government protection from five to 15 years, and licensing is also protected with an extension of ten years.

3. Advertising and Promotion, Price Regulation

All advertising and promotion of drugs is regulated by the government through the Ministry of Health and includes promotion through newspapers, television, radio, and road board.

All prices of pharmaceutical products are regulated by the Ministry of Commerce through the Office of Price Regulation, with the following margins: from the CIF price of a product, the Office allowed an increased price of 30% to the retail pharmacy or drugstore and an increase of between 30% and 33% of that price to the public.

4. Drug Approval and Government Regulation

Government regulation of drug approval is covered by Decree 93 of 1962, with the following registration procedures: a registered distributor of drugs has to present a petition to the Ministry of Health, countersigned by a lawyer and a registered pharmacist, accompanied by:

a. A free sales certificate from the health authorities of the country where the drug is manufactured, legalized by the Panamanian consulate
b. Four original samples of each product
c. Four labels of each product
d. Four scientific publications on the product (pharmaceutical studies of the product)
e. The qualitative and quantitative formula of each product
f. The analytical assay
g. A brief description of the manufacturing process.

All the documents have to be presented in Spanish. The Department of Pharmacy and Drugs of the Ministry of Health studies the petition and sends the samples to a Food and Drug Laboratory at the University of Panama, where an analytical assay is performed. If the product passes the assay according to the U.S.P., the product will receive a free sale certificate for five years.

5. Competition: Brand and Generic

The pharmaceutical product competition in the country is rough and stormy. Branded and generic products are marketed by Panamanian companies, as well as the majority of pharmaceutical laboratories from the U.S.A., Germany, France, Italy, England, Switzerland, Spain, Canada, Central America, South America, Korea, Taiwan, and Japan. This rough competition makes the containment activities very expensive, especially in the field of promotion by the detail men (scientific salesmen) and the distribution activities.

C. PHARMACY PRACTICE

1. Dispensing a Prescription

Based on government regulation, the pharmacy is the only place able to dispense a prescription. Since the country has a price regulation law, every pharmaceutical product has a price printed on the label, and extra charges for the pharmaceutical professional service are not permitted.

There is no national formulary, and the official books are the U.S. Pharmacopeia, the U.S. National Formulary, and the International Pharmacopeia from WHO.

2. Record Keeping

The pharmacy or drugstore must have a book where all prescriptions must be transcribed for record keeping as well as a file to keep the prescriptions. The pharmacy is allowed, by law, to give a copy of the prescription to the client if it is required.

3. Patient Education

During the last decade, patient education has been developed in all the pharmacies that belong to the Social Security system, with a good acceptance, and in the private pharmacies where the owner is a pharmacist.

4. OTCs and Other Classes and Schedules of Drugs, Nonpharmacy Outlets

There is no regulation of OTC drugs, but there is one that covers popular drugs that are similar to OTC. During the last 15 years, the professional pharmacy has changed in the majority of cases to a drugstore very similar to U.S. drugstores because of the price regulation law. In the near future, OTC regulation must be present in the country.

A few of the supermarkets in the capital city have sold popular drugs and other pharmaceutical products, in a wildly open violation of the law.

Popular drugs are products that do not need prescriptions, and their sale is permitted in nonpharmacy outlets.

5. The Use of Technicians and Others

The technician in pharmacy is not common and is not present in any future educational program. The pharmaceutical assistant is present in all public and private pharmacies and normally is a pharmacy student.

In a few private pharmacies, a person called a practical pharmacist is present. The practical pharmacist normally is a person without university studies in pharmacy but who has developed a great knowledge of the pharmaceutical profession by working and learning behind a pharmacy counter.

6. Pharmacy in Primary Care

The contribution made by pharmacy in the primary care system became significant in 1978, when the integration of the National Care System (Social Security and the Ministry of Health) was completed in eight of nine provinces. The first step in that direction was made by the Social Security Committee of the Drug Formulary when it divided the pharmaceutical products present in the formulary by drugs used in the three stages of health care. The second step was when pharmacists started the patient education program.

7. Drug Insurance and Third Parties

The Social Security system provides, throughout its pharmacies, all prescriptions made to associates and their dependents without any charges. It also has another drug program, called subrogation pharmacy, with the private pharmacies in the country, which provide drugs to associates when they are not present in the Social Security pharmacies but are present in the Formulary.

The three private insurance programs present in the country provide drugs only to associates when they are in the hospital.

D. CONCLUSION AND ASSESSMENT OF ACHIEVEMENTS

All data for this evaluation has been obtained from the Panamanian statistical system and from the reports of the Social Security system and Ministry of Health at the end of 1987.

The country has been subjected to a social and economic crisis during the last decade, with the greatest impact in the last two years. Panama has one of the highest external debts in Latin America and a high rate of unemployment. However, Panamanian health indicators and the national health program, on a national average, have exceeded the proposed regional goals for attaining health for all by the year 2000.

The service network, functionally integrated throughout the Social Security fund and Ministry of Health, has provided a degree of health coverage, particularly in drinking water supply services, immunization, and sanitary educational programs, which may be said to represent the health sector's contribution to the indicators now achieved: life expectancy of 74.13 years in the cities and 70 years in rural areas, infant mortality of 19.1 per 1,000 live births, child mortality (1-4 years) of less than 10.6 per

1,000, drinking water supply coverage of around 92%, and good immunization coverage under the EPI.

The significant advances made by the country in health coverage in the last two decades are the main benefits of the national health program started in 1973, called Health for Everyone, which was supported by the Social Security fund, the Ministry of Health, and the Pan American Health Organization (PAHO). Nevertheless, serious studies carried out under the project for adaptation of the service network have identified previously overlooked groups in need of priority attention in future operational plans. The country has analyzed its population's access to the first level of care in each population center and has proposed changes needed for fulfilling the government's objective of guaranteeing access by the entire population to the service network. This has increased dramatically the number of health centers, from 65 in 1973 to 183 in 1978. The secondary level health centers grew from 96 in 1973 to 178 in 1987, and health posts grew from a few in 1973 to 435 in 1987.

SELECTED READINGS

Panama en cifras, publicado por la Direccion de Estadistica y Censo de la Contraloria General. Panama, Noviembre de 1987.

Situacion Social y Servicios de Salud, publicado por la Direccion de Estadistica y Censo de la Contraloria General de Panama, Noviembre de 1987.

Indicadores Economicos y Sociales de Panama, 1978-1987, publicado por la Direccion de Estadistica y Censo. Panama, Noviembre de 1988.

Indicadores Economicos y Sociales de Panama, publicado por la Direccion de Estadistica y Censo. Panama, Noviembre de 1987.

IMS. Third World Health Insights, 1978-1979; 121-126.

Region of the Americas. Panama. 1984; 170-174.

Farmaco, organo del Colegio Nacional de Farmaceuticos de Panama. Ano 1985, Nos. 1 y 4; ano 1986, Nos. 2, 9 y 12.

INCAP, CDC. Nutritional evaluation of the population of Central American and Panama: regional summary. Washington, DC: DHEW, 1972.

Ministerio de Salud. Defunciones de menores de 1 ano y de 1-4 anos y tasas de mortalidad de la Republica de Panama, segun provincia y distrito Panama ciudad. Minsterio de Salud, 1982.

Memoria de la Caja de Seguro Social, publicada por Caseco en Enero de 1988.

Memoria del Minsterio de Salud, publicada por Caseco, en Enero de 1988.

Chapter 19

Spain

Joaquima Serradell
Eugeni Sedano
Joan Serra

A. THE NATIONAL HEALTH CARE SYSTEM

1. History and Background

The history and development of the health care system in Spain is complicated, and it is linked to political changes as well as economic and industrial development.

The 1822 Health Code (Codigo Sanitario de las Cortes de Cadiz) was the first governmental attempt to organize public and community health. This first attempt failed because a disagreement existed about which scientific measures were to be used to support collective health. In 1855, the General Health Board (Direccion General de Sanidad) was established, and its framework lasted for about a century. Efforts to create first social and then health insurance were in place in 1904 with the creation of the General Instruction of Public Health legislation (Instruccion General de Salud Publica).[1] As a way of summarizing the history of health care provision in Spain until the 1930s, it can be said that there were a few stages of private medical care and charitable public assistance without coordination and with great limitations to access.

From 1931 to 1939, there were serious attempts to reform the health sector by the government of the Spanish Republic to develop a system of social and health insurance. The Republican Constitution recognized the right to health insurance. In Catalonia, because of its autonomous government during that period, the first design for a regionalization of hospitals and health services was made.[2] However, due to the consequences of the

Spanish Civil War, many of the progressive advances during the 1930s were never realized.

The Law of Bases for Health (Ley de Bases de Sanidad) in 1944 organized the embryo of what would become Social Security (Seguridad Social) and the Compulsory Sickness Insurance Program (Seguro Obligatorio de Enfermedad). Originally it was intended only to cover the lower-paid workers in industry, excluding those earning above a stipulated minimum income. At that time, only one-fourth of the Spanish population was covered by public assistance.

The government took responsibility for developing a system of disease prevention, basically taking care of child vaccination programs. The function of provision of care was its primary goal. This function was developed through compulsory sickness insurance (Seguro Obligatorio de Enfermedad). It was the first attempt to create a centralized state structure for the delivery of health care services. However, health assistance was also provided by other public structures, such as hospitals, which depended on the church or the provincial government. Overall, though, the delivery of health care was still very fragmented and unorganized.

With the return of democratic institutions and the new Constitution in 1978, Spain was divided into 17 autonomous regional governments. A transfer of control and authority by means of transferring money and manpower from the Central Administration to the autonomous governments is now underway.

Specifically in health issues, the ultimate goal of the Health Law (Ley de Sanidad, 1986) is the creation of a National Health Service system. The National Health Service requires the integration of all of the old health delivery structures so that a more coordinated and even distribution of care will result. The National Health Service coordinates all autonomous regions. The General Law of Health (Ley General de Sanidad) is being phased in.

As of 1988, only four regions had had transferences granted regarding health issues: Catalonia, 1981; Andalucia, 1984; Basque Provinces, 1988; and Valencia, also in 1988.

The financing of the health system is multiple and mixed. During the transfer period, financing comes from the Central Administration. Once the transfer is a fact, the regional administration or autonomous government has direct resonsibility for the financing through taxes, funds from the central government in Madrid assigned to each region, and resources coming from the Fondo de Compensacion Interterritorial. This national government agency guarantees an even distribution among the 17 regions.

The basic characteristics of the National Health Service are:

— Health insurance covering 100% of the population (foreigners as well as Spaniards)
— Comprehensive health services; that is, disease prevention, health promotion, and delivery of care
— Integration and coordination of public health resources in a single network
— Health education and evaluation of health initiatives.

The application of the law facilitating the National Health Service relies on the creation of Health Areas (Areas de Salud). These are the basic administrative health structures, and they are responsible for providing health programs in a given area to a population of between 200,000 and 250,000. Great emphasis is placed on the prevention of diseases and environmental protection.

The integration of the private sector is by contract. The integrated private sector will guarantee hospital services free of charge.

The financing of the National Health Service was one of the problems faced during the reform. The financing will not only be through taxes, but through:

— Copayment from workers and employers
— Government budgets guaranteeing equal distribution among regions
— Copayments for certain services, for instance, copayments for drugs (ticket moderador)
— Money from autonomous (local) governments and local corporations.

2. Social and Political Environment

Geography

Spain is a country situated in the southwest corner of the European continent. Its surface is 504,703 square kilometers. To the north is France, to the east and south is the Mediterranean Sea, and to the west is Portugal. The total coastal perimeter is 3,904 kilometers. Climate is varied: humid to the north, extreme or continental in the center, and moderate along the Mediterranean region.

Demographic Characteristics

Spain has a population of nearly 40 million inhabitants. Population density is 76.6 inhabitants per square kilometer.[3] There are great demographic differences between the most developed regions (Madrid, Basque Provinces, Catalonia, Valencia, etc.) and those less developed regions such as Castilla, Aragon, Extremadura, Leon, etc.

Health Indicators

Infant mortality rates are shown in Table 1.[4]

Morbidity data are shown in Table 2.[5] The figures are from hospital diagnoses for discharged patients.

The mortality rate was 7.8 per 1,000 inhabitants in 1983.[5] In 1980, the most frequent causes of death for both genders were related to the circulatory system, tumors, and the respiratory system.

Spanish Political Context

After a long period of dictatorial government, Spain has become a democratic state with rights defended by the Constitution. These rights are freedom, justice, equality, and political pluralism. The form of government is a parliamentary monarchy, where the chief of state is a king who represents the unity of Spain. The Spanish Constitution directs a decentralization of the state into 17 autonomous regions. An important move for Spain was its integration into the European Economic Community (EEC) as a full member in 1987.

Economic Context

In 1983, the gross national product (GNP) was Pts. 22,682,761 (pesetas) or Pts. 596,130 per person.[6]

TABLE 1. Infant mortality rate evolution (1978-1982; per 1,000 live births).

Year	Catalonia	Spain
1978	12.3	15.3
1979	11.3	14.3
1980	10.5	12.3
1981	11.3	12.5
1982	9.5	11.5

3. Public and Private Financing, Health Costs

In the European Community, 22% of the social protection expenditures go for health. In Spain in 1985, 22.5% was spent on health.[7] According to data from the Ministry of Health and Consumption (Ministerio de Sanidad y Consumo), the public expenditures for health for 1988 totaled Pts. 1.8 billion, which represents around 5% of the GNP and 73% of all the expenditures for health.[8] Table 3 includes the expenditures for health by the public sector organization in 1987.[9]

Table 4 shows the distribution of the Organization for Economic Cooperation and Development (OECD) countries' expenditures on health according to their GNP.[10] The table includes GNP expenditures for the public as well as the private sectors.

Private Health Expenditures

Private Spanish health expenditures in 1987 were 1.8% of the GNP, or 27% of total health expenditures. In Spain, the private sector does not limit its activity to the delivery of privately financed health care. The

TABLE 2. Morbidity data 1984, Spain.

Diagnosis	Men	Women
Infectious Diseases	52,464	35,498
Tumors	111,881	109,890
Endocrine and Immune System	20,477	36,008
Blood Formation	8,056	8,731
Mental Diseases	52,814	30,350
Nervous System	71,856	66,450
Circulatory System	142,767	110,311
Respiratory System	172,266	104,369
Digestive System	221,266	160,369
Genito-Urinary System	94,211	136,100
Pregnancy & Complications in Delivery	---	574,677
Skin Disorders	32,335	22,679
Skeleto-Muscular	63,984	55,405
Congenital Disorder	25,848	12,016
Perinatal	21,511	16,263
Not Well Defined	138,348	119,057
Traumas, Poisons	194,032	109,679
Unknown Causes	174,428	183,870

TABLE 3. Expenditures for health in the public sector, 1987.

Organization	Expenditures (in millions pesetas)	Total (%)
A. Central Administration	1,490,283	84.68
B. Territorial Administration	269,498	15.31
Autonomous Regions	131,295	7.46
"Disputaciones"	105,969	6.02
City Governments	32,234	1.83
Total Public Expenditures	1,759,781	100.00

TABLE 4. OECD countries and their health expenditures per GNP, 1984.

Country	Public Sector Expenditures as a % of GNP	Total Health Expenditures as a % of GNP	Public Expenditures/ Total Health Expenditures (%)
Australia	6.6	7.8	8 5
Austria	4.4	7.2	6 1
Belgium	5.7	6.2	9 2
Canada	6.2	8.4	7 4
Denmark	5.3	6.3	8 4
Finland	5.4	6.6	8 2
France	6.5	9.1	7 1
West Germany	6.4	8.1	7 9
Greece	3.6	4.6	7 8
Iceland	6.5	7.9	8 2
Ireland	6.9	8.0	8 6
Italy	6.1	7.2	8 5
Japan	4.8	6.6	7 3
Luxemburg	6.8	6.4	8 2
Holland	4.4	8.6	7 9
New Zealand	---	5.6	- - -
Norway	5.6	6.3	8 9
Portugal	3.9	5.5	7 1
Spain	4.9	6.7	7 3
Sweden	8.6	9.4	9 1
Switzerland	---	7.8	8 2
United Kingdom	5.3	5.9	9 0
United States	4.4	10.7	4 1

private health sector has a close relationship with the public sector. First, it has a role collaborating in the management of care for the insured population through the public system (Seguridad Social). This activity is achieved by means of contracting with clinics, hospitals, and corporations. Second, the private sector insures groups of people who choose private insurers. Private financing is principally used to cover the expense of drugs, private medical visits, dentistry, and prothesis.

4. Facilities

Hospitals

Hospitals and the number of beds in 1986 in Spain, according to their ownership, are presented in Table 5.[11]

More than one-half of private hospitals provide care to Social Security patients, by contracting with the public system. Private hospitals generally specialize in minor surgery with short stays, while public hospitals provide diagnostics and surgery for more serious pathologies.

The number of public hospital beds per 1,000 inhabitants is 3.3, while there are 1.7 private beds per 1,000 inhabitants. The index of utilization is lower in public hospitals if acute pathologies are considered. Utilization of hospital services is regulated by the number of centers providing care and by the financing of those services through contracts. Not all hospitals provide free access to their services. Some provide services only to patients who have prepaid and joined their plan.

Private hospitals that wish to offer services to patients insured by Social Security (public assistance) contract with the public system through the Social Security management agencies (ICS, Institut Catala de la Salud; Osakidetza, Servicio Basco de Salud; Insalud, etc.).

Psychiatric care is interesting in Spain. The Catholic Church, in its

TABLE 5. Hospitals and beds according to ownership.

	Hospitals		Beds	
	Number	%	Number	%
Public	387	43	128,721	69
Private	512	57	57,330	31
Total	899	100	186,051	100

provision of care for the poorest groups of the population, has had a role in the care of the mentally ill. In 1987, 25% of the psychiatric beds belonged to Church-owned hospitals. In addition to the Church, provincial institutions have traditionally taken care of psychiatric patients. In Spain, psychiatric care is generally not covered by Social Security (only acute cases, but not chronically ill patients). The Health Law (Ley de Sanidad) foresees the inclusion of all psychiatric centers in the Social Security network.

Ambulatory Care

In Spain, ambulatory care in the public sector is practiced in health centers (centros ambulatorios and consultorios publicos), where several professionals varying in number and qualifications guarantee primary care to the publicly insured population. The definitions for the two modalities of primary care centers are:

1. Centros Ambulatorios (ambulatory centers): Centers where provision of primary care, general medicine, pediatrics, and other medical specialties and surgery facilities in a nonhospital regimen are offered. In the near future, after restructuring the public sector, these centers will be the health centers (centros de salud) where curative as well as preventive medicine will take place for a population of between 50,000 and 250,000. Health education will also be a function of these centers.
2. Consultorios (consultories): Centers where primary care, that is, general medicine, pediatrics, and nursing consulting is provided for a population of between 5,000 and 25,000. (See Table 6.)

Private Sector

While there are 232 entities of health insurance in Spain in the private sector, a great concentration exists. In 1986, 2 main entities (SANITAS and ASISA) represented 35% of the total, and the 10 top entities have 65% of the total volume. Of the top 10 medical insurance organizations, 4 have links with medical associations. Foreign capital also is present in these 10 top medical insurance companies.

In 1987, the number of persons covered by private health insurance was 6,001,566 with a premiums' volume of Pts. 90,469,000. Health insurance expenses are reimbursed in the private sector by capitation and fee-per-service. Capitation represents 15% of the total premiums and does not allow freedom in selecting physicians. The fee-per-service system allows

TABLE 6

Autonomous Communities	Centros Ambulatorios (Ambulatory Centers)		Consultorios (Aonsultories)	
	Number	Pop./center	Number	Pop./center
Andalucia	68	98,411	95	70,422
Aragon	14	86,719	41	29,611
Balearas	8	84,067	12	56,045
Canarias	15	95,422	12	119,278
Cantabria	4	131,441	23	22,859
Castilla-La Mancha	19	87,939	17	98,284
Castilla-Leon	29	89,758	71	36,662
Catalonia	51	118,414	113	53,444
Valencia community	28	134,528	150	25,112
Extremadura	11	98,285	38	28,451
Galicia	21	136,415	14	204,622
Madrid	26	187,282	133	36,611
Murcia	11	90,723	33	30,241
Navarra	6	86,740	23	22,628
Pais Vasco	31	73,232	107	21,217
Asturias	6	189,747	78	14,596
La Rioja	5	52,294	5	52,294
Total	355		965	

freedom of choice of physicians and the patient has a copayment for service of Pts. 50 to 150.

5. Manpower

Health personnel include physicians, dentists, pharmacists, veterinarians, and nurses. Table 7 shows the changes in health personnel 1982-1986.[14]

6. Social Security System (Seguridad Social)

In Spain the first attempts to structure different types of social insurance began early in the twentieth century following some of the social reforms in Bismark's Germany.

TABLE 7. Health professionals in Spain.

	1982	1983	1984	1985
Physicians	104,759	115,251	121,362	127,195
Dentists	4,065	4,458	4,682	5,135
Pharmacists	26,274	27,646	28,748	30,569
Veterinarians	8,037	8,404	8,660	8,705
Nurses	136,992	139,846	142,542	143,508

Massive opposition from organized medicine slowed down these first initiatives. In 1908, the Instituto Nacional de Previsión was created to manage the system of social programs and after a few years, to manage the Social Security system. In 1948, this organization started to build its own hospitals and ambulatory care centers.

Within the Spanish Republic (1931), work accidents were covered through the social insurance, and in 1936, sickness insurance for workers and professionals was also mandatory. In 1942, the Obligatory Sickness Insurance Program (Seguro Obligatorio de Enfermedad) was under way. The sickness fund was compulsory for industrial workers with lower than average incomes, which amounted to one-quarter of the work force.

With time, coverage was granted to larger groups of people. In 1953, it covered 30%; in 1964, 54%; and in 1988, 97%. The sickness fund evolved from a system of heterogeneous and isolated social insurances to a true system of Social Security covering the vast majority of Spanish residents.

The Spanish Social Security program is a quasi-governmental organization in charge of the social protection of its covered population. The health organization of the Social Security program is responsible for the provision of health care. It supports hospitals and primary health centers.

Three autonomous institutes constitute the structure of the Social Security program. These three organizations were created in 1978 from the original managerial system of the Instituto Nacional de Previsión. The three autonomous institutes are:

1. Instituto Nacional de la Salud (INSALUD) (National Health Institute), created in 1978, is responsible for the delivery of health care. Its organization and management are under the control of the Ministry of Health and Consumption. INSALUD is the organization responsible for managing health care and health promotion policies. Every Spanish province has a different organization for management of health care delivery at the local level.

2. Instituto Nacional de la Seguridad Social (INSS) (National Institute of Social Security), was created in 1978 to manage and administer the economic provision of Social Security (retired pensions and workers compensation). It is regulated by the Ministry of Work and Social Security. It is responsible for the affiliation and regulation of business corporations and employees in the Social Security system.
3. Instituto Nacional de los Servicios Sociales (INSERSO) (National Institute of Social Welfare) was also created in 1978. It manages all the other complementary services covered by Social Security, such as services for the physically and mentally handicapped and the elderly. It is regulated by the Ministry of Work and Social Security.

The three institutes have two levels in their organizational framework: national and provincial.

To better control the economy of the three organizations, there is a single budget for the three institutes: Tesoreria General de la Seguridad Social (Treasury of the Social Security). This treasury collects the different premiums from industry, farmers, miners, students, artists, etc., and then distributes the money to the different institutes.

The population covered by Social Security includes card holders (workers and retired people) and beneficiaries (family and dependents of the card holders).

Parallel management structures for health and Social Security delivery exist for special groups of employees. These are, for example, state civil servants, local civil servants, and the military. In addition, some organizations sponsor delivery systems for their workers or affiliates, e.g., Telefonica (telephone company).

7. The Medical Establishment:
Entry, Specialists, Referrals

In the public sector, the insured population is distributed among family practice physicians and pediatricians working in primary health centers. Each physician is assigned a minimum and a maximum quota. People are assigned to physicians according to the neighborhood or town in which they live. People have freedom to change physicians if they are not satisfied with the assigned ones.

Entry into the public sector is, therefore, at the primary care level at the area health center. The health center has a team of health professionals: family physicians, pediatricians, nurses, and social workers. The primary care team takes care of the acute cases and also provides follow-up care for chronically ill patients. The primary care team is also responsible for emergency calls during night and weekend hours. They even make calls when patients cannot travel to the clinic.

In emergency situations, patients also have the option of the hospital emergency room. Whether they are taken as inpatients or not, they will receive a report of that event to present to their assigned primary physicians to be added to their charts.

At a second level of health assistance, there are the medical specialists in health centers (ambulatorios). If a visit with a specialist is needed, the primary physicians will sign a referral form that the patient takes to the specialist or hospital office. Once the specialist has seen the patient, the specialist will provide the patient with a patient report, including the proposed treatment. These documents are then given to the primary physician. If the patient needs surgery or a complex diagnostic procedure, she or he is referred to the hospital that could provide such treatments. Subsequently, the patient receives a discharge report for the primary physician, who will eventually monitor the patient after the hospital stay.

8. Payment and Reimbursement

In Spain, all people covered by Social Security (97%) have the right to receive free outpatient health care as well as hospital care. The only out-of-pocket payment by the insured is in the partial payment of drugs and pharmaceutical products (like gauze). In 1980, a change was introduced in the payments for drugs (ticket moderador). The percentage paid by patients was increased to 40% of drug cost. This policy was an attempt to reduce Social Security expenditures. Exempt from drug copayments are retired people and payment for certain drugs for chronic conditions.

Private insurance covers medical assistance but does not cover pharmaceuticals.

B. PHARMACEUTICAL INDUSTRY
AND DRUG DISTRIBUTION

1. Structure of the Industry

The pharmaceutical industry produces drug products as well as raw materials for pharmaceutical use. The pharmaceutical laboratories must be registered as pharmaceutical industry companies and are included in the General Subdirection of Chemical, Textile and Pharmaceutical Industries as part of the Ministry of Industry and Energy.

Most of pharmaceutical firms are enrolled in the manufacturer employer's organization, Farmaindustria, created in 1963 as a technical service inside the Chemical Industries Labor Union. In 1977, under the law that establishes freedom to select from among labor unions, Farmaindustria obtained independence as the Entrepreneur's Association of the Pharmaceutical Industry. Since 1979, it has represented the interests of companies related to drug products.

In 1980, a Royal Decree was passed that allows total freedom for the installation, enlargement, and moving of the manufacture of pharmaceutical raw materials as well as drug products, excluding only the manufacture of narcotics.

As a brief summary of what Farmaindustria represents today, it should be pointed out that in 1987, the number of enrolled laboratories was 249. Catalonia and Madrid include the largest number of laboratories (51% and 36.9%, respectively). The companies with annual sales higher than Pts. 1,300 million represent 26.1% (classified as "large companies" by the association criteria), and 46.2% are small laboratories, with sales not higher than Pts. 250 million. Foreign companies represent 52% of the big and medium laboratories and are relatively more concentrated in the area of Madrid.

Table 8 shows the distribution by autonomous communities of laboratories enrolled in Farmaindustria in 1987.[15]

TABLE 8. Distribution of pharmaceutical manufacturers by location associated with Farmaindustria, 1987.

Autonomous Communities	Number of Laboratories
Andalucia	7
Aragon	4
Asturias	1
Baleares	1
Canarias	--
Cantabria	1
Castilla-La Mancha	--
Castilla-Leon	--
Catalonia	127
Galicia	2
Madrid	92
Murcia	1
Navarra	2
Pais Vasco	2
Rioja	--
Valencia	9
TOTAL	249

In addition to the information from Farmaindustria, data provided by the Ministry of Industry are shown in the following tables: the number of pharmaceutical laboratories (Table 9), pharmaceutical industry personnel (Table 10), pharmaceutical production (Table 11), and European Economic Community (EEC) pharmaceutical industry basic data for 1986 (Table 12).[16,17]

2. Research and Development, Imports, Exports, Patents, Licensing

Research and Development

The percentage of the Spanish internal gross national product used in research and development is 0.48%; the average percentage for the EEC countries is 2.45%. The U.S. has the highest investment in this field with 2.83%, followed by Germany (2.66%) and Japan (2.61%). A similar situation is found when considering the number of researchers per 100,000 people. Spain has one-tenth that of Japan and half of what Italy has.

It has been estimated that the private sector finances 20% of the total research in Spain, while this financing reaches 55% in more developed

TABLE 9. Number of pharmaceutical laboratories in Spain.

Year	Total
1980	401
1981	380
1982	375
1983	370
1984	368
1985	365
1986	362

TABLE 10. Persons employed in the pharmaceutical industry.

Year	Raw Materials Production	Pharmaceuticals Production	Total
1984	4,700	36,000	40,700
1985	4,800	35,900	40,700
1986	---	---	---
1987	1,965	33,055	35,020

TABLE 11. Pharmaceutical production (millions of pesetas).

Year	Raw Materials	Drug Products	Total Sales
1984	71,050	244,600	315,650
1985	84,500	272,000	356,500
1986	98,100	297,600	395,700
1987	108,560	304,175	413,735

countries and 75% in Japan. On the other hand, Spain is one of the countries that has had a higher percentage of the national budget assigned to research and development in the past ten years. According to the National Plan of Scientific Research and Technologic Development, it is expected that at the end of this decade the portion of the national budget assigned to biomedical and health sciences research will reach Pts. 250,000 million (around U.S. $227 million). This amount is undoubtedly modest when compared with the budgets assigned to medical research in the United States in 1986 ($14,338 million), and in Japan in 1985 ($14,111 million), but it is close to the amounts seen in the EEC countries.

Focusing the analysis on the pharmaceutical industry research in Spain, it is estimated that at present there are more than 30 research-intensive pharmaceutical companies and that their total investment in research and development (R&D) is around Pts. 8,000 million. The percentage of investment varies greatly among the 30 different companies and can be estimated at between 8% and 10% of the sales for the four companies that, in absolute value, invest the most in this field. These four companies have significantly increased their investment in R&D during the past ten years, from 307 million pesetas in 1974 to 2,300 million pesetas in 1985.

Imports, Exports

From data provided by the General Direction of Customs, in 1986 the exports of the pharmaceutical sector (raw materials and drug products) to the EEC countries totaled Pts. 19,501 million, while the level of imports were Pts. 32,578 million.

In the raw materials sector, the negative balance with the EEC increased from Pts. 8,757.8 million in 1985 to Pts. 12,161 million in 1986. In the drug products sector, the deficit in 1986 was Pts. 915 million while in 1985 it was Pts. 630 million.

Table 13 shows the total Spanish commercial trade (imports and exports of pharmaceutical raw materials and drug products) with the rest of the world.[18]

TABLE 12. Pharmaceutical industry of the EC, 1986 Data.

Country	Population (000)	Per capita (ECUS)	No. of Manufacturers	Employed Personnel	Pharmaceutical Production (mill. ECUS)
West Germany	61,050	14,978	1,000	72,500	9,573
France	55,390	13,042	345	82,500	8,824
Italy	57,220	8,966	340	64,582	6,523
Holland	14,560	12,125	294	12,500	943
Belgium/Luxemburg	9,900	11,520	240	15,853	1,077
United Kingdom	56,760	9,781	301	87,100	6,606
Ireland	3,540	7,432	120	4,800	1,056
Denmark	5,120	14,102	35	9,311	724
Greece	9,970	3,974	55	5,000	296
Portugal	10,290	2,853	83	9,000	288
Spain	38,670	6,012	345	35,020	2,879
TOTAL	322,470	11,551	3,158	398,166	38,789

TABLE 13. Spanish pharmaceutical imports and exports.

Year	RAW-MATERIALS*		DRUG-PRODUCTS*		TOTAL*	
	Imports	Exports	Imports	Exports	Imports	Exports
1983	28,699	13,588	11,658	15,526	40,357	29,084
1984	28,821	15,945	13,670	17,975	42,491	33,920
1985	26,049	18,095	17,828	22,480	43,877	40,575
1986	30,934	25,322	24,993	25,544	55,927	50,866

* in millions of pesetas

473

Patents

The Spanish law of patents, dated March 20, 1986, is directly adopted from the European Treaty of Patents and contains, in general, an equivalent regulation to the ones used in any other countries of the European Community. The law regulates the patentability of chemical, pharmaceutical, and food products, although implementation on the two former areas is postponed until October 7, 1992, when the government will pass the law as a Decree.

The introduction patents, which were ten years old and required only that the drug had national newness, have been eliminated. They no longer contribute effectively to technological development, and are incompatible with the European regulation of patents. At the EEC meetings, Spain agreed formally to their suppression. The new regulation requires that if a patent includes a manufacturing process with a new product, then every other similar product must be produced with the same patented procedure. The patent has a duration of 20 years, starting from the application date, and cannot be extended. It is effective from the day in which the concession is published.

The Treaty for European Patents in Munich (CPE) unifies the patents concession procedures of the different countries, but once granted, the patents will be regulated by individual national legislation, with some exceptions. The CPE does not exclude the patentability of chemical and pharmaceutical products, and this is why Spain did not originally sign the treaty, although it participated in the conference. As agreed in the CPE, which took effect in 1977, Spain will accept the patentability of chemical and pharmaceutical products starting July 10, 1992.

Licensing

The application for the patent as well as the granted patent itself are transferable and can be licensed to others. Contractual licenses (exclusive and nonexclusive) can be granted anywhere in the nation during the patent's validity. The licenses can be contractual, full right, or mandatory.

Contractual Licenses

Contractual licenses are regulated by the free enterprise laws. They should be inscribed in the Industrial Property Registration so that the licensee can state on the manufactured products that the patent is either applied for or granted. The contracts will be included in a public document, and the registration will indicate the legality, validity, and efficacy of the contracts that should be inscribed.

Full Right Licenses

The owner of a patent can offer full right licenses to anybody interested in using the registration. This will allow the owner to pay half of the annual fees for the patent. The licenses will be nonexclusive and if the parties do not reach agreement, the registration office will set the amount to be paid by the licensee. When the agreement is reached and the contract takes effect, the licensee will pay the fees quarterly and will inform the patent holder about the use of the invention. If the patent holder withdraws the full right offer, whenever there is no contract in effect, the patent holder will have to pay the registration office for the reduction of fees.

Mandatory Licenses

The patent holder has the obligation of exploiting the patented invention by the commercialization of the obtained results to satisfy the national market demand. The exploitation must be done within four years from the application date or within three years from the granting publication date, whichever is later.

The government, through a royal decree, can transfer a patent to the mandatory licenses office. This will be done when the export market is not being supplied sufficiently due to insufficient production using the patent, which can lead to damage of the economic and technological development of the country. By royal decree and for public interest reasons, the government can transfer a patent to the mandatory licenses agency. It is considered that public interest exists when the invention (its initiation, increase, generalization, or improvement) is necessary for the public health or national defense.

Mandatory licenses will not be exclusive, except in cases of the national interest. In the same manner, the licensee will not be allowed to import the patented object, except in cases of national interest.

3. Manufacturing

When the order of April 18, 1985 took effect, establishing good manufacturing practices and quality control of drugs, Spain started a period in which the laboratories established in the country would have one year to adjust their installations, means, and procedures to good manufacturing practices and quality control of drugs.

The present practices are mandatory and have been designed to systematize the general, fundamental, and minimum principles that should control the manufacture and quality control of drug products in Spain. In the future, they will be developed and adjusted to the new needs that the progressive evolution of science and technology will impose. They are based on the World Health Organization's practices (WHO 28.64 Resolution of 1975) and on the Pharmaceutical Inspection Convention (Document PH3/83 of June 1983). These practices will define the job of the Inspection Services of the General Direction of Pharmacy and Drug Products at the drug products laboratories. The fulfillment of these practices will contribute toward guaranteeing a high and uniform quality of drug products, as well as ensuring that safety and effectiveness is established in the permanent updated registration.

The practices consist of: definitions, staff, installations, machinery, hygiene, documentation, manufacture, quality control, finished products, and rejected materials.

Drug Wholesale Distribution System

The 1964 law regulates the wholesale drug industry.[26] The law considers drug wholesalers as those warehouses in which the main goal is the distribution of pharmaceutical products to community pharmacies. The wholesale distribution system is established under the authorization of the General Board of Health (Direccion General de Sanidad). In addition to distributing pharmaceuticals, they can distribute any other products to be sold or dispensed in pharmacy settings. A pharmacist has to be the technical director of the wholesaler.

There are several types of drug wholesaling or drug distribution in Spain:

1. The pharmaceutical centers (centros farmaceuticos) are drug warehouses whose owners are pharmacists. They supply 35% of the Spanish market. They formed an association, ACFESA (Asociacion de Centros Farmaceuticos de Espana S.A.). Administratively, it operates as a limited liability company.
2. The pharmaceutical cooperatives (cooperativas farmaceuticas) supply 47% of the pharmaceutical market. They have more than 10,000 pharmacists as cooperators. They also formed a group, ACO-FARMA (Agrupacion de Cooperativas Farmaceuticas).
3. Lastly, there are warehouses with nonpharmacist owners. They supply 18% of the Spanish market.

In 1980, a super association including the three types emerged: FEDIFAR

(Federacion Nacional de Asociaciones de Mayoristas Distribuidores de Especialidades Farmaceuticas y Productos Farmaceuticos).

4. Advertising and Promotion, Price Regulation

In 1980, a disposition about the "promotion, information and advertising of drug products" was passed, which includes and updates all the previous ones.[19] This disposition applies to "all the information or advertising that in any way, direct or indirect, is done about drug products for human use, and about the installations or companies that manufacture, transform, store, deliver, dispense, sell or apply drug products, as well as the objects, apparatus and methods designed for improvement and prevention of treatment and diagnostics." It extends also to any formula or product subject to registration.

In the same year, the General Direction of Pharmacy and Drug Products was created, as well as the Information and Advertising Pharmaceutical Commission, with representation of the government, the General Councils of Medical and Pharmaceutical Colleges, the INSALUD, and the companies. It has consulting functions, and the purpose of knowing and informing about application of the present rules.

Information About Drug Products

Information about drug products can be distributed only by the manufacturing company and must be adjusted to the technically approved label or to the approved package insert. It cannot induce error and must indicate the retail price.

The routes of information can be only the following:

1. Medical visit. The present rules do not allow doctors or dentists in practice to be drug company representatives, and neither can pharmacists with pharmacies or pharmacists who are pharmaceutical technical directors in pharmaceutical distribution or in other staff positions depending on hospitals.
2. Written instructions such as the package insert
3. Written information to the laboratory directors, directly or by medical visit
4. Authorized free samples, which must follow several requirements. The sample must be identical in appearance to the item for sale, except for the Social Security coupon, and must have "Free sample. Sale forbidden." written on it. Free samples of narcotics that may cause dependence are prohibited.

5. Advertisements in scientific, technical, and professional publications. Authorization by the General Direction of Pharmacy and Drug Products is required. The scientific content may not be less than 70% of the approved original text.
6. Audiovisual techniques, which follow the same rules as advertisements.
7. Scientific conferences and research meetings.

Advertisement of Drug Products

Advertisement of drug products is any information about a drug product addressed to the general public. Advertisement, either direct or indirect, is prohibited and considered illegal, unless the drug product has been classified as "advertising." For these advertising drug products, any route of advertising is allowed, with previous approval by the General Direction of Pharmacy and Drug Products, for both the drug product and the related company. The advertisement cannot induce error and must be based on the approved label record or package insert.

In 1985, the content of the advertising messages for drug products, toothpaste, medicinal plant formulations, physical contraceptives, gauzes, and sterile applications became regulated.[20]

The following criteria for the regulation of advertising message content have been established:

1. Identification of the drug product and the manufacturing company
2. Truthfulness, limited to the contents of the registration, without exaggerating or deforming it
3. Professional integrity requires that advertising:
   - Should not mention, when alluding to components of other authorized drug products, their absence or that they are less safe.
   - Should not attribute as exclusive a general characteristic that should also be included in other drug products
   - Should not mention recommendations or testimonials from health professionals or encourage consultation with them
   - Should not have messages or images that guarantee healing, inspire a threat of worse sickness if not taken, or affirm that the drug can lead to improved normal health
   - Should not be addressed to children or suggest that it should be applied without being authorized by the registration. Toothpastes are excluded.
   - Should not suggest that its value comes from being a "natural product."

4. To ensure correct use of products, advertising is required to:
   – Provide clear texts, avoiding terms that require specialization
   – Recommend the reading of the label or package insert
   – Promote responsible self-medication
   – Refrain from any references to the Health Administration or the public health services.

The following specific criteria for advertising drug products prepared with medicinal plants have been established:

1. When there are one or two components, the international common name (WHO) or in its absence, the generic name (INN), should appear, together with the commercial name. If there are more than two components, together with the commercial brand name, one of the following should be mentioned, depending on what would lead to better indentification:
   – The international nonproprietary names (INN) or, in its absence, the generic or scientific names, of at least the two most important components
   – The most important pharmacological properties
   – A general expression indicating the nature or activity (multivitamin, analgesic, etc.)
   – The expression "see the composition"
2. Advertising should:
   – State the most important therapeutic action
   – Provide warnings and precautions about the possible side effects that should be highlighted as well as any other aspect of the message, except the name
   – Encourage public confidence in drug products and in their advertising
   – Recommend consultation with a physician or a pharmacist.

The advertising messages will be kept with the General Direction of Pharmacy until six months after the end of the advertising campaign.

Prices of Drug Products

Prices are regulated in a Decree of 1963, which requires a preliminary report to the Superior Assembly of Prices and Ministry approval. This procedure must be followed at the registration as well as in later price modifications.

Since 1964, pharmaceutical laboratories must present documentation of their total costs. The professional profit of the laboratory is estimated as 15% of the total cost of the drug product manufacture, that of the wholesaler as 12% of the amount obtained from subtracting the public sale price from the pharmacist profit, and that of the pharmacist as 30% of the public sale price with the taxes included.[21]

For the clinical containers, the wholesaler profit is 5% of the pharmacist's sale price and that of the pharmacist is 10% of the public sale price, with the taxes included.

Food drugs are by definition drug products and therefore must be registered as such. They have always had a special scheme with respect to prices and commercial benefits.

In 1965, the industrial benefits and those corresponding to the wholesaler and the professional pharmacist were fixed for the drug products enrolled in the Drug Products Registration, and were established as follows:

— Industrial profit, 12.5% of the manufacture cost
— Wholesaler profit, 5% of the sale price to the pharmacy
— Pharmacy profit, 10% of the public sale price[22]

5. Drug Approval and Government Regulation

In Spain, access to the registration was not free. The Decree of December 7, 1977, specified three access routes: drug products that represented absolute newness, relative newness, and repeated. This was contrary to the governing tendencies of the EEC. A preliminary review of the records to have access to registration was later added. Finally, once the records have been approved, the laboratories must manufacture a first lot of 2,500 units to obtain the commercialization authorization.

Until the Drug Law was created, the adaptation of the Spanish legislation to the EEC directions had started with Royal Decree 424/88 April 29 by which the registration procedure of drug products had been changed. This Royal Decree 424/88, as stated in its preamble, tries to be a first adaptation to the EEC laws with respect to drug products, specifically the contents of the Directive 65/65/EEC January 26 and its later modifications.

The innovations of the Royal Decree are:

1. The elimination of the previous classification for registration
2. The omission of the first lot mandatory manufacture.
3. A 120 work days period is established for negotiation.
4. The use of international common denominations of drug products as

monodrugs is enforced to facilitate international uniformity, as well as to have better, clearer, and rational information.

The coordination of the Directive Centers of the Health Ministry was also accomplished. The requirement of annual renewal of the commercialization authorization of drug products was also established, as a necessary step for the information of the Registration and to avoid the existence of drug products in the market without a sponsor.

When the Product Patent Law in 1992 takes effect, copies of drug products will disappear from the Spanish Registration, in which they have been abundant, because in Spain the manufacture patent is protected. The Registration has allowed copies of drug products by laboratories which were not the discoverer of the product. The legislation with respect to this is contained essentially in four directives:

1. Directive 65/65 sets forth the administrative requirements for commerialization in member countries, including which data and documents are required. Among these requirements is that of the drug product denomination. The authorization can be denied only if it is harmful, has no therapeutic effect, or is not justified (it cannot be denied because of a price dispute).
2. Directive 73/318 completes Directive 65/65, specifying the information that must be submitted with the registration application regarding composition, description, posology, control, and results of the registered clinical trials.
3. Directive 75/319 is the first step in the reciprocal recognition of the authorization and registration of drug products. A drug product committee is created through which the authorization application is sent to different countries' members when a member has already authorized a drug product and the laboratory wants the application to be sent to at least five other members.
4. Directive 83/570 modifies different aspects of the former directives.
5. Directive 86/609 deals with animal protection in experimental research and other scientific goals.
6. Directive 87/18 deals with legal and administrative regulations related to application of good manufacturing practice and applications control for chemical substances testing.

Directive 87/21 modifies Directive 65/65 with respect to the lack of a requirement for showing the results of the pharmacologic, toxicologic, and clinical tests in specific cases. Directive 87/19 deals with the rules and protocols for the analytic, toxicologic, pharmacologic, and clinical testing of the drug product. Directive 87/22 is related to the commercialization of high-technology drug products, especially the ones obtained by biotechnology.

Although the time limit for the incorporation of these directives into the Spanish legislation has expired, a first step was taken in the most urgent aspects (as those related to the registration), and the rest will come with the Drug Product Law.

The principles related to the drug product free market are included in the EEC treaty, and they could be restricted due to health or public security reasons. In spite of the legislation's harmonization, a drug product does not freely circulate in Europe. At present, it seems difficult to indicate what its future could be.

6. Competition: Brand and Generic

In 1981, only 107 generics with brand name (the manufacturer's name) existed in spite of the great number of drug products on the market under brand names. In Spain, the consumption of generics by 1988 was still incipient in relation to the brands. This is due to the fact that at present, generics do not yet have characteristics that differentiate them from the rest of the drug products, and there is not yet specific legislation for them.

C. PHARMACY PRACTICE

1. Dispensing a Prescription: Where, What, Payment, Formulary

Dispensing of prescriptions in the primary setting takes place in community pharmacies. The current General Law of Health (Ley de Sanidad 16/86) establishes that "the safeguard, conservation, and dispensing of medications shall correspond" to the legally authorized community pharmacies and to the pharmacy services of hospitals, health centers, and the primary care level structures of the National Health Service. The law also establishes that community pharmacies open to the public shall be considered "health establishments" and "will be regulated by health planning authorities." Finally, the law specifies that "pharmacies can only be owned by pharmacists and the pharmacist owner has to place her/his license in the pharmacy."

In 1988, there were 17,415 pharmacies in Spain, or 1 for every 2,244 inhabitants. (The rate of increase, on the other hand, has decreased in recent years.[23]) There are no community pharmacies that are government owned. Pharmacies in Spain are privately owned. Only one exception exists: the Ministry of Defense has its own pharmacies for dispensing medications to the military.

Fifty-six percent of the community pharmacies are located in towns with more than 25,000 inhabitants, 17% in towns with between 5,000 and 25,000 inhabitants, and the remaining 23% in populations of less than 5,000 inhabitants.[23] To practice as a pharmacist in a community pharmacy, a pharmacist must meet the following conditions:[24]

- The pharmacist must have been granted either a B.S. in Pharmacy (Licenciado), which in Spain is five years of university training, or have a Ph.D. after additional doctoral courses and completion of a doctoral thesis.
- The pharmacist must be a member of the pharmacy association in the province in which he or she intends to practice.
- The pharmacist cannot own more than one pharmacy. To avoid possible duplication in pharmacy ownership, the pharmacy association stamps the back of the license to indicate ownership of a given pharmacy.
- The pharmacist must not practice in another exclusive activity. There are specific practices that are incompatible with pharmacy ownership. For example, a hospital pharmacist cannot own a community pharmacy.
- The owner of the pharmacy must be the manager as well. Several pharmacists may be coproprietors of a pharmacy, but a pharmacist cannot be the proprietor of several pharmacies.
- The establishment of pharmacies is not limited, but certain regulations do apply.
- Each municipality may arrange to have a community pharmacy, but the total number of pharmacies per municipality may not exceed 1 for each 4,000 inhabitants.
- The distance between pharmacies is regulated according to population density.

Hospital Pharmacy

Dispensing of medications in the hospital setting takes place in the hospital pharmacy services or in the hospital's drug stock areas.

In general, all hospitals for the acutely ill having more than 200 beds have the obligation to provide a pharmacy service within the general services of clinical support of the hospital center. In Catalonia, the health planning criteria require the implementation of a hospital pharmacy service in all centers for the acutely ill having more than 100 beds, in the centers for the chronically ill having more than 200 beds, and in the psychiatric centers having more than 300 beds. The upcoming Drug Law (Ley del Medicamento) establishes that all centers having more than 100 beds must provide a pharmacy service.

To authorize a pharmacy service, the proprietary entity of the hospital is required to send to the health authorities the following documentation:

— Type of hospital specialty, number of beds, and level of assistance provided
— Building plans with details of the sections
— Report of each section's manpower and facilities.

The pharmacy service area must be divided into:

— General warehouses (storage areas) and special warehouses (narcotics, inhalers, radiopharmaceuticals, etc.)
— Administration
— Library and drug information center
— Laboratory analysis and control of medications
— Compounding area
— Dispensing.

At the forefront of the pharmacy service is the chief pharmacist, who has complete responsibility. She or he must be a member of the pharmacy association and is subject to the incompatibility law with other pharmacy functions. A hospital pharmacist cannot be either an owner or a manager of a community pharmacy, nor can she or he work in the pharmaceutical industry.

The number of beds determines the number of additional hospital pharmacists who must be hired. They also must be members of the pharmacy association and subject to the law of incompatibility.

The accreditation for hospitals in Catalonia establishes as a minimum of two pharmacists for the first 200 beds plus a pharmacist for each additional 200 beds, or fraction thereof above 100. Hospital pharmacists must have finished their hospital residency and received the degree of Hospital Pharmacy Specialist granted by the Ministry of Science and Education.

Each hospital has a Pharmacy and Therapeutics Committee in charge of the drug policies for that institution. These policies include: selection of drugs for the formulary (Guia Farmacoterapeutica), drug utilization studies, quality control, and follow-up of treatments.

At present, there are 340 hospital pharmacy services in existence in Spain, representing 38% of all Spanish hospitals (899). In Catalonia, the number of pharmacy services is 80, which represents 49% of the hospital centers (164).[25]

Hospitals that lack pharmacy services must have a drug stock or depository (deposito de medicamentos). This stock is supplied by a community pharmacist in town who is also responsible for it. In Catalonia, the autonomous government has established that these drug stocks can be linked either to a community pharmacy in town or to another hospital pharmacy. The pharmacist in charge of these drug stocks must meet the same requirements as other hospital pharmacists and have some experience in hospital pharmacy. The time spent for activities related to these drug stocks shall be 10 hours per week in centers of less than 50 beds and 20 hours in those of more than 50 beds.

Drug Prices and Coverage in the Community Pharmacy

In the dispensing of drugs to the population, the community pharmacy has a profit margin of 30% above cost when the medications are supplied by a wholesaler and a margin of 42% when the pharmacy is supplied directly by the manufacturing laboratories.

For the population covered by the Social Security system, dispensing in pharmacies is fulfilled by official Social Security prescription forms, and the medications are free for retirees, while the actively working people who have not reached retirement age (65 years) must pay in the pharmacy at the time of dispensing 40% of the value of the medication. This applies to a branded product and to extemporaneously compounded products. The pharmacy sends these prescription forms to its pharmacy association, which submits them for reimbursement to the managerial entity of Social Security (INSALUD, ICS, SAS, SVS, Osakidetic) so that the following month each pharmacy may be reimbursed. Drugs used for certain chronic conditions, such as tuberculosis, and some antihypertensives, are reviewed by a national commission from the Health Ministry which may grant exemption from payment of the 40% patient's portion.

Generally, when prescriptions are written by private physicians, the insured must pay the entire price of the drug when it is dispensed.

Pharmaceutical consumption under the Spanish Social Security system was Pts. 347,970 million in 1988, up by 17.9% from 1987. The number of prescriptions issued under Social Security was 463.4 million, an increase of 4.3% over the previous year. The average price per prescription was Pts. 750.8, up by 13% in current terms.

Of the total consumption under Spanish Social Security in 1988, patients contributed Pts. 44,148 million, an increase of 14.2% over 1987. The increase in patient contributions was due primarily to the fact that new measures were introduced at the beginning of the year, under which patients also had to pay 40% of the cost of extemporaneous products instead of a fixed sum of Pts. 75.[27]

2. Record Keeping

Community pharmacies do not keep individualized patient records of dispensed drugs in order to know a patient's drug history, possible interactions, etc. In Catalonia, however, a pilot study is underway. A notebook of patients' personal medication histories for a few pathologies (hypertension, tuberculosis) or for certain medications (growth hormone) was designed. The physician writes the drug regimen in the first page of the notebook and gives it to the patient, together with the prescription. When this reaches the pharmacy, the pharmacist writes in the notebook the dispensed medication and the dates of the prescriptions, so when the patient returns to the physician for the follow-up visit, the physician knows if the medication has been obtained in the recommended time intervals and may thus verify initial patient compliance with treatment.

At the hospital level, with a unit dose distribution system, the pharmacy service keeps drug records for each patient from the moment of admission discharge. The goal is to verify that the doses and drug regimen are correct and that there are no pharmacological interactions.

3. Patient Education

In Spain, the central government, as well as many of the autonomous governments, have developed health education initiatives aimed at the general population. Pamphlets and health books on information about drugs are distributed, principally from the community pharmacies, hospital pharmacies, and primary health centers. In Catalonia, the Department of Health, in collaboration with the Department of Education, has started a program about health education and disease prevention in schools for children from 4 to 14 years old. Information about use of medicines is also provided. Moreover, the Catalonian Institute of Health (Institut Catala de la Salud) and its administrative office for the delivery of health care through Social Security developed several videos about health education and use of medicines. These videos are shown in the waiting rooms of the primary health centers.

4. OTCs and Other Classes and Schedules of Drugs

Drugs that do not require prescription (OTC) are those drugs for pathologies that do not need a precise diagnosis. OTC drugs are mostly sold in the community pharmacies. To a lesser extent than in the United States, OTC drugs can also be found in supermarkets.

OTC drugs must meet the following criteria:

- They are designed for the prevention, alleviation, or treatment of minor symptoms.
- They are not designed for the prevention or curing of pathologies that require a diagnosis and a prescription.
- They have been demonstrated to be safe and efficacious for a given indication.
- In their application, they may not be used by the parenteral route.
- They are formulated with substances from the list issued by the Minister of Health and Consumption.
- OTC drug advertising can be aimed at the general public. However, it should not include any wording that furnishes a guarantee for cure nor testimony of health professionals that may persuade consumers, and it should include information about side effects and recommendations for appropriate use.

5. The Use of Technicians and Others

Formal education and certification for pharmacy technicians does not exist in Spain. However, pharmacy technicians who work in community pharmacies (auxiliares de farmacia) receive their training through experience (by working with a pharmacist) or through participation in a six-month course organized by the pharmacy associations. This six-month course is not accredited or recognized by the Ministry of Education. In hospital pharmacies, in addition to auxiliares de farmacia, nursing personnel are hired and trained for preparation of sterile products such as intravenous drugs, parenteral nutrition and reconstitution of anticancer agents.

6. Pharmacy and Primary Care

At the primary level, community pharmacies have a role not only in the provision of care but in the areas of health promotion, disease prevention, and health education of the population in the area of drugs.

In the provision of care, pharmacists are in charge of the production, distribution, custody, and dispensing of drugs for diagnostics, disease treatment, and rehabilitation functions. In addition to dispensing a drug, a pharmacist provides information to the patient to improve the rational use of medicines. Information may include dosage and administration guidance, as well as education about following physicians' prescriptive directions.

Community pharmacists today collaborate with other health professionals and institutions in activities to promote health and prevent diseases. For example, they can collaborate in the early detection of hypertensive patients or in providing information about sexually transmitted diseases, vaccination schedules, dietary recommendations, and prevention of poisoning in the household.

The Health Reform Act establishes two levels of community pharmacies at the primary care level. At the basic level, each community pharmacy belonging to a primary care area (area basica de salud) will collaborate directly with the primary health team of that area. The community pharmacist will act like a family pharmacist for a given town or neighborhood. The pharmacist can help in the detection of early disease states and other similar activities. At a second level, a new unit of pharmaceutical care (unidad de atencion farmaceutica) is created. This unit serves as a support for all of the activities of care by the primary health team. These drug-related support activities are in the field of health promotion and health education. The unit of pharmaceutical care structures qualitative and quantitative drug utilization studies. The outcome of these studies is intended to create better and more rational drug use.

The integration of the pharmacist as a drug advisor at the primary level includes two functions. The first occurs internally at the primary care center and concerns the use of drugs at that center. The second function uses its relationship with other community pharmacies and the hospital pharmacy as a reference to coordinate drug activities programmed by the primary health team.

D. CONCLUSIONS

Pharmaceutical services in Spain are under a free market economic system. However, regulations by the government include the definition and function of pharmacy services. In the specific case of community pharmacies, the government regulates them by limiting their number by geographic distance and population served. At the community level, pharmacies are privately owned and, therefore, function in a free market system. All pharmacies, through their pharmacy associations, have a contract with the public sector (or Social Security) to dispense drugs to the insured population (97%).

With the implementation of a National Health Service (NHS), every citizen (100%) will be insured through the public sector. The NHS will coordinate the role of community pharmacists and integrate them with the primary care physicians. The objective is to rationalize the use of drugs at the primary level. The global objective of the relationship between community pharmacies and primary health centers is that, in addition to the pharmacist's role as a dispenser and provider of information, the pharmacist will collaborate in other health initiatives, such as health promotion, disease prevention, and health education. The community pharmacy is like a community health center.

The objective of integrating the pharmacist into activities other than dispensing has an obvious conflict: reimbursement is only based on the products sold. Therefore, collaborating in all the other activities might initially be on a voluntary basis. However, it would be reasonable for the Spanish authorities to examine the provision of complete pharmaceutical services offered in Holland, Sweden, and the United States to obtain some ideas about how different nations approached this dilemma to make pharmacy services more professional and valuable to the public.

At the hospital pharmacy level great gains have occurred in recent years. Pharmacists play an indispensable role in the provision of rationalized therapies. In recent years, more pharmacies have opened in hospitals and more hospital pharmacists are employed. The Pharmacy and Therapeutics Committee is enhanced as the basic organization to provide quality assurance.

Finally, one of the principal objectives of public administration regarding pharmacy is the promulgation of the Drug Law (Ley del Medicamento). The law ensures that Spain will have drugs that are safe, efficacious, high quality, easily and correctly identifiable, and which follow all of the health laws of the EEC. The law also regulates policies for manufacturing, research, supervision of a center for postmarketing surveillance, and norms for distributing information and advertising.

REFERENCES

1. Ordeig Fox JM. El sistema Espanol de la Seguridad Social. Madrid: Edersa, 1982.

2. de Miguel J. La sociedad enferma. Madrid: Akal, 1979.

3. Boletin Estadistico 471, Mayo/Junio (INE), Madrid, 1988.

4. Instituto Nacional de Estadistica (INE). Movimiento natural de la poblacion, Madrid: 1983.

5. Anuario Estadistico de Espana (INE), Madrid, 1984.

6. Boletin Estadistico 546 (INE), Madrid, 1985.

7. Barea J. Gasto publico en asistencia sanitaria. Papeles de Economia Espanola, Madrid, 1988; 37.

8. Artells JJ. Asistencia sanitaria: Financiacion y Rentabilidad. Papeles de Economia Espanola, Madrid, 1988.

9. Presupuestos Generales del Estado. Madrid, 1988.

10. Financing and Delivering health care. Paris: OCDE, 1987.

11. Consejo Generales de los respectivos Colegios Professionales, Madrid, 1988.

12. Memoria Estadistica. Ministerior de Sanidad y Consumo, Madrid, 1985.

13. Benalbas L. El seguro de asistencia sanitaria. Papeles de Gestion Sanitaria, Madrid, 1988.

14. Catalogo Nacional de Hospitales. Ministerior de Sanidad y Consumo, Madrid, 1987.

15. Farmaindustria. Madrid, 1988.

16. Ministerior de Industria y Energia, Madrid, 1988.

17. Instituto Nacional de Estadistica, Federacion Europea de Asociaciones de la Industria Farmaceutica, Banco de Bilbao, Bilbao, 1988.

18. Direccion General de Aduanas. Estadistica de Comercio Exterior de Espana, Madrid, 1987.

19. Royal Decree 3451/1977 de 1 Diciembre. BOE de 24 de Enero 1988 y Orden de 30 de Mayo BOE de 16 Junio de 1980.

20. Orden de 10 Diciembre, BOE de 18 Diciembre de 1985, Madrid.

21. Real Decreto 86/1982 de 15 de Enero. BOE del 18 Enero, y Orden de 22 de Enero, BOE de 25 de Enero 1982.

22. Orden de 12 de Diciembre, COE del 27 de Febrero de 1965.

23. Consejo General de Colegios de Farmaceuticos, Madrid, 1987.

24. Sune Arbussa JM. Legislacion Farmaceutica Espanola. 8a ed. Barcelona, 1987.

25. Sociedad Espanola de Farmaceuticos de Hospital, Madrid, 1989.

26. Orden de 7 de Abril, BOE del 27 de Abril de 1964.

27. SCRIP No. 1429, London July 14, 1989; p. 4.

Chapter 20

Sweden

Tommy Westerlund

A. THE NATIONAL HEALTH CARE SYSTEM

1. History and Background

Sweden covers 450,000 square kilometers (173,700 square miles) and in terms of land area, is the fourth largest country in Europe. The number of inhabitants is 8.5 million.

The need in Sweden for government-sponsored social welfare policies grew with the full emergence of industrialism toward the end of the nineteenth century. It may be said that the history of Swedish social welfare began at the start of the 1880s with industrial and social reforms aimed at improving conditions for working people.

Early in the twentieth century, Sweden was passing through a period of very rapid public health improvement. Infant mortality was halved in a few decades, and average life expectancy increased by more than five years. It was considered axiomatic that the government had a duty of working to improve the state of public health, but a great deal remained to be achieved. Average life expectancy in those days was 57 years, compared to 77 years today. Infant mortality was 70 per 1,000 births, 10 times what it is today.

The causes of concern in the 1930s were the housing shortage and a declining birth rate. During the following years, the Parliament passed many reforms to improve the living conditions and health of young families. Child allowance, maintenance advances, free school meals, and an expanded system of free mother and child health care were introduced. As a result, infant mortality declined sharply.

During the 1940s, there was widespread criticism of outpatient medical care. Outpatient care away from the hospitals existed on a modest scale. In an official report, the Director General of the National Board of Health reiterated the proposal already put forward in 1934 of an expansion of free outpatient care, with special emphasis on preventive screening. This proposal, however, came under such heavy fire that the government never introduced legislation on the subject. For the next 25 years, efforts were instead concentrated on a tremendous expansion of institutional medical care. The number of beds rose from 77,000 in 1945 to 120,000 in 1970.

Starting in the second half of the 1960s, however, the development of health and care underwent several important changes. The national economic growth rate began to decline. Despite the deceleration of economic growth, medical care continued its rapid expansion. A fixed low charge for medical consultations (7 Swedish crowns) was introduced in 1970. Social and economic factors no longer played such an important part in determining people's access to care.[1]

2. Social and Political Environment

Present-day technologies increasingly affect social and productive patterns. In employment terms, fewer and fewer people are engaged on the shop floor while ever larger numbers work in front of computer screens. Thirty percent of the labor force is employed in industry, mainly in manufacturing. A similar proportion is engaged in the service industries (e.g., commerce and transportation) while just over 30% work in the public sector in fields such as child day care, care of the elderly, and health care and medical services.

Today, almost as many women as men are in gainful employment, with 75% between the ages of 16 and 64 (65% in the case of women with children under 7) earning a regular income. A much larger proportion of women than men are in part-time work. Parents with children under the age of eight are legally entitled to work shorter hours with a corresponding reduction in pay. This entitlement is extended to men and women and covers all forms of employment, both in the public and private sectors.

Average unemployment in Sweden, at 3% or less for the last few years, is low in relation to that of many other countries. The rate varies within the country, however, and is considerably higher in certain areas and for certain population groups.

Until the Second World War, Sweden's population was unusually homogeneous. But war refugees, large-scale immigration in response to the new recruitment policy, and in recent years, a steady influx of political refugees, have led to a substantial representation from a wide variety of ethnic origins. Today, 10% of the population (around 775,000) are foreign nationals or were born abroad.

The majority of households consist of only one or two members. Most families are made up of one or two adults and one or two children.

Although a million new homes were built between 1964 and 1973, there is an acute shortage of living accommodations in places like Stockholm. In urban areas today, almost as many people live in detached or semidetached houses as in blocks of flats. The same pattern is found in the country as a whole. Most flats are rented on an indefinite leasehold. Rent is normally payable on a monthly basis.

Sweden has a larger public sector than any other comparable industrialized country. Public authorities are responsible for education; health and medical care; the social services; national social insurance; the provision and maintenance of living accommodations and power supplies; transport; the correctional system; the postal, wire, and telephone services; and much more. Public support is also extended to the arts and leisure activities (Figure 1).[2]

The rapid increase in the number of elderly (65 years and over) that Sweden experienced during the 1970s and early 1980s will decrease somewhat by the year 2000. It is estimated, however, that there will be an additional increase of 125,000 in the 80-years-and-over age group by that time. Sweden has a high standard of education. This implies that the elderly population of the future will be better educated than the current population.

The rapid structural change of the working environment and the large proportion of people receiving long-term sickness benefits have accentuated recent debates concerning problems of a psychosocial nature. A general change in lifestyle is undoubtedly a key explanation behind the fact that Swedes live longer and feel healthier. Alcoholism decreased during the early 1980s but is now showing signs of increasing again. Since the 1960s, approximately one million Swedes have given up smoking. In 1977, 40% of the men aged 16-74 smoked daily, but 5 to 6 years later, only 30% of them smoked daily. The number of women who smoke also decreased, although to a lesser degree. This, however, does not apply to the younger age groups.

Eating habits have also improved to some extent. Moreover, many studies now demonstrate that it is possible to influence living habits, and this can directly affect health and mortality.[4]

Sweden is a constitutional monarchy as well as a parliamentary democracy. Apart from a period between 1976 and 1982 when the non-Socialist block was able to form a succession of governments, the Swedish Social Democratic Party has been the party of government since the 1930s.

Sweden is divided into 24 regional units or counties, each the seat of a

FIGURE 1. Public sector share of GNP (GNP 1986 = SEK 933,686 million) in percentage, 1975-1986.[3]

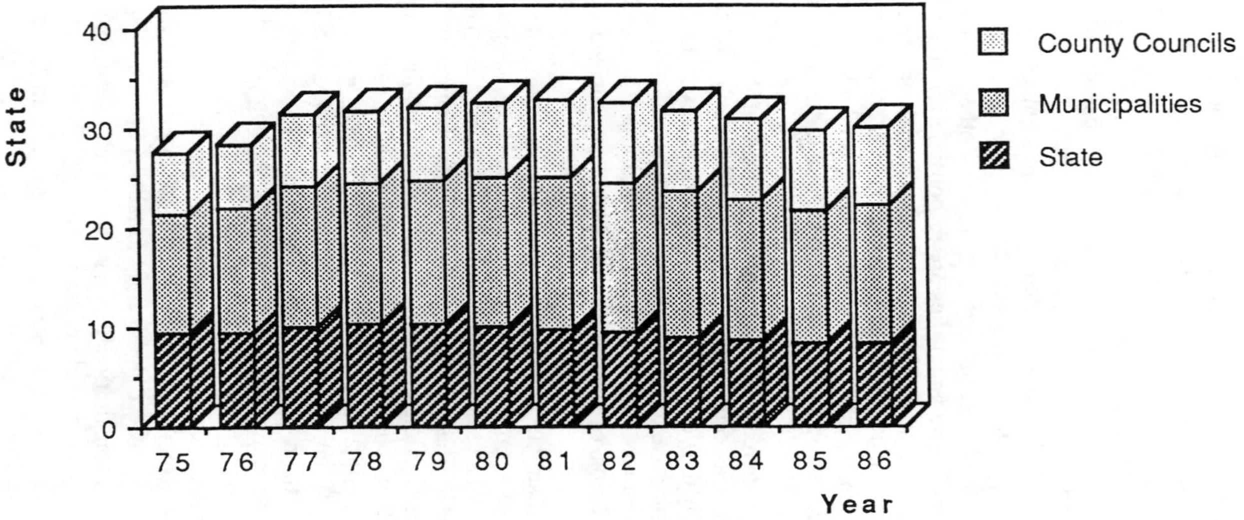

regional administration headed by a governor appointed by the central government. These county administrative boards are responsible for ensuring that central and local government activities conform to policy guidelines for the use and conservation of land and water. The county administrative boards also exercise a supervisory function with respect to the municipally-run social services.

Local government at the regional level is the responsibility of 23 county councils. Like their municipal counterparts, these are elected bodies with powers to levy taxes from within their jurisdictional areas, which, in general, coincide geographically with the counties. Each county council area has within its boundaries a number of municipalities.

The County Council's principal concern is the provision of health and medical care. Other responsibilities are the National Dental Service and certain types of care for the disabled and intellectually handicapped.

The country is divided into 284 municipalities, each governed by a locally elected council. Municipal authorities are vested with broad powers and maintain a wide range of services. These include social welfare facilities, schools, public libraries, leisure amenities, water and power supplies, and sewage disposal.

3. Public and Private Financing, Health Costs

Health care is regarded in Sweden as being clearly a task for the public sector. Like social welfare services, it is provided mainly by local authorities. Responsibility for both individual- and population-oriented health services and for medical care, both outpatient and inpatient, rests with 23 county councils and three large municipalities that are not part of county council areas: Goteborg (Gothenburg), Malmo, and the island of Gotland.

Of the county councils' total expenditure, public health and care of the sick accounts for some 75%. The county councils receive approximately 60% of their finances from county revenue district taxes. One-third of direct taxes go to the county councils. The average county revenue district tax is just over 13% of taxable income. Approximately 400,000 people (13% of all employed people in Sweden) work for the county councils (Figure 2).[5]

In 1980, both the U.S. and Sweden allocated 9.5% of GNP for health and medical care. Since then, the U.S. has raised its percentage to 11 (1986), while in Sweden the figure has fallen to 9.1%. State grants to the county councils have fallen in relative terms since 1982, from 28% of county council expenditure in 1982 to 20% in 1987. In addition, Parliament has passed legislation reducing the county council tax (Table 1).[1]

The conflict between the actual financial situation on one hand and

FIGURE 2. County Council revenue 1986.

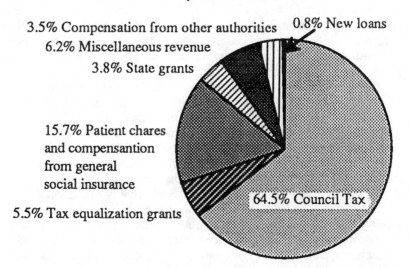

3.5% Compensation from other authorities   0.8% New loans
6.2% Miscellaneous revenue
3.8% State grants

15.7% Patient chares
and compensantion
from general
social insurance

5.5% Tax equalization grants

64.5% Council Tax

demands for high quality health care on the other, has intensified over the past few years. Health service staff frequently complain that budget cuts prevent them from providing full quality care. At the same time, the general public places high expectations on the quality and accessibility of health care.

The economic situation within the county councils has increased interest in the return on financial investments in terms of quality and results in health care. The American diagnosis-related groups (DRG) system has created worldwide interest, and even in Sweden it is hoped that DRGs will lead to a greater cost awareness and eventually better utilization of resources. The quality of health care is being discussed more and more. Recently, Sweden undertook development of a system of quality asssurance in the Swedish health service, in accordance with WHO's objectives for the European region.

Although in many respects the county councils operate their health care systems independently, the Swedish state has supervisory powers over these activities. The state is responsible for ensuring that the health care system develops efficiently and in keeping with its overall objectives based on the goals and constraints of social welfare policy and macroeconomic factors. The state is also responsible for ensuring everyone equal access to health care and guaranteeing them equal protection before the law with regard to safety and high quality medical care. Central govern-

TABLE 1. Approximate distribution of costs between various fields of activity in public medical care in percentages.[1]

	1970	1983	1986
Short-term somatic care	60	50	48
Long-term somatic care (inc. local nursing homes)	17	24	27
Psychiatric care	17	13	12
Non-hospital outpatient care	6	13	13

Source: Hur urnyttjas halso-och sjukvardens resurser? Spri rapport 193, 1985 and Spri three-year programme for 1987-1989. (Spri: The Planning and Rationlisation Institute for the Health and Social Services.)

ment administration below Parliament and Cabinet is divided into two levels. In the case of health care these are the Ministry of Health and Social Affairs (Socialdepartementet) and relatively independent administrative agencies, primarily the National Board of Health and Welfare (Socialstyrelsen).

The Ministry of Health and Social Affairs is part of the government offices. It prepares Cabinet business and draws up general guidelines in such fields as health care, social welfare services, and health insurance.

The National Board of Health and Welfare is a central administrative agency for matters concerning health care, supply of pharmaceutical products, and social welfare services. Its tasks include:

— Handling planning questions (providing background material for planning to the Cabinet, Parliament, and the county councils and otherwise monitoring and assisting in the planning activities of the county councils)
— Supervising outpatient and inpatient health care and the professional performance of health care personnel
— Carrying out health information programs.

The Swedish Planning and Rationalization Institute of the Health and Social Services (Spri) is jointly owned by the central government and the county councils. The county councils provide most of its funds. Spri works in collaboration with the county councils and central government agencies on planning and efficiency measures, as well as undertaking special investigative tasks. In various ways, Spri supports research and development work in health care administration (Figure 3).

4. Facilities (Hospitals, Clinics, etc.)

The number of hospital beds in Sweden is relatively high, equivalent to about 16 per 1,000 inhabitants (5 general, 5 long-term, 3.5 psychiatric, and 2.5 for the intellectually handicapped). There are also about 7 places per 1,000 inhabitants in municipal homes for the elderly. In contrast, the number of outpatient visits to physicians is comparatively low, about 2.5 medical treatment visits per inhabitant per year. On top of this are about 1.2 public health care visits. Of visits to doctors, 53% take place at hospitals, 30% are to district physicians within the primary care system, and 17% are to doctors in private practice.

Outpatient care is organized into primary care districts, each with 5,000 to 50,000 inhabitants. Their task is to assume primary responsibility for

FIGURE 3. The organization of health services in Sweden.[6]

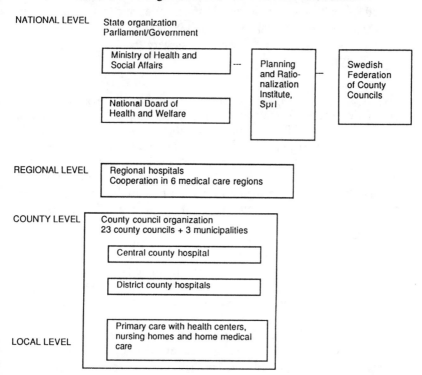

NATIONAL LEVEL   State organization
Parliament/Government

Ministry of Health and Social Affairs

National Board of Health and Welfare

Planning and Rationalization Institute, Spri

Swedish Federation of County Councils

REGIONAL LEVEL   Regional hospitals
Cooperation in 6 medical care regions

COUNTY LEVEL   County council organization
23 county councils + 3 municipalities

Central county hospital

District county hospitals

LOCAL LEVEL   Primary care with health centers, nursing homes and home medical care

the health of the population in their areas. Each district has one or more local health care centers plus one or more nursing homes for long-term care. At the health centers, district physicians, both general practitioners and specialists, provide medical treatment, advisory services, and preventive care. The outpatient system also includes district nurses and district midwives. Ordinarily there are also special centers that provide child health and maternity health care services.

The responsibilities of county medical care cover patients with critical conditions or other illnesses requiring access to staff or technical resources which, for various reasons, must be concentrated to one or a few hospitals within each county council area.

These county-operated hospitals are divided according to their size and degree of specialization into:

—Central county hospitals for 200,000-300,000 inhabitants, with 15-20 specialties, ordinarily one hospital for each county council area
—District county hospitals for 60,000-90,000 inhabitants, with at least four specialties (internal medicine, surgery, radiology, and anesthesiology).

The regional medical care system is responsible for those few patients who offer especially formidable problems, requiring collaboration among a large number of highly trained specialists and perhaps also special equipment. For these purposes, Sweden is currently divided into six medical care regions, each serving a population averaging more than 1 million inhabitants. Their activities are regulated by agreements among the county councils included in each of the respective regions. Within each region there is one (sometimes more) regional hospital which possesses the higher degree of specialization required by the regional medical care system. These hospitals are found in Umea, Uppsala-Orebro, Stockholm, Linkoping, Goteborg, and Lund-Malmo. All are affiliated with medical schools and also function as research and teaching hospitals. In addition, there are currently a number of special purpose hospitals, mainly for psychiatric care. Currently, however, the aim is to merge psychiatric with somatic care institutions as much as possible.[6]

Table 2 shows the patient capacity, etc., of institutions run by the county councils and the municipalities of Malmo, Gothenburg, and Gotland. The county councils availed themselves of approximately 6,500 additional places in private nursing homes. Of these places, just over 40% were intended for psychiatric care and almost 50% for long-term care. In addition, places in sheltered accommodations with full board (former old people's homes) were used for long-term care. The number of places for long-term care has increased sharply in recent years, while places for short-term care and, in particular, psychiatric care, have decreased. In 1986, the number of visits to doctors in private practice was 3.6 million, including visits to doctors under health care contract.[3]

A heavy expansion of outpatient care amenities was inaugurated in the mid-1960s. In 1951, home help services were being provided for 16,000 senior citizens. By 1985, the figure was 313,000. This is a multiple increase, even when compared with the growing number of elderly persons. Personnel strength in primary care has tripled in 20 years, up to the year 1980, and viewed in relation to the expansion of inpatient care, primary care also increased slightly during the 1970s. However, under the current spending restrictions it has not been able to hold its ground.[1]

TABLE 2. Patient capacity, patient days and visits to doctors by main category, 1986.

Category	Patient capacity	No of patient days (thousands)	Visits to doctors (thousands) total	Inc primary care
Short-term care, med.	16775	5104	3632	328
Short-term care, surg.	19068	5074	5566	467
Long-term care	51196	17598	165	28
Psychiatric care	19189	5947	773	88
General med care	-	-	7866	7866
Other	4863	627	873	17
TOTAL	111091	34349	18876	8794

5. Manpower

The Swedish health care system underwent rapid expansion during the 1970s, at the same time as major changes took place in its organizational structure and areas of emphasis. The increase in health care personnel has been very great during the past few decades. In 1985, about 450,000 people were employed in this sector, equivalent to almost 10% of all employees in Sweden. Comparable figures for 1960 were 115,000 and about 3%. The 1985 figure is roughly equivalent to 310,000 full-time jobs. The rapid growth in the number of employees in the health care sector is, to some extent, due to shorter working hours and to the growing proportion of part-time employees.

There is currently a shortage of some categories of trained personnel, for instance, physicians — particularly in general practice, psychiatry, long-term medical care, and certain laboratory specialties — but there is also a shortage of physiotherapists. The situation is expected to improve gradually, however[6] (see Table 3).

6. The Social Security System

Social insurance forms the heaviest element of Swedish social welfare policy, which includes the health and medical services, family benefits, and other benefits of various kinds. Social insurance is defined to include medical, dental, and parental insurance, and partial, basic, and supplementary pensions — all of which come under the National Insurance Act — as well as compulsory work injuries insurance and voluntary unemployment insurance (see Figure 4).

National insurance is administered by the regional social insurance offices. There are 26 such offices covering different geographic areas, each with numerous branch offices. The activities of these offices are centrally administered and regulated by a government agency, the National Social Insurance Board. Anyone dissatisfied with the decision of a regional office can appeal to an insurance court. Further appeal can be made to the Supreme Social Insurance Court, which is the highest adjudicating authority in such cases.

All insured persons resident in Sweden are automatically registered at a regional social insurance office within a month of reaching the age of 16. Foreign nationals resident in Sweden may also be covered by the National Insurance Act if their country of origin has signed an agreement — a convention — with Sweden. These conventions put the nationals of such countries on the same footing with Swedish citizens, which means they are insured while in Sweden even if they are not domiciled here. Swedish

TABLE 3. County Council employees by profession and form of employment, 1987.[3]

Profession	Under County Council and Municipality contract			Total	Under other contracts[1]	Total
	Full-time	part-time	hourly paid			
Total	218 334	167 898	34 345	420 577	16 169	436746
Men	52 572	6 558	6 758	65 888	5 508	71396
Women	165 762	161 340	27 587	354 689	10 661	365350
including						
Physicians	16 384	831	178	17 393	-	-
Adm.nurses	9 775	1 955	264	11 994	-	-
Nurses/ass./ midwives	28 654	25 558	2 038	56 250	-	-
Occ. therapists	2 765	1 062	36	3 863	-	-
Lic. prac. nurses	15 119	6 990	2 400	24 509	-	-
Nursing aux.	22 310	21 616	2 179	46 105	-	-
Welfare officers	29 676	39 305	11 658	80 639	-	-
Physiotherapists	1 954	1 477	66	3 497	-	-

[1] Salaried according to State contracts, emergency measure contracts, etc.

FIGURE 4. Swedish social insurance and other benefits.[7]

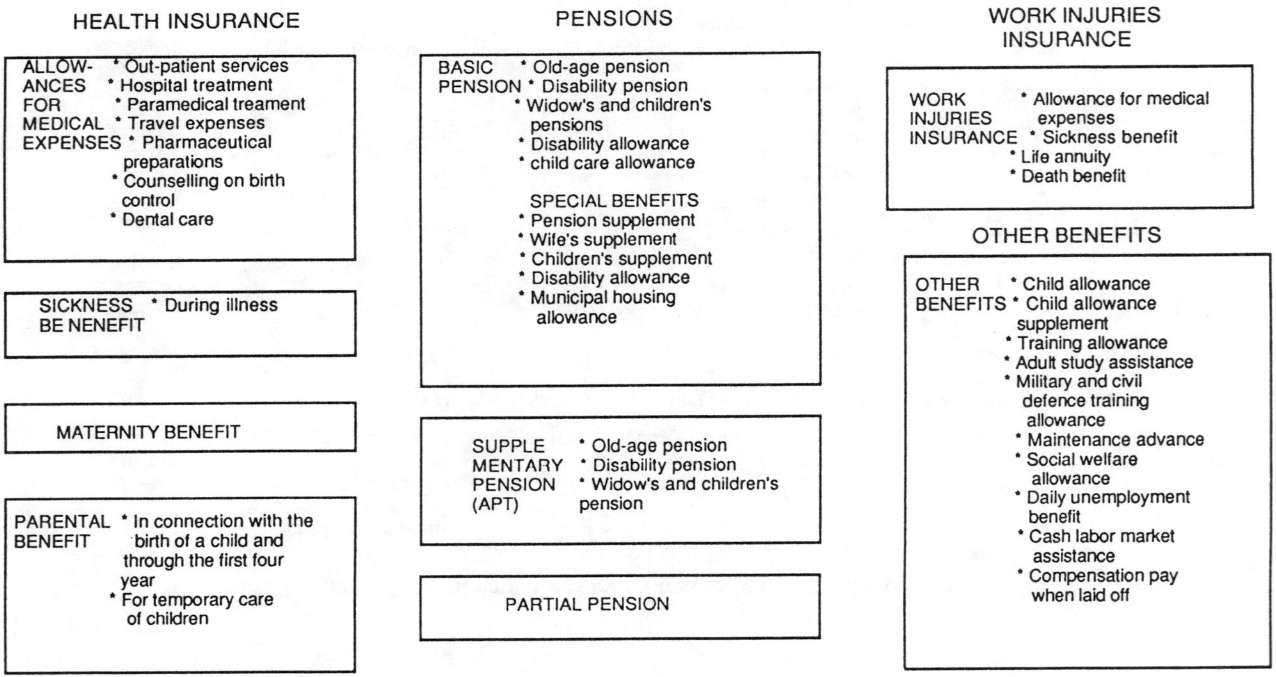

HEALTH INSURANCE

ALLOW-
ANCES
FOR
MEDICAL
EXPENSES
* Out-patient services
* Hospital treatment
* Paramedical treatment
* Travel expenses
* Pharmaceutical preparations
* Counselling on birth control
* Dental care

SICKNESS
BE NENEFIT
* During illness

MATERNITY BENEFIT

PARENTAL
BENEFIT
* In connection with the birth of a child and through the first four year
* For temporary care of children

PENSIONS

BASIC
PENSION
* Old-age pension
* Disability pension
* Widow's and children's pensions
* Disability allowance
* child care allowance

SPECIAL BENEFITS
* Pension supplement
* Wife's supplement
* Children's supplement
* Disability allowance
* Municipal housing allowance

SUPPLE
MENTARY
PENSION
(APT)
* Old-age pension
* Disability pension
* Widow's and children's pension

PARTIAL PENSION

WORK INJURIES INSURANCE

WORK
INJURIES
INSURANCE
* Allowance for medical expenses
* Sickness benefit
* Life annuity
* Death benefit

OTHER BENEFITS

OTHER
BENEFITS
* Child allowance
* Child allowance supplement
* Training allowance
* Adult study assistance
* Military and civil defence training allowance
* Maintenance advance
* Social welfare allowance
* Daily unemployment benefit
* Cash labor market assistance
* Compensation pay when laid off

citizens who do not live in Sweden are entitled to medical benefits only if the need for care has arisen in the course of a temporary sojourn in their native country.

The total cost of Swedish social welfare policy amounted to around SEK 155,000 million in 1985. The greater part of these costs related to the national social insurance scheme. The insurance is financed out of government subsidies, charges levied upon employers and the self-employed, and — in the case of unemployment insurance — contributions from the insured themselves.[7]

7. The Medical Establishment: Entry, Specialists, Referrals

Training of doctors takes place at six medical schools, which in most cases are part of a university. They are located in Umea, Uppsala, Stockholm (the Karolinska Institute), Linkoping, Goteborg, and Lund. Teaching hospitals, which also serve as regional hospitals, are at the disposal of the study programs. Admissions are restricted. Approximately 850 medical students are accepted each year. The basic study program requires five years, divided into preclinical, preparatory, and clinical periods. After receiving the medical degree, all doctors must complete a 21-month period of general internship to obtain a license to practice. After that, most physicians complete postgraduate training programs of 4-5 years which qualify them as specialists or general practitioners. The National Board of Health and Welfare allocates the blocks for postgraduate training courses in accordance with a physician distribution program set up by the health planning committee of the Ministry of Health and Social Affairs. The allocation of places among the various specialties is intended to reflect long-term national needs.

In 1985, there were over 25,000 physicians in Sweden, or more than one doctor per 400 inhabitants (see Table 4). There is still a considerable shortage of doctors, but their number is expected to grow to more than 28,600 by the year 2000.

Nurses in Sweden are trained at one of the 40 county council-administered colleges of nursing. Applicants choose from any six specialized disciplines: general nursing, operating room nursing, psychiatric nursing, ophthalmic nursing, oncology, and diagnostic radiology. Training courses last, as a rule, for two years. Prior to nursing training, candidates must have successfully completed the two-year nursing line at upper secondary school.[6]

TABLE 4. Number of physicians by medical specialty.

	1985	1986	1987	1988[1]
GENERAL PHYSICIANS	4 164	4 502	4 682	4 886
MEDICINE	4 233	4 395	4 416	4 496
General medicine	1 931	2 024	1 996	1 973
Pulmonary diseases	180	177	186	188
Renal diseases	39	47	78	67
Rheumatology	95	102	114	129
Child medicine	1 086	1 105	1 106	1 138
Chronic diseases	611	656	628	647
Medical rehabilitation	119	113	127	132
INFECTIOUS DISEASES	306	326	340	362
DERMATOLOGY	283	286	309	316
RADIOLOGY	249	266	280	279
NEUROLOGY	184	189	203	213
SURGERY	2 464	2 529	2 574	2 633
General surgery	1 365	1 377	1 373	1 381
Neurosurgery	93	85	76	82
Orthopedic surgery	612	642	681	713
Plastic surgery	87	78	87	93
Thoracic surgery	62	73	78	78
Urologicical surgery	143	155	159	166
Child surgery	61	72	70	71
GYNECOLOGY and OBSTETRICS	1 088	1 099	1 105	1 154
ENT	616	646	643	661
OPHTHALMOLOGY	530	552	569	580
PSYCHIATRY	1 506	1 609	1 687	1 717
Child psychiatry	276	287	285	299
General psychiatry	1 088	1 172	1 250	1 267
SERVICE	2 649	2 794	2 897	2 970
Anesthetics	863	940	981	1 008
Roenthenology	818	856	877	909
Laboratory work	968	998	1 039	1 053
OTHER	3 904	4 066	4 491	4 403
TOTAL	22 176	23 259	24 196	24 670

[1]February

8. Payment and Reimbursement

Allowance for medical care is the collective term covering the various payments that are made in connection with medical attendance, hospital treatment, dental treatment, etc. As a rule, these allowances are paid directly by the social insurance office to the health care administration or the individual practitioner responsible for the treatment. Usually, the patient is charged a modest fee at the consultation.

A uniform tariff applies to public outpatient services (as provided by district medical doctors and at hospitals). This means that the patient pays SEK 60* (April 1989) for visiting a doctor. Visits to private practitioners come under a separate reimbursement list fixed by the government. As a rule, the patient pays SEK 60 out of his own pocket for each visit, in some cases SEK 70, seldom more. The fees paid by patients cover not only a visit to the doctor, whether the doctor is a public employee or a private practitioner, but also include the following items: issuance of a prescription and a doctor's certificate to qualify for sickness benefit; X-ray and laboratory examinations to which the patient is referred; X-ray, radium, and other treatments to which the patient is referred; and referral to a specialist with no charge made for a first visit.

Allowances for hospital treatment in connection with illness or maternity are paid directly by the social insurance office to the county council. Hospital treatment is thus free of charge for the patient, except for a daily fee of SEK 55 that is deducted from the sickness benefit.

Rebates are given on pharmaceutical preparations (officially registered drugs, which are discussed in Section C.).

Health insurance also extends to dental care, defined to include not only all forms of treatment but also prophylaxis. Its benefits cover the treatment given by members of the public dental service as well as the vast majority of dentists in private practice. The dentists must adhere to an established tariff whose rates may not be exceeded. A patient pays 60% of the costs to the dental service not exceeding SEK 2,500, and 25% of the cost above this amount, in the course of any single treatment period. The dentist is directly reimbursed for the rest from the social insurance office. Children and young people are entitled to free care of their teeth from the public dental service until they reach the age of 19.

Sickness benefit is the compensation paid for earned income lost due to illness. It amounts to 90% of the income normally earned. No sickness benefit is payable on that portion of gross annual income that exceeds SEK 209,250. Sickness benefit is taxable just like other forms of income, which means that it also qualifies for future payments from the national supplementary pension scheme (ATP) in the same way as other earned income. Any registered person who is not entitled to a sickness benefit due to lack of income or who is entitled to a benefit lower than the guaranteed daily minimum of SEK 60 can contract a voluntary insurance, thus receiving SEK 60 per day.

---

*SEK 1 = US $0.15 = £0.09

9. Coverage

All residents in Sweden, irrespective of nationality or occupation and whether in or out of employment, are covered automatically by the scheme. Foreign citizens may, however, be required to meet certain residence qualifications to become eligible for some benefits, such as retirement pension.

B. PHARMACEUTICAL INDUSTRY
AND DRUG DISTRIBUTION

1. Structure of the Industry

The development of the Swedish pharmaceutical companies did not occur, in the main, until after World War II, although most of them were founded during the early portion of the twentieth century. A decisive factor in the rapid growth of the drug industry has been the high level of research in medicine and chemistry in Sweden. Cooperation in research between the universities and the industry has long been a matter of course, and this cooperation now has a global scope. It was an early ambition of the industry to operate in clearly demarcated areas demanding a high degree of competence where the necessary conditions existed to become leaders by virtue of highly qualified work.

More than 18,000 people are now employed in the Swedish pharmaceutical industry. Due to a number of mergers, there are now two major concerns; Kabi Pharmacia (a merger of KabiVitrum and Pharmacia), which includes ACO and Leo-Ferrosan, and Astra, which includes Draco and Hässle. Pharmacia is involved in pharmaceuticals, diagnostics, and products for biochemical separation and ophthalmology. In 1986, an agreement was reached for Pharmacia to purchase the shares of Leo-Ferrosan. Astra has specialized in diseases of the central nervous system, local anesthesia, infectious diseases, respiratory diseases, and allergic disorders, as well as in cardiovascular and gastrointestinal diseases. KabiVitrum has concentrated primarily in the fields of parenteral nutrition, hematology, and growth factors. ACO's operations lie within five different business areas: hospital care, primary care, self-care, preventive care, and skin care. Ferring has specialized since the early 1950s in products based on peptides.[9]

The Association of the Swedish Pharmaceutical Industry (LIF), established in 1951, is the trade association of the corporate groups which together form the Swedish pharmaceutical industry and which account for

42% of the sales of pharmaceutical products on the domestic market. The member companies are very international in outlook, and 80% of their pharmaceutical output is sold abroad. LIF cooperates closely with RUFI, the Association of Representatives of Foreign Pharmaceutical Industries in Sweden.

A large number of pharmacists are employed in the pharmaceutical companies. There are some years when half of the graduating pharmacists go to the industry.

2. Research and Development, Imports, Exports, Patents

Sweden, where sales of medicines amounted to approximately SEK 7 billion in 1987, answers for about 0.5% of the total annual worldwide consumption of medicines (see Figure 5). Sweden's pharmaceutical industry has undergone rapid expansion since World War II. Expressed in current prices, sales increased by a factor of about 50, and more than three-quarters of the production volume is exported.[10]

Over the years, the product profiles of the Swedish pharmaceutical companies have undergone marked development, with continued concentration on different specialized fields of medicine. Since the 1930s, Pharmacia has had collaborative research with universities and hospitals in Sweden and abroad. Nearly one-quarter of the employees of Astra are engaged in research and development, and one-fifth of the sales revenue is invested in research. Research and development are undertaken mainly in Sweden, while 85% of sales are made abroad. The extensive research of KabiVitrum has resulted in nutrient solutions for intravenous administration, proteins obtained from human blood plasma, agents for the treatment of thrombotic diseases, and human growth hormone. Research and development for Ferring is carried out in Sweden and abroad, mainly in the field of peptides.[9]

The world market was not achieved by high-level product development alone. It is also the result of the development of an extensive network of contacts for research and marketing in every part of the world. These activities are being continued and further developed. In some cases, for legal, technical, or economic reasons, Swedish pharmaceutical companies have also assigned some of their manufacturing operations to foreign subsidiaries.

The Swedish pharmaceutical industry has an unconventional approach to mutual cooperation in foreign markets. As a rule, the most important export products of the companies belong to different fields of therapy, and not infrequently it is therefore more natural to cooperate than to compete

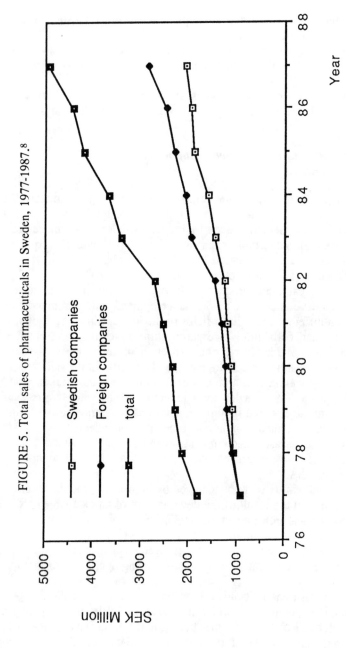

FIGURE 5. Total sales of pharmaceuticals in Sweden, 1977-1987.8

510

in a joint effort to ensure an appropriate place for Swedish pharmaceutical preparations in international medical care[11](see Figures 6 and 7).

Patent protection is an important prerequisite if companies are to finance the development of a new original preparation. If products are protected by patent, other companies cannot make generic products. As far as pharmaceuticals are concerned, Sweden's participation in the European Patent Convention meant the products became patentable in Sweden from June 1, 1978. At the same time, the patent term was extended from 17 to 20 years. The effective period of protection offered by a patent has, however, been reduced in recent years, partly because the time that elapses from the approval of a patent to the introduction of a new medicine on the market is now longer.

3. Manufacturing

The manufacturer and the control of manufacturing processes are governed by a number of different regulations and conventions. They include the important GMP rules developed by WHO and in the Pharmaceutical Inspection Convention (PIC). These rules were adopted in the form of a convention in 1971. Sweden is one of the 14 signatory countries.

4. Advertising and Promotion, Price Regulation

The pharmaceutical industry's marketing operations and information dissemination are subject to a number of different rules, i.e., statutory regulations and the industry's own rules. The statutory regulations are found in the Marketing Practices Act, the Drugs Act, and rules in the regulations of government agencies.

LIF, the Association of the Swedish Pharmaceutical Industry, and RUFI have issued detailed "Rules Governing Drug Information" with which their members have agreed to comply. The rules cover all the pharmaceutical information produced by the companies for doctors and other medical personnel. The rules apply to both verbal and written information. The pharmaceutical industry founded the Information Practices Committee (NBL) to monitor compliance with the rules or contravention of equitable business practices. The NBL independently processes complaints about inappropriate pharmaceutical information. LIF and RUFI have also appointed a special information examiner whose main task is to monitor the pharmaceutical industry's printed information.

The demand for certification of the industry's pharmaceutical information officers can also be regarded as part of the system of rules. To ensure that these officers possess good knowledge of their product and therapy

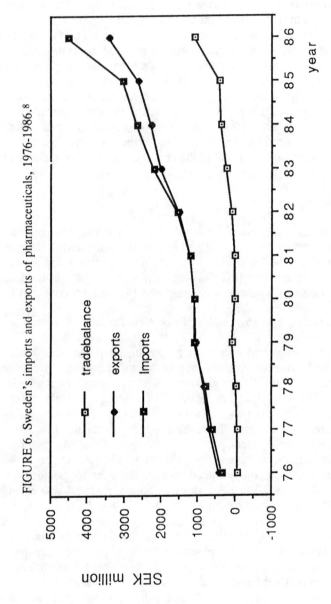

FIGURE 6. Sweden's imports and exports of pharmaceuticals, 1976-1986.8.

FIGURE 7. Sales of pharmaceuticals in Sweden and abroad by LIF companies, 1987. Price to pharmacy, SEK million.[8]

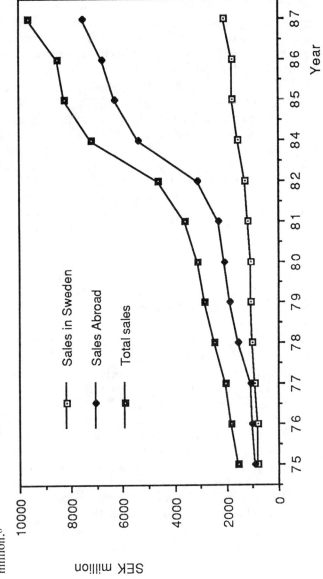

fields, the trade associations have adopted relative comprehensive training requirements.

There are two agreements, in addition to the various rules, that affect the industry's behavior in information matters. The pharmaceutical industry has agreements on information matters with the Federation of Swedish County Councils and the Swedish Medical Association.[12]

Promotion of OTCs is based on the Marketing Act and supervised by the National Board of Consumer Policies (NBCP). All product claims for conditions that may be suitable for self-medication must be substantiated, and no product claims may be made about diseases or symptoms that require a physician's diagnosis (as determined by the National Board of Health and Welfare). In practice, such claims are rarely examined because of the lack of resources and the internal priorities of the NBCP.

Advertisements are not intended as an educational media for the public. They are used to catch attention and to communicate a short message about a medication. Consequently, advertisements are used intensively when new products are introduced or when prescription medicines become available over-the-counter. This was the case when hydrocortisone preparations became available for topical use without prescription in 1983. Consumer education is expected to take place in pharmacies, predominantly through written and oral information given by pharmacy personnel.[13]

Every pharmaceutical registered in Sweden is subject to price control. Apoteksbolaget (The National Corporation of Swedish Pharmacies) is the main price control agency. The point of departure for the corporation's assessments is laid down in the Drugs Act, which stipulates that the price of a pharmaceutical must be "reasonable." A pharmaceutical must be approved by the National Board of Health and Welfare if it is to be sold in Sweden. Prior to approval, a pharmaceutical must be assigned a "reasonable price," in addition to being of good quality and medically efficacious. Since 1971, the price reached after negotiations between the corporation and the producer is generally regarded as a reasonable price. If the two parties fail to agree on the price, the pharmaceutical company can refer the matter to the National Board of Health and Welfare.

The corporation makes its assessments according to the following factors, irrespective of whether a drug is new or a registered product:

— Therapeutic importance and impact on treatment costs
— Price of any synonymous products
— International price comparisons

— Product cost calculations
— Sales volume.

Swedish companies must add company costs to this list (see Figures 8 and 9).

5. Drug Approval and Government Regulation

The pharmaceutical sector is governed by strict regulations. The basic regulations for drug control are found in the Drug Ordinance and in supplementary instructions on the application of this ordinance. The government strives, "by supervising the different stages in drug supply, to guarantee the consumers of drugs adequate drugs at reasonable prices, and to promote the proper use of drugs."

The Drug Ordinance specifies which goods are considered to be pharmaceuticals and the standards that apply to them. The pharmaceutical specialties are the principal type of medicine. They are defined as standardized medicines intended to be supplied to the user in the manufacturer's original packing. Before they can be sold in Sweden, pharmaceutical specialties must be registered with the National Board of Health and Welfare. This process requires the manufacturer to produce documentation to show that the medicine is an appropriate one, that its quality is satisfactory, and that its side effects are acceptable in light of the intended effect. In addition, only products with prices set at a reasonable level may be registered (the prices are subject to negotiation).[10]

A department of the National Board of Health and Welfare, the SLA, employs some 180 people, this being the equivalent of 150 annual employees. The total annual turnover of the SLA is about SEK 60 million. The work of the SLA is financed entirely by charges. These normally consist of a fee for applying for registration and an annual fee. The work of the SLA is grouped into seven subprograms, two of which are pharmaceuticals before registration and pharmaceuticals in use.

The pharmaceuticals-before-registration program deals with the work done before a pharmaceutical specialty is registered, including the work on applications for selling unregistered preparations under license, applying for the clinical testing of a pharmaceutical, approval of new indications, and registering new pharmaceuticals. About 29,000 licenses were issued in 1986-1987, some 1,300 applications for clinical tests were processed, and about 250 new pharmaceutical specialties were registered (see Tables 5 and 6).

The pharmaceuticals-in-use program involves the control of registered pharmaceutical specialties or radiopharmaceutical specialties. This program also deals with the inspection of manufacturers, wholesalers, phar-

FIGURE 8. The price construction of pharmaceuticals sold in Sweden.[8]

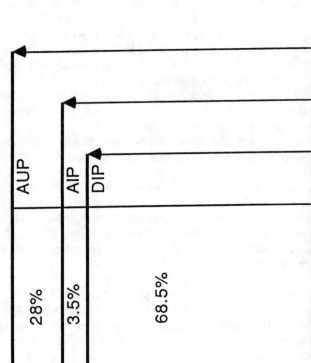

AUP, AIP and DIP 1987

AUP = Pharmacy selling price
The pharmacies add just under 39% to their purchase price(AIP), which gives them a margin of 28%

AIP = Pharmacy purchase price
The pharmaceutical trade adds some 5.3% to the manufacturers prices, which gives the trade a margin of 5%. This accounts for an average of 3.5% of the pharmacy selling price (AUP)

DIP = Pharmaceutical trade purchase price
The manufacturers' prices to the trade (DIP) account on average for 68.5% of the pharmacy selling price (AUP)

FIGURE 9. The relative share of the distribution trade in 1986 in percentage of the price of registered drugs in various countries (outpatient care).

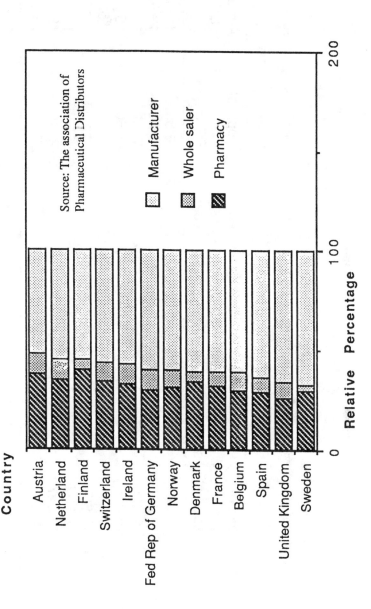

TABLE 5. Number of pharmaceutical specialties (approved drugs) registered on January 1, change during the year and annual registration fees.[8]

	No. of registered specialities	Newly registered during year	Removed from register during year	Application fee SEK	Annual fee SEK
1979	2 461	85	114	16 500	11 000
1980	2 433	123	96	20 800	13 900
1981	2 460	143	158	23 600	15 900
1982	2 445	171	69	23 600	15 900
1983	2 547	120	111	23 600	15900
1984	2 556	118	107	24 900	16 750
1985	2 567	175	115	26 100	17 700
1986	2 627	168	100	27 400	18 600
1987	2 696	210	137	27 400	18 600

TABLE 6. Time needed for examination of registration applications for pharmaceutical specialties.[8]

Year	No of pharm. specs registered	Mean value	Median	Range
1981	143	27	22	1-111
1982	171	25	15	1-145
1983	120	23	16	1-140
1984	118	20	10	1-174
1985	175	19	15	1-122
1986	169	20	11	1-180
1987	210	18	11	1-187

macies, and medical care, and also with checks on side effects and information issued to doctors and pharmacists. One objective of this program is to ensure that the manufacture and application of drugs correspond to the specification under which they were registered. Several thousand side effect reports, rejects, and cases where changes are required are handled every year. A large number of inspections, prescription controls, and various laboratory analyses, etc., are also carried out.[10]

6. Competition: Brand and Generic

The total number of registered drugs (pharmaceutical specialities) in Sweden, including different package sizes, strengths, dosage forms, etc., is approximately 2,800 (April 1989).

After the patent of a licensed pharmaceutical expires, a number of generics are usually introduced by competing industries. There are, however, so far only a few companies in Sweden that are strictly generic. Their market shares are still very small, but may be growing. The total market share of generics in Sweden is estimated at around 10%-15%. In 1987, The 1983 Pharmaceutical Inquiry Commission's report (LU83) was submitted to the Minister of Public Health.[20] The main task of the Commission was to review existing pharmaceutical legislation. As a result, the Commission has proposed a new Drugs Act.

The Commission did not propose any major changes, with one important exception. The majority of the Commission members proposed the introduction of a completely new system for designating pharmaceuticals in Sweden, i.e., generic names. This means that pharmaceuticals would henceforth be registered under generic names combined with the manufacturer's name. The Commission also suggests that generic names be introduced for many pharmaceuticals that are already registered in Sweden. According to the proposal, a doctor would be required to specify either a generic name alone, such as ceftazidime, or the generic name plus the manufacturer's name, such as ceftazidime-Glaxo, in prescriptions. However, the doctor would not be allowed to indicate the product's trade name (Fortum, in the example).

The Commission's proposed change to generic names for pharmaceuticals in Sweden has thus still not been adopted. Due to the immense number of objections, it is not very likely that it ever will.

7. Cost-Containment Activities

Outpatient care, assessed in terms of visits to physicians, has increased in scope. However, this has not resulted in increased prescribing of drugs in outpatient care. On the contrary, assessed in terms of the number of prescription items per visit to the physician, prescribing has decreased. However, the costs per prescription have increased.

The ratio of drug costs to total health care costs has fallen from 10.7% to 8.1% between the years 1970 and 1985 (Figure 10). The pharmaceutical cost price index increased less than the net price index during the period December 1970-December 1987 (Figure 11). The same applies to the pharmaceutical retail price index, compared with the net price index.[14]

The need for well-developed research in health economics, as a basis for more effective use of resources in health care, was also emphasized strongly by the 1983 Pharmaceutical Inquiry (LU83).[20] The concept behind the proposal for generic prescribing or, as it is called by the Commission, the use of substance name, is to decrease the drug costs for society and to improve the use of drugs. (See Section B.6). Other examples of proposals aimed at achieving those goals are introducing a health record card and improving drug information both to the health care staff and to the patients. In the latter case, the pharmacies are expected to take a greater part than they do now. The Commission realizes that the pharmacies are the only body that has resources which may be utilized and that has an existing organization spread over the country.[15,20]

It is difficult, though, to make relevant comparisons between pharmaceuticals and other costs. However, for a certain drug and a certain diagnosis or group of diagnoses, it is possible to perform a profitability calculation for the drug in question. According to LU83, this is an important area for development. By studying alternative methods for treating dis-

FIGURE 10. Ratio of drug sales (retail price) to total health care costs during 1970-1985 in Sweden.

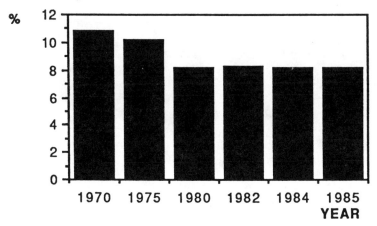

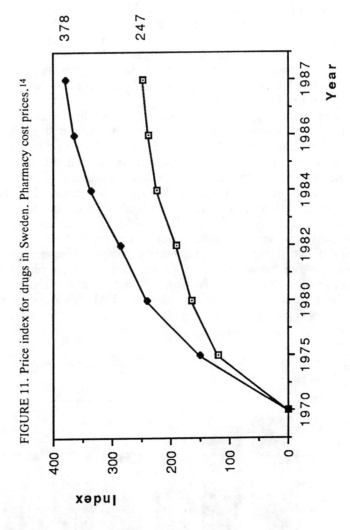

FIGURE 11. Price index for drugs in Sweden. Pharmacy cost prices.[14]

ease – drug therapy, surgery, change in lifestyle, etc. – it is possible to arrive at the treatment that is most suitable economically and medically. For example, in treating a certain disease, costly but potent drugs may prove to be a cheap alternative if the reduced costs of inpatient care, absence from work through illness, etc., are taken into account.

In 1987, Apoteksbolaget increased its commitment to research in health economics and analysis through its decision to acquire the Swedish Institute for Health Economics (IHE) and to make allocations to Apotcksbolaget's fund for research and studies in the fields of health economics and social pharmacy.

Some cost-containment activities are also tried through local formularies, issued by the Pharmacy and Therapeutics Committees.

C. PHARMACY PRACTICE

1. Dispensing a Prescription: Where, What, Payment, Formulary

In 1963, a new government committee, the Inquiry on Drug Supply, was appointed to consider a new system of drug supply on the retail level. The committee's work, and the negotiations carried on under its auspices with the pharmacy owners' organization, resulted in a recommendation that the pharmacy system should be reorganized as an incorporated company under the joint ownership of the government and Apotekarsocieteten (the then organization of the pharmacy owners). In May 1970, Parliament decided accordingly. On August 24, 1970, Apoteksbolaget AB was established, and on September 18 an agreement was reached between the newly founded corporation and the government, creating the framework for company operations.[16]

Apoteksbolaget, the National Corporation of Swedish Pharmacies, handles purchases and sales of pharmaceuticals in Sweden. The company has the sole right and responsibility to engage in retail trade in drugs. Apoteksbolaget has two main tasks: to provide a pharmaceutical service, that is, to supply drugs and information about drugs, and in free competition, to satisfy the requirements of customers for goods and services within the sectors of medical care and body care. The guidelines for Apoteksbolaget's activities are stated in an agreement between the company and the Swedish government. Apoteksbolaget is required, among other things:

- To be responsible for a good supply of drugs in Sweden at the lowest possible cost
- To determine where and to what extent pharmacies and other sales outlets for drugs shall be located
- To negotiate the company's purchase prices for drugs with producers
- To ensure that drug prices are uniform throughout the country
- To negotiate with the principals for the medical care services on the conditions governing the company's supplies to public health and medical care services
- To seek close cooperation with the principals for medical care services
- To supply such high quality goods within the health care field as are naturally linked to the company's activities
- To promote the dissemination of sufficiently comprehensive factual information in the pharmaceutical field
- To assist in the continuous production of statistics on the consumption of pharmaceuticals
- To ensure that personnel engaged in the distribution of pharmaceuticals satisfy the qualifications necessary from a safety viewpoint.

The company's operations are managed by the president, reporting directly to the board. The president's management group consists of the president, vice president, board secretary, and managers of the sectors into which operations are divided. The president's office includes resources for providing external information and performing internal inspection (Figure 12).

Operationally, Sweden's 840 pharmacies are organized in 42 pharmacy groups. Each group has a manager who is also pharmacy manager at one of the group's pharmacies and who has complete charge of pharmacy operations within the group. Activities between and among the groups include coordination of internal and external services and information, training, and accounting operations.

The aim of the work organization at the pharmacies is, with due observance of stringent safety requirements, to give customers their medications as speedily as possible and thereby contribute to more effective medical care. Waiting times should normally be short enough so that customers do not have to make a return trip to collect their medicines.

Modern technology is employed to improve service. When, on arrival from the doctor's office, the patient presents his or her prescription, several operations must be performed at the pharmacy before the medicine is ready to be handed to him or her. For example, the medicine has to be priced and supplied with a label bearing the patient's and doctor's names

FIGURE 12. Apoteksbolaget, the National Corporation of Swedish Pharmacies, was recently reorganized. This is the current organization chart, valid since January 1, 1989.

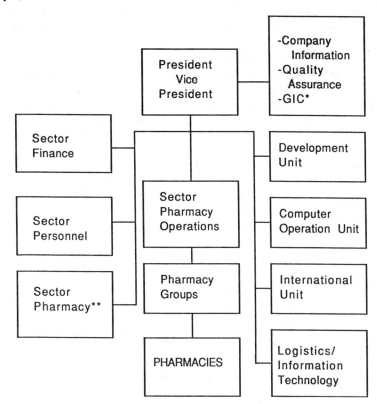

and dosage instructions. A prescriptionist or pharmacist checks the drug, dosage, etc., before the customer receives his or her medicine.[17] During the 1980s there has also been an increasing emphasis on patient drug information in the Swedish pharmacies.

Rebates are given on pharmaceutical preparations (officially registered drugs). The maximum sum payable by the prescription holder at any one time is SEK 65. Pharmacies receive the difference from the National Social Insurance Board. No charge is made for drugs in the lifesaving category, i.e., pharmaceutical preparations needed to treat chronic and serious diseases. The pharmaceutical benefits scheme also covers oral contraceptives which are prescribed for birth-control purposes only. For persons with considerable costs for medical treatment and pharmaceutical preparations, there is a "15-card" which entitles them to free treatment and/or preparations after the fifteenth visit or purchase and during the 12 months following the date on which the card was issued.

Formularies are made up by Pharmacy and Therapeutics Committees, which usually have at least one representative from primary care, in most cases a district general practitioner.

On the wholesale side, Sweden has consistently adopted the practice of single-channel distribution. This means that manufacturers of pharmaceuticals distribute their entire range through a single wholesale company. This system has not been adopted fully by any other comparable country. As a result of structural rationalization within the wholesale trade, there are now two wholesale companies in Sweden Apoteksbolaget's ADA subsidiary and Kronans Droghandel (KD). Regarding pharmaceuticals, ADA accounts, in terms of monetary value, for about 78% and KD for the remainder of approximately 22%. ADA and KD have concluded an agreement on common transportation of goods to pharmacies. ADA administers the coordination of transportation arrangements.[14]

2. Record Keeping

All dispensing of prescriptions in Sweden is conducted by aid of computers. They are, however, used as a dispensing aid only, for pricing, printing labels, and giving codes on what drug information to deliver. For confidential reasons, personal data related to the individual patient may not be stored. Computerized patient records or profiles, like the ones used in the Netherlands and the U.S., have not yet been tried. Some depersonalized information is saved, though, for statistical purposes.

In Sweden, the introduction of a "patient card" (smart card), similar to those used in automatic teller machines, has been discussed. Information could then be added to the card both by the physician and the pharmacy. This system eliminates any violation of confidentiality, since the patient has possession of the card.

3. Patient Education

"The Company shall promote the development of good information in the drug field," according to the agreement between the government and Apoteksbolaget.[16] Specific, printed "Guidelines for Drug Information at the Pharmacies" were introduced by Apoteksbolaget in 1978. Providing information on drugs and other products in stock is thus an important part of the pharmacies' operations. An important channel of information to the public is the customer journal *Apoteket* (*The Pharmacy*), published five times a year. This journal, together with a number of informational brochures published by Apoteksbolaget, is available free of charge at pharmacies.

The pharmacies cooperate with the local medical care services in providing advice and instructions concerning suitable treatment of minor ailments. This is done, for example, by publishing booklets on self-care that are distributed free of charge to the public.

In Sweden, an investigation carried out toward the end of the 1970s showed that the general introduction of package inserts would result in high costs. As an alternative, Apoteksbolaget then compiled written information on drugs. This was done in cooperation with the pharmaceutical industry and experienced clinicians. Each information sheet covers a group of drugs and is handed to patients receiving prescription drugs at pharmacies. The written information is supplemented with oral information.

Since 1987, the information sheets cover 90% of prescribed drugs. The Swedish pharmacies' computer system contains information telling pharmacy personnel which information sheets should be handed out and which aspects should be emphasized orally.

In addition to their main informational activity for the general public in the course of customer service, the pharmacies have — especially in recent years — assisted in providing general drug information for special target groups. An example of this activity is a two-hour instructional program on drugs for use in the ninth grade of comprehensive school. Drug information programs are presented for the elderly as well. The pharmacies have also taken an increasingly active role in providing municipal home health care personnel with suitable information and training concerning drugs and drug handling. A special educational package has been produced for this purpose.

There are one or two information pharmacists employed in each pharmacy group. They conduct information and training programs internally and externally. This internationally unique position was introduced in Sweden in 1979-1980 and fosters the development of patient drug information.

4. OTCs and Nonpharmacy Outlets

Large- and medium-sized pharmacies are established with a self-selection department where the customer can obtain a good view of the range of self-care products offered. The range of drugs for the self-selection departments has been selected (in light of, for example, safety aspects of certain group of drugs) consultation with the National Board of Health and Welfare and the National Board for Consumer Policies. Within the framework of this selection, the individual pharmacy, in consultation with the medical care personnel in the community, adapts its own self-selection range to the local therapeutic tradition.

In recent decades, a number of pharmaceuticals that were formerly obtainable only by prescription have now become available for self-care. In Sweden, such drugs include paracetamol and ibuprofen, hydrocortisone preparations for use in mild dermatitis, topical lidocaine, and decongestant nose drops and nasal sprays. In connection with the removal of pharmaceuticals from the list of prescription drugs, pharmacy personnel have received extra training relating to the diseases involved and their treatment. Brochures describing the disorders and stating when the nonprescription medication can be used have been prepared for customers.

The number of OTCs available in Sweden is smaller than in most Western countries. The National Board of Health and Welfare seems, however, to have changed attitude lately, and more Rx-OTC switches are therefore expected (Figure 13).

The pharmacy network is supplemented by a large number of pharmacy representatives in localities lacking a pharmacy. The work of the representatives, usually general food stores, may be limited to accepting orders and distributing packages or it may also be combined with sales from a stock with a very limited range of goods direct from the pharmaceutical storeroom. The total number of pharmacy representatives is about 1,500.

Otherwise, OTCs may not be sold outside pharmacies. The grocery store merchants would, of course, like to see a change.

5. The Use of Technicians and Others

There are two training alternatives for pharmacy personnel. One is the four-year pharmacist's course held in the Faculty of Pharmacy at the University of Uppsala. The other alternative, a two-year course of prescriptionist's training, is also available in the same faculty and includes one term of practical training at a pharmacy. The new prescriptionist's course is a result of a merger of two earlier courses, those for pharmacy technicians and prescriptionists. The two-year pharmacy technician training at high school level has thus ceased. The proportion of technicians in Swedish pharmacies will consequently gradually decrease. For those who have been previously trained as pharmacy technicians, there is now a special supplementary course which confers a prescriptionist's qualification. Both pharmacists and prescriptionists have the right to dispense prescriptions under their own responsibility.

It is estimated that due to a large number of retirements in the 1990s, there will be a severe shortage of pharmacists and prescriptionists in Sweden (see Table 7).

FIGURE 13. The proportion (%) of OTC-sales of total sales in the Swedish pharmacies is approximately 8%, but should grow due to Rx-OTC switches.[17]

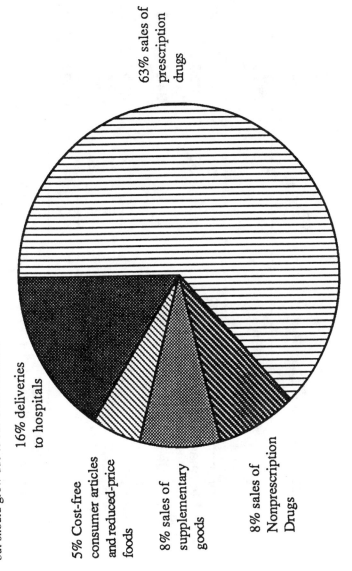

63% sales of prescription drugs

16% deliveries to hospitals

5% Cost-free consumer articles and reduced-price foods

8% sales of supplementary goods

8% sales of Nonprescription Drugs

TABLE 7. Breakdown of Apoteksbolaget employees by category and sex, December 1988.

	Women No.	%	Men No.	%	Total No.	%
Pharmacy managers*	603	72	223	27	826	7
Other Pharmacists	292	70	128	30	420	3
Other Prescriptionists	3382	99	42	1	3424	28
Pharmacy Technicians	5039	98	120	2	5159	43
Cleaners	1031	99	9	1	1040	9
Others	899	74	324	26	1223	10
Total	11246	93	846	7	12092	100

Source: Apoteksbolaget AB
* Both pharmacists and prescriptionists can become pharmacy managers.

6. Pharmacy in Primary Care

As a result of the cooperation between Apoteksbolaget and the principals for medical care services, new pharmacies have increasingly been localized to hospitals and medical care centers, where prescriptions are issued. Today, there are about 150 such pharmacies in Sweden. The main operation of these pharmacies is dispensing prescriptions. Close cooperation with doctors and other medical care personnel has resulted in an adaptation of the range of goods at these pharmacies to suit local requirements. Thus, despite their small size, these pharmacies can provide good service.

Sweden has progressed comparatively far in the integration of outpatient and inpatient care. The outpatient care sector is now represented in the majority of Pharmacy and Therapeutics Committees. Recommendations from the overall committees are discussed and adapted in local groups in which medical and pharmaceutical expertise is represented. However, much work remains to be done before the same compliance with the committees' recommendations is attained in outpatient care as in inpatient care.

Apoteksbolaget is also more and more engaged in drug-related informational and training activities directed to physicians and other health care personnel.

The journal *DistriktsLakaren* (*District Medical Doctor*) is a new channel used by Apoteksbolaget to supply information to physicians. Each issue has a pharmaceuticals page, dealing with matters of interest.

A new edition (the seventh in the series) of the therapy handbook *Lakemedelsboken* (*LB*) was published in 1989. A drug handbook employing a problem-solving approach, it is intended primarily for primary care and trainee physicians.

The pharmaceutical statistics produced and published by Apoteksbolaget constitute a well-established aid in studies of drug consumption. *Svensk lakemedelsstatistik* (*Swedish Drug Statistics*), published annually, gives sales figures at national and county level. Sales information related to a particular health care district, municipality, or pharmacy can be ordered.

7. Drug Insurance and Third Parties

The idea of refunding the patients' costs of drugs, either partially or totally, became more widely accepted and formalized as a composite scheme when sickness or health insurance was nationalized and made compulsory to the whole population in 1955 in Sweden.

The social insurance usually pays most of the cost of prescribed medicine. Individuals pay a patient's fee. If a patient has been given several prescriptions at the same time by the same doctor, the patient need only to pay one patient's fee (SEK 65 in April 1989) for all the medicines together, provided he or she buys all the medicine on the same occasion. But if the doctor has prescribed medicine for six months, the patient is allowed to buy only enough medicine for 90 days' use for one patient's fee. The patient then has to pay a patient's fee again when getting the refill.

For certain long-term and serious illnesses, medicine is free. Patients can also obtain certain other medical articles free of charge. For contraceptive pills, there is one patient's fee for each three-month supply. If the doctor telephones the pharmacy with a prescription, the patient has to pay an extra charge.

Persons under 16 years of age who need specially composed foodstuffs (because of illness) can obtain them at a lower cost.

D. UNIQUE OR INTERESTING FEATURES OR SPECIAL SITUATIONS IN SWEDEN

One distinctive feature of Sweden is its social welfare policy, which includes health and medical services, family benefits, and other benefits. Even if it is not internationally unique, the Social Security system in Sweden is probably the most comprehensive one existing. However, it is not without problems. Some current trends and subjects of discussion will be described.

The Swedish pharmacy system, previously described, with all pharmacies belonging to a single government-owned corporation (Apoteksbolaget), is thought to be unique in the Western part of the world. Even Swedish pharmacists not in favor of socialized systems are generally very positive toward Apoteksbolaget. Since the corporation was established in 1970, there has been quite a development of pharmacy premises and services, as well as of educational activities both for the pharmacy personnel and external target groups.

The pharmacologically-oriented research of Apoteksbolaget is conducted at the central laboratory, as well as at hospital pharmacies, mainly by pharmacists at regional hospitals (regional information pharmacists with Ph.D. degrees).

The internationally unique information pharmacists in all the pharmacy groups of Sweden were previously mentioned.

The Swedish pharmacy system has also enhanced the comprehensive pharmaceutical statistics continuously produced and published by Apoteksbolaget.

The Swedish Pharmaceutical Society, Apotekarsocieteten, and the

Swedish Pharmaceutical Association, Sveriges Farmacevtförbund, are also important factors in the development of pharmacy practice in Sweden.

E. CONCLUSIONS

The Swedish Social Security program has been rather successful as a whole. There are problems, though, of increasing costs, leading to a conflict between the actual financial situation and the demands for high quality care. The long waiting times are another problem, especially for specialist referral or for surgery.

There has also been a growing dissatisfaction in the 1980s among hospital nurses and midwives due to hard working conditions and low pay. In many hospitals there is a shortage of nurses.

Another topic being discussed is a possible change in the division of responsibilities in the health care field between the city councils and the municipalities.

The Health and Medical Services Act, HS 90, and a Parliamentary Directive resulting from a Government Health Proposition (1984/85:181) reflect the government's views on the essential guidelines for future health services. This includes the changing structure of health care, for example a reorientation of psychiatric care away from its concentration on psychiatric hospitals toward more open, community-based initiatives. Likewise, long-term care must expand in the form of neighborhood care centers to the greatest possible extent. Hospital-based long-term care would be limited to those requiring special facilities for examination and rehabilitation.

The role of preventive care within both the health care sector and the community at large is emphasized by HS 90 and the Government Health Proposition. HS 90 emphasizes that the health services should give priority to equalizing access to care and that the county councils should develop more need-related planning methods and more effective follow-up and evaluation methods.[4,18]

The general public is not prepared to accept national cutbacks in the health care sector. There has been an ongoing debate between the political parties concerning private and public health care delivery and opposition from specialized medical practitioners concerning the probable consequences of reducing resources in these areas. There is a gradual increase in private medical care. But even among those who recommend a large public sector, there is a desire for greater freedom of choice and better performance. There is also a general willingness and tendency to try alternative organizational forms.[4]

The mass media frequently calls attention to issues concerning patient satisfaction with health services. The shortcomings of the Swedish health system are mentioned more often than its merits.

Much has happened in the Swedish pharmaceutical industry during the past few years. Many organizational changes have taken place, especially involving pharmacy. There is no reason to believe that the Swedish companies will not continue their growth and success. In fact, the manufacturers of pharmaceuticals are, together with the car industries, the most prosperous ones in Sweden.

In light of the problems observed in drug controls, the Swedish National Audit Bureau (RRV), in its audit in 1987, considered ways of making the present organization more effective. The main proposal is to reconstruct the National Board of Health and Welfare Department of Drugs (SLA) into an independent agency with overall operative responsibility for the work of drug controls.[10]

As previously mentioned, the 1983 Pharmaceutical Inquiry Commission (LU83) was appointed to investigate the regulation of pharmaceuticals and some related matters. According to its instructions, the Commission should as a first task consider if, in light of present development and experience acquired from the activities of Apoteksbolaget (The National Corporation of Swedish Pharmacies), there would be any reason for changes for the time period after 1985 in the two existing agreements between the government and Apoteksbolaget on the activities of the corporation. The commission's study of the corporation shows that in all essentials the corporation has well carried out its task of being responsible for a good drug supply in Sweden. This has not prevented the Commission from making certain remarks on a few points. Among the viewpoints on the future are some commentaries concerning the problem of keeping a monopoly business vital. The strong decentralization is considered valuable.[19,20] Apoteksbolaget remains vital, despite the monopoly, through continuous work on improving both quantity and quality of services.

There is a growing need in Apoteksbolaget for developing more methods for measuring the quality of services. There is also a need for studies of the pharmacies in health economic terms. Productivity, for example, the number of prescriptions filled per hour, etc., has been measured continuously in the Swedish pharmacies for some time.

Pharmacy personnel sometimes feel a conflict between providing fast, productive service on one hand, and fulfilling an increasing number of high quality tasks, including patient drug information, on the other. Interestingly, according to a Gallup Poll in Sweden in 1988, the public rated the pharmacies first among 15 different government agencies and service companies regarding their function of services.[21]

A new trend is the establishment of drug information centers, mainly at regional hospital pharmacies in cooperation with clinical pharmacologists. These centers reply in depth to advanced inquiries, mostly from physicians. The National Poison Information Center is now part of Apoteksbolaget as well.

Another trend is the increase in home health care as a pharmacy service, ranging from mere care of the elderly to advanced services. The delivery of individually packaged dosages of drugs is one step in this development. This new outpatient pharmacy service is now provided to around 15,000 patients. Up to 100,000 might be in need of the service.

While it could be said that most services are readily available in Swedish pharmacies, there is room for improvement. One example is the pharmacies' computer system, which could be more developed. This has also been proposed by the Swedish National Audit Bureau (RRV), and there are a number of computer projects planned.

There is some uncertainty regarding the future. One source of unrest is the question of how the Common Market will influence Sweden and Swedish pharmacy practice. Another is the expected shortage of pharmacists and prescriptionists, due to a large number of retirements and fewer young people in the 1990s (there will be 20%-25% fewer 20-year-old persons). Swedish pharmacies might thus face recruitment problems.

REFERENCES

1. The National Board of Health and Welfare, Public Health Report, Stockholm, 1988.

2. The National Board of Health and Welfare, The Social Services in Sweden, Stockholm, 1988.

3. Federation of County Councils, Swedish County Councils, Stockholm, 1988.

4. The Swedish Planning and Rationalization Institute For the Health and Social Services (Spri), SPRI 3 years ahead 1987-89, Stockholm, 1986.

5. Federation of County Councils, A Presentation, Stockholm, 1985.

6. The Swedish Institute (SI), The Health Care System in Sweden, Stockholm, 1987.

7. The Swedish Institute (SI), Social Insurance in Sweden, Stockholm, 1987.

8. LIF/RUFI, FAKTA 88, Stockholm, 1988.

9. LIF, The Swedish Pharmaceutical Industry, Stockholm, 1988.

10. The Swedish National Audit Bureau (RRV), The State System of Drug Controls, (Audit report, summary, Dnr. 1986:58), Stockholm, 1986.

11. LIF, The Swedish Pharmaceutical Industry, Stockholm, 1982.

12. RUFI, Information on pharmaceuticals — rules and agreements, RUFI Newsletter, Stockholm, 1986;3.

13. Levin LS, Beske F, Fry J, Self-Medication in Europe, WHO, Copenhagen, 1988.

14. Apoteksbolaget AB, Annual Report, Stockholm, 1987.

15. Eklund LH, Drug Cost Containment — The Case for Generics, JSAP, Stockholm, 1989;1.

16. Apoteksbolaget AB, The first 10 years, Stockholm, 1981.

17. Apoteksbolaget AB, Pharmacies in Sweden, Stockholm, 1986.

18. The National Board of Health and Welfare, HS 90, The Swedish Health Services in the 1990's, Stockholm, 1985.

19. Apoteksbolaget Towards the Turn of the Century (SOU 1984:82), Stockholm, 1984.

20. The 1983 Pharmaceutical Inquiry Commission's Report, LU 83 (SOU 1987:20), Stockholm, 1987.

21. Statens Institut for personalutveckling (Sipu), Servicebarometern, Stockholm, 1988.

Chapter 21

Taiwan, Republic of China

Weng F. Huang

A. THE NATIONAL HEALTH CARE SYSTEM

1. History and Background

Long before Chinese immigration to the island in the thirteenth century, Malay-Indonesian aborigines had inhabited Taiwan for hundreds or even thousands of years. Portuguese sailors in 1557 unintentionally viewed the island and called it Ilha Formosa — Portugese for "beautiful island." Formosa has been synonymous with Taiwan since then and has been an island of international involvements through the Spanish (1626 to 1642), the Dutch (1624 to 1661), the Ming Dynasty (1661 to 1683), and the Ching Dynasty (1683 to 1895) of China. Taiwan was ceded to Japan in 1895 when the Ching Dynasty lost the Sino-Japanese War. Japanese administration ended in 1947 when Japan was defeated in World War II and abandoned sovereignty of Taiwan in accordance with the Potsdam Declaration of 1945. Taiwan then became a province of the Republic of China when the People's Republic of China was founded in 1949.

The country has been better known as Taiwan than by its official name, the Republic of China, due to international politics when both the People's Republic and the Republic of China declared that there is only one China and each represents the sole legitimate government of China. The Republic of China (Taiwan) became more isolated from the international political arena when its legitimacy in the United Nations was challenged by the People's Republic in 1972 and the United States changed official recognition to the People's Republic in 1978. However, Taiwan survived despite international political setbacks during the 1960s and 1970s, and has begun to attract more attention internationally, mainly because of its impressive economic development and strong foreign reserves of U.S. $76 billion (second only to Japan as of December 1988). The influx of foreign reserves is mainly attributed to a continuous international trade surplus since the late 1970s. Per capita income in Taiwan exceeded U.S. $6,000 in 1987 and is projected to reach U.S. $10,000 in five years.

2. Social and Political Environment

An island of 36,000 square kilometers (14,000 square miles), Taiwan has a population of 20 million. Taiwan's successful economic development can be attributed to a high literacy rate; diligent workers in a relatively stable political environment; and an effective, planned free economic system. The political system was basically a controlled democracy under an authoritarian leadership (i.e., Generalissimo Chiang Kai-Shek and his son Chiang Chin-Kuo) for 40 years, until the death of Chiang Chin-Kuo in January 1988 and the lifting of martial law in July 1988. Opposition parties can now organize legally, and they actively represent increasingly diversified social strata on various issues.

The health status of the people of Taiwan has improved greatly with the progress of economic development. For instance, life expectancy increased by 17.87 years, from 53.10 years in 1950-1951 to 70.97 years in 1986, for males and 18.56 years, from 57.32 years to 75.88 years in the same period, for females. Some economic and social welfare indexes are presented in Table 1 and Table 2.

3. Public and Private Financing, Health Costs

Health expenditures include those of the national government, as well as health and medical institutions at various levels of provincial, municipal, county, and city government. The government health budget increased from 1.34% of total government expenditures in 1951 to 3.92% in 1987. During the same period, individual health and medical expenditures also increased, from 2.58% to 5.23% of total individual expenditures. Because only 45% of the population is covered by various health insurance programs, total expenditure on health care figures is not available. It is estimated that about 3.5% of GNP was spent on health care in 1988, and this figure is expected to reach 7% in 10 years. The average increment of medical care cost between 1970-1984 is estimated at 11.4%, but in the two largest insurance programs — Labor Insurance and Government Employees Insurance — the average increments during the same period are 16.9% and 19.0%, respectively.

TABLE 1. Economic development in Taiwan area, R.O.C. (1981 value).

Year	Mid-Year Population		GNP		Per Capita income		Per Capita Expenditure		Industrial Product
	1000	1952=100	NT$	1952=100	NT$	1952=100	NT$	1952=100	(1981=100)
1956	9,823	115.0	20,377	118.8	19,783	117.3	13,949	111.3	3.83
1966	13,283	155.5	33,26	197.2	32,984	195.6	21,338	170.3	13.27
1976	16,329	191.2	69,964	407.8	68,382	405.4	39,857	318.1	61.21
1986	19,357	226.6	132,019	769.6	126,328	749.0	64,253	512.9	148.20
1987	19,564	229.1	146,111	851.7	142,733	846.2	70,039	559.1	166.63

Source: Health and Medicine in Taiwan Area, Department of Health, Executive Yuan December 1988.

TABLE 2. Social indices in Taiwan area, R.O.C.

Year	Per Capita Medical Expenditure		Education (6 year +)		No. of Telephone/ 100	Kcal Intake	Protein Intake (9)
	NT$	1052=100	College & Above %	Literacy %			
1956	383	8.19	1.7	62.9	0.56	2,262.0	53.9
1966	897	153.38	2.5	76.9	1.47	2,432.8	62.3
1976	1,963	454.51	5.6	87.9	8.46	2,770.5	75.9
1986	3,237	814.40	9.4	92.0	31.24	2,969.1	85.1
1987	3,442	872.31	9.8	92.2	32.78	2,999.4	88.4

(Note: per capita medical expenditure at 1981 value)

Source: Health and Medicine in Taiwan Area, Department of Health, Executive Yuan December 1988.

4. Facilities (Hospitals, Clinics, etc.)

At the end of 1986, there were 12,533 public and private hospitals and clinics in Taiwan. The public hospitals and clinics include 85 hospitals, 487 clinics, and 496 health stations at various administrative levels (medical centers, provincial, municipal, county, etc.). Private hospitals and clinics number 750 and 10,715 respectively, including private medical schools affiliated with teaching hospitals, nonprofit hospitals, proprietary hospitals, corporate hospitals, and clinics. The total number of hospitals and clinics in 1986 is 34% more than the 9,344 facilities in 1976. At the end of 1986, there were 81,502 hospital beds: 29,792 in public hospitals, 2,332 in public clinics, 41,381 in private hospitals, and 7,997 in private clinics. This supply represents an average of 42 beds per 10,000 persons, while the number of persons per hospital bed was 237 in 1986, compared with 503 in 1976.

5. Manpower

There are 8 medical schools, 10 paramedical junior colleges, and 14 paramedical vocational schools. Each year, they produce about 800 physicians, 400 dentists, 500 medical technology professionals and technicians, 1,000 pharmacists, and 5,000 nurses and midwives. The number of medical students has increased recently, and more than 1,000 medical graduates per year are expected in the 1990s.

At the end of 1987, there were 77,246 health professionals of all kinds practicing at either public or private hospitals, clinics, and pharmacies (see Table 3).

6. The Social Security System

The Social Security system in Taiwan is segregated into various programs administered by different government ministries. The premium of these Social Security programs varies from 3% to 12%, depending on the scope of coverage, while the percentage of government subsidization also varies from 25% to 65% of the premium.

Approximately 45% of the population of Taiwan is covered by various health insurance schemes which provide free medical care services and certain cash benefits. The largest insurance plan is the Labor Insurance Program (LIP) which covers 6 million people. Government Employees Insurance (GEIP) covers another 850,000 people, Military Insurance (MIP) 810,000, and Farmers Insurance (FIP) initially 400,000 (1987) but expanded to 1.2 million farmers in 1989. A national health insurance pro-

TABLE 3. Health manpower in Taiwan area, December 1987.

Category	Number in Service	Pop / Person in Service	Person in Service Per 10,000
1.Physician Chinese medicine Doctor	17,045 2,324	1,016	9.85
2. Dentist Dental assistant	4,150 119	4,608	2.17
3. Pharmacist Assistant Pharmacist	9,259 8,340	1,118	8.95
4. Medical technologist &technician	2,540	7,745	1.292
5. Nurse	30,174	652	15.34
6. Midwife	2,380	8,266	1.21
7. Radiological Technologist $technician	915	21,500	0.47
Total	77,246	255	39.27

Source: Health and Medicine in Taiwan Area, Republic of China, Department of Health, December 1988.

542

gram for the year 2000 was proposed by the government earlier, but the proposed schedule has been advanced to 1995 due to escalating public expectation and pressures.

7. Medical Establishment: Entry, Specialists, Referrals

According to the Medical Care Law, institutes are classified into medical centers, regional hospitals, district hospitals, and clinics. A medical care referral system has been drafted and is intended to be enforced in accordance with the establishment of the Medical Network Program (Figure 1).

8. Payment and Reimbursement

The Labor Insurance Program is the largest insurance program in terms of population insured. Approximately six million people are covered under this insurance scheme, which includes capital payment of birth, disability, pension, and death benefits, as well as capital payment and medical care for injuries and diseases. The LIP is administered by the Council for Labor Affairs, Executive Yuan.

The Government Employee Insurance Program (GEIP) covers approximately 850,000 government employees, including teachers in public schools and workers in public enterprises. Insurance coverage includes: capital payment for disability, pension, death, and funerals of spouses; and medical care for birth, injuries, and diseases. The GEIP is administered by the Ministry of Personnel Qualifications under the Examination Yuan.

The Insurance Program for Teachers and Staff in Private Schools covers about 28,000 people. The insurance premium and scope of coverage are the same as for GEIP and so is the Ministry in charge.

Health Insurance of Government Employees Spouses is an extension of GEIP and covers about 180,000 spouses (wife or husband) and parents of insured government employees. It covers only medical care reimbursement for injuries and diseases.

Farmers Health Insurance Program (FHIP) is the second largest insurance program and covers about 1.2 million farmers. The scope of coverage includes capital payment for birth and funerals, and medical services for injuries and diseases.

FIGURE 1

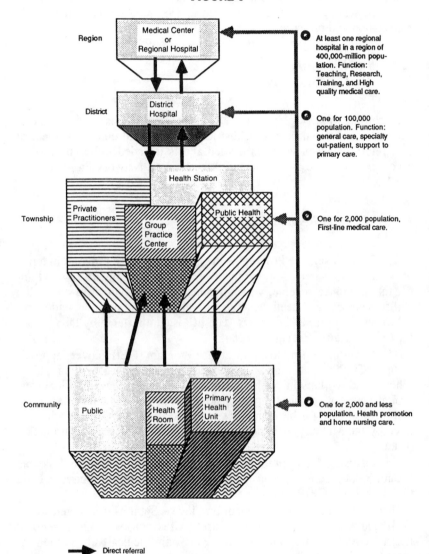

9. Coverage

All persons covered by any of the social insurance schemes are entitled to inpatient and outpatient health services, including curative dental care. Inpatient and certain outpatient health services are provided by hospitals having a contract with the insurance agencies. Drug coverage in these programs is quite comprehensive, as nearly all insurance programs cover prescription medicines, even outpatient visits.

B. PHARMACEUTICAL INDUSTRY
AND DRUG DISTRIBUTION

1. Structure of the Industry

Taiwan's pharmaceutical industry is composed predominantly of small-scale generic manufacturers, except for some 28 foreign-investor plants which formulate and produce drug products for their multinational owners. There are 406 registered manufacturers of Western pharmaceuticals and an additional 261 producing traditional Chinese herb medicines as of June 1989.

In a recent survey of 240 pharmaceutical establishments by the Industrial Technology and Research Institute (ITRI), only 40 had a registered capital of at least NT $50 million (e.g., approximately U.S. $2,000,000), while 22 had 100 or more employees (Table 4 and Table 5). It should be pointed out that ITRI's survey focused on GMP manufacturers and a limited number of bulk pharmaceutical manufacturers.

TABLE 4. Registered capitals and market share of Taiwan's pharmaceutical establishments, ITRI survey, 1988.

Registered Capital (million NT$)	Number of Establishment	Sales Revenue (million NT$)	Market Share (%)
>100	19	7500	40.7
50 -100	21	3300	18.0
10 -49	100	7000	38.0
<10	35	600	3.3
Total	175	18400	100.0

Source: Taiwan Economic Research Monthly, December 1988, p. 40.

TABLE 5. Number of employee and market share of Taiwan's pharmaceutical establishments, ITRI Survey, 1988.

Number of Employee	Number of Establishment	Sales Revenue (million NT$)	Market Share (%)
200 - 400	6	4400	23.9
100 - 199	16	4000	21.7
50 - 99	43	6500	35.3
<50	110	3500	19.1
Total	175	18400	100.0

Source: Taiwan Economic Research Monthly, December 1988, p. 39.

2. Research and Development, Imports and Exports, Patents, Licensing

Research and development (R&D) activities in the industry are limited to formulation studies; it was estimated that only about 0.5% of annual production revenue is devoted to R&D expenditures. Government research grants are a major source of research funding in Taiwan. A gradual increase in pharmaceutical R&D has been advocated by various government ministries, such as the Ministry of Economic Affairs, Department of Health, and National Council of Science. Biotechnology has been selected as one of the major strategic industries to be promoted, and a government-sponsored Development Center for Biotechnology was established in 1985 for such a purpose.

Imports and Exports

Most bulk pharmaceuticals are imported, and the top suppliers are Japan, Italy, the U.S., Spain, and West Germany. Imported finished products represent approximately 50% of the market, mainly from Italy, West Germany, the U.S., Switzerland, and the United Kingdom. In contrast to the mounting trade surplus against other countries, Taiwan has been continuously relying on imported pharmaceuticals, especially new chemical entity products. Statistics indicate that the proportion of import value of pharmaceuticals versus exports has been kept steadily at 5-6:1, and this level may be sustained in the foreseeable future (see Table 6).

TABLE 6. Pharmaceutical import and export of Taiwan in NT$, 1983-1987.

	1983	1984	1985	1986	1987
Import	6,457M	7,196M	7,271M	9,007M	8,215M
Export	1,463M	1,395M	1,280M	1,560M	1,489M

Source: Import and Export Statistics Monthly.

Patents and Licensing

The Patent Law of Taiwan was revised to grant product protection per se on December 24, 1986. Pharmaceutical products were not patentable prior to that revision, and only process patents are granted currently. Because the revision was not retroactive, new chemical entities will not be practically protected by the patent law until the 1990s due to the time lag between product research and regulatory approval.

Licensing of products to local manufacturers is not mandatory, since it can be arranged through a technical cooperation agreement among the manufacturers involved.

3. Manufacturing

Pharmaceutical establishment registration is required for any manufacturing activity related to pharmaceuticals. Compliance with GMP guidelines has been required since 1982, when a 5-year program was instituted for manufacturers to fully comply with the regulation. There are 222 certified GMP manufacturers.

4. Advertising and Promotion, Price Regulation

Pharmaceutical advertisements in the public media require prior approval by provincial or municipal health authorities, while prescription drug product ads are limited to professional journals or magazines.

Promotional activities to the medical and pharmacy professions have not been regulated, although the pharmaceutical law stipulates that professional sales representatives shall register with local health authorities.

Promotional materials such as pamphlets, brochures, etc., are not very effectively controlled due to insufficient enforcement manpower employed by local health authorities. There is no regulation on drug pricing except for drugs reimbursed under various insurance programs.

5. Drug Approval and Government Regulation

Taiwan's pharmaceutical industry is basically a generic industry, and most of the product applications are for generics of local and foreign origin. The Bureau of Drugs in the Department of Health receives an average of 1,000 generic applications per year. Registration records indicate diversified sources of supply from nearly 2,000 manufacturers in 50 countries. Such flooding of imported generics has been better contained since the implementation of GMP. The Bureau of Drugs also processes approximately 50 new drug applications (NDA) each year, mostly from multinational drug companies. This application requires documentation of pharmacological and toxicological effects, physical and chemical characteristics, and testing specifications and methods, together with clinical documentation of properly controlled double-blind clinical studies with explicit criteria for judgment and statistical analyses. Local clinical trials are not mandatory; instead, at least two official approvals from ten NDA reference countries are required for NDA review in lieu of local clinical trials. The ten NDA reference countries are: the U.S., Canada, West Germany, the United Kingdom, France, Switzerland, Belgium, Sweden, Australia, and Japan. A drug advisory committee consisting of clinicians, pharmacologists, pharmacy scientists, etc., is organized to coordinate the NDA reviews. An average approval of 25-30 NDAs per year illustrates implementation of the NDA country reference system.

6. Competition: Brand and Generic

Prior to the 1986 Amendment of Patent Law, product patents per se did not effectively exist because only process patents were legally protected. Generic versions of new chemical entities were limited because of the GMP program and the establishment of a postmarketing surveillance system in 1983.

Because hospital revenue has been closely interwoven with the utilization of drugs, markup on drug procurement represents approximately 50% of revenue derived from this source. Public hospitals have been induced to prescribe brand products more frequently than generics, while private sector practitioners prescribe more generics than brands, mainly due to hospital revenue and operating cost concerns and the fact that drug reimbursement is not billed independently.

The pharmacist law permits dispensing pharmacists to substitute other products of the same dosage form and effective ingredient quantity for prescribed products.

7. Cost-Containment Activities

Nearly all current insurance programs have been criticized for under-paying services rendered which, in turn, results in over-utilization of drugs because the large markup on procured drugs becomes the major source of revenue for the hospital. Cost containment for drug reimbursement in insurance programs focuses on classifying drug products into original brands, branded generics, and other generics, with price ceilings set for each class of drugs. Drugs will soon be reimbursed based on the acquisition cost plus a fixed dispensing and management fee instead of a flat percentage markup on procurement cost.

C. PHARMACY PRACTICE

1. Community Pharmacy Practice

The pharmacists or assistant pharmacists in practice do not reflect the actual output of the pharmacy education system in Taiwan because of the very high percentage of manpower loss due to the existence of an unfavorable practice environment. The discrepancy between pharmacy manpower licensed and in practice does not take into account pharmacy practitioners in public health services, hospital settings, and manufacturing facilities. It is estimated that there are approximately 3,000 additional pharmacists practicing in public health services, hospitals, and manufacturing plants, with at least two-thirds of them in public service and the hospital sector.

The number of persons in community practice can be classified further into two types of practice: those who own and operate pharmacies and those who are employed by drugstores (Table 7).

TABLE 7. Pharmacy manpower distribution in community practice.

	Pharmacy	Drugstore
Pharmacist	2,945 (21.0%)	3,389 (24.2%)
Assistant Pharmacist	3,958 (28.2%)	3,736 (26.6%)
Subtotal	6,903 (49.2%)	7,125 (50.8%)
Total	14,028 (100.0%)	

Source: Department of Health, June 1989.

2. Hospital Pharmacy Practice

Since the community practice environment has been deteriorating, hospital pharmacy practice has become a land of escape for pharmacy professionals. In 1987, a hospital accreditation program was initiated by the Department of Health on a voluntary basis, following the Effectiveness of Medical Care Law, which classified medical institutions into four categories: medical centers, district hospitals, regional hospitals, and clinics. The first phase of hospital accreditation, which excludes clinics, involves a total of 170 hospitals, 11 medical centers, 38 district hospitals, and 111 regional hospitals. According to the Department of Health, 85% of these hospitals complied with the standards.

Pharmacy practice represents 10% of the accreditation weight. It includes pharmacy personnel in terms of quantity and quality, pharmacy operation, pharmacy management, teaching, and continuing education. It was found that hospital pharmacy services remained relatively traditional and could not achieve a satisfactory professional level in view of the rapid advancement of health care sciences and technology and improvement of hospital facilities over the past decades. The failure to promote competent professional services in pharmacy and to develop clinical pharmacy programs could be attributed to an overall negligence of hospital management to improve the pharmacy department amid an ever-increasingly complicated medical care system.

A survey of 79 major hospitals in October 1988 by the Hospital Pharmacist Association presents an overall picture of hospital pharmacy practice in Taiwan (Table 8).

3. Dispensing a Prescription: Where, What, Payment, Formulary

Pharmacy practice in hospital settings has remained static in dispensing prescriptions by disregarding the rapid progress of medical care technology and the increasing demands of patient-oriented practice. Current pharmaceutical laws permit medical doctors to dispense their own prescriptions under designated conditions, which leaves community pharmacy practice with nearly no prescriptions to dispense. Prescription dispensing in most hospitals involves conventional counting or compounding, but medical centers and some district hospitals have established unit dose dispensing. Only recently have medicines dispensed by certain hospital pharmacies been labeled with product names and dosage, and medical practitioners have been very reluctant to inform patients about the medicines prescribed for them.

TABLE 8

Scope of Practice	Percentage of Hospital Pharmacy Involved
1. Prescription Dispensing	
1) Unit-Dose Delivery	24% (Partial); 7.5% (Complete)
2) Computer-Aided Dispensing	20% (Partial); 25% (Complete)
3) Out-Patient Prescriptions	70%
4) Patient Consulting	38%
2. Total Parenteral Nutrition Dispensing	10%
3. Cancer Chemotherapy Dispensing	5%
4. Drug Information Services	66%
5. Therapeutic Drug Monitoring	8% (by Pharmacy Department); and 33% (by Medical Technology Dept.)
6. Drug Utilization Review	12.5%

Insurance reimbursement does not include a professional pharmacy service fee, as patients are required to have their prescriptions dispensed in the same hospital or clinic where they received physician service. Indeed, current insurance reimbursement mechanisms prohibit payment if the prescription is dispensed outside the hospital or clinic setting. Most medical centers and district hospitals have their own hospital formulary, while others have drug listings to a certain extent. There is no national formulary except for the third edition of the Chinese Pharmacopeia, which was officially published in 1980.

4. Record Keeping

Individual patient drug profiles are not available in most hospitals or pharmacies. Moreover, in most clinics and some hospitals, prescription dispensing is performed according to medical orders written on patient medical records without separate issuance of a prescription.

5. Patient Education

Patient education on drug-related knowledge is limited to instruction for drugs dispensed. Very few hospitals allocate pharmacists for general patient education because the heavy load of prescription filling prohibits the pharmacy department from rendering additional professional services. In most cases, patients are not even informed of the medicines prescribed and dispensed, while information on drug bags or bottles is not adequately presented.

6. OTCs and Other Classes and Schedules of Drugs, Other Nonpharmacy Outlets

Drug products are classified into three categories: prescription-only medicine, medicine under professional instruction, and OTCs. Medicine under professional instruction is available in pharmacies and drugstores without a physician's prescription. There is no controlled substance scheduling in addition to the regulation on narcotics. Psychotropic drugs such as secobarbital, amobarbital, triazolam, flunitrazepam, lormetazepam, estazolam, alprazolam, temazepam, tramadol, brotizolam, ephedrine bulk, and phenylbutazone bulk have been closely monitored in their import, manufacture, and distribution on a batch-by-batch basis to help prevent abuse.

7. The Use of Technicians and Others

Poor planning of the pharmacy education system during the 1950s has resulted in the production of pharmacy personnel on two different levels: pharmacists and pharmacy technicians (or assistant pharmacists). Because the latter are not required to practice under a pharmacist's supervision, the real difference exists only in the public image of practice (assistant pharmacists are not permitted to dispense prescriptions containing narcotics or to sell poisonous drugs). The actual environment of professional practice, though, does not have much impact, as physicians seldom release their prescriptions to patients since current law permits physicians to dispense their own prescriptions. There is also concern that computer-aided dispensing could further downplay the role of pharmacists in hospital settings and, in turn, limit the potential development of clinical pharmacy services.

8. Pharmacy and Primary Care

Pharmacist participation in primary care is minimal; blood pressure measurement has been strictly interpreted by the health authority as a semimedical practice, which usually leads to diagnosis and selling of antihypertensive agents without a physician's prescription.

9. Drug Insurance and Third Parties

Nearly all insurance programs are administered by government agencies. Drug coverage has been quite comprehensive, as prescriptions for both outpatient visits and hospitalization are reimbursed without much containment, monitoring for fraud and abuse, or control of administrative expense. This is one of the major factors that results in a relatively high percentage of drug expenditures in overall medical bills.

D. UNIQUE OR INTERESTING FEATURES OR SPECIAL SITUATIONS IN TAIWAN

Traditional Chinese medicine plays a significant role in health maintenance for the public in Taiwan. However, professionalism related to Chinese medicine has been complicated by a lack of established education for professional training. Patients seek Chinese medicine services as a complement to Western medicine. Regulation of Chinese herbalists involves a direct conflict of interest, as pharmacists regard the retailing and dispensing of Chinese herbal medicines as part of the professional pharmacy domain, but traditional Chinese herbalists are trained through apprenticeship without formal pharmacy education. As of June 1989, there were 1,470 registered pharmacists practicing Chinese herbal medicine, including dispensing of granules made from concentrated herbal extracts, in addition to Western pharmacy practice. There are also 6,027 Chinese herbal medicine retailers who mostly engage in Chinese herbal material transactions; however, 999 are under the supervision of registered Chinese medicine doctors. The coexistence of professional and commercial practice has become a hotly debated issue as reimbursement for traditional Chinese medicines is discussed for the upcoming national health insurance program.

E. CONCLUSIONS

The pharmacy profession has been living in an unfavorable environment for professional practice. There are four barriers which the pharmacy profession has to tackle with great effort to establish the professionalism of pharmacy.

1. Physician's Right to Dispense

Although prescription dispensing by physicians has been an integral part of medical practice for decades — long before the establishment of the pharmacy profession in this island country — it has become a vested interest since the revenue generated from prescription dispensing compensates for the underpayment of medical services. Worst of all, physicians in private clinics seldom perform dispensing by themselves as required by the law; this function in clinic settings is usually carried out by unqualified personnel. This is detrimental to the professionalism of pharmacists and increases the risk to patients.

2. Poorly Planned Pharmacy Manpower Supply

Poor planning of the pharmacy education system during the 1950s has resulted in an excessive manpower supply and poor practice environment for pharmacists. There are four levels of pharmacy education that have produced pharmacists or assistant pharmacists of questionable quality in numbers far exceeding the demand until the year 2000. Graduates from junior pharmacy colleges, e.g., 6 years after ninth grade, have become the major source of pharmacists since 1980. For example, in 1989, graduates from schools of pharmacy numbered only 250 per year versus 600 from junior colleges; there are an additional 130 pharmacy graduates from night schools of pharmacy. There were 12,000 assistant pharmacists graduated from pharmacy vocational high schools during 1960-1980. Pharmacy seems doomed to an oversupply of manpower; only time can tell how to balance this excess with the less than 1,000 medical doctors graduated per year.

3. Nightmare of Drugstores

Legally speaking, a pharmacy or drugstore owner is not necessarily a pharmacist or assistant pharmacist. He or she may employ one as required by pharmacy law, but an employer of a drugstore can very easily function as a de facto pharmacist on site, while the pharmacist in charge is nothing but a ghost who collects the fee for renting a license. This inherited practice of a drugstore can be shown in the statistics of Table 8, where 50.8% of drugstores are of this type. Regulatory efforts to overturn such vested realism or strictly enforce the presence of a pharmacist on site have met strong resistance from drugstore owners. As there has not been a supportive environment for professional pharmacy practice for four decades, pharmacists are more inclined to practice in hospital settings and forsake the prestige of serving the public in community pharmacy. It is a Catch-22 dilemma that no one likes.

4. Dilemma of Chinese Herbalists: Professionals or Businessmen?

Pharmacists have long been educated in the utilization of natural herbal materials. Moreover, traditional Chinese medicine has many features in common with the principles of pharmacognosy, yet the way these agents are prescribed by Chinese medicine doctors is substantially different from modern medical practice. In fact, natural herbal materials are seldom regulated by most Western countries, or even in Hong Kong and Singapore where the majority of the constituents are of Chinese origin. Here, Chinese herbal materials are regulated as general agricultural commodities.

Pharmacists in Taiwan regard the handling of Chinese herbal medicines as their responsibility, partly due to professional motives and, undoubtedly, partly because of commercial interests. This issue has become more controversial concerning the demarcation between professionalism and commercialism. Moreover, the transaction of Chinese herbal materials has been complicated by cultural and folk concepts about Chinese herbal materials.

* * *

With the approaching implementation of a national health insurance program in 1995, the redistribution of health care resources is entering a turning point. This is perhaps the last opportunity to reestablish a more reasonable medical practice model in which physicians diagnose and prescribe and pharmacists dispense. Pharmacist organizations strongly support the return to pharmacy professionalism, but the pharmacy profession has been long blocked by the four barriers mentioned above. Yet there seems no consensus among pharmacists as to how to bring down these barriers. The pharmacy profession in Taiwan does not remember that the role of this profession is to provide professional knowledge for the betterment of the public in the use of drugs. Unless pharmacists in Taiwan can justify their expertise in drug knowledge to protect the public, these barriers surrounding the pharmacy profession will continue to isolate them from the public.

SELECTED READINGS

M. C. Chang, "Summary Report of Pharmacy Practices in Hospital Accreditation Program, 1988," Hospital Pharmacy 1989; 6: 1-35.

Y. Y. Hsieh and W. F. Huang, "The Drug Regulatory Process of the Republic of China," Journal of Clinical Pharmacology 1988; 28: 200-203.

C. H. Hsieh, et al., Trends and Strategy of Medical Care and Financial Operations in Labour Insurance and Government Employee Insurance, Research, Development and Evaluation Commission, Executive Yuan, December 1985.

Weng F. Huang, "Opinion Survey of Pharmacists About On-Site Practice," Research Project of Bureau of Drugs, Department of Health, December 1987.

Peter Kaim-Caudle and Tom Yee-Huei Chin, Cost and Provisions of Health Services in the Republic of China on Taiwan 1979-1981, Research, Development and Evaluation Commission, Executive Yuan, February 1983.

F. S. Lee and W. N. Yu, "Current Status and Assessment of Pharmacy Administration in Taipei County," Research Project DOH 78-65, Department of Health, Executive Yuan, Taipei, Taiwan, July 1989.

Y. H. Lin and H. Y. Chen, "Competition of Pharmaceutical Industry Between Taiwan and South Korea," Taiwan Economic Research Monthly, December 1988, p. 39-46.

W. N. Yu, W. F. Huang, and Y. T. Hong, "Survey on Quality of Services in Taiwan's Community Pharmacies," Public Health 1988; 15: 139-156.

Health and Medicine in Taiwan Area, Republic of China, Department of Health, Executive Yuan, Taipei, Taiwan, December 1988.

Workshop in Impact of National Health Insurance on Pharmacy, Kaoshiung Medical College, October 15, 1989, Kaoshiung, Taiwan.

Establish the System of Clinical Pharmacy — Report of the Hospital Pharmacist Association of R.O.C., Research Project DOH 78-35, June 1989.

Chapter 22

United Kingdom

Peter R. Noyce
Jeannette A. Howe

A. NATIONAL HEALTH CARE SYSTEM

Britain was the first country in the world to provide universal coverage of free health care, i.e., to the whole population. The National Health Service (NHS) started on July 5, 1948. It not only made services free to patients, but also allowed a more even distribution of health care. Several separate strands of provision and resource were brought together.

1. History and Background

In 1911, a National Insurance Act was adopted, which required lower paid workers to have compulsory medical insurance for primary medical and pharmaceutical services. The scheme covered only employees, *not* their families or others. These were obliged to belong to a separate insurance scheme on a voluntary basis or pay private fees for medical care. For the poor, the choice was either to seek attention at the casualty department of a charitable hospital, if one was in the vicinity, or to submit to a means test for admittance to care by the poor law doctor.

In 1919, the Ministry of Health was established, and from 1929 local municipal authorities were given the authority to provide the full range of hospital services. However, up until 1939, the Ministry of Health had no direct command over the development of health services, and local authorities were largely free to accord their own priority to health care; health provision across the country was patchy. Much of the acute hospital care was provided through charitable hospitals, relying on donations and the free services of specialist medical staff who derived their income from treating private patients elsewhere. Local authorities provided the vast majority of hospital care for the chronically sick, and those with infectious diseases, mental illness, and mental handicaps. Hospital obstetric services were also generally provided by local authorities who had responsibility for ante- and post-natal clinics and welfare centers and employed health visitors and midwives. Nationally 80% of beds were under local authority responsibility.

Dental services, where much of the activity was directed towards tooth extraction rather than maintenance, were limited to nursing and expectant mothers and young children. Some other services (e.g., developmental screening) were provided through the School Health Service. Ambulances were operated by a variety of organizations, including local authorities.

World War II led to the establishment of the Emergency Medical Service, which involved the government paying voluntary hospitals to provide services. From this arrangement, the advantages of rationalization and coordination of both private and public sector hospitals soon became clear to all.

Following the Beveridge Report in 1942 and a simultaneous report from the medical profession urging the establishment of a centrally planned public health service, the government produced a White Paper (policy statement) proposing a National Health Service. In 1946, the necessary legislation was laid by Aneurin Bevan, the Minister of Health in the post-war government.

*The Inception of the National
Health Service (NHS)*

The institution of the NHS saw a tripartite structure covering primary care, hospitals, and community services. Primary care was provided by independent professional contractors to give family doctor, dental, pharmaceutical, and ophthalmic services. The entire population was entitled to free care through registration with a family doctor. Primary care contractors were administered through 138 Executive Councils across England and Wales, much as Insurance Committees had operated under the 1911 Act. Primary care accounted for approximately one-third of the cost of the NHS in 1948.

The vast majority of hospitals were transferred to national ownership, totaling some 2,700 hospitals in England and Wales which collectively contained nearly 500,000 beds, of which 190,000 were devoted to the care of the mentally ill and mentally handicapped. Fourteen Regional Health Boards (RHBs) were established and appointed by the Minister of Health in England to administer hospital services through approximately 400 Hospital Management Committees (HMCs). The 36 teaching hospitals in England and Wales had separate boards of governors, directly responsible to the Minister of Health. Usually, separate HMCs were established for hospitals for the mentally ill and mentally handicapped. In the care of the elderly, separate facilities were arranged for the chronically sick, under the NHS, and those requiring residential facilities, which remained the responsibility of municipal authorities. The administrative structure of NHS in 1948 is shown in Figure 1, and early NHS statistics for the United Kingdom are given in Table 1.

Within hospitals, a tripartite administrative structure generally evolved, involving a clerk (administrator), matron (chief nurse), and chairman of the medical staff committee (senior medical specialist), all having direct access to the HMC itself. Approximately 50% of NHS costs in 1949 were expended in hospitals, to which referral, except in emergencies, was through family doctors.

Community services, such as health visiting and midwifery, and ambulance services remained under the direct control of local municipal authorities.

The First 25 Years

In 1951, as part of a policy for applying ceilings to NHS expenditure, charges were introduced for spectacles and dentures. During 1952, prescription charges and charges for some forms of dental treatment were also levied. Prescription charges were abolished in 1965, reintroduced in 1968, and increased in 1971. Throughout the 1960s, substantial resources went into building programs for hospitals, alongside efforts to rationalize design and improve the economics of construction and use of hospitals. There were also major increases in NHS manpower, particularly of nurses. The proportion of the U.K. gross national product spent on the NHS increased from 3.8% in 1960 to 5.3% in 1974, much of it going toward the hospital sector.

In 1968, the Ministries of Health and Pensions were joined to form the Department of Health and Social Security (DHSS). In the late 1960s, significant problems were being encountered in the coordination of the various components and bodies within the NHS. The Labor government of this time produced two policy discussion documents on restructuring the NHS. It was left, however, to the next Conservative Government to devise a new administrative structure for the NHS. But it was actually a Labor government, in 1974, which implemented the first major reorganization of the NHS. By this time, the balance of expenditure had swung toward the hospital sector, with 72% being spent on hospital and community health services and 20% on primary care.

FIGURE 1

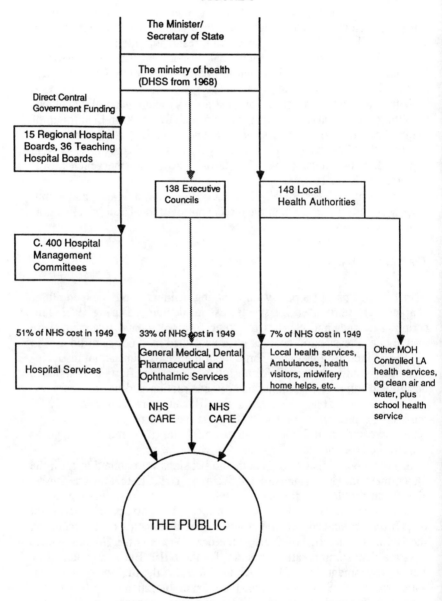

TABLE 1. Statistics for early years of NHS.

UK Population	49.5m	(1949)
Total expenditure on NHS	£437m	(1949)
Gross expenditure on NHS (as % GNP)	3.92	(1949)
Total No. of hospital beds	542,000	(1951)
Total No. of staff employed in NHS hospitals	406,750	(1951)
Hospital Activity:		
Discharges and deaths	3.81m	(1951)
Out-patient attendances	45,810	(1952)
Primary Care:		
Total cost of FPS service	£145m	(1949)
Total No. of prescriptions	225.1m	(1949)
Total cost of pharmaceutical services	£35m	(1949)

The 1974 Reorganization

The main effects of the 1974 reorganization were to convert RHBs into Regional Health Authorities (RHAs) with executive authority, and to establish under them a further executive tier of 90 Area Health Authorities (AHAs). These Health Authorities (HAs) adopted the responsibility for community services from Municipal Authorities and combined these with hospital services, including those provided by teaching hospitals. Primary care services (known more commonly as family practitioner services) were also reorganized under 90 Family Practitioner Committees (FPCs), each coterminus with the respective Health Authority. A chain of corporate responsibility was introduced from AHA through RHA to the Secretary of State for Health (Senior Minister). Areas were further subdivided operationally into districts, each with a District Management Team.

This reorganization was also characterized by the standardization in the structure of management teams of officers and the consensus style of management expected of these teams. At each tier, the core membership of these teams was composed of an administrator and finance, medical, and nursing officers. The medical representation was large at the district level, and professional advisory machinery existed at both AHA and RHA levels. Pharmaceutical officers were appointed at regional, area, and district levels.

Health Authority membership included a substantial municipal authority representation, together with professional representatives, and Joint

Consultative Committees were established to coordinate health and social care between Health and the local Municipal Authorities. Community Health Councils were statutorily established to represent the interests of consumers at the district level. The administrative arrangements for NHS following the 1974 reorganization are shown in Figure 2.

Planning As a Priority

Planning became a priority in the system of delegated authority and consensus management in the mid-1970s. An integrated system for strategic and operational plans allowed HAs to plan and manage the demand for services locally within the policies and priorities set by government. The proportion of gross national product spent on health services rose from 5.3% in 1974 to 5.7% in 1976.

In 1976-1977, the main DHSS themes were greater emphasis on preventative services and community care and priority for the development of services for the mentally ill, mentally handicapped, physically handicapped, elderly, and children. Also in the late 1970s, a system of resource allocation, based essentially on population and comparative deprivation data, was introduced.

The Need for Stronger Management

The particular concern of the Conservative government elected in 1979 was to improve NHS management. DHSS set about developing a standard management information system; introduced annual review meetings between Ministers and RHAs; and established a monitoring system based on performance indicators in clinical, financial, and personnel functions.

A further reorganization of the administrative arrangements for the NHS was undertaken in 1982. The dominant themes were simplicity, devolution, and flexibility, but involving a strengthening of accountability. AHAs were scrapped, and executive authority was devolved to District Health Authorities (DHAs) with the emphasis on operational management being moved to still smaller units. The intention was to create a system for health care more sensitive to consumer needs and, at the same time, to cut the cost of administration. DHAs were largely left to establish their own managerial arrangements, encompassing the philosophy that managerial decisions should be made at the lowest level practicable.

In 1983, the first tentative moves toward a business culture in the NHS were made by encouraging Health Service Managers to seek competitive tenders for hospital hotel services, which had previously always been provided by direct labor. Much more significantly, an independent NHS

FIGURE 2

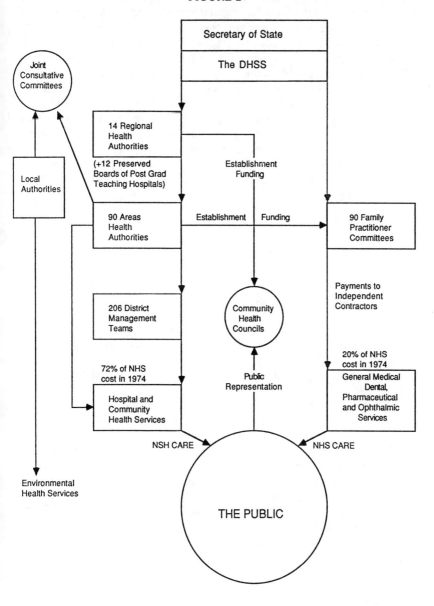

Management Inquiry was undertaken by Sir Roy Griffiths which was to have far-reaching effects. It highlighted the intrinsic weaknesses of management by consensus and stressed the need for competent, individually accountable managers to become responsible for running NHS. It led to the appointment of general managers at all levels in the NHS in 1985, and within DHSS, the establishment of an NHS Management Board chaired by a Chief Executive. The annual review process, set against DHSS service objectives and policy aims, was extended down to unit level. The inquiry also stressed the need for tight budgetary and management information systems and for involving clinicians in the management process. Indeed, a number of clinicians became general managers.

General managers established directorate teams with financial, personnel, estate, and quality assurance responsibilities. Performance-related pay has become the basis of remuneration for senior NHS managers.

Health Care Outside Hospitals

During this process of strengthening management, quiet but fundamental change had been occurring in the care of priority groups (i.e., the mentally ill, learning impaired, and elderly) identified for attention in 1977. Throughout the time of the NHS, and even before that, the care of these clients had centered on long-term institutions. Now, the shift was to care within the community, with the development of residential facilities and social care under local authorities, underpinned on the health side by primary care services and day-care facilities.

Most of the changes in the NHS have focused on the hospital side. However, in 1987, the government published a White Paper entitled "Promoting Better Health," following the first comprehensive review of primary care services in the NHS. Gross cost of these services in 1987-1988 had reached £5,080 million. The policy aims were to make services more responsive to the consumer and to provide a wider range of choice, promote health and prevent illness, raise standards of care, and improve value for money. Patients, as a result, will receive more information about family doctors and the services they provide, and it will be easier for them to change doctors. To increase the consumer sensitivity of family doctors, the proportion of their income derived from capitation will increase. Fees will also be paid when specified levels of vaccination and screening are achieved, and remuneration structures will encourage the raising of standards and range of care provided. Community pharmacists will be paid for performing a wider role than simply dispensing. Dentists will be encouraged, through their remuneration system, to concentrate on preventative dental services. These developments are being financed by requiring those who can pay to bear the cost of sight tests and to make a contribution toward dental examinations and treatment. All these changes will occur alongside a strengthening of management of primary care services by FPCs.

In 1988, DHSS split into separate Departments of Health (DH) and Social Security (DSS). The DH Ministerial Team is now composed of the Secretary of State for Health, who is of Cabinet rank, supported by two other Ministers, the Minister of State for Health and the Parliamentary Under Secretary for Health.

The Future

Following a comprehensive review of all aspects of the NHS during 1988 and building on the reforms in primary care already begun, the latest White Paper, entitled "Working for Patients," was unveiled in January 1989. This was followed the next month by eight Working Papers covering the detailed aspects of the reforms. The major reforms planned, mainly for implementation in 1991, include the following:

a. The introduction of competition between hospitals within the NHS, while services remain free at the point of delivery to the patient. This will be achieved, in part, through the establishment of hospitals run by self-governing trusts, which will be free to determine the range of specialist services they provide, the staff employed, and the salaries paid. HAs will be able to purchase services from NHS hospitals outside their locality, from private hospitals, as well as their own NHS hospitals. They will also have a duty to ensure the collective provision of a full range of services that meet acceptable quality criteria.

b. The introduction of common management and scope for full integration of all aspects of the NHS in 1990, through making FPCs, as well as HAs accountable to RHAs. Budget ceilings will also be introduced for primary care services for the first time in the history of the NHS.

c. Large group practices of family doctors, who will remain NHS contractors, will be allocated individual budgets under the NHS to provide and purchase services.

d. All family doctor practices will be accorded indicative drug budgets, within which they will be expected to contain their prescribing costs. Computerized individual prescribing statistics are already routinely made available to all family doctors. Both the development of practice formularies and medical audits are being encouraged. These initiatives — jointly — are expected to exert a downward pressure on drug expenditures in primary care.

e. The membership of HAs is to be reconstituted to comprise executive and nonexecutive members, rather than being largely representative bodies.

2. Constitutional and Political Environment

The U.K. is a political and economic union of England, Scotland, Wales, and Northern Ireland (Ulster). The Channel Islands and the Isle of Man retain their own legislative assemblies. Each of the home countries has a separate administrative arrangement, and in Scotland, an independent legal system. A Senior Minister represents each home country, government, ministry and department in the British government at the Cabinet level.

The U.K. is a constitutional monarchy and since 1972 has been a member of the European Economic Community (EEC). By 1992, barriers to this internal market are intended to be removed to facilitate the free movement of individuals, goods, and services across the 12 member states (Belgium, Denmark, France, Greece, Holland, Ireland, Italy, Luxembourg, Portugal, Spain, the U.K., and West Germany).

For most of this century, Britain has been governed by one of the two main political parties: Conservative (Tory) or Labor (Socialist). Recently, there has been some reemergence of a central, Liberal and Democrat political grouping. Until recently, the government was Conservative, under the premiership of Margaret Thatcher.

Social Trends

The U.K. population has remained stable over the last ten years at around 56 million, but the proportion of elderly and ethnic minority groups has increased, which has particular implications for the burden and delivery of health care. In the mid-1990s, the number of school leavers troughs, which is of particular concern in staffing the NHS.

Social trends over the last 25 years suggest an increasing need for social support. The number of single-person households has doubled in 25 years (now 25%), the number of single-parent families has doubled in 15 years (14%), and the number of illegitimate births has doubled in 6 years (23%). The U.K. has the highest marriage and divorce rates in the EEC.

The social aim continues toward equal opportunities for males and females in educational achievement, job opportunities, and pay.

There has been a significant shift away from manufacturing to service industry over the last ten years. The unemployment level is steadily and consistently dropping, even though significant levels may remain in particular localities or ethnic groups. The latest Organization for Economic Cooperation and Development (OECD) level is 8.0% (up to August 1988) against a 1983 annual average of 12.5%.

There has been a move to lower the personal income tax. Currently, a married man is allowed an annual income, after allowable expenses of £4,095 before taxation, then the lower tax rate of 25% applies for the next £19,300 and a higher rate of 40% for the remainder. The standard rate of purchase tax (value-added tax — VAT) on goods and services is 15%. Real household disposable income is, on average, estimated to have risen 3% per year over the last five years (i.e., up to 1987).

3. Public and Private Financing, Health Costs

Expenditure on health care under the NHS in the U.K. is given in Table 2, and Table 3 indicates the proportional sources of NHS funds.

The Consolidated Fund is derived from income tax, excise duties, VAT, and other tax revenue sources. National Insurance, which is payable by all employed persons and their employers (e.g., at an annual salary of £10,000, both the employer and employee contributions are 9%), has made a consistent contribution to NHS costs over the last 20 years. Three percent of NHS finance originates from charges to patients for prescriptions, dental treatment, and ophthalmic services.

Expenditure upon the FPS represents 24% of total NHS expenditure and

TABLE 2. Public expenditure on health, United Kingdom (million).

	1982-83 outturn	1987-88 estimated outturn	1988-89 plans
Current expenditure			
Hospital and community			
health services	10,228	14,109	14,980
Family practitioner services	3,181	4,724	5,067
Central health and other			
services	407	705	720
Capital expenditure			
Hpspital and community			
health services	837	969	993
Family Practitioner services	9	22	15
Central health and other services	22	40	43
Total Public expenditure on health	14,685	20,569	21,818

1 Figures do not include subsequent in-year additions in1987-88 and 1988-89.
Source: The Government's Expenditure Plans (Cm 288), HM Treasury

Figure 3 demonstrates how these funds are shared among the component services and medicines. Medicines account for 37.6% of NHS primary care expenditure, and the provision of pharmaceutical services a further 10.4%.

Private sector health remains a relatively small component in the U.K. — around 15%. Although the rate of growth of the number of avail-

TABLE 3. National Health Service expenditure by source of finance, United Kingdom.

	1966	1976	1986*
		Million (per cent)	
Consolidated Fund	1102 (77)	5546 (88)	16292 (86)
National Insurance Contributions	166 (12)	605 (10)	2252 (11)
Charges to patients	30 (2)	128 (2)	620 (3)
Local Authorities	134 (9)	-	-
Total	1432	6279	19801

*estimate, Source: Office of Health Economics (1987).

FIGURE 3. Gross cost of the family practitioner services in the United Kingdom 1987/1988.

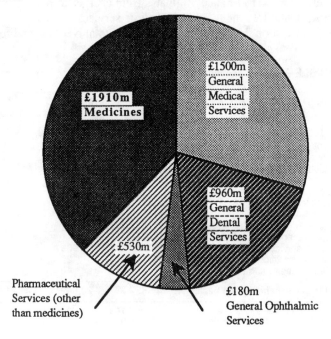

able beds and outpatients has increased substantially (28% and 9%, respectively) between 1985 and 1986, the number of private beds still represents only 16% of total available beds. At the primary care level, there is a much lower provision and demand on the private sector. While all British nationals have access to free NHS facilities, 5.3 million (1987) are now also covered by private medical insurance.

4. Facilities

Total NHS capital expenditure for the U.K. in 1985-1986 was £1,084 million, which represents £19 per head of population. In 1986-1987, 14 major hospital building schemes over £5 million were approved, and 480 schemes over £1 million were in various stages of development in England.

There is a steady decrease in the total number of hospitals, with an annual reduction of about 30 hospitals per year across the U.K. In 1985, there were 2,423 NHS hospitals in the U.K., with a total of 404,000 beds available for inpatient admissions. The greatest reduction has been in the number of beds in large psychiatric hospitals, reflecting the trend toward community care. Shorter inpatient stays and increases in day care have reduced demand for acute care hospital beds.

The Department of Health evaluates expensive items of equipment, and funding for these is allocated on a regional basis. For other equipment, decisions are left to individual HAs.

5. Manpower

Over 1 million people are employed or contracted to provide professional services in the NHS, as detailed in Table 4. Approximately 50% of these are nursing and midwifery staff; 10% are medical and dental practitioners, split almost equally between primary care and hospital services; and other professional and technical staff account for a further 10%.

In 1988, there were 36,779 members of the Royal Pharmaceutical Society of Great Britain (M.R.Pharm.S.) registered to practice pharmacy in the U.K. (There is a separate Pharmaceutical Society of Northern Ireland.) The distribution of registered pharmacists is given in Figure 4, in which the overall male/female ratio of the register is 60:40, but women currently constitute approximately 60% of entrants to the Register. Of pharmacists living in Great Britain (30,717 in 1986), 62% are involved in community pharmacy, 14% in hospital pharmacy, 4.5% in industry, and 2% in academic pharmacy. (Since the free movement of pharmacists between EEC member states came about at the end of 1987, some 50 pharmacists have established reciprocal registration in and out of Britain.)

TABLE 4. Manpower in health (from Social Trend 19, 1989, London: HMSO).

Thousands

United Kingdom	1976	1981	1983	1984	1985	1986
Regional and District Health Authorities						
Medical and dental (excluding locums)	43.8	49.7	51.2	51.4	52.1	52.4
Nursing and midwifery (excluding agency Staff)	429.6	492.8	502.0	500.8	505.2	506.0
Other Staff 2	454.4	485.4	484.7	473.3	462.7	448.8
Total Regional and District Health Authorities	927.8	1,027.9	1,037.9	1,025.5	1,020.0	1,007.2
Family practitioner service	48.7	54.3	56.8	58.2	59.3	57.3
Dental Estimates Board and Prescription pricing Authority / Prescription Pricing Division 3	4.4	4.7	4.6	4.6	4.5	4.0

1: figures for family practitioner services are numbers, all other figures are whole-time equivalents. See Appendix, Part 7: Manpower.
2: Includes professional and technical, administrative and clerical, ancillary, works, maintenance and ambulance staff.
3: In Northern Ireland the Central Service Agency.
(Source: *Department of health*; *Scottish Health Service*, *Common Services Agency*; *Welsh Office*, *Department of Health and Social Services*, *Northern Ireland*)

FIGURE 4. Distribution of membership by sex and age (unadjusted).

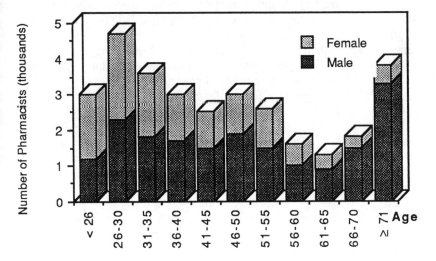

There is a polarization in the populations of pharmacists involved in community practice (older and male) and hospital practice (younger and female). Within pharmacy, as in most health disciplines, there is a considerable component of part-time workers. Recent studies have shown that women pharmacists work professionally for 40%-45% of their potential working hours, and the 4,200 pharmacists engaged in hospital work constitute 2,700 full-time equivalents. Many health disciplines are now looking to modify work practices to increase the contribution of female professionals.

Pharmacy is an all graduate-entry profession in the U.K. There are 16 schools of pharmacy in universities and polytechnics, which collectively graduate approximately 1,000 students annually from 3-year (England) and 4-year (Scotland) courses. Graduates then undergo a 12-month period of supervised practice prior to registration.

6. The Social Security System

The Social Security system provides for pensions on retirement, support when unemployed or incapacitated through ill health or disability, and other situations of financial hardship. Retirement pension, sickness benefit, and unemployment benefit all depend upon an adequate payment of National Insurance (NI) contribution while employed.

In the U.K., state retirement pension has two main components and is payable to women at 60 and men at 65. The basic pension is paid at a single rate of £43.60 per week or £69.80 per couple (1989). The state also requires that an additional pension provision be made on an earnings-related basis, either through NI payments, a formally recognized employer scheme, or an agreed personal pension plan. In addition, individuals may also make their own personal pension arrangements.

A child benefit is paid for each child under 16, at a current rate of £7.25 per week, regardless of parental income or NI contribution.

If a person is unable to work, through illness or disability, for more than four days and up to six months, he or she is entitled either to statutory sick pay from his or her employer, up to a rate of £52.10 per week, or to sickness benefit from the state of £33.20 (assuming adequate NI contributions have been paid). Longer term invalidity allows a basic entitlement of £43.20 per week.

The unemployment benefit is a weekly cash payment of £34.70 for up to a year for people who have lost their jobs and are available to work immediately.

There is a range of benefits to people on low incomes, whether they are working or not, including income support, housing benefits, and the social fund, from which payments can be made to cover exceptional expenses (e.g., fuel during periods of cold weather, or funeral payments).

Also, there are maternity allowances payable, additional support for raising children, and pensions and benefits payable as a result of industrial injuries sustained. All pensioners, children under 16, pregnant women, nursing mothers, and those receiving income support are relieved of paying NHS charges for prescriptions and dental treatment.

7. The Medical Establishment: Entry, Specialists, Referrals

There are 28 schools of medicine in U.K., and each year some 3,600 medical students graduate, usually with the double degree of Bachelor of Medicine (M.B.) and Bachelor of Surgery (Ch.B.). All medical graduates then have to undertake a house year (internship) in hospital clinical practice, under supervision, prior to registration as medical practitioners. The registration body is the General Medical Council, and the professional representative body is the British Medical Association (BMA). The basic choice on registration is to decide whether to go into general practice (in primary care) or to begin a specialist career in hospital medicine.

For entry into general practice, there is a requirement to undertake a three-year vocational training scheme, which involves experience in both hospital medicine and attachment to a general practice as a trainee. National arrangements exist for provision of vocational training and continuing education for general medical practice. The Royal College of General Practitioners (RCGP) takes a lead in setting standards in practice and education.

Hospital doctors are organized into specialist clinical firms, each with a senior specialist (consultant) (e.g., psychiatrist, cardiologist, urologist, etc., in charge). House officers or senior house officers are the most junior grades, attached to firms for periods of six months as part of a process of gaining experience in a variety of specialties related to the individual's career intentions. The next grade is registrar, in which the incumbent undertakes a two- to three-year structured program within a specialty, perhaps working in several hospitals. The most senior "junior" hospital doctor is the senior registrar, who understudies and deputizes for the consultant. Advancement is via a combination of personal qualities, experience, and performance in specialist examinations, usually organized by the specialist Royal College (e.g., surgeons, psychiatrists, etc.). Success in the membership exams may be a requirement for middle range posts, and the attainment of a fellowship a prerequisite for a consultant's post. There is no guarantee of advancement or of ultimately obtaining a consultant's post.

Consultants within a hospital unit are usually organized into broader divisions (e.g., medicine), and the chairman of each of the divisions constitutes the Medical Executive Committee, the chairman of which is the corporate representative of medical opinion in a hospital or unit.

There are 53 recognized medical specialties in the NHS, and RHAs currently determine the allocation and appointment of consultant posts. Consultants are usually appointed on a salaried sessional basis to NHS hospitals, which may either collectively result in a full-time commitment to being wholly employed by the NHS or allow scope for private medicine.

Doctors who have taken the GP route and completed their vocational training have to find vacancy in a general medical practice as an assistant. At this point, the doctor becomes a general medical contractor to an FPC. Increasingly, general medical practice is becoming a group activity, and the number of single-handed practitioners is decreasing. Overall, an FPC controls the number of general medical practitioners it has in each location.

All NHS patients are expected to register with a general medical practitioner who is responsible for their primary medical care. If the patient requires specialist medical advice or care, a general practitioner has the freedom to refer the patient to a consultant of his or her choice. While in the hospital, a patient remains the clinical responsibility of that particular consultant, in a bed or on a ward, unless the responsibility is transferred to another consultant. It is the mutual decision of the patient's general practitioner and the respective consultant as to when clinical responsibility is reassumed by the GP after a period of hospital treatment. In emergencies, patients go without referral to accident and emergency departments of NHS hospitals, from which they may be admitted.

8. Payment and Reimbursement

As a general principle, all hospital staff are direct employees of respective Health Authorities and are paid a wage or salary on a national scale, negotiated through the central Whitley arrangements for determining NHS pay. Recently, rolling contracts and performance-related pay have been introduced for senior hospital managers. There is a long established system of merit awards for medical and dental consultants, who normally hold tenured appointments. Recently, for other professional groups (e.g., nursing, pharmacy) a system of flexible grading based on national pay schemes has been established to allow local determination of pay.

Typical NHS salary bands are:

Staff Group	Minimum Pay £	Maximum Pay £
Medical Consultant	29,700	38,340 (*plus* merit awards of 6,760- 36,420)
Director of Nursing	14,470	25,300
Director of Pharmacy	18,500	28,700
Medical House Officer	10,280	11,600 (with overtime average salary of approx. 16,670)
Third-Year Student Nurse		5,950
Preregistration Pharmacist	5,968	

In primary care, professional services (i.e., medical, dental, optical, and pharmaceutical) are all provided on a contractual basis against a system of fees and expenses agreed nationally. Community nurses are employed by Health Authorities. Of course, many pharmacists working in primary care are employees of NHS pharmacy contractors, and their salaries are individually agreed between the pharmacy and the employee pharmacist.

About half of a general practitioner's income is derived from the capitation fees of patients registered with him or her, and this proportion is to be increased against acknowledged practice standards, including defined levels of vaccination and screening. Practice fees will be allowed a financial ceiling, while specialist procedures will attract individual item of service fees. In the future, practice budgets are to be introduced in large practices and indicative budgets developed for all practices to cover prescribed drugs.

A capitation component is to be introduced into the remuneration package for contracted dentists, with fees for courses of treatment and individual procedures accounting for the main sources of income.

Community pharmacy contractors are paid a professional dispensing fee and reimbursement of drug and container costs for each prescription, as well as for other agreed items of service.

9. Coverage

Apart from a limited range of charges (e.g., prescriptions), health care under the NHS is free at the point of delivery to all U.K./EEC nationals. There is, of course, a finite resource available to the NHS, which results in limits being placed on high-tech equipment and procedures (e.g., transplant surgery) and waiting lists for "cold" surgery (e.g., hernias, hemorrhoids) can be more than a year. Hospitals in the private sector relieve some of the demand in these areas of routine surgery and in techniques such as fertility enhancement which may not receive a high priority under the NHS.

In NHS primary health care in 1987, there were, on average, 1,970 patients per doctor, 3,200 per dentist, and 4,540 per pharmacy.

B. PHARMACEUTICAL INDUSTRY
AND DRUG DISTRIBUTION

Following the discovery of penicillin and the early success of the sulfonamide drugs, Britain has a significant modern history of drug research and innovation, with the development of semisynthetic penicillins by Beecham, discovery of cephalosporins at Oxford, and beta blockers by Sir James Black at Imperial Chemical Industries (ICI). More recently, both cimetidine and ranitidine, the outstandingly successful H^2 antagonists, were discovered in British laboratories (respectively by Smith Kline & French, and Glaxo). Acyclovir, the first effective antiviral, and zidovudine, the AIDS-treatment drug, have both been discovered by the British pharmaceutical company Wellcome.

Many economic parameters also demonstrate the prominence of the U.K.-based pharmaceutical industry. In 1987-1988 of the 20 top-selling medicines worldwide, five were discovered in Britain, including the top three. In 1987, it was third in the league of medicine-exporting nations, after West Germany and the U.S., and just ahead of Switzerland. Medicinal exports totaled £1,621 million and are expected to rise to £1,730 million in 1988. Within the U.K., pharmaceuticals became the second highest contributor to the balance of payments for manufactured goods, after oil. The income from exports over imports for pharmaceuticals for 1987 was £834 million estimated to rise to £890 million in 1988. The U.K. pharmaceutical industry employs 87,000 of the 450,000 people employed in this sector in Europe, which is 5.5% more than France. Pharmaceuticals, followed by electrical engineering, have been the country's fastest growing manufacturing sector during the period 1977-1987.

1. The Structure of the Industry

There are approximately 300 companies involved in pharmaceutical manufacture in the U.K. The industry is dominated by a relatively small number of very large producers, with five companies accounting for one-third of the output. Currently, there are 63 companies covered by the Pharmaceutical Price Regulation Scheme (PPRS), for which the current criteria for participation is an annual turnover of more than £4 million in NHS medicines. The nationality of ownership of companies covered by PPRS is given in Table 5. The large proportion of very small companies includes: generic and OTC medicine manufacturers, which require a relatively low level of research and development (R&D); and foreign companies that have established British subsidiaries.

In 1987, total gross output of the U.K. pharmaceutical industry was £4,877 million, which included imports valued at £787 million. (The gross figure includes intercompany trading, veterinary sales, dressings and appliances, functional services, and other miscellaneous items.) Total home market sales were £2,757 million of which 595 million (22%) were OTC medicines and £2,138 million were prescription medicines. (The latter represent 9.9% of NHS total costs, a proportion that has varied only slightly for the previous 6 years.) Financial statistics for the British pharmaceutical industry in the 1980s are given in Table 6.

TABLE 5. Ownership of companies covered by PPRS scheme.

UK	9
USA	25
German	6
French	6
Swiss	5
Scandinavian	7
Other	5

	63

The total NHS generic market is approximately of £190 million, with 80% supplied from 30 generic manufacturers.

Three to five percent of branded products sold to the NHS are parallel imports, licensed as such in the U.K. from other EEC member states. The current U.K. NHS market for parallel imports is valued at around £70 million.

Some manufacturers distribute their own products directly to retail pharmacies, and many more do so to hospitals. Most distribution, however, is undertaken by pharmaceutical wholesalers. In the U.K., there are approximately 40 full-range wholesale distributors, operating from around 100 depots in the U.K., most of which serve only regional or local areas. There are also a number of short-line wholesalers who sell only a restricted range of fast-moving products. Under PPRS, 12.5% is allowed for the wholesaler distribution margin, and recently, action has been directed toward improving the service benefits from the distribution network to NHS hospitals. Wholesaler discounting is a routine component of trade with retailers, and for medicines dispensed under the NHS, there is a discount recovery component in the remuneration package of NHS pharmaceutical contractors.

2. Research and Development, Imports, Exports

Twelve percent of world pharmaceutical R&D activity is undertaken in the U.K., while the U.K. medicine sales represent only 4% of the world market. In 1987, £668 million was spent on R&D in the U.K., similar to the amounts spent in the aerospace industry. This sum is roughly split: 50% on basic and applied research, 30% on pharmaceutical development

TABLE 6. From "The Pharmaceutical Industry and the Nation's Health," 1988, London: ABPI (in £million).

Year	Gross Uk Pharm Output	NHS Sales	Household (OTC medicine)	Exports	Imports	Balance of Trade
1980	2,422	1,034	246	745	222	523
81	-	1,202	274	852	298	554
82	3,050	1,399	291	978	375	603
83	3,274	1,579	354	1074	470	604
84	3,667	1,688	373	1222	542	680
85	4,030	1,786	451	1426	590	835
86	4,426	1,947	536	1533	679	853
87	4,877	2,138	595	1621	787	834

and 20% on capital investment. It represents just under 14% of output, a stable proportion over the last five years. Eighteen percent of the work force of the U.K. pharmaceutical industry is employed in R&D activities.

The major therapeutic classes for research expenditure are cardiovascular, anti-infective, and central nervous system agents. Twelve new chemical entities (NCEs), were launched in the U.K. in 1987, maintaining the level of innovation of the previous three years. In the period 1985-1987, the most productive areas have been in the anti-infective market and gastrointestinal, cytostatic, and thrombolytic fields, each having four NCEs, closely followed by cardiovascular, musculoskeletal, dermatological, and genitourinary areas.

With the increasing complexity of pharmaceutical research, the rate of appearance of NCEs has attenuated in the last ten years, and the cost and time of bringing a drug to market has increased. This led, in January 1989, to the removal of the 4-year licenses-of-right provision in the current 20-year U.K. patent life for pharmaceuticals. However, the Association of British Pharmaceutical Industry (ABPI) has estimated that the average on-the-market patent life of medicines marketed since 1980 is seven years, but an option exists under an EEC directive for high-tech products to be given market exclusivity for ten years. This option has already been seized upon for cimetidine.

Exports and Imports

The U.K. contributes 12% of the world trade in pharmaceuticals, and around 40% of medicines produced by the U.K. pharmaceutical industry are exported. Pharmaceutical exports totaled £1,621 million, in 1987, with antibiotic products as the largest group. The major export market was the EEC, particularly West Germany, Holland, Ireland, and France, although the U.K. export market is generally broad-based, with no single country being responsible for as much as 10% of exports. Exports to the U.S. were £155 million in 1987 (9.6% of the U.K. export trade) and £91 million to Japan (5.6%).

Imports contribute to 16% of the home market in pharmaceuticals. By far the largest source of imports is West Germany, contributing just under 25% of the total, with a value in 1987 of £195 million. Altogether, EEC member states contribute two-thirds of the British pharmaceutical import totaling £529 million. Imports from the U.S. were £70 million in 1987 (8.9% of U.K. imports) and, from Japan £6 million (0.8%). Although the trade with EEC partners has increased markedly, that with the U.S. has changed modestly in favor of the U.K. in recent years.

There is also a parallel import market in the EEC which accounts for less than 10% of pharmaceutical imports to the U.K. and involves currently about 200 branded products, with a significant difference between U.K. prices and those in other member states.

3. Manufacturing

Capital investment in the pharmaceutical industry is buoyant in the U.K., with a program of £400 million for 1988. Production output in the U.K. is less than that in West Germany and France, but this is primarily due to the smaller home market.

The manufacture of medicines in the U.K. is strictly controlled under the Medicines Act, 1968. Before a medicine can be manufactured or marketed, it must be the subject of a product license, which is issued after the government regulatory authorities are satisfied of a product's safety, efficacy, and quality. A manufacturer's license is also required to authorize manufacture of an approved product. The award of a manufacturer's license is based on the suitability of premises and equipment and the competence of personnel engaged in pharmaceutical manufacture. The British Medicines Control Agency also maintains a Medicines Inspectorate which regularly inspects manufacturing sites and, through statutory powers, maintains production standards. Britain is a signatory to the Pharmaceutical Inspection Convention, which provides for the sharing of information on pharmaceutical manufacturing facilities gained by authorized inspectors in different countries.

Special licensing arrangements cover the preparation and assembly of nonlicensed medicinal products produced to meet specific clinical needs.

The distribution of medicines is also subject to control under medicines legislation. A wholesaler dealer's license is required for the sale of medicinal products to anyone other than the ultimate user.

4. Advertising and Promotion, Price Regulation

Controls on advertising and promotion are also exercised under the Medicines Act of 1968, through a voluntary code of practice operated by the Association of British Pharmaceutical Industry (ABPI), and through a ceiling on promotion costs allowed under the PPRS scheme. Both the medical and pharmaceutical professions in the U.K. also have reservations about advertising of their own professional services.

Under the Medicines Act, the reference statement for promotion of medicinal products is the data sheet. This is a factual and balanced description of the product and its actions that has been approved, on the basis of evidence submitted, as part of the product license application. The information contained in the data sheet includes the name of the product, active constituent, presentation, clinical indication, dose and route of administration, contraindications, side effects, and product license number and holder. (The ABPI publishes a Data Sheet Compendium annually, which is distributed to medical practitioners and pharmacists as a reference text.)

Before an advertisement (e.g., a mailing) is sent to a medical practitioner or a representation (i.e., through a medical representative) is made on a particular product, the company has a responsibility to ensure that the clinician has had an opportunity to see the respective data sheet.

The content of pharmaceutical advertisements in professional publications must represent the product fairly, and the claims made need to be consistent with the data sheet of the product. In particular, a product cannot be promoted for clinical indications other than those contained in the data sheet and, therefore, for which the licensing authority is satisfied of the safety and efficacy in the use of the product.

The ABPI also operates its own code of practice, and claims made against member companies concerning the marketing of prescription medicines are formally investigated. The code is concerned with mitigating misleading representation of medicinal products and guarding against unsubstantiated comparisons between products of different companies. It is also keen to preserve the overall standard of advertising of medicines and draws a clear distinction between promotional material and that for ensuring awareness of important factual information (e.g., on the safety of a product or product recall). Further, particular guidance is given in the code on the acceptable size of promotional gifts and aids, hospitality, and marketing research.

The advertising of prescription-only medicines (POM) to the public is prohibited, along with promotion of the use of medicines in certain serious diseases (e.g., cancer). An EEC directive was passed in 1988 requiring all packs of medicine to contain, in the future, information to patients on the dose and indications of the product. Patient package inserts are now regularly screened by the British regulatory authorities as part of the product licensing process.

Strict product liability also became a legislative reality in 1988. Through this legislation, a plaintiff can sue for damages the supplier of a medicinal product that can be demonstrated to have damaged the patient without evidence of negligence or culpability on the part of the supplier.

The level of original pack dispensing (OPD) is increasing rapidly for branded products, with new products invariably in OPD. OPD is also being extended to generic products, but this is yet to be reflected in the Drug Tariff.

Price Regulation

The main mechanism for price regulations of prescription medicines in the U.K. is the Pharmaceutical Price Regulation Scheme (PPRS) operated by the Department of Health. The first version of the scheme was introduced in 1957, and the present scheme came into operation in October 1986 for a six-year period, with a midpoint review opportunity.

In December 1988, EEC adopted a directive dealing with the transparency of pharmaceutical price control arrangements. National authorities within the EEC have a right to control medicine prices for use in national health care systems, but by 1992 the EEC, as part of a commitment to ensure the free movement of goods, has an obligation to remove impediments between member states with respect to pharmaceuticals and to move toward the closer harmonization of medicine prices. It is also committed to establishing a database on medicine prices in EEC member states. The initial aims of the directive are to identify the various types of control arrangements practiced by national governments, to lay down ground rules on operating price control machinery, and, to require comprehensible explanation of national price control systems.

PPRS controls branded medicine prices indirectly by limiting the amount of profit a company is allowed to make from selling medicines to NHS. Although the aim is to ensure a reasonable price to the NHS, PPRS also incorporates a recognition of the R&D costs of new medicines. A target profit is set for each company, expressed as a percentage return on capital employed, and is currently within the range of 17%-21%. If a company fails to achieve its target profit, it bears the shortfall; there is no guarantee. If a company exceeds its approved profit, it may keep whole or part, providing that it can demonstrate this is due to innovation or increased efficiency. If not, then excess profit is relieved by a reduction in prices or direct repayment to the Department of Health. Apart from the target profit levels, which were increased as above in the present scheme (dating from 1986), other features include the exclusion of generic products, an agreement on efficiency in pharmaceutical operating costs, a specific government undertaking on R&D recognition, a strict limit on promotional spending, introduction of rules on transfer pricing, and agreements on incentives for innovation and efficiency.

The limit on promotional spending has been set at 9% of sales income, and for the industry as a whole, 20% of sales income is allowed for R&D activity. On a worldwide scale, this latter figure is on the high side, but it reflects the very high commitment by the pharmaceutical industry in the U.K.

Since October 1986, generic medicines have fallen outside the PPRS, and price control measures are exacted through the reimbursement levels to dispensing pharmacists for particular products under the NHS. These are listed in the Drug Tariff. The NHS generic market in the primary care sector was approximately £150 million in 1987, and the Drug Tariff contains slightly under 800 preparations. These products are divided into five categories; their allocation is a reflection of the level of use and number of competitive suppliers, and takes account of the market position of products newly out of patent as well as those of declining importance. The basis of pricing, allocation to category, and price of individual products is agreed between the Department of Health and the Pharmaceutical Services Negotiating Committee (PSNC).

5. Drug Approval and Government Regulation

The manufacture, distribution, and marketing of medicines is regulated by a comprehensive system of controls, generally exercised under the Medicines Act of 1968, which became fully operational in 1971. In the run up to 1992, when free trade within EEC is planned to become fully operational, the coverage of directives is being extended to harmonize the regulation of medicines across the EEC and to make arrangements for facilitating the mutual recognition of drug approval across EEC member states.

Detailed guidance on the operational aspects of medicines legislation are provided in MAL (Medicine Act Leaflet) publications, and MAL 99 describes the overall control of medicines in the U.K. Amendments are published in the monthly *MAIL (Medicine Act Information Letter)* series.

The Medicines Commission, comprised of a panel of experts, has been established under the Medicines Act to advise Health and Agriculture Ministers on the issue of licenses for the production, importation, and marketing of human and veterinary medicines. From April 1989, these functions are being discharged by the Medicines Control Agency (MCA) of the Department of Health and by the Ministry of Agriculture Fisheries and Foods (MAFF).

Under the auspices of the Medicines Commission and particularly concerned with the licensing of medicines are the Committee on the Safety of Medicines (CSM) and the Committee on the Review of Medicines (CRM). The CSM is responsible for advising on the efficacy, safety, and quality of new medicines for human use, and to aid it in its task, it has established a number of standing subcommittees (e.g., Chemistry, Pharmacy, and Standards Subcommittee; Adverse Reaction Subcommittee; etc.). The CRM advises on the review, again from the aspects of safety, efficacy, and quality of products already on the market that have not previously been the subject of scrutiny of the U.K. regulatory authorities. Other committees established under medicines legislation deal with dental and surgical materials and veterinary products.

A medicine may not generally be imported, marketed, or manufactured in the U.K. unless it is the subject of a product license. This is normally held by the company marketing the product.

Human medicines may be classified as either prescription-only (POM) or general sales list (GSL). Otherwise, they are designated pharmacy (P) medicines. Both POM and P medicines may only be sold or supplied from a registered pharmacy, under the supervision of a pharmacist. Additionally, POM medicines can only be supplied in accordance with a prescription issued by a medical or dental practitioner. GSL medicines can be obtained not only from pharmacies, but also in certain packs from other retail outlets.

The U.K. Regulatory Authority (i.e., MCA) is formally involved in drug approval in three areas: clinical trial approval, product licensing, and postmarketing surveillance. Before a drug substance can be tested on human patients in clinical trials, it has to be approved for this purpose by the issue of a clinical trial certificate (CTC) or be formally exempted under the clinical trial exemption (CTX) scheme. Both procedures require the submission of basic data on the chemistry, pharmacy, pharmacology, and toxicology of the drug, but the CTX route achieves, by a process of "negative vetting" within appropriate safeguards, an accelerated pathway.

Animal pharmacology and toxicology studies are usually adequate for CTC or CTX approval, but human volunteer studies are usually undertaken prior to clinical trials to obtain human pharmacodynamic parameters for the drug and optimal administration profiles. Recently, the Royal College of Physicians and APBI have issued ethical guidelines on volunteer studies to assist local ethical committees when considering submissions for human experimental studies on drugs to be undertaken in NHS hospitals.

At the product license stage, a full submission is required on the chemistry, pharmacy, pharmacology, toxicology, and clinical assessment of the product in terms of safety, efficacy, and quality of the product. At the same time, presentational material has to be submitted, including the proposed packaging and labeling, for approval, together with the text of the data sheet and any patient information to be provided with the product by the supplier. Separate expert summaries are also required on the pharmacy, pharmacology, and toxicology of new products. The average time period between receipt of submission and granting of a license for a new product is about 20 months (1989).

For generic medicinal products (i.e., for which a product license already exists for a comparable branded product), a system of abridged submissions exists. Within these there is generally no requirement to submit clinical or in vivo data. Machinery also exists to deal with minor variations on product licenses. In the period April 1988-March 1989, product licenses were granted for 833 medicinal products involving 623 different active agents.

As a further and increasingly critical stage of involvement in drug licensing, the MCA maintains a register of adverse reactions on all licensed medicinal products. Nationally, adverse drug reaction (ADR) monitoring is achieved through the wide distribution of characteristic ADR "yellow cards," which clinicians are encouraged to complete and forward to the MCA upon diagnosis of a suspected ADR. In a bid to improve the reporting of ADRs, local collation schemes have been established in several major centers, and experimental approaches (e.g., event monitoring) have been pursued to strengthen postmarketing surveillance.

The MCA publishes an occasional bulletin, *Current Problems*, to raise the awareness of prescribers of emerging concerns about the safety of particular products. In more serious circumstances, the MCA has the power to suspend or revoke product licenses.

The British Pharmacopoeia Commission has also been established under the Medicines Commission and is responsible for the publication of the British Pharmacopoeia (BP) and the British Pharmacopoeia (Veterinary) and for assisting in the elaboration of the European Pharmacopoeia (Ph Eur). It also selects and publishes British Approved Names (BAN) of medicinal substances. The BP provides a comprehensive reference source of published standards and test methods on medicinal materials.

6. Competition: Brand and Generic

U.K. spending on medicines, including both NHS prescription medicines and OTC sales, was an average of £48 per head in 1987, equivalent to 0.8% of gross national product (GNP). (In 1986, 6.2% of GNP was devoted to health expenditure, less than 15% of which was through the private sector.)

In 1987, the U.K. drug market share was split almost evenly between companies of British (34%), European (36%), and American (30%) ownership. Over 60 companies have prescription medicine sales exceeding £4 million annually in the U.K.

The generic substitution of prescriptions for branded products is not allowed in primary care by NHS pharmacy contractors. Thirty-nine percent of general practice prescriptions are written generically and 29% can be, and are, dispensed generically, which represents 9% of the overall FPS medicine bill. The generic market of the FPS sector has a value of £150 million. Generic substitution is the norm in NHS hospitals, and generic products account for about 10% of the hospital drug bill.

Eighty percent of the NHS generic demand is met by generic manufacturers and the remaining 20% by research-based companies. Eight generic manufacturers, three of which are owned by research companies, account for two-thirds of the overall market, and some twenty others compete for the remainder.

As mentioned earlier, the price of branded products is controlled through the PPRS mechanism and the price for individual generic products through the Drug Tariff.

Competition continues through the supply chain, with wholesalers competing on the basis of service and discounts.

7. Cost-Containment Activities

The cost-containment machinery of PPRS and the Drug Tariff have already been mentioned. In April 1985, a selected list of medicinal products that could be prescribed on the NHS was introduced in seven therapeutic groups: minor analgesics, antacids, benzodiazepines, cough and cold remedies, laxatives, tonics and bitters, and vitamins. Within these groups, doctors can prescribe only the generic products listed. However, for most of the preparations, other than the benzodiazepines, branded products are available OTC. An expert committee, the Advisory Committee on NHS Drugs (ACD) is maintained to consider additions, deletions, or amendments of products on the selected list, which in its first year of operation reduced drug expenditure in the listed categories by £75 million. There are no plans to extend the list.

In 1988, all NHS hospitals were instructed to introduce formulary management systems, which have already been demonstrated in the U.K. to exert a significant check on drug expenditure.

In FPS, there is a long-standing arrangement by which high cost prescribing medical practitioners are identified and counseled. More recently, improved and detailed information of individual prescribing behavior has been produced under the PACT (Prescription Analysis and Cost) initiative by the Prescription Pricing Authority (PPA) and circulated on a quarterly basis to prescribers. From 1991, all general medical practices are to be allocated indicative drug budgets within which they will be expected to contain their prescribing costs. They are intended to exert a downward pressure on drug expenditure.

C. PHARMACY PRACTICE

1. Dispensing a Prescription: Where, What, Payment, Formulary

Where

Prescriptions for medicines may be dispensed under the NHS (1) in community pharmacies, which have a contract to provide NHS pharmaceutical services, (2) by doctors (only in rural areas) that have dispensing contracts, and (3) at NHS hospital pharmacies. Private prescriptions (i.e., those for which the patient pays the complete cost) and which may or may not be ultimately reimbursable from private medical insurance, may be dispensed from all community pharmacies, private hospitals, and NHS hospitals for private inpatients and, exceptionally, outpatients.

Over 95% of all prescriptions dispensed in the U.K. are NHS prescriptions, and over 98% of community pharmacies have a contract to provide NHS pharmaceutical services, specified in Schedule 4 of the National Health Service (General Medical and Pharmaceutical Services) Regulations of 1974.

The terms of service for an NHS chemist (pharmaceutical) contract include the following:

> A chemist [pharmacist] shall supply with reasonable promptness to any person who presents an order for drugs, appliances or listed drugs or medicines on a [NHS] prescription form signed by a doctor . . . or a dentist . . .[and]
>
> The dispensing of medicines shall be performed either by or under the direct supervision of a registered pharmaceutical chemist . . .

For prescriptions to be dispensed under the NHS, they must be written on standard NHS prescription forms. FP10 forms are issued to all NHS general medical practitioners, and FP14 forms are issued to all dentists providing NHS services.

Currently, both the hours of service and location of the individual pharmacy are specified in the contract.

Since 1987, legislation has been introduced to limit the control of entry to NHS pharmaceutical contracts, and FPCs are required to apply a "necessary or desirable" test to any new application for the provision of community pharmaceutical services prior to awarding an NHS contract. In rural areas, controls are more complex and involve an independent statutory body called the Rural Dispensing Committee (RDC). Apart from the application of the "necessary or desirable" test, any new application to provide a dispensing service, either by a family doctor or a community pharmacy, has to be considered in the context of whether the new arrangement proposed is prejudicial to the existing provision of NHS medical and pharmaceutical services.

The total number of pharmacies in Great Britain (i.e., in England, Scotland, and Wales) shows an overall decline from 15,313 in 1954 to 10,623 in 1980. Since then, there had been a steady increase which peaked at 12,021 in the autumn of 1987, but with the control of entry provisions, the number had dropped to 11,761 by the end of 1988.

Population survey figures for 1987 give an average figure for the U.K. of one community pharmacy per 4,540 people. The distribution of community pharmacies in England and Wales, based on the average number of NHS prescriptions dispensed monthly, is given in Table 7.

In England alone in 1986, there were 3,162 dispensing doctors (13% of all NHS family doctors), compared with 9,763 community pharmacies. But the former dispensed only 7% of the total number of NHS prescriptions.

In 1987, a total of 413 million prescriptions were dispensed in the

TABLE 7. Frequency distribution of chemists: by monthly Rx band: 1987, E & W.

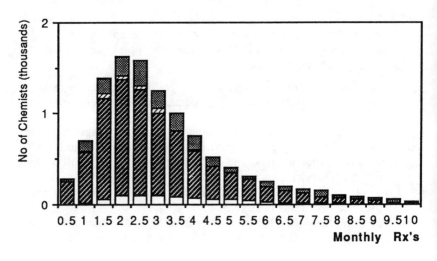

U.K.—a rise of 4% over 1986—through NHS community pharmacies. This represents an average of 7.3 prescriptions per person per annum, ranging from 6.2 in the Oxford region to 8.9 in the Mersey (Liverpool) region.

What

In principle, there is total freedom in prescribing choice under FPS, except for the modest limitations of the Drug Tariff (discussion to follow). Normally, prescribed medicines have a product license. The main exceptions to this are medicines still under clinical trials, those manufactured or imported for a named patient, or products extemporaneously prepared and dispensed.

In April 1985, the selected list was introduced, limiting the prescribing of medicines under the NHS in seven categories, namely, minor analgesics, antacids, cold and cough remedies, benzodiazepines, laxatives, tonics, and vitamins, to listed generic preparations. Of course, a full range of proprietary products remains available on private prescription, and most of the blacklisted products, except benzodiazepines, can be purchased as OTC products from community pharmacies. The British National Formulary (BNF) indicates which products in each of these groups are not available on NHS prescriptions. The list of prescribable products in these categories is provided in the Drug Tariff.

In the community pharmacist's contract there is no provision for generic substitution, and the brand prescribed must be supplied, although this may be a parallel imported product from within the EEC. Although 39% of prescriptions are written generically, and theoretically the choice of products selected is open to the pharmacist, generic equivalents are available for only 29% of FPS prescriptions. The reimbursement prices of generic products are fixed and listed in the Drug Tariff. Generic substitution is routine in NHS hospitals, which have access to their own quality control facilities.

The Drug Tariff, published monthly by the Department of Health, and issued to all NHS medical and pharmaceutical contractors, also lists the prescribable items allowed in other restricted groups (i.e., dressings, appliances, and chemical reagents) and foods prescribable against certain diagnoses or clinical conditions (borderline substances).

Payment

Patient Charges

Medication received as a hospital inpatient or on discharge from the hospital does not raise a levy. For all other NHS prescriptions, either arising from a hospital or the community, a standard prescription charge per item is made, regardless of the unit cost of the medicine or quantity dispensed. This charge is normally reviewed annually and in 1989 was £2.80 per item. (For high users, prepayment certificates offer financial advantages.) However, in England in 1987, 77% of prescriptions were for patients in exempt categories (Table 8), who receive all their NHS medication free of charge. Included among these are prescriptions for contraceptives, which are also free. Pharmacies are free to determine the cost of private prescriptions, although recommended scales are issued by professional organizations.

National Statistics

In 1987 in the U.K., the total ingredient cost of prescriptions dispensed by community pharmacists under the NHS was £2,263 million, which gives an average ingredient cost of £4.47 per prescription item, and a total ingredient cost per person annually of £39.86. It is interesting to note that the average ingredient cost for prescriptions dispensed in England to exempt patients is £4.35, and for those dispensed where charges are levied it is £5.43. The annual total prescription ingredient cost per head in England ranges from £30.18 in Oxford to £38.99 in North West (Manchester). Approximately £300 million is spent on medicines in hospitals.

Payment to Pharmaceutical Contractors

The contract is between the owner of the pharmacy and the FPC, rather than the pharmacist himself or herself. Each pharmacy has a separate contract, even though it may be part of a group or chain. NHS income is primarily determined by NHS prescription volume, and therefore it is this level that determines the income to an individual pharmacy, regardless of whether it is part of a chain or an independent pharmacy. Fifty-six percent of pharmacies in England are owned by independent proprietor pharmacists, and the rest are split between small multiple groups (2-4 pharmacies; 17%) and multiples or chains (above 5 outlets; 27%). The largest is Boots, with over 1,000 pharmacies in the U.K. The NHS income of pharmacies varies between those who represent 5% or less of turnover to those where

TABLE 8. Exemptions

Provided that the appropriate declaration is received, a charge is not payable to the contractor or dispensing doctor for drugs or appliances, including elastic hosiery, supplied for:

- children under 16
- students under 19 in full-time education
- men age 65 and over
- women aged 60 and over

People holding Family Practitioner Committee exemption certificates, which are issued to:

- expectant mothers
- women who have borne a child in the last 12 months
- people suffering from the following specified medical conditions:

  i.   Permanent fistula (including caecostomy, colostomy, ileostomy or laryngostomy) requiring continuous surgical dressing or an appliance.

  ii.   The following disorders for which specific substitution therapy is essential:

  - Addison's disease and other forms of hypoadrenalism
  - Diabetes insipidus and other forms of hypopituitarism
  - Diabetes mellitus
  - hypoparathyroidism
  - Myasthenia gravis
  - myxoedema (hypothyroidism)

  iii.   Epilepsy requiring continuous anti-convulsive therapy.

  iv.   A continuing physical disability which prevents the patient leaving his residence except with the help of another person (this does not mean a temporary disability even if it is likely to last a few months).

People holding DHSS exemption certificates, which are issued to:

- war and service pensioners (for prescriptions needed for treating their accepted disablement)
- people receiving income support or family credit*
- Anyone aged 16 or over (except students under 19 in full-time education) whose income is not much above income support level)
- (Young people can apply for an exemption certificate on their own income, whether or not they are in work and irrespective of their parents' circumstances).

People who have purchased a prepayment certificate (FP96) from the Family Practitioner Committee.

Full details of the exemption arrangements are contained in Leaflet P11, stocks of which are obtainable from the Family Practitioner Committee. The leaflet includes claim forms and guidance for patients to apply for exemption certificates.

*These certificates also cover dependents

it is virtually 100%. For the average pharmacy in the U.K., NHS dispensing income represents 67% of turnover, with an average of 70% for proprietor-run pharmacies and 53% in managed pharmacies.

The overall remuneration package for community pharmacies in England and Wales — or chemist contractors, to use the legalistic term — is determined annually by the Department of Health and Pharmaceutical Services Negotiating Committee (PSNC). Separate negotiation arrangements apply in Scotland. Individual contractors forward their prescriptions monthly to the Prescription Pricing Authority (PPA), which processes them for pricing and information purposes, and then the respective FPCs pay pharmacy owners. Payment for dispensing medicines under the NHS falls into two main components: dispensing fee and reimbursement of the drug costs. There is also an allowance to cover containers.

Dispensing fee. Currently, there are three levels of dispensing fees related to the volume of prescriptions dispensed. From April 1989, these are, respectively, £1.34 per prescription for the first 1,400 prescription items dispensed per month, 61.5p per prescription for the next 5,250 prescriptions dispensed per month, and 68p per prescription for prescriptions thereafter. There are separate scales of fees for special dispensing operations, fitting surgical hosiery and appliances, and the supply of oxygen. In recent years, several changes to the contractual arrangements have been introduced to improve the cost efficiency of pharmacies and to simplify the financial administration. In 1987, a package was introduced to encourage pharmacies that dispensed less than 16,000 NHS prescriptions annually to relinquish their NHS contracts. In 1989, a cost-plus contract, based on regular surveys of labor, overhead, and capital costs, was replaced by a straightforward fee system.

Reimbursement of drug costs. This is calculated against a formula of total cost (determined in accordance with Drug Tariff conditions) minus apportioned discount (currently based on a sliding deduction scale from 2.29% to 9.79% related to prescription volume and based on discount surveys) plus an on-cost allowance of 5% (on total cost before deduction). Prices of proprietary branded drugs are set by the individual manufacturers or suppliers within the overall controls of the Prescription Price Regulation Scheme (PPRS). The pricing of generic products is outside the PPRS, and the allowable reimbursable costs for these drugs are listed in the Drug Tariff. These reimbursement prices are paid generally regardless of the brand supplied against an NHS prescription for a generic product. Regular surveys are conducted to determine levels of wholesaler discount and of parallel-import use, and the graduated discount scale is applied to drug reimbursement costs, as previously detailed. An allowance is paid to community pharmacies that provide internship facilities, and the expenses of pharmacists providing NHS pharmaceutical services are paid for them to undertake approved continuing education programs. Also, special financial arrangements exist to maintain essential small pharmacies in rural areas.

Currently, new components to the community pharmaceutical service arc being negotiated to provide a fee structure to establish and maintain patient medication record systems, and for the provision of comprehensive pharmaceutical services to residential homes for the elderly. These were part of the packages announced in the Primary Care White Paper issued in November 1987 and noted earlier.

Formularies

Formularies first began to appear in NHS hospitals in the late 1970s as products of active Drug and Therapeutics Committees. In 1980, a Coordinating Center for Drug and Therapeutics Committees was established in Southampton and funded by the Department of Health for a period of five years. The intention was to raise awareness of good prescribing practices and to encourage the adoption of cost-effective prescribing across the NHS. By 1985, 40% of hospitals claimed to have an operating formulary, and a further 30% operated a partial formulary or policies for individual drug groups. In 1988, the Department of Health issued a circular requiring hospitals to establish full formulary management systems that involved Drug and Therapeutics Committees, clinical pharmacists, and computerized information.

On the primary care side, for many years, prescribing statistics have been produced by the PPA for individual medical practitioners, and high cost prescribers have been visited by medical officers, operating in an educational mode, employed by the Department of Health with a view to improving prescribing behavior. Also, cost comparison charts have been routinely distributed for particular therapeutic groups of drugs.

Since August 1988, the quality of information provided to doctors, through the PACT (Prescription Analysis and Cost) initiative, has improved dramatically. The information is at three levels. Level 1 provides an overall comparison of prescribing volume and costs against FPC and national averages of total prescribing and within major therapeutic groups (e.g., cardiovascular drugs). Level 2 extends the detail of the comparative analysis to allow high-cost areas of prescribing to be readily identified, for example, ACE inhibitors. Level 3 provides a complete breakdown of prescribing so that the number of prescriptions and quantity prescribed of each preparation can be pinpointed. Level 1 is sent out on a routine quarterly basis, and Levels 2 and 3 are available on request. Practices with overall prescribing costs 25% above the FPC coverage or costs within a therapeutic group 75% above the FPC average receive Level 2 automatically.

The British National Formulary (BNF), prepared under the joint auspices of the British Medical Association, Royal Pharmaceutical Society of Great Britain, and the Department of Health, is revised and issued on a six-month basis as a comprehensive prescribing guide, but this is not restrictive. Its therapeutic classification is used as the basis for PACT. Two publications, the *Drug and Therapeutics Bulletin* and *Prescribers Journal*, are routinely circulated nationally with a view to improving prescribing. Regional Drug Information Centers also provide evaluated information on medicines on a more local basis.

Initial interest in practice formularies occurred in the early 1980s and the Royal College of General (Medical) Practitioners appointed a prescribing fellow several years later to act as a focus for the development of prescribing initiatives and the production of a model practice formulary. There are, however, probably still less than 100 practices in the U.K. with practice formularies.

A key component of the NHS review is the establishment of budgets in primary care in 1991. Large practices (i.e., over 11,000 patients) will be provided with actual practice budgets and all other practices will be allocated indicative drug budgets. In support of the budgetary system for drugs, the establishment of formularies and independent professional advice for FPCs is being encouraged, and a drug evaluation resource is being funded within the Mersey Regional Drug Information Service.

2. Record Keeping

There is no general provision under the NHS for prescriptions to be dispensed more than once (except under special arrangements for the daily dispensing of narcotic prescriptions for addicts), and NHS prescriptions once dispensed are forwarded to the PPA for payment. There is no requirement, except for narcotics, for community pharmacists to keep records of NHS prescriptions. In contrast, dispensing records of private prescriptions for all types of medicines need to be retained for two years as records of bona fide sales of prescription medicines. In NHS hospitals, prescribing and drug administration charts are retained for a minimum of five years.

Under the Misuse of Drugs Act, all legitimate transactions of narcotics and similar medicines (i.e., involving receipt or supply in both hospital and community pharmacy) have to be recorded in standard Controlled Drugs Registers according to a specified format.

Most hospital pharmacies are now computerized, as are probably 70% of community pharmacies. The primary uses of computers in the community are stock management and prescription labeling; the development of patient medication records (PMRs) will require extended computerized capacity. From 1989, community pharmacists will be paid an allowance under the NHS to maintain the PMRs for elderly and confused patients on long-term medication. Adequate record keeping is also a key component of efficient supervision of the use of medicines in residential homes, another service for which community pharmacists will be paid as of 1989.

Interest is also developing in patient-held medication records — "smart cards" — and collaborative trials involving general practitioners, pharmacists, dentists, and hospitals are already in progress.

3. Patient Education

At a basic level of patient information, supplementary administration and cautionary statements are routinely included on medicine labels. The BNF contains the text for 28 sets of such material for routine application. Apart from warning cards supplied for patients prescribed lithium, monoamine oxidase inhibitors, steroids, and aspirin, there is no general move within community pharmacy to provide information to patients on medicines. Neither are counseling areas in pharmacies a universal feature.

Recent surveys (Shafford and Sharpe) have shown that community pharmacists have a priority order for their advisory function, which runs in descending order from advice on prescription medicines, through that on OTC medicines, to advice on symptoms, and then to general health education. About one hour a day is spent by the average community pharmacist on giving OTC advice, and the majority of the public value the advice given. In contrast, only half that time is spent on general health education; less than half of community pharmacists consider this to be of major importance, and the majority of the public do not look to the community pharmacy for general health education.

Against this, two particular ventures have been successfully conducted in community pharmacy. First, the National Pharmaceutical Association (NPA), the trade association of community pharmacy, has run several media campaigns starting in 1983, based on the theme of "Ask your pharmacist." Second, a consortium composed of pharmaceutical, health educational, and family planning bodies launched the "Health Care in the High Street" campaign in 1986. This is based on health education leaflets. Ninety percent of community pharmacists are said to have participated, and 60% fielded display stands. Several specialist health charities, ranging from family planning to coronary prevention, have impressive statistics to demonstrate the effectiveness of this scheme. From 1989, the scheme has been relaunched as "Pharmacy Healthcare," with Department of Health sponsorship.

Several local health promotion initiatives involving community pharmacists in the areas of smoking cessation and blood pressure monitoring have shown encouraging results. The value of community pharmacists participating in needle exchange schemes, as part of AIDS prevention programs among drug misusers, has recently been demonstrated. Currently, a pilot study is being undertaken on the scope for a cholesterol testing and counseling service from community pharmacies. This section naturally provides only a broad treatment of the subject of patient education. For further details, readers are referred to *The Pharmacist as a Health Educator*, A. Shafford and K. Sharpe, Research Report No. 24. London: Health Education Authority, 1989.

4. OTC Medicines, Nonpharmacy Outlets

There are two classes of OTC medicines in the U.K.: pharmacy (P) and general sales list (GSL). P medicines are limited to sale or supply through pharmacies, while those in the GSL classification may have a wider retail distribution. In 1988, the U.K. market for nonprescription medicines was £560 million; the major components are listed in Table 9.

Some 75% of OTC medicines are bought through pharmacies and 10% through grocers, including supermarkets. This proportion may vary with the product range; for example, in minor analgesics, pharmacies hold 69% of the market share and the grocery outlets 22%.

5. Use of Technicians and Others

Pharmacy technicians are a recognized staff group within NHS hospitals, with their own national qualifications and career and salary structures. They represent 40% of the total pharmacy work force in NHS hospitals, and unqualified assistants make up a further 13%. Technicians are employed in both technical and supervisory levels, particularly in preparative, analytical, purchasing, and dispensing areas.

In community pharmacy, the picture is less clear-cut, being currently

TABLE 9. Major components of the UIH non-prescription medicine market.

| | £m | | | |
	1985	1986	1987	1988
Cold, Cough and throat remedies	133	141	147	156
Minor analgesics	95	103	110	118
Vitamins	59	63	64	72
Indigestion and stomach remodies	37	41	44	47
Skin preparations	37	39	42	46
Laxatives	12	13	14	15
Topical analgesics	-	-	-	12
Eye care products	8.5	9	10	10
Antidiarrhea	-	-	-	7.5
Hayfever	-	-	-	6.5

complicated by an apparent polarization within the profession over the interpretation of supervision of dispensing. It is estimated that 30%-40% of community pharmacies employ a dispensing assistant, the majority of which are nonqualified. The average pharmacy also employs two to four counter assistants.

Interestingly, many of the OTC manufacturers are now targeting courses and other training material at pharmacy assistants. Both of the two main British pharmacy weeklies, *The Pharmaceutical Journal*, and *Chemist and Druggist*, now publish pharmacy assistants' supplements. Despite this, the uptake of the NPAs distance-learning course for pharmacy assistants remains low, although Boots, the largest pharmacy chain, requires the dispensing staff to undergo a set training course.

6. Pharmacy in Primary Care

The Nuffield Report on Pharmacy, published in 1986, having endorsed the role of clinical pharmacy in hospitals, considered the extension of the principles to community pharmacy. Currently, administrative arrangements within the NHS frustrate this evolution. Under current NHS legislation, a community pharmacist can only be reimbursed under the NHS for services associated with the dispensing of prescriptions. Prescriptions must not be "directed" to a particular pharmacy; the choice must remain open to the patient. Nevertheless, building on the work of progressive community pharmacists, payment under the NHS for the maintenance of PMRs and the supervision of the administration of drugs in residential homes will extend the pharmaceutical influence and involvement in primary care.

The current success of the health care leaflet campaign under the "Pharmacy Healthcare" initiative demonstrates the potential for health education from community pharmacies. This potential also applies to some screening services (e.g., blood pressure and cholesterol monitoring) which have been piloted.

Successful primary health care depends on an integrated team approach, and community pharmacy in the U.K. does not actively project itself as part of that team. Most patients and health professionals have yet to make the link between the activities of community pharmacy and other components of primary care services.

SELECTED READINGS

National Health Care System

Abel-Smith, B. National Health Service – The first thirty years. London: HMSO, 1978.

Central Statistical Office (Eds. Griffin, T. and Church, J.). Social Trends 19. London: HMSO, 1989.

Compendium of Health Statistics. (6th ed. London: Office of Health Economics, 1987.

Department of Health. Health and Social Services Statistics for England. (1988 ed.) London: HMSO, 1988.

Department of Health. Working for Patients. White Paper and 8 NHS Review Working Papers. London: HMSO, 1989.

Promoting Better Health – The Government's Programme for Improving Primary Care. Cm 249. London: HMSO, 1987.

Taylor, D. Understanding the NHS in the 1980s. London: Office of Health Economics, 1984.

Pharmaceutical Industry

Association of British Pharmaceutical Industry. Annual Report 1987-88. London: ABPI, 1988.

Association of British Pharmaceutical Industry. The Pharmaceutical Industry and the Nation's Health. London: ABPI, 1988.

Britain's Pharmaceutical Industry. London: Central Office of Information (COI), 1989.

Department of Health and Social Security. Medicines Division. The Control of Medicines in the United Kingdom. MAL 99. London: DHSS.

Department of Health and Social Security. Medicines Division. Guide to the Licensing System. London: DHSS, 1984.

Chemicals Economic Development Committee (Barnes, D.). New Focus on Pharmaceuticals. London: NEDO, 1986.

Pharmacy Practice

Committee of Inquiry (Chairman: Sir Kenneth Clucas). Pharmacy—A Report to the Nuffield Foundation. London: The Nuffield Foundation, 1986.
Dale, J.R. and Appelbe, G.E. Pharmacy Law and Ethics. 4th ed. London: The Pharmaceutical Press, 1989.
Department of Health. Drug Tariff. London: HMSO, 1989 (published monthly).
Department of Health. Statistical Bulletin. Prescriptions Dispensed by Pharmacy and Appliance Contractors—England 1977-1989. London: Department of Health, 1989.
Joint Formulary Committee. British National Formulary. London: British Medical Association and Royal Pharmaceutical Society of Great Britain, 1989 (published every six months).
Nielson Marketing Research. The Retail Pocket Book 1989. Oxford-NTC Publications, 1989.

APPENDIX. General statistics – United Kingdom.

Demography

Population (million) - Total	56.9	1987	SC 1.2
Males	27.7	1987	
Females	29.2	1987	
Annual average population growth rate (%)	0.16	1981-87	SC 1.8
Population aged under 15 (%)	19	1987	SC 1.2
Urban population (% of total)			
Birth rate (per 1,000 per annum)	13.6	1987	SC 1.9
Death rate (per 1,000 per annum			
overall	11.3	1987	SC 1.13
Males	11.5	1987	
Females	11.2	1987	

Economy

Gross Domestic Product (at current prices) (billion £ sterling)	413.7	1987	Nielsen
GDP growth rate % change (at current prices)	+9.3	1986-87	Nielsen
Imports (billion US)			
Exports (billion US)			
Rate of inflation			
Agriculture as £ of GDP			
US conversion factor (average value per £)	1.639	1987	Nielsen

Health Expenditure

Public health expenditure as % of GDP NHS expenditure as % GNP	6.17	1987	OHE 2.16
% of private incomes spent on health	1.3	1986	
Pharmaceutical market valuation pharm. Indus. gross output (£m)	4,877	1987 +Nations	Pharm. Ind. Health, fig. 4
Average annual increase in pharma market (%) Average increase in gross output (%)	9.9	1983-87	Calculated
		from Pharm. Ind.+	Nat. Health Fig. 4

Health Care Availability

Number of Doctors (1) No. of GPs (including restricted + unrestricted principals, assistants and trainees)	32,847	1986	OHE 4.11

(2) No. of GPs (unrestricted principals only) (thousands)		30.7	1987	SC 7.27
Population per doctors (GPs)				
(No.of GPs taken from (1) above) (000s)		1.7	1986	
(No. of GPs taken from (2) above) (000s)		1.97	1987	SC 7.27
Number of hospitals		2,423	1985	OHE 3.20
Av. number of beds available daily (thousands)		392	1987	SC 7.24
Population per hospital bed		145	1987	Calculated
Number of pharmacies				
Great Britain only - Registered pharmacies only		11,974	1987	Reg. Rpt
UK Chemist + Appliance contrtrs		12,190	1986	OHE 4.32

<u>Health Factors</u>

Life expectancy, Male (years)		71.5	1985	SC 7.2
Females (years)		77.4	1985	
Infant mortality per 1,000 live births		9.4	1986	SC 7.3
Defined causes of death (% of recorded cases)				R 3.14
Heart + hypertensive disease				
	Males	35	1986	
	Females	30	1986	
	Overall	33	1986	
Neoplasms				
	Males	25	1986	
	Females	22	1986	
	Overall	24	1986	
Cerebrovascular disease				
	Males	10	1986	
	Females	15	1986	
	Overall	13	1986	
Pneumonia				
	Males	4	1986	
	Females	6	1986	
	Overall	5	1986	
Bronchitis, emphysema, asthma				
	Males	3	1986	
	Females	2	1986	
	Overall	3	1986	

Notifiable communicable diseases (000's of recorded cases)			
Infective Jaundice	4.4	1987	SC 7.8
Whooping cough	17.4	1987	
Measles	46.1	1987	

SC = Social Trends 1989
RT = Regional Trends 1988
OHE = OHE Compendium 1987

Chapter 23

United States

Kathleen A. Johnson

A. THE NATIONAL HEALTH CARE SYSTEM

1. History and Background

The current health care system in the United States has evolved over a period of time beginning with the mid-1800s. Around 1850, the institutionalization of health care began with the development of large hospitals as primary units around which health care services were organized. In the early 1900s, the scientific method was introduced into medicine and medical schools, and a scientific basis for teaching and practice was developed.

Beginning in the 1940s, there was a great deal of interest in the social and organizational structure of health care in the United States. Scientific advances flourished during this period, and there was concern that financing for newly developed health care technology be available. Private health insurance as a mechanism for paying for health care expanded for the employed, and the federal government took on an increased role in the construction of hospital facilities and research. In 1965, with the passage of the amendments to the Social Security Act, Medicare and Medicaid were instituted as mechanisms for the federal government to finance health care for the aged and poor.

Beginning in the late 1970s and early 1980s, there has been interest in developing strategies to control the growth and cost of the health care system and reorganizing the financing mechanisms and methods used to deliver care. Health maintenance organizations (HMOs) were organized in the 1970s as a method of organizing health services into a system of care to help control spiraling costs. New government legislation was developed in the early 1980s, changing hospital reimbursement methods to help foster efficiency incentives and to control costs within the health sector. Private employers and insurance companies have also instituted cost control measures. The health care system in the United States today is still largely based in the private sector, with private insurance as the principle mechanism for financing care. However, the federal government has played an increasingly important role in financing and regulating health care in the United States.

2. Social and Political Environment

Historically, the United States health care system has changed, from one which required individual responsibility or charity to provide for care, to one in which there is a larger federal government role in organizing and financing health care. Still, there is no single system of health care that provides services to the entire population. Health care is made up of many different subsystems which serve different populations. The employed segment of the population and their dependents generally have private health insurance which is provided by the employer and which pays for health care provided through the private sector. Employees may select the insurance plan and method of organizing health care services from a number of options, including HMOs and the fee-for-service sector. This system, serving the majority of the United States population, probably provides some of the best health care available in the world.

For the poor, the unemployed population, and others with no private insurance coverage, the federal government gives responsibility to state and local governments for providing health care. State Medicaid programs, which obtain some funds from the federal government, provide health care financing for a segment of the poor and unemployed, such as the disabled and families with dependent children meeting poverty level eligibility requirements. Medicaid recipients, depending on the state and locale, may use the private sector for obtaining health care services, with payment provided by the government, or they may use local public hospitals and clinic facilities. State Medicaid programs, depending on the wealth of the state and comprehensiveness of services offered, provide access for eligibles to the private health care sector, which provides outstanding care. However, many state programs, in attempts to control rising costs, have placed limits on payments to providers of care, thus reducing the availability of some services. Additionally, many of the poor remain ineligible for the Medicaid program and must obtain health care services from poorly organized and overburdened local government hospitals and clinics.

Medicare provides the elderly over 65 years of age with federally-sponsored health insurance coverage through the same private system of health care as the employed population. Other populations of Americans, such as

Native Americans and the military and their dependents, have their own systems for obtaining health care services.

Although these apparently separate subsystems exist, there is some overlapping of services. Public health services are provided for all by local or state governments. The private system is sometimes used by government programs through contractual arrangements. However, in general, the whole system lacks an organizational scheme to coordinate services and ensure proper utilization. Many aspects of the system arc inadequate for those with no insurance and who are unable to pay for care. There is also a general problem of ineffective and inefficient utilization of resources. Many people feel that more effort should be made toward health planning.[1]

3. Public and Private Financing, Health Costs

Health Costs

In 1987, $500 billion was spent on health care in the United States, an average of $1,987 per capita.[2] This amount represents spending on health care that is 11.1% of the United States gross national product (GNP). Health care spending in the United States has been increasing steadily since 1965, after the introduction of Medicare. Figure 1 shows the rise in health care spending in the United States as a percentage of the GNP since 1965.

Table 1 shows the relative amount of the United States health dollar spent for various components of health care in 1987. Hospitals consumed the highest percentage, and the amount has been rising yearly. New government payment regulations for hospitals were adopted in 1983, using diagnostic classifications to determine payment in an attempt to control hospital costs for Medicare patients. However, each year since 1984, hospital spending has risen faster than the previous year. It appears that hospital spending increases in recent years may be primarily due to increases in outpatient service spending rather than to growth of inpatient costs. Payments to physicians, which had been stable, and to nursing homes are also increasing.[2]

In 1987, spending on drugs and medical sundries was $34 billion, almost 6.8% of the health care dollar, a number that has been fairly stable since 1980. Spending on drugs and medical sundries, including over-the-counter, non-prescription drugs, has declined as a percentage of total health expenditures since 1965, when it was 12.4%. In 1981 approximately 57% of expenditures in this category were for outpatient prescription drugs, 31% for OTC drugs, and 12% for drug sundries.[3]

FIGURE 1. Health care expenditures (United States).

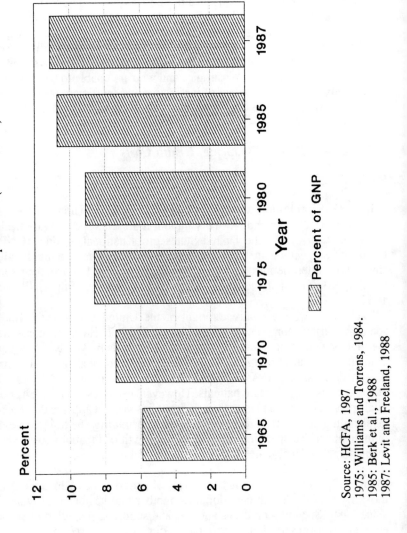

Source: HCFA, 1987
1975: Williams and Torrens, 1984.
1985: Berk et al., 1988
1987: Levit and Freeland, 1988

TABLE 1. Health care expenditures, 1987 (United States).

EXPENDITURE TYPE	AMOUNT BILLIONS $	PERCENT OF HEALTH CARE EXPENDITURE
Hospital Services	194.7	39.0%
Physicians' Services	102.7	21.0%
Dentists' Services	32.8	6.5%
Other Professionals	16.2	3.0%
Drugs and Medical Sundries	34.0	6.8%
Durable Medical Products	9.5	2.0%
Nursing Homes	40.6	8.0%
Other Personal Health Care	12.0	2.4%
Program Admin. + Cost of Private Health Insurance	14.7	3.0%
Research + Construction	17.1	3.4%
TOTAL	$500.3	100%*

*not exact due to rounding

Source: Levit and Freeland, 1988

Recent (1986) causes of increased health care expenditures have been: general inflation in the economy (32% of health care expenditure growth), increased services and intensity per capita (35%), population growth (11%), and excess health care sector inflation (22%).[4]

4. Facilities (Hospitals, Clinics, etc.)

Hospitals

Health care in the U.S. operates in a hospital-dominated system in which the hospital is central to patient care as well as to the training and education of health care personnel and research. As of 1987, there were 6,821 short- and long-term care hospitals in the U.S. with 1,266,700 beds.[5] Short-term hospitals have an average length of stay of less than seven days. Long-term hospitals have an average length of stay of three to six months. The average size is 125 beds. Table 2 shows that most hospitals in the United States are nongovernment, private, not-for-profit facilities.

Federal hospitals primarily serve special groups such as veterans and military personnel. State hospitals are generally long-term hospitals serving the mentally ill. Local hospitals are short-term general hospitals and often disproportionately serve the poor, especially those located in large

TABLE 2. Health care facilities (United States).

FACILITY	NUMBER	PERCENT
HOSPITALS(1)	6821	100%
Private Not-For-Profit	51%	
Private For-Profit	14%	
Total Non-Government Private Hospitals		65%
City/County	24%	
State	6%	
Federal	5%	
Total Government Hospitals		35%
NURSING HOMES(2)	23,065	100%
For-Profit	81%	
Not-For-Profit	15%	
Other	4%	
HOME HEALTH CARE AGENCIES(3)		
Medicare Certified	5317	

(1) 1987
   American Hospital Association, 1988

(2) 1980
   Williams and Torrens, 1988

(3) 1985
   Pharmaceutical Manufacturers Association, 1988

urban cities. Private, nongovernment hospitals are usually short-term general hospitals serving patients with insurance or government sources of financing health care.

Increased competition for patients and cost-containment efforts are changing the organization and operations of hospitals in the U.S. After many years of emphasis on building new hospitals, refurbishing older ones, and increasing the number of beds during the 1950s and 1960s, the concern developed that there was an oversupply of hospitals and too many beds. Since 1978, the number of hospitals and beds have actually decreased.[5] Occupancy rates of hospitals in the U.S. average 70%; however, there are local variations. Still, many hospitals today operate at only 69%

capacity. Many are looking at new sources of revenue through expansion of services offered, such as home health care and other outpatient programs.

Physicians Offices/Clinics

Ambulatory, out-of-hospital, patient care services are provided by physicians and other providers in private physician offices, clinics associated with hospitals, ambulatory surgery centers, urgent care centers, and specialized nongovernment and government clinics. Still, most ambulatory care services are provided in office-based settings, including solo and group physician practices.[1]

Nursing Homes

In the United States, nursing homes are licensed by the state and provide long-term health care services to the elderly and chronically ill. There are several different types, each providing a different level of care. A nursing home must meet special criteria to provide care for patients whose care is paid for by federal (Medicare) or state (Medicaid) programs.

A skilled nursing facility, a type of nursing home where at least one registered nurse or licensed practical nurse is employed, provides 24-hour a day nursing care as well as other services. An intermediate care facility is a nursing home where the emphasis is on providing regular, but not 24-hour a day, nursing care and social and rehabilitative services for people not fully capable of independent living. Residential care homes, domiciliary care facilities, and board and care homes provide sheltered living and some meals to people who do not need nursing care or other therapies. People who live in these facilities are given assistance with some activities, but there are no health care professionals on staff. Some nursing homes provide different levels of care within one facility.[1]

In 1980, there were 23,065 nursing and related care homes in the U.S. with 1,537,338 beds.[1] Homes range in size from 3 to 1,200 beds, with 95% having less than 200 beds. Typically, nursing home beds are in demand and have high occupancy rates (90%-98%). Nursing homes in the U.S. tend to be privately owned (see Table 2).

Home Health Care

Home health care grew rapidly in the United States during the 1980s. The growth of care for patients in their own homes is probably the result of financial pressures on hospitals to discharge patients earlier, the lower cost of caring for patients at home, and patient acceptance. Home health care agencies provide many types of services to patients in their homes, from occupational therapy to pharmacy services, respiratory care, and chemotherapy. Some agencies may also provide 24-hour care, housekeeping, and other personal care services. Many services of home health care agencies are paid for by private insurance coverage and/or the federal government for patients served by Medicare. The demand for home health care services is increasing. In 1985, there were 5,317 Medicare-certified home health care agencies in the United States, up from 1,019 in 1966.[6]

5. Manpower

The health care industry, the largest single employer all of industries monitored by the Department of Labor in the U.S., comprises more than 5% of the U.S. labor force and is employing an increasing number of people.[1] Recently, there has been an increase in new categories of personnel, such as physicians assistants, nurse practitioners, nurses aides, and other technicians, probably due to technical innovations, increasing specialization, and growth of the hospital as a center for care.

The numbers of health personnel in all fields have continued to increase in the 1980s, although at a slower pace than in previous years.[7] In the future, it is expected that the supplies of health personnel will continue to grow, and some have predicted surpluses in some professions. But with the continued growth and aging of the population, it is expected that the demand for health personnel will also increase. Some specialized areas in the fields of medicine, nursing, and public health may still be in short supply, and there are shortages of some health personnel in many rural areas. In fact, it is felt that 6% of the U.S. population remain underserved.[7] There are more women entering the fields that have traditionally been dominated by men. This trend is expected to continue. Tables 3 and 4 list the numbers of selected health care personnel in the United States.

From 1965 to 1982, there was an increase of 62% in the supply of physicians in the United States.[1] Although some reports predicted a surplus of physicians by 1990, there are still indications that there are shortages of some primary care specialties in rural regions and an oversupply of surgical and medical specialties in large metropolitan areas. The supply of dentists has increased in the past ten years, resulting in a surplus and a decrease in dental school enrollments during the past few years. Still, dentists are variably distributed across the U.S., with decreased numbers in rural areas. Almost 45% of the U.S. population had some form of dental insurance coverage in 1983.[1]

The number of nursing personnel can be seen in Table 3. Approxi-

TABLE 3. Health care personnel (United States).

CATEGORY	NUMBER	(YEAR)	PER 100,000 POPULATION
PHYSICIANS	501,200	(1983)	210.7
DENTISTS	137,950	(1984)	58.0
NURSING			
REGISTERED NURSES	1,453,000	(1984)	613.0
PRACTICAL NURSES	402,000	(1985)(2)	
NURSING AIDES	1,242,000	(1985)(2)	
PHYSICIAN ASSISTANTS	15,100	(1983)(1)	
HEALTH TECHNOLOGISTS AND TECHNICIANS	1,112,000	(1985)(2)	

Source: USDHHS, 1986
(1) USDHHS, May 1984
(2) PMA, 1988

TABLE 4. Pharmacy personnel (United States).

			PER 100,000 POPULATION
PHARMACISTS	157,000	(1984)	66.0

Employment Site(1)
Independent Community Pharmacy	39.1%
Chain Community Pharmacy	28.9%
Clinic/Medical Building Community Pharmacy	4.0%
Hospital	19.9%
Nursing Home Pharmacy	1.6%
Pharmaceutical Industry	2.5%
Pharmacy Education	1.3%
Other	2.6%

Source: USDHHS, 1986

(1) USDHHS, January, 1984

mately two-thirds of the registered nurses work in hospitals.[1] During the past years, there has been a perceived shortage of nurses, and many causes have been discussed, one of which has been that there is an increased demand for nursing services. There have also been increased roles for nurses in areas such as quality assurance. Nurse practitioners have more training and provide primary care in ambulatory care clinics and physicians offices in such specialties as pediatrics, obstetrics and gynecology, and adult medicine. Additionally, there have been increasing numbers of practical nurses and nursing aides in recent years.

Physicians' assistants, like nurse practitioners, provide services to patients under the direction and supervision of physicians. They have the ability to diagnose and treat common medical problems, provide preventive services, and in general, they provide more patient counseling and education than physicians. They, along with nurse practitioners, were developed when there was a shortage of physicians. Most work in primary care areas and serve in nonurban areas more often than physicians.

Table 4 shows the numbers and employment patterns of pharmacists in the United States. Most pharmacists work in community and hospital pharmacies. Projections to the year 2000 indicate a balance in the supply and requirements for pharmacists, although some states are experiencing shortages of qualified pharmacists at present.[7] Education of pharmacists occurs at 74 schools of pharmacy in the U.S. There are two entry-level degrees, the five-year baccalaureate (B.S. and B.Pharm.) and the six-year Pharm.D. Some schools also offer the postbaccalaureate Pharm.D. In 1987, there were 5,854 entry-level degrees awarded.[9] The entry level Pharm.D. made up 11.8% of the degrees awarded and has been steadily increasing for almost every year since 1960.[9] There is still disagreement among pharmacy educators about whether there should be a single entry-level degree, such as the Pharm.D.[10] Roles for pharmacists in the U.S. are changing such that more emphasis is placed on clinical, patient care functions. Educational programs are continually changing to prepare future pharmacists for these new roles.

A trend in recent years has been the growth of technologists and technicians in various health care professions who assist health care professionals with routine functions. For example, in dentistry, dental hygienists are licensed and provide oral prophylaxis services and dental health education. Dental assistants help dentists with patients, and dental laboratory technicians make oral appliances. Because the auxiliary personnel have taken over increasing functions, they have increased the productivity of dental practices.[1] In pharmacy practice, technicians, although not an official or licensed group in most states, provide support for pharmacists, especially in the hospital setting.

6. The Social Security System

Working persons contribute 7.5% of their monthly income, based on incomes of up to $48,000 per year in 1988, toward payment of Social Security taxes. After the age of 65 (60 in some instances), monthly Social Security income is available. Spouses who have not worked are also eligible for Social Security benefits. Monthly Social Security income amounts are dependent upon the length of time contributions were made to the fund and the amount contributed.

Health benefits are provided to all persons over the age of 65 through Medicare. Part A of Medicare provides hospital benefits. Part B, which is a voluntary program, provides payment for physician and some other services for a monthly premium.

7. The Medical Establishment: Entry, Specialists, Referrals

Because there is a multiplicity of medical care systems in the U.S. and no single financing mechanism, entry into the health care system depends upon the form of health insurance coverage or the system of health care being utilized. Most HMOs and many insurance plans require a patient to enter the system for first evaluation via a primary care provider, such as a physician in family or general practice, an OB/GYN specialist, an internist, or a pediatrician. If the problem requires the care of a specialist, the patient is referred for further care. Some private insurance plans and government programs allow the patient to choose the medical practitioner. One problem with this system is that a patient may go directly to a specialist if he or she believes there is a problem requiring that specialty. The visit to a specialist is usually more expensive than a visit to a primary care practitioner and is often unnecessary.

There are a large number of medical specialties in the U.S. In fact, for some specialties there is an oversupply of practitioners. For example, some studies have found that a higher supply of surgeons leads to more surgeries when compared to a similar population in a matched area with fewer surgeons. Quality of care and cost ramifications for these situations are being studied.

8. Payment and Reimbursement

Table 5 shows the sources of health care financing in the United States for 1987. These percentages have not changed appreciably in the last five years. Prior to the development of government Medicare and Medicaid programs in 1965, private sources played an even greater role in financing health care.

Public sources provided less than 30% of financing in 1965, but have increased to 40% since 1975.[11] With the bulk of financing for health care coming from private health insurance and government sources rather than direct consumer payments, these payers have been more influential in implementing cost control measures.

Funding for various health care services is dependent upon the payer of care. For example, Medicare provides health care financing primarily for hospital care. Medicaid is almost the only source of third-party financing for nursing home care (there is almost no private insurance coverage for long-term care). Most financing for nursing home care comes from direct patient payment. Table 6 shows the extent of payment that private insurance provides for various personal health care expenditures. Insurance provides the best coverage for hospital and physician services and the worst for drugs, eyeglasses/appliances, and nursing home expenses. Consumers pay for physician services and drugs primarily out-of-pocket.

TABLE 5. Sources of financing health care (United States).

	PERCENT OF FUNDING	
	1987(1)	1965(2)
SOURCE OF HEALTH CARE FUNDS		
Private Insurance	32%	
Direct Consumer Payment	25%	
Other Private and Charity	2%	
Total Non-Governmental Sources	59%	78%
Medicare	17%	
Medicaid	10%	
Other Government	14%	
Total Governmental Sources	41%	22%

(1) Levit and Freeland, 1988
(2) HIAA, 1987

TABLE 6. Private insurance coverage, personal health care expenditures.[8]

TYPE OF EXPENDITURE	PERCENT PAID BY INSURANCE
Total Personal Health Care Expenditures	51%
Hospital Services	77%
Physicians' Services	63%
Dentists' Services	35%
Other Professional	32%
Drugs and Sundries	<16%
Eyeglasses and Appliances	<15%
Nursing Home	< 2%

Source: Waldo, 1986

Fairly recently, private insurance has begun to provide payment for out-of-hospital drug use. However, direct patient payments still account for 75% of the funding for drugs and medical sundries.[2]

9. Coverage

Approximately 18% of persons less than 65 years old do not have health insurance coverage. Studies find that the uninsured are poor, minorities, young adults, rural residents, and employees of small companies.[12] Although care is available through government clinics and hospitals for those not eligible for government-sponsored programs and without insurance coverage, studies have shown that the uninsured utilize preventive services less and have no usual source of care.[12] Although Medicare and Medicaid programs begun in 1965 have filled the gaps in providing health care for all, there are still many people who do not qualify for government programs and remain uninsured or underinsured.[12] This is a problem that is currently being discussed and will hopefully be resolved in the near future.

B. PHARMACEUTICAL INDUSTRY AND DRUG DISTRIBUTION

1. Structure of the Industry

In 1986, there were approximately 680 corporations and companies in the United States engaged in the manufacture and sale of prescription and nonprescription pharmaceuticals. About 150 firms conduct research and development and produce patented prescription drug products.[3] The U.S. Department of Commerce estimated the value of pharmaceutical shipments by all manufacturers to be $25.3 billion in 1986.[6]

The pharmaceutical industry in the United States began its development into a major industry more than 40 years ago. Before drugs were generally available from manufacturers, most pharmaceutical products were compounded by pharmacists in local pharmacies. Prior to the 1930s, although there were many small pharmaceutical firms, only a few conducted research.[6]

According to the Pharmaceutical Manufacturers Association (PMA), the United States pharmaceutical industry is the largest of any nation. In 1987, total worldwide sales of PMA member companies were $41 billion, and foreign sales of PMA member companies represented 35% of total worldwide sales. Some companies have jointly owned affiliates in other countries.

2. Research and Development, Imports, Exports, Patents, Licensing

The pharmaceutical industry in the United States has had and continues to have a major commitment to research and development of pharmaceutical products.[6] Total worldwide spending for research and development by PMA member companies was $4.7 billion in 1986. Research and development as a percentage of total pharmaceutical sales was 15.1% in 1987 and has been above 10% since 1965.[6] Investment in research and development is necessary to assure continued profitability in the increasingly competitive industry.[3]

Both private industry and the federal government play key roles in pharmaceutical research and development in the United States. In 1987, 42% of total spending for biomedical research and development (much of which is pharmaceutical) in the United States was invested by industry, and 34% was spent by the National Institutes of Health (NIH), a federal government agency. Almost all research and development conducted by pharmaceutical companies is financed by the pharmaceutical companies themselves, with a small amount (less than 1% in 1986) of research being financed under contract with the U.S. federal government.

The decision of the U.S. pharmaceutical industry to pursue research and development has been linked to consideration of four major factors: medical need; probability of success and time and funds required; a balanced portfolio; and profitability.[3] Recent development efforts of U.S. pharmaceutical companies have focused on specific diseases. In 1986,

72% of research and development expenditures were toward finding new drugs to treat four major disease categories: cardiovascular disease (25.8%), infections (13.9%), neoplasms (18.1%), and central nervous system problems (14.2%).[6]

Recently, the Food and Drug Administration (FDA) review process has generated controversy due to the lengthy period of time required to complete it and the resultant high costs involved, especially since there is increased competition in the worldwide market. A total of 1,771 drugs were introduced into the United States market from 1961 through 1987. United States companies were responsible for 421 (24%). Only 7.5% of the 1,771 new drugs were marketed first in the United States, which probably reflects the long time required for the Food and Drug Administration review process.[6] In addition, in 1986 approximately 18% of U.S. company-funded research was conducted abroad.[6]

Imports and Exports

Since 1984, the United States has been the largest importer and exporter of pharmaceutical products. Japan is the largest recipient of U.S. pharmaceutical exports, receiving more than 20% of the total. Almost half of U.S. exports are to Europe, and 20% to other countries.[6] In 1987, exports from the U.S. totaled more than $3.2 billion, and imports totaled a little more than $2.8 billion, for a net export of $394.3 million. Most exports consist of ingredients and intermediate products which are then processed and packaged abroad.

Patents and Licensing

Patents in the United States may be obtained for new chemical entities, for novel manufacturing processes, and new uses of existing compounds. A new drug may have several different patents. In the United States, patents provide inventors with an exclusive right for 17 years to sell these inventions in the U.S.[6] A patent application is usually filed as soon as a promising chemical compound is discovered. Testing and Food and Drug Administration approval of a drug may take anywhere from seven to ten years. Thus the patent protected marketing time is substantially reduced. This reduction in the time in which a manufacturer may recover investment costs by being the sole producer of a new drug has led to discussion about how these decreased incentives for pharmaceutical companies adversely affect new drug research and development activities in the United States. In 1986, 2,504 new United States patents for drugs and medicines were recorded. Of these, 1,198 (48%) were U.S. in origin, while 1,306 (52%) were of foreign origin.[6]

3. Manufacturing

Drug manufacturing facilities are concentrated in nine states (New Jersey, New York, Pennsylvania, Indiana, Illinois, Michigan, Missouri, Ohio, California) and Puerto Rico. Products are distributed from pharmaceutical companies to consumers via networks of distribution centers and wholesale distributors.

In the United States, most drugs are purchased by hospitals and community pharmacies through wholesalers and directly from manufacturers. Recently, there seems to be a trend toward increased distribution through wholesalers and decreased manufacturer direct sales distribution. In 1972, about 54% of the drugs sold in the United States were distributed through direct sales from manufacturers to pharmacies and hospitals. In 1986, approximately 68% of prescription drugs were distributed through wholesalers, and 32% were distributed directly from the manufacturer to hospitals, pharmacies, and other outlets.[6]

4. Advertising and Promotion, Price Regulation

There are strict controls over advertising and promotion of pharmaceutical products in the United States. The 1962 amendments to the Food, Drug, and Cosmetics Act regulate the advertising and promotion of prescription drugs. They require companies to include generic names on drug labels along with brand (trade) name, and restrict all advertising claims in labeling and packaging inserts to those approved by the FDA. A recent trend, which has generated some controversy, has been the advertising of prescription drugs directly to consumers. Additionally, the influence of the drug industry's promotion (estimated to be 20% of sales dollars) of products to physicians has been the subject of much discussion.[3]

While there are no uniform price regulations, many payers are introducing programs to limit or control drug costs. Federal and state regulations control the amount the government pays for drugs in the Medicare and Medicaid programs, through various mechanisms. State Medicaid programs have controls over the prices paid to pharmacists for drugs dispensed to Medicaid recipients. Most states have MAC (maximum allowable cost) programs and fixed professional fees, which are paid to pharmacists when a drug is dispensed to a patient who is covered by a government-sponsored program.

5. Drug Approval and Government Regulation

Many laws, both state and federal, govern the production, distribution, and marketing of prescription drugs in the United States. Federal regulation began in 1906 with the Pure Food and Drugs Act, which was intended to protect consumers by requiring truthful labeling of medicinal ingredients. The 1938 Food, Drug, and Cosmetic Act strengthened the 1906 law and also required that manufacturers demonstrate drug safety for labeled use. The 1962 amendments to the Food, Drug, and Cosmetic Act require that manufacturers show that a new drug is effective for its intended use.[6]

The process for the approval of a prescription drug to be used in the United States is lengthy and highly regulated. All drugs must be determined to be safe and effective for intended uses. In general, required laboratory and animal studies on a newly discovered compound take one to two years. If a drug shows promise and is determined to be safe for limited controlled clinical trials in humans, the company must file an Investigational New Drug (IND) application with the Food and Drug Administration. This informs the FDA that human efficacy and safety studies will be initiated within 30 days unless the FDA objects. Approximately one out of ten drugs for which INDs are issued goes through all phases of testing.[6] From 1963 to 1986, between 671 and 2,112 INDs were submitted annually, about half of which were abandoned by the sponsor.[6]

After the IND is filed, clinical drug studies in humans begin and are divided into three phases. Phase I trials are initial studies of a new drug in a few healthy individuals. These studies are carried out to determine the pharmacological profile of the drug and take, on average, one year to perform. Phase II trials study the drug in patients who have the disease for which the drug is intended to be effective. This phase, on average, takes approximately two years. Phase III trials include more extensive and longer testing to determine efficacy and generally require three years to conduct. After the clinical trials are completed, a New Drug Application (NDA) is filed with the FDA. The NDA contains extensive documentation of safety and effectiveness studies and manufacturing control procedures to be used in preparing the product. The FDA review of all the available data generally takes two to three years. On average, the time from discovery of a new drug to marketing is ten years in the United States.

Between 1984 and 1987, 93 new molecular entity drugs were approved by the FDA in the United States, 21 in 1987.[6] Seventy-two of these were first approved in a foreign country, a trend which seems to be more common. This trend appears to be a result of the fact that the average development and review time for drug approval in many foreign markets is shorter. It is also expensive to bring a new drug through the approval process in the United States (over $100 million from discovery to approval).[6] These two issues have led to concerns. about the continued growth and leadership of the United States pharmaceutical industry in the future.

6. Competition: Brand and Generic

To market a generic drug in the United States, a company must submit an Abbreviated New Drug Application (ANDA). Generic drugs may be marketed in the U.S. only after patent expiration of the original product. As long as the generic sponsor of the drug plans to sell the drug that is identical in potency, dosage form, and product labeling, the FDA does not require any preclinical or clinical testing. Prior to marketing of the generic versions, bioequivalency to the innovator's drug product must be demonstrated and should not differ from that of the innovator's product by more than 20% with regard to the mean extent of absorption.[13] The time for the approval of an ANDA can vary from approximately nine months to two years and is significantly less expensive than filing an NDA.[13] Many of the research-based pharmaceutical companies and others feel that the availability of generics negatively affects the research and development of new drugs.

In 1980, there were 2,400 basic pharmaceutical chemical entities on the market in many different dosage forms.[14] Multiple-source drugs, supplied by several manufacturers under their own brand names, comprise more than 44% of total prescriptions.[15] Currently, most states have generic substitution laws which permit the pharmacist to substitute a generic drug for a brand name drug specified in a prescription without the prescribing physician's permission. Drug product substitution is an increasingly important role for the pharmacist.[3] Generic substitution usually leads to reduced drug costs.

There has been much controversy about the true bioequivalence of generically equivalent drugs. In an attempt to provide information to prescribers and pharmacists about generic drugs, the FDA publishes a list called the *Approved Drug Products with Therapeutic Equivalence Evaluation*, also known as the *Orange Book*, which lists different therapeutic ratings for marketed generic products. There are approximately 6,000 marketed generic drugs listed in the *Orange Book*. According to the Pharmaceutical Manufacturers Association, approximately 20% have not been determined by the FDA to be therapeutically equivalent to the pioneer product.[16]

7. Cost-Containment Activities

The amount of money spent on prescription drugs is relatively small compared to the expenditures for other segments of health care. Sales of drugs and sundries (prescription and nonprescription drugs and related products) in community pharmacies accounted for 6.7% of the total national health expenditures of 1986.[6] Community pharmacy prescriptions alone accounted for approximately 3.9% of the national health dollar.[6]

Although the amount of the GNP spent on drugs is relatively low compared to other sources of expenditures, such as hospitals, payers of care continually attempt to reduce drug costs even further. There are many regulations pharmacists must follow to correctly submit claims for reimbursement, and each payer may be different. Pharmacists are encouraged to use generics, some plans have formularies, and pharmacists' fees may be lowered to reduce overall drug costs. Although pharmacists are under pressure to expand their role in drug product selection and patient counseling, there is no provision for reimbursing pharmacists for patient care services provided.[3]

Many large operations—hospitals, HMOs, and community pharmacy chains—negotiate lower purchase prices for drugs from manufacturers in attempts to stay competitive and increase revenues. Many small pharmacies have formed buying groups through professional associations to take advantage of volume purchase prices. Still, profits have been declining for independent community pharmacies and chains.[3]

Federal and state governments have also tried to control drug costs. Through the Medicaid program, the federal government has adopted regulations to encourage generic substitution and thus reduce drug expenditures. Maximum allowable cost (MAC) regulations have also been developed. The MAC regulations encourage pharmacists to dispense the lowest cost generic product because they restrict the amount of payment to pharmacists providing drugs to patients participating in government sponsored programs such as Medicaid. Some states have adopted restrictive lists or formularies of drugs that may be prescribed for Medicaid recipients as another mechanism to reduce government drug program costs.

C. PHARMACY PRACTICE

1. Dispensing a Prescription

All drugs designated as legend drugs require a prescription. Prescriptions can be dispensed only by a licensed pharmacist or an authorized health care practitioner, such as a physician or a dentist. Over 1.5 billion prescriptions are dispensed in the United States per year from community pharmacies, clinics, and outpatient pharmacies in hospitals.[10] A few prescriptions are dispensed directly to patients by physicians or dentists from their offices. Although not a large percentage of the total prescriptions are dispensed in this manner, there are indications that the practice of physician dispensing may be growing as a result of alleged financial rewards, and it has generated considerable controversy.

In 1985, there were 68,304 community pharmacies in the United States.[6] Pharmacies may be privately owned, part of a large privately owned chain, or part of a government facility or clinic. Most community pharmacies are located in medical buildings with physician offices, in small shopping centers, or are "stand alone" operations. Pharmacies may be privately owned by persons other than pharmacists, but there are restrictions on the percentages that can be owned by physicians due to conflict of interest. All pharmacies must employ a licensed pharmacist who is designated "in-charge" and responsible for ensuring that laws are followed. More than 5,000 of 5,732 community short-term hospitals in the United States have their own pharmacy.[6] These supply medications for patient use in the hospital and a limited quantity of medicine upon discharge.

Formularies

Most HMOs and many state Medicaid programs maintain a formulary of drugs that may be prescribed and dispensed to enrollees. Under most circumstances in the "fee-for-service" health care sector, any prescribed drug may be dispensed to a patient without formulary restrictions. Many hospitals also have a restricted drug list which is agreed upon by a Pharmacy and Therapeutics Committee. The drugs selected to be on the formulary or "restricted list" are chosen because of therapeutic and economic considerations.

2. Record Keeping

Records of all prescriptions written by an authorized health practitioner and dispensed to a patient must be stored in the pharmacy for a minimum of three years. Most pharmacies assign a number to prescriptions dispensed and then file them numerically. Certain information must be recorded on the actual prescription at the time the prescription is dispensed and for each refill. All drugs must be dispensed in child-proof containers (except nitroglycerin, which must be dispensed in the original bottle) unless the pharmacist is authorized by the patient to do otherwise. Pharmacies that use a computer to perform labeling and record keeping functions must also keep a log of prescriptions dispensed on a daily basis, as well as keep a file of the original prescriptions.

An inventory of all controlled substances must be performed every two years and records kept in the pharmacy. Purchasing records of all prescription drugs must also be maintained. Hospitals must keep records of all drugs administered to patients, as well as purchasing and inventory records.

3. Patient Education

Patient education is an important function of the pharmacist and will assume an ever-increasing role with the development of newer and more sophisticated drug therapy. There are a variety of resources available to assist pharmacists in the U.S. with patient education. The United States Pharmacopeial Convention (USP), which is a branch of the FDA, provides several resources to help the pharmacist fulfill this important role. *Drug Information for the Health Care Profession* and *Advice for the Patient* are two books that provide excellent drug information. The USP also publishes one-page patient education leaflets that can be used by pharmacists when discussing drug therapy with patients. In addition, there are a variety of sources of information available on prescription drugs, including publications produced by the FDA, professional journals and other professional publications, and computerized drug information systems. Most computer systems provide at least a brief message on the prescription label to the patient regarding drug precautions. Many computer systems used by pharmacists today print a short synopsis of information about the drug that can be given to the patient when the prescription is dispensed. Many states now require that pharmacists obtain continuing education to maintain a current pharmacist license. Frequently, these programs emphasize the role pharmacists can play in clinical pharmacy services and patient education. When certain drugs are dispensed, patient education is mandated by law. The consumer must be warned when a drug combined with alcohol may impair the ability to drive or operate machinery. Also, when birth control pills, estrogens, progestational agents, and certain other drugs are dispensed, the patient must be given a patient package insert provided by the drug company.

Many community pharmacies are using other resources to educate patients on drugs. Some have developed patient education centers within the pharmacy, with brochures and other patient-oriented literature. Others have videos for patient information on drugs or disease states, which may be viewed while in the pharmacy or checked out and taken home. In the hospital, many pharmacists provide counseling on all medications the patient takes home upon discharge. Pharmacists working for home health agencies may go to the patient's home to monitor and educate the patient and family members about drug administration and use, especially for intravenous drug therapy and medication administered via pumps. Some pharmacists give patient education lectures to community groups and schools. Others write columns in local newspapers and patient- or physician-oriented newsletters.

4. OTCs and Other Classes and Schedules of Drugs, Other Nonpharmacy Outlets

There are essentially two classes of drugs in the United States, prescription and nonprescription (OTC) drugs. All prescription drugs must be prescribed by a physician, osteopath, dentist, or veterinarian and dispensed by the prescriber or a pharmacist. Controlled substances are prescription drugs that have been further classified into Schedules I-V because they have abuse potential. Schedule I drugs have very high abuse potential, have no medical purpose, or are for investigational use only. Schedules II-V include drugs with varying levels of abuse potential. Special laws govern the prescribing, dispensing, and record keeping of these medications.

OTC drugs can be used freely by the consumer and are labeled for proper use so as to be self-administered safely. OTC drugs are sold in pharmacies but are also available in nonpharmacy outlets such as grocery stores. Although there is controversy concerning the availability of OTC drugs in nonpharmacy outlets, many feel the price would increase and availability would be limited if they were sold only in pharmacies. Some would argue that many pharmacists do not consult with the consumer about these purchases even in pharmacies, since OTCs are available on shelves and are not behind the prescription counter. In the United States, the volume of self-prescribed OTC drugs used tends to outnumber physician-prescribed drugs approximately 3:2, but the total cost of OTC drugs bought by consumers is about one-half or less the cost of drugs prescribed by physicians.[3] OTC drugs are heavily advertised to the consumer.

There are exceptions to the rule that prescription drugs must be dispensed only on the prescription order of an authorized prescriber. In Florida, a third class of drugs may be prescribed and dispensed by pharmacists. All of these drugs are available by prescription-only in other states; they include drugs for which there is an OTC with the same ingredient and some other prescription drugs. The second exception to the rule is that in some, but not all states, Schedule V controlled substances which require a prescription and have low abuse potential may be dispensed by a pharmacist without a prescription. Examples of these include some cough syrups containing codeine. Insulin and insulin syringes may also be dispensed without a prescription by a pharmacist.

5. The Use of Technicians and Others

Technicians are a widely used and accepted part of hospital pharmacy practice. They perform many important functions under the one-to-one direction and supervision of a pharmacist (e.g., make intravenous solutions and fill/check patient cassettes with 24-hour unit dose medications). The use of technicians allows pharmacists to provide many clinical patient care services.

Some junior colleges provide technician training. However, most technicians learn from on-the-job experience. Although some people have advocated that technicians be certified or licensed, there is so far no formal licensing or certification procedure required.

While technicians are an accepted part of hospital pharmacy practice, there is considerable controversy about the use of technicians in the ambulatory care, community pharmacy setting. Although clerks can help pharmacists by typing labels and putting away stock, in most cases only pharmacists or intern pharmacists are allowed to fill the prescription. Because the pharmacist must still perform a considerable amount of the dispensing functions, clinical pharmacy in this setting has not progressed as far as in the hospital. Several states are experimenting with the use of technicians in community settings. For example, Washington state has legislation that permits the use of technicians to help pharmacists fill prescriptions. In exchange, pharmacists are required to provide patient counseling on all new prescriptions. Opposition to the use of technicians is focused on quality of care and employment concerns.

6. Pharmacy and Primary Care

There are many pharmacists in the United States who perform primary care services. Pharmacist practitioners in the Public Health Service may work in clinics and provide routine primary care in the Indian Health Service as well as in other settings. In many HMOs and other health care environments, pharmacists are running anticoagulation and hypertension clinics. Pharmacists in community pharmacies help patients select the proper OTC medication in the treatment of self-limited medical problems.

Pharmacists may also screen patients for hypertension or diabetes in the ambulatory care setting, such as in a community pharmacy. Many pharmacies provide ostomy fitting and supply, durable medical equipment, and diabetic teaching. The federal government requires that a pharmacist review the medication records of all patients in skilled nursing facilities monthly. Pharmacists working as consultants to skilled nursing facilities monitor drug therapy, make recommendations on changing medication regimens, and order laboratory tests to monitor patient progress.

Clinical pharmacy, which began in the United States around 1965, has provided opportunities for pharmacists to expand their roles in direct patient care within different health care settings, especially the hospital. Documentation of the value of the services provided by pharmacists has been demonstrated in the hospital, in the skilled nursing facility, and in other institutional settings. The development of clinical pharmacy services in community practice and documentation of the value of these services is beginning to occur and is one area of future pharmacy practice emphasis and research in the United States.

7. Drug Insurance and Third Parties

Almost all health insurance plans provide payment for drugs administered to hospitalized patients. On the other hand, most prescriptions dispensed for patients not in the hospital are paid for out-of-pocket directly by the patient. Although the amount is growing, only about 30% of the cost of outpatient prescription drugs are paid for by third-party prescription insurance plans.[9] Somewhat higher amounts are paid out-of-pocket by the elderly.[3] Sources of funding for drugs and medical sundries in 1987 can be seen in Table 7.

Patients enrolled in HMOs generally have some outpatient prescription

TABLE 7. Sources of funding for outpatient prescription drugs, OTCs and medical sundries (United States).

SOURCE OF FUNDING	DOLLARS (BILLIONS)	PERCENT
Private Insurance	4.0	13.8%
Federal Government	1.9	5.6%
State / Local Government	2.0	5.9%
Direct Consumer Payment	25.5	75.0%
Total	$ 34.0 billion	

Source: Levit and Freeland, 1988.

drug coverage which pays the total cost of prescription or requires a modest patient copayment per prescription, usually $1 to $5. In this case, the quantity of a chronic medication dispensed may be limited to a one- to three-month supply, depending on the drug and the health plan. Some HMOs have their own pharmacies. Others contract with existing pharmacies in the community to provide outpatient prescription drugs to enrolled members. Patients over 65 years old who have Medicare also have all inpatient drugs covered.

D. CONCLUSIONS

The American health care system provides some of the most sophisticated and highly technical care available anywhere in the world. New innovations in medical treatments are researched, developed, and rapidly integrated into mainstream medicine. Technology assessment is now evaluating the medical effectiveness of various treatment modalities (drug treatment versus surgery, hospitalization versus home care). Quality assurance activities, especially in the hospital, provide support to continually evaluate and improve existing services. Highly trained and specialized professionals and technical staff provide some of the highest level of skill possible.

But there are limits to the resources available that can be spent on health. While a high level of care is available, it is not accessible to those without insurance or the resources to pay for care. Many in the United States remain uninsured or underinsured, a problem that is currently being examined.

There has been a tremendous amount of discussion about whether newly developed technical innovations actually improve the quality of life, are cost-effective, or are worth the enormous expense. Many relatively inexpensive programs, such as standard prenatal care for the poor and uninsured populations, may have been underfunded in favor of more expensive and highly sophisticated hospital services, such as neonatal intensive care units. These issues still must be addressed.

The future of health care in the United States will see more emphasis on cost control and efficiency measures. There will be less emphasis on hospital care and more on outpatient services and other cost-effective treatments. There will be continued growth of innovations, but with careful attention to assessment of costs and benefits. The government will have an increasingly important role in assuring access to care, controlling costs, and regulating health care in the United States.

REFERENCES

1. Williams, Stephen J., and Torrens, Paul R. *Introduction to Health Services*. 3rd ed. Wiley and Sons, Inc., New York, 1988.

2. Levit, Katherine R., and Freeland, Mark S. "National Medical Care Spending." *Health Affairs*. Winter 1988, pp. 124-136.

3. Goldberg, Theodore, DeVito, Carolee A., and Raskin, Ira E., eds. *Generic Drug Laws: A Decade of Trial—A Prescription for Progress.*" National Center for Health Services Research, Public Health Service, U.S. Department of Health and Human Services, June 1986.

4. Gibson, Robert M., and Waldo, Daniel R. "National Health Expenditure, 1981." *Health Care Financing Review*. Vol. 4(1), September 1982, pp. 8-9.

5. American Hospital Association. *Hospital Statistics*. 1988 Edition. Chicago, 1989.

6. Pharmaceutical Manufacturers Association. *Statistical Fact Book*. PMA, Washington, DC, August 1988.

7. *Fifth Report to The President and Congress on the Status of Health Personnel in the United States*. Bureau of Health Professions, Health Research Services Administration, Public Health Service, U.S. Department of Health and Human Services, 1986.

8. Waldo, Daniel R., Levit, Katherine R., and Lazenby, Helen. "National health expenditures, 1985." *Health Care Financing Review*. Vol 8(1), Fall 1986.

9. Penna, Richard P. and Sherman, Michael S. "Degrees Conferred by Schools and Colleges of Pharmacy, 1986-1987." *American Journal of Pharmaceutical Education*. Vol. 52 (3), Fall 1988, pp. 276-292.

10. Smith, Mickey C. and Knapp, David A. *Pharmacy Drugs and Medical Care*. 4th ed. Williams and Wilkins, Baltimore, 1987.

11. Health Insurance Association of America. *1986-1987 Source Book of Health Insurance Data*. Washington, DC, 1987.

12. Davis, K., and Roland, D. "Uninsured and Underserved: Inequities in Health Care in the United States." *Millbank Memorial Fund Quarterly*. 61, 1983, pp. 149-177.

13. Troetel, William M. "How New Drugs Win FDA Approval." *U.S. Pharmacist*, November 1986, pp. 54-68.

14. Pharmaceutical Manufacturers Association. *1980 Fact Book*. PMA, Washington, DC, 1980.

15. Feldstein, Paul J. *Health Care Economics*. 2nd Ed. Wiley and Sons, Inc., New York, 1983.

16. Pharmaceutical Manufacturers Association. *Generic Drugs*. PMA, Washington, DC, 1987.

SELECTED READINGS

Pharmaceutical Industry

Berk, Marc L., Monheit, Alan C., and Hagen, Michael M., "How the United States Spent its Health Care Dollar: 1929-1980." *Health Affairs*. Fall 1988, pp. 46-60.
Bezold, Clement. *The Future of Pharmaceuticals: The Changing Environment for New Drugs*. John Wiley and Sons, New York, 1981.
Bezold, Clement, ed. *Pharmaceuticals in the Year 2000: The Changing Context for Drug R&D*. Institute for Alternative Futures, Alexandria, VA, 1983.
Comanor, William S. "The Political Economy of the Pharmaceutical Industry." *Journal of Economic Literature*. Vol. XXIV, September 1986, pp. 1178-1217.
Grabowski, Henry, G. and Vernon, John M. *The Regulation of Pharmaceuticals: Balancing the Benefits and Risks*. American Enterprise Institute for Public Policy Research, Washington, DC, 1983.

Health Services in the United States

Donabedian, Avedis, Axelrod, Solomon J., Wyszewianski, Leon, and Lichtenstein, Richard L., eds. *Medical Care Chartbook*. 8th ed. Health Administration Press, Ann Arbor, Michigan, 1986.
Health Care Financing Administration. "National Health Expenditures, 1986-2000." *Health Care Financing Review*. Vol. 8(4) Summer 1987, pp. 1-36.
Star, Paul. *The Social Transformation of American Medicine*. Basic Books, Inc., New York, 1982.
Wilson, Florence A. and Neuhauser, Duncan. *Health Services in the United States*. 2nd ed. Ballinger, New York, August 1987.

Pharmacy Practice in the United States

American Pharmaceutical Association. *The APhA Pharmacy Commission on Third Party Programs Final Report*. Washington, DC, APhA, 1986.
Bezold, Clement, Halperin, Jerome A., Binkley, Howard L., and Ashbaugh, Richard R., eds. *Pharmacy in the 21st Century. Planning for an Uncertain Future*. Institute for Alternative Futures and Project HOPE, American Association of Colleges of Pharmacy, Washington, DC, 1985.
Lipton, H. L. and Lee, P. R. *Drugs and the Elderly: Clinical, Social, and Policy Perspectives*. Stanford University Press, Stanford, 1988.
Smith, Mickey C. and Brown, Thomas R. *Handbook of Institutional Pharmacy Practice*. Williams and Wilkins Co., Baltimore, 1979.

Index

Page numbers followed by "t" indicate tables or figures.